Kernel Methods for Remote Sensing Data Analysis

Kernel Methods for Remote Sensing Data Analysis

Edited by

Gustavo Camps-Valls
University of València, Spain

Lorenzo Bruzzone
University of Trento, Italy

WILEY
A John Wiley and Sons, Ltd., Publication

This edition first published 2009

Registered office
John Wiley & Sons Ltd, The Atrium, Southern Gate, Chichester, West Sussex, PO19 8SQ, United Kingdom

For details of our global editorial offices, for customer services and for information about how to apply for permission to reuse the copyright material in this book please see our website at www.wiley.com.

Library of Congress Cataloguing-in-Publication Data

Camps-Valls, Gustavo, 1972-
Kernel methods for remote sensing 1 : data analysis 2 / Gustavo Camps-Valls and Lorenzo Bruzzone.
p. cm.
Includes bibliographical references and index.
ISBN 978-0-470-72211-4 (cloth)
1. Remote sensing. I. Bruzzone, Lorenzo. II. Title. III. Title: Kernel methods for remote sensing one.
G70.4.C36 2009
621.36'780285631–dc22

2009015958

A catalogue record for this book is available from the British Library.

ISBN: 978-0-470-72211-4 (Hbk)

Set in 10/12pt Times by Thomson Digital, Noida, India
Printed in Singapore by Markono Print Media Pte Ltd

Contents

About the editors

Gustavo Camps-Valls was born in València, Spain in 1972, and received a B.Sc. degree in Physics (1996), a B.Sc. degree in Electronics Engineering (1998), and a Ph.D. degree in Physics (2002) from the Universitat de València. He is currently an associate professor in the Department of Electronics Engineering at the Universitat de València, where he teaches electronics, advanced time series processing, machine learning for remote sensing and digital signal processing. His research interests are tied to the development of machine learning algorithms for signal and image processing, with special attention to adaptive systems, neural networks and kernel methods. He conducts and supervises research on the application of these methods to remote sensing image analysis and recognition, and image denoising and coding. Dr Camps-Valls is the author (or co-author) of 50 papers in refereed international journals, more than 70 international conference papers, 15 book chapters, and is editor of other related books, such as *Kernel Methods in Bioengineering, Signal and Image Processing* (IGI, 2007). He has served as reviewer to many international journals, and on the Program Committees of SPIE Europe, IGARSS, IWANN and ICIP. Dr Camps-Valls was a member of the EUropean Network on Intelligent TEchnologies for Smart Adaptive Systems (EUNITE), and the Spanish Thematic Networks on 'Pattern Recogntion' and 'Biomedical Engineering'. He is active in the R+D sector through a large number of projects funded by both public and industrial partners, both at national and international levels. He is an Evaluator of project proposals and scientific organizations. Since 2003 he has been a member of the IEEE and SPIE. Since 2009 he has been a member of the Machine Learning for Signal Processing (MLSP) Technical Committee of the IEEE Signal Processing Society. Visit http://www.uv.es/gcamps for more information.

Lorenzo Bruzzone received a laurea (M.S.) degree in electronic engineering (summa cum laude) and a Ph.D. degree in telecommunications from the University of Genoa, Italy, in 1993 and 1998, respectively. From 1998 to 2000 he was a Postdoctoral Researcher at the University of Genoa. In 2000 he joined the University of Trento, Italy, where he is currently a Full Professor of telecommunications. He teaches remote sensing, pattern recognition, radar and electrical communications. Dr Bruzzone is the Head of the Remote Sensing Laboratory in the Department of Information Engineering and Computer Science, University of Trento. His current research interests are in the area of remote-sensing image processing and recognition (analysis of multitemporal data, feature extraction and selection, classification, regression and estimation, data fusion and machine learning). He conducts and supervises research on these topics within the frameworks of several national and international projects. He is an Evaluator of project proposals for many different governments (including the European Commission) and scientific organizations. He is the author (or co-author) of 74 scientific publications in referred international journals, more than 140 papers in conference proceedings and 7 book chapters. He is a Referee for many international journals and has served on the Scientific Committees of several international conferences. He is a member of the Managing Committee of the Italian Inter-University Consortium on Telecommunications and a member of the Scientific Committee of the India–Italy Center for Advanced Research. Since 2009 he has been a member of the Administrative Committee of the IEEE Geoscience and Remote Sensing Society. Dr Bruzzone gained first place in the Student Prize Paper Competition of the 1998 IEEE International Geoscience and Remote Sensing Symposium (Seattle, July 1998). He was a recipient of the Recognition of IEEE Transactions on Geoscience and Remote Sensing Best Reviewers in 1999 and was a Guest Editor of a Special Issue of the IEEE Transactions on Geoscience and Remote Sensing on the subject of the analysis of multitemporal remote-sensing images (November 2003). He was the General Chair and Co-chair of the First and Second IEEE International Workshop on the Analysis of Multi-temporal Remote-Sensing Images (MultiTemp), and is currently a member of the Permanent Steering Committee of this series of workshops. Since 2003, he has been the Chair of the SPIE Conference on Image and Signal Processing for Remote Sensing. From 2004 to 2006 he served as an Associate Editor for the IEEE Geoscience and Remote Sensing Letters, and currently is an Associate Editor for the IEEE Transactions on Geoscience and Remote Sensing, and the Canadian Journal of Remote Sensing. He is a Senior member of IEEE, and also a member of the International Association for Pattern Recognition and of the Italian Association for Remote Sensing (AIT).

List of authors

Jerónimo Arenas-García
Dept. Signal Theory and Communications, Univ. Carlos III de Madrid, Spain

Amit Banerjee
Applied Physics Laboratory, The Johns Hopkins University, USA

Jon Atli Benediktsson
Faculty Electrical and Computer Engineering, Univ. of Iceland, Iceland

Joshua Broadwater
Applied Physics Laboratory, The Johns Hopkins University, USA

Lorenzo Bruzzone
Dept. Information Engineering and Computer Science, University of Trento, Italy

Philippe Burlina
Applied Physics Laboratory, The Johns Hopkins University, USA

Javier Calpe-Maravilla
Image Processing Laboratory (IPL) & Dept. Enginyeria Electrònica, Universitat de València, Spain

Gustavo Camps-Valls
Image Processing Laboratory (IPL) & Dept. Enginyeria Electrónica, Universitat de València, Spain

Jocelyn Chanussot
Signal & Image Department, Grenoble Institute of Technology, France

Chris Dichl
Applied Physics Laboratory, The Johns Hopkins University, USA

M. Murat Dundar
Indiana University-Purdue University, Indianapolis, USA

Mathieu Fauvel
Signal & Image Department, Grenoble Institute of Technology, France, Faculty Electrical and Computer Engineering, University of Iceland, Iceland

Giles M. Foody
School of Geography, University of Nottingham, UK

Glenn Fung
Siemens Medical Solutions Inc., Malvern, USA

Peter V. Gehler
Max Planck Institute for Biological Cybernetics

Luis Gómez-Chova
Image Processing Laboratory (IPL) & Dept. Enginyeria Electrònica, Universitat de València, Spain

Yanfeng Gu
College of Electronics and Information Engineering, Harbin Institute of Technology, Harbin, P.R. China

J. Anthony Gualtieri
NASA/GSFC, CISTO & Global Science and Technology

Bor-Chen Kuo
Graduate Institute of Educational Measurement and Statistics, National Taichung University, Taiwan

Cheng-Hsuan Li
Graduate Institute of Educational Measurement and Statistics, National Taichung University, Taiwan

Xiaowen Li
Research Center for Remote Sensing and GIS, Beijing Normal University, Beijing, P.R.China

Mattia Marconcini
Dept. Information Engineering and Computer Science, University of Trento, Italy

Manel Martínez-Ramón
Dept. Signal Theory and Communications, Univ. Carlos III de Madrid, Spain

Gabriele Moser
Dept. of Biophysical and Electronic Engineering (DIBE) and Interuniversity Research Center in Environmental Monitoring (CIMA) University of Genoa, Genoa, Italy

Jordi Muñoz-Marí
Image Processing Laboratory (IPL) & Dept. Enginyeria Electrònica, Universitat de València, Spain

Nasser M. Nasrabadi
US Army Research Laboratory

Kaare Brandt Petersen
Epital ApS, Denmark

José Luis Rojo-Álvarez
Dept. Signal Theory and Communications, Univ. Rey Juan Carlos, Madrid, Spain

Bernard Schölkopf
Max Planck Institute for Biological Cybernetics

Sebastiano B. Serpico
Dept. of Biophysical and Electronic Engineering (DIBE) and Interuniversity Research Center in Environmental Monitoring (CIMA) University of Genoa, Genoa, Italy

Yanfei Wang
Key Laboratory of Petroleum Geophysics, Institute of Geology and Geophysics, Chinese Academy of Sciences, Beijing, P.R. China

Björn Waske
Faculty Electrical and Computer Engineering, Univ. of Iceland, Iceland

Changchun Yang
Key Laboratory of Petroleum Geophysics, Institute of Geology and Geophysics, Chinese Academy of Sciences, Beijing, P.R. China

Jinn-Min Yang
Department of Mathematics Education, National Taichung University, Taiwan

Preface

Machine learning experienced a great advance in the 1980s and 1990s due to the active research in artificial neural networks, adaptive schemes and fuzzy systems. These methodologies demonstrated good results in many real applications, especially for classification and regression tasks, since neither a priori knowledge about the model of the distribution of the available data nor the relationships among the independent variables should be necessarily assumed. These desirable properties are at the basis of the success of these methods in the field of the analysis of remote sensing images, where a wide literature refers to the definition of classifiers and estimation algorithms based on neural networks and fuzzy systems.

In the 1990s, *Kernel methods* emerged as innovative techniques in the framework of machine learning. Kernel methods can be simply viewed as a two-fold methodology. The first step consists of mapping the data from the original input space into a kernel feature space of higher dimensionality through a nonlinear function. The second step solves a linear problem in the transformed kernel space. These methods allow us to design and interpret learning algorithms geometrically in the kernel space (which is nonlinearly related to the input space), thus combining statistics and geometry in an effective way, and all of this while obtaining solutions with that desirable property that is uniqueness. A few sets of free parameters are commonly needed to make the algorithms work properly. In addition, the inclusion of regularization in the function to be optimized becomes a natural and theoretically well-founded task. This theoretical elegance is also matched by their practical performance. Interestingly, this framework allows us to create non-linear methods from linear well-established ones.

The recent application of machine learning and pattern recognition approaches based on kernel methods (KMs) to the field of remote sensing data analysis provided excellent results in many different application domains. Kernel methods proved effective in the analysis of images of the Earth acquired from airborne and satellite sensors, by improving the results offered by traditional statistical and neural networks methods in real-life applications (e.g. natural resource control, detection and monitoring of antrophic infrastructures (e.g. urban areas), agriculture inventorying, disaster prevention and damage assessment, anomaly and target detection, biophysical parameter estimation, etc.). Nowadays, KMs are standard techniques for many remote sensing applications.

The book *Kernel Methods for Remote Sensing Data Analysis* presents research related to remote sensing based on the recent advances in kernel methods. The book is organized into five parts. The first part of the book presents two background chapters on the key aspects of machine learning for remote sensing, and the theoretical and practical foundations of kernel methods.

The remaining four parts address the most recent research in developing kernel methods in remote sensing for supervised classification, semi-supervised classification, regression, and feature extraction.

The four main parts of the book are preceded with introductory chapters. Chapter 1, written by Waske, Fauvel, Benediktsson and Chanussot, reviews the main concepts of machine learning in the context of remote sensing, and focuses on the supervised classification case. Written by Gehler and Schölkopf, Chapter 2 provides the reader with an exhaustive introduction to the framework of kernel methods. The authors address the main concepts and theoretical derivations, discuss theoretical and practical issues, and give pointers to useful references. These introductory chapters can be useful both for the non-expert reader in remote sensing and kernel methods, and for practitioners interested in the theoretical and practical issues of kernel machines. Then, three parts are devoted to the application and development of kernel methods in different learning paradigms, i.e., in supervised classification, semi-supervised classification, function approximation and regression, and feature extraction. The fourteen chapters of these blocks include a literature review of the specific application, a critical discussion of the needs and demands of each one, novel research contributions, and experimental results to demonstrate method capabilities. Chapters are written by leading experts in their respective fields. A brief description on the specific contribution of each of the chapters follows.

The first block of chapters is devoted to supervised image classification. This is the most active research area for kernel methods in remote sensing due to the special characteristics of the problems and the advantages offered by kernel machines. In Chapter 3, Gualtieri reviews the formulation and foundations of the classical support vector machine (SVM), and shows results in hyperspectral image classification. Also, special attention is paid to the parallelization of the algorithm. In all cases, theory is supported with experimental results. In Chapter 4, Foody analyses in detail the design of training and testing stages for image classification with SVM. Next, in Chapter 5, Dundar and Fung analyse another kernel method, the Kernel Fisher's Discriminant. After reviewing the formulation, a novel version with heterogeneous kernels is presented, and results in multi and hyperspectral image classification are presented. Chapter 6 by Muñoz-Marí, Gómez-Chova, Martínez-Ramón, Rojo-Álvarez, Calpe-Maravilla and Camps-Valls presents a general framework based on composite kernels for dealing with multitemporal and multisource image classification and change detection. Two kernel classifiers are used to illustrate the properties of the framework: the standard SVM and the one-class SVM. The latter is useful for target detection and anomaly detection problems. Related to this, in Chapter 7, Nasrabadi provides a performance comparison of various linear and nonlinear anomaly detection techniques based on kernels. In particular, subspace anomaly detectors are reviewed and their kernel versions discussed. Also, the kernel Reed–Xiaoli (RX) anomaly detector is presented and benchmarked. This book part is concluded with Chapter 8 by Banerjee, Burlina and Diehl, in which one-class SVMs for hyperspectral anomaly detection is proposed as an alternative to the RX method. Performance is illustrated in airborne mine and hyperspectral imagery detection.

The third part of the book is devoted to the emerging field of semi-supervised image classification. In particular, depending on whether training and test data are drawn from the same remote-sensing image or not, one should work under a domain adaptation problem or under a sample selection bias problem. The first case is treated in Chapter 9 by Marconcini and Bruzzone, where a novel domain adaptation SVM technique that extends SVMs to the domain adaptation framework is presented, and a novel validation strategy for domain adaptation is proposed. The second case is treated in Chapter 10, by Gómez-Chova, Calpe-Maravilla,

Bruzzone, Camps-Valls, which presents a semi-supervised SVM classifier based on the combination of the expectation-maximization (EM) algorithm for Gaussian mixture models (GMM) and the *mean map* kernel.

The fourth part of this book is devoted to the field of regression and model inversion. Several interesting applications are included. First, in Chapter 11, written by Broadwater, Banerjee, and Burlina, the concept of a kernel unmixing algorithm for hyperspectral imagery is introduced. This new model generalizes the linear mixing model allowing for both linear and nonlinear mixture estimation. Then, in Chapter 12 by Wang, Yang and Li, the theory and methods for quantitative remote sensing inverse problems with kernel-based operator equations is developed. The authors focus on the important bidirectional reflectance distribution function (BRDF) model inverse problem and the distribution function of atmospheric aerosols. Finally, in Chapter 13, written by Moser and Serpico, land and sea surface temperature estimation with support vector regression is treated in detail, and the span bound is proposed for free parameter tuning.

Finally, the book finishes with a part dealing with kernel-based feature extraction. Three chapters are included. Chapter 14 by Arenas-García and Brandt Petersen reviews the principles of several multivariate analysis methods and their kernel extensions. Application to contextual multispectral image classification and pixel-based hyperspectral classification are illustrated. Then, in Chapter 15, by Gu, kernel PCA is revised in detail for target and anomaly detection. An improved selective KPCA is proposed in hyperspectral images. The book ends with Chapter 16, by Kuo, Yang and Li which deals with the kernel nonparametric weighted feature extraction (KNWFE). The experimental results show that KNWFE outperforms KPCA and GDA remote sensing image classification.

Gustavo Camps-Valls and Lorenzo Bruzzone
València and Trento, 2009

Acknowledgments

The editors would like to acknowledge the help of all involved in the collation and review process of the book, without whose support the project could not have been satisfactorily completed. A further special note of thanks goes also to all the staff at John Wiley & Sons, Ltd, whose contributions throughout the whole process, from inception of the initial idea to final publication, have been valuable. Special thanks also go to the publishing team at John Wiley & Sons, Ltd, and in particular to Alex King and Nicky Skinner, who continuously prodded via e-mail keeping the project on schedule.

We wish to thank all of the authors for their insights and excellent contributions to this book. Most of the authors of chapters included in this book also served as referees for articles written by other authors. Thanks go to all those who provided the constructive and comprehensive reviews.

In addition, a qualified set of 'external' reviewers participated actively in the revision process. We would like to express our deepest gratitude to them: Francesca Bovolo, University of Trento (Italy); Marco Diani, University of Pisa (Italy); Gregoire Mercier, Institut Telecom / Telecom Bretagne (France); Jose Nascimento, Instituto Superior de Engenharia de Lisboa (Portugal); Allan Aasbjerg Nielsen, Technical University of Denmark (Denmark); Emilio Parrado, University Carlos III de Madrid (Spain); Fernando Pérez-Cruz, Princeton University (USA); Antonio Plaza, University of Extramadura (Spain); Frédéric Ratle, University of Lausanne (Switzerland) and Devis Tuia, University of Lausanne (Switzerland).

Gustavo Camps-Valls and Lorenzo Bruzzone
València and Trento, 2009

List of symbols

Physical symbols			
Symbol	**Description**	**Symbol**	**Description**
λ	wavelength	ν	frequency
c	speed of light	$\hbar$	Planck's constant
F_0	solar constant	$F_0(\lambda)$	solar spectral irradiance
$B(\lambda, T)$	Planck's distribution	k_B	Boltzmann's constant
Φ	radiant power	I	radiant intensity
A	area	ω	solid angle
F	irradiance	L	radiance
θ	zenith angle	ψ	azimuth
μ	$\cos(\theta)$	$1/\mu$	optical mass
β_e	volume extinction coefficient	τ	optical thickness or depth
log	natural logarithm	e	base of the natural log
T	transmittance	ρ	reflectance
y	surface along-track dimension	x	surface across-track dimension
l	image-lines dimension	p	line-pixels dimension
N_l	number of image lines	N_p	number of pixels per line
b	image-bands dimension	N_b	number of spectral bands
$H(x, \lambda)$	optical system response	$S(p, b)$	CCD sensitivity
$I(l, p, b)$	image values	$a(p, b)$	calibration coefficients
$v(p, b)$	vertical striping factors	$\alpha(p, b)$	integrated line profile
f	profile frequency	A	profile amplitude
D	spectral distance	R	spatial filter response
λ_b	channel mid-wavelength	Λ_b	channel bandwidth

Dimensions, indexes and basic algebra			
$\mathbb{R}$	real numbers	$\mathbb{N}$	natural numbers
$\mathbf{x} \in \mathcal{X}$	input and input space	$y \in \mathcal{Y}$	output and output space
$f(\mathbf{x})$	real-valued function	$\mathcal{F}$	class of real-valued functions
$\mathbf{x}^\top, \mathbf{X}^\top$	transpose of a vector/matrix		
n	number of samples	$\|\mathbf{x}\|_p$	p-norm
ℓ	labelled training samples	d	number of input features
δ	confidence	u	unlabelled training samples
$\mathbf{C}$	covariance matrix	ε	error probability
$(x)_+$	equals x, if $x \geq 0$ else 0	$\mathbf{I}$	identity matrix
$\text{sgn}(x)$	equals 1, if $x \geq 0$ else -1	e	base of the natural log
		#	cardinality of a set
Clustering			
i	sample index	k	cluster index
ω_k	cluster k	c	number of clusters
$\boldsymbol{\mu}_k$	mean feature vector	$\boldsymbol{\Sigma}_k$	covariance matrix
$p(\mathbf{x}\|\boldsymbol{\Theta})$	probability density function	$\boldsymbol{\Theta}$	vector of parameters
$p(\mathbf{x}\|\omega_k, \boldsymbol{\Theta})$	conditional pdf	$P(\omega_k)$ or π_k	prior probability
$\mathcal{L}$	log likelihood	$\mathcal{N}_d(\boldsymbol{\mu}_k, \boldsymbol{\Sigma}_k)$	d-variate Gaussian
$P(\omega_k\|\mathbf{x}_i, \boldsymbol{\Theta})$	posterior probability	h_{ik}	posterior or membership
Endmembers and unmixing			
$\mathbf{a}_i$	vector of abundance fractions	$\mathbf{m}_q$	endmember
		Q	number of endmembers
q	endmember index		
SVM and kernels			
$\mathcal{L}_p$	Lagrangian primal problem	$\mathcal{L}_d$	Lagrangian dual problem
$R_{\text{emp}}(f)$	empirical risk	$V(f(\mathbf{x}), y)$	cost function
$\mathcal{H}$	Hilbert space	$\boldsymbol{\phi}, \boldsymbol{\varphi}$	mapping functions to $\mathcal{H}$
$\mathbf{w}$	weight vector	b	bias term
ξ	slack variables	λ, C	regularization parameter
ϵ	insensitivity zone	$\boldsymbol{\alpha}, \boldsymbol{\beta}, \boldsymbol{\eta}$	Lagrange multipliers
$\langle \mathbf{x}, \mathbf{z} \rangle$	inner product of $\mathbf{x}$ and $\mathbf{z}$	$K(\mathbf{x}, \mathbf{z})$	kernel function $\langle \boldsymbol{\phi}(\mathbf{x}), \boldsymbol{\phi}(\mathbf{z}) \rangle$
$\mathbf{K}$	kernel matrix	$\mathbf{L}$	graph Laplacian
S	subset of training set	$\boldsymbol{\Phi}(S)$	S in kernel space
$\boldsymbol{\phi}_\mu(S)$	mean of S in kernel space	$\boldsymbol{\phi}_\mu$	mean map
σ	Gaussian width	R	radius containing the data

List of abbreviations

AATSR	Advanced Along Track Scanning Radiometer
AIC	Akaike's Information Criterion
ATGP	Automated Target Generation Process
ATLAS	ATmospheric Laboratory for Applications and Science mission
AVHRR	Advanced Very High Resolution Radiometer
AVIRIS	Airborne Visible/Infrared Imaging Spectrometer
BRDF	Bidirectional Reflectance Distribution Function
BRF	Bidirectional Reflectance Factor
CASI	Compact Airborne Spectrographic Imager
CCA	Canonical Correlation Analysis
CCD	Charge Coupled Device
CHRIS	Compact High Resolution Imaging Spectrometer
COT	Cloud Optical Thickness
CTP	Cloud Top Pressure
CVA	Change Vector Analysis
DB	Davies–Bouldin index
DEM	Digital Elevation Model
DN	Digital Number
DOY	Day Of Year
EM	Expectation-Maximization
EMR	ElectroMagnetic Radiation
EnMAP	Environmental Mapping and Analysis Program
ENVISAT	ENVIronmental SATellite
EO	Earth Observation
ERM	Empirical Risk Minimization
ERS	European Remote Sensing satellite
ESA	European Space Agency
EURECA	EUropean Retrieval CArrier mission
FCLSU	Fully Constrained Linear Spectral Unmixing
FOV	Field Of View
FR	Full Resolution
GLRT	Generalized Likelihood Ratio Test

GMES	Global Monitoring for Environment and Security
GMM	Gaussian Mixture Model
ICA	Independent Component Analysis
ISMD	Invariant Subpixel Material Detection
IR	InfraRed
KISD	Kernel Invariant Subpixel Detection
KM	Kernel Method
KPCA	Kernel Principal Component Analysis
LAI	Leaf Area Index
LapSVM	Laplacian Support Vector Machine
LCA	Land Cover Abundances
LCC	Land Cover Classification
LDA	Linear Discriminant Analysis
LPF	Low Pass Filter
LSU	Linear Spectral Unmixing
LSMM	Linear Spectral Mixture Model
LUT	Look-Up Table
MAE	Mean Absolute Error
MAP	Maximum *A posteriori* Probability
MDL	Minimum Description Length criterion
ME	Mean Error
MERIS	MEdium Resolution Imaging Spectrometer
MFD	Matched Filtering Detector
MIR	Middle-InfraRed
MLC	Maximum Likelihood Classifier
MODIS	MOderate Resolution Imaging Spectroradiometer
MODTRAN	MODerate resolution TRANsmittance
NASA	National Aeronautics and Space Administration
NDVI	Normalize Difference Vegetation Index
NDOI	Normalized Difference at Oxygen-A absorption Index
NIR	Near InfraRed
NN	Neural Network
OA	Overall Accuracy
ODH	Optimal Decision Hyperplane
OLCI	Ocean and Land Colour Instrument
OSP	Orthogonal Subspace Projection
PCA	Principal Component Analysis
PLS	Partial Least Squares
pdf	Probability Density Function
POLDER	POLarization and Directionality of the Earth's Reflectances
PROBA	PRoject for On-Board Autonomy
RAA	Relative Azimuth Angle
RAZ	Relative Azimuth Angle
RBF	Radial Basis Function
RGB	Red-Green-Blue
RKHS	Reproducing Kernel in Hilbert Space
RMSE	Root Mean Square Error

ROC	Receiver Operating Characteristic
ROI	Regions Of Interest
RR	Reduced Resolution
RT	Radiative Transfer
RTM	Radiative Transfer Model
SAA	Sun Azimuth Angle
SAD	Spectral Angle Distance
SLST	Sea and Land Surface Temperature
SNR	Signal-to-Noise Ratio
SPECTRA	Surface Processes and Ecosystem Changes Through Response Analysis
SRF	Spectral Response Function
SRM	Structural Risk Minimization
SSL	Semi-Supervised Learning
SV	Support Vector
SVDD	Support Vector Domain Data Description
SVM	Support Vector Machine
SVR	Support Vector Regression
SWIR	ShortWave InfraRed
SZA	Sun Zenith Angle
TM	Thematic Mapper
TIR	Thermal InfraRed
TOA	Top Of Atmosphere
TSVM	Transductive Support Vector Machine
UV	Ultra Violet
VIS	Visible
VISAT	VIsualization and Analysis Tool
VNIR	Visible and Near Infra Red

I

Introduction

1

Machine learning techniques in remote sensing data analysis

Björn Waske[1], Mathieu Fauvel[1,2], Jon Atli Benediktsson[1] and Jocelyn Chanussot[2]

[1] *Faculty Electrical and Computer Engineering, Univ. of Iceland, Iceland*
[2] *Signal & Image Department, Grenoble Institute of Technology, France*

Several applications have been developed in the field of remote sensing image analysis during the last decades. Besides well-known statistical approaches, many recent methods are based on techniques taken from the field of machine learning. A major aim of machine learning algorithms in remote sensing is supervised classification, which is perhaps the most widely used image classification approach. In this chapter a brief introduction to machine learning and the different paradigms in remote sensing is given. Moreover this chapter briefly discusses the use of recent developments in supervised classification techniques such as neural networks, support vector machines and multiple classifier systems.

1.1 Introduction

1.1.1 Challenges in remote sensing

Owing to the recent development of different Earth observation platforms with increased spatial and spectral resolution as well as higher revisit times, remote sensing provides more detailed information on land cover and the environmental state than ever before. Moreover, different Earth-observation systems, such as multi-spectral and SAR systems operate in different wavelengths, ranging from visible to microwave. The data sets consequently provide different, but complementary information. The classification of such data might be considered

Kernel Methods for Remote Sensing Data Analysis Edited by Gustavo Camps-Valls and Lorenzo Bruzzone

complex on the one hand, but with regard to recent and upcoming missions, remote sensing applications become even more attractive, on the other.

Many early techniques have been taken directly from signal processing and these methods are often based on simple data models and approaches. However, when dealing with recent data sets these well-known classifiers can be limited (Richards 2005). In addition, increased performance requirements such as speed (e.g., for operational monitoring systems and near-real time applications) and accuracy, further demand the development of more sophisticated analysis concepts (Jain *et al.* 2000). Thus, the development of adequate methods for various data sets is an important ongoing research topic in the field of remote sensing. This chapter is organized as follows. In Section 1.1.2 a general introduction to machine learning is given, followed by a discussion on different paradigms in remote sensing. In Section 1.2 various supervised classifiers are introduced. A conclusion is given in Section 1.3.

1.1.2 General concepts of machine learning

Machine Learning is an area of artificial intelligence and generally refers to the development of methods that optimize their performance iteratively by *learning from the data*. Such methods can be predictive (e.g., a regression model) and make a prediction of a specific phenomenon or descriptive (e.g., a classification model) and distinguish for example between different classes of patterns. In the field of remote sensing, descriptive machine learning algorithms often focus on land cover classifications, and thus provide important input information in several environmental monitoring systems, as for instance in the area of flood forecast, urban sprawl and land degradation. In this context, we are dealing with the differentiation between several land cover classes, and the algorithm learns to differentiate between different types of patterns (i.e. land cover classes).

Let us assume the detection of a specific land cover class, e.g., buildings, to be a general machine learning problem. The corresponding machine learning formulation will be:

$$\text{Find } f : \mathbf{x} \mapsto y = f(\mathbf{x}), \tag{1.1}$$

where $\mathbf{x}$ is a feature vector from the image and y is a scalar that indicates the presence ($y = 1$) or the absence ($y = 0$) of a building $\mathbf{x}$. Note that the values chosen for y are arbitrary. Machine learning theory aims at estimating the functional f from some prior data. Several techniques have been introduced to achieve this goal. They can be separated into several categories, e.g., depending on the type of the training data, the form of function $f(\mathbf{x})$ and potential assumptions on the underlying density function (Figure 1.1). Two main categories, *unsupervised* and *supervised*, based on whether they include a priori knowledge during the decision process by using labelled training samples or not. In contrast to this, *semi-supervised* concepts are a special type that rely on both labelled and unlabelled samples.

1. *Supervised algorithms*: These correspond to the situation where a set of labelled samples, $\{(\mathbf{x}_i, y_i)\}_{i=1}^{\ell}$, our training data, is available and the objective is to predict the value y corresponding to a new sample $\mathbf{x}$, i.e. determining the class-memberhsip of x.

2. *Unsupervised algorithms*: These correspond to the situation where only the data, *i.e.*, $(\mathbf{x}_j)_{j=1}^{\ell}$, are known and the objective is to describe how the data are organized or clustered.

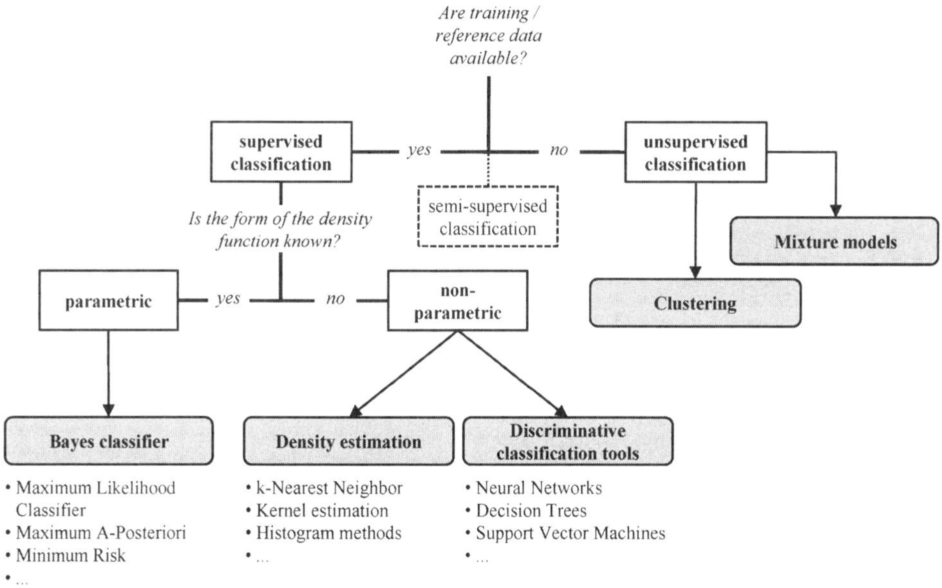

Figure 1.1 Overview of different classifier categories (after Jain et al. 2000).

3. *Semi-supervised algorithms*: In this case, the two previous approaches are combined. The learning is based not only on the available set of labelled samples $\{(\mathbf{x}_i, y_i)\}_{i=1}^{\ell}$, but also on some additional data $(\mathbf{x}_j)_{j=\ell+1}^{\ell+u}$ for which no prior knowledge y_j is available.

In a supervised setting, (1.1) can be reformulated as

$$\text{using } (\mathbf{x}_i, y_i)_{i=1}^{\ell}, \text{ find } f : \mathbf{x} \mapsto y = f(\mathbf{x}). \tag{1.2}$$

In order to find the best functional f, and consequently to optimize the detection performances, an additional constraint needs to be included in the formulation of the problem. Denoting $E[f]$ as a measure of error, i.e. the prediction $f(\mathbf{x})$ differs from the truth y, (1.2) should be rewritten as:

$$\text{Using } (\mathbf{x}_i, y_i)_{i=1}^{\ell}, \text{ find } f : \mathbf{x} \mapsto y = f(\mathbf{x}) \text{ for which } E[f] \text{ is minimal.} \tag{1.3}$$

Depending on $f(\mathbf{x})$ the machine learning algorithms can be further differentiated into two subgroups:

1. *Linear algorithms*: The variables $\mathbf{x}$ and the output y are assumed to be linearly related and f is an affine function $f(\mathbf{x}) = \langle \mathbf{w}, \mathbf{x} \rangle + b$.

2. *Nonlinear algorithms*: The relation between $\mathbf{x}$ and the output y is nonlinear and f can be any function. On example is the quadratic function $f(\mathbf{x}) = \mathbf{x}^\top \mathbf{A}\mathbf{x} + \mathbf{b}^\top \mathbf{x} + c$.

When using nonlinear algorithms, the user hence needs to choose the set of functions that will be considered. Depending on the application and on the level of the user's knowledge, this choice may be difficult. For instance, if it is assumed that grey-level of building pixels follow a Gaussian distribution $\mathcal{N}_1(\boldsymbol{\mu}_1, \boldsymbol{\Sigma}_1)$ and background pixels follow $\mathcal{N}_2(\boldsymbol{\mu}_2, \boldsymbol{\Sigma}_2)$, where Σ and μ denote the variance–covariance matrix and mean vector, respectively. Following a Bayesian framework (see Section 1.2.1), the decision function can be derived as follows (Duda *et al.* 2001):

$$f(\mathbf{x}) = \begin{cases} 1 \text{ if } -\frac{1}{2}(\mathbf{x} - \boldsymbol{\mu}_1)^\top \boldsymbol{\Sigma}_1(\mathbf{x} - \boldsymbol{\mu}_1) + \frac{1}{2}(\mathbf{x} - \boldsymbol{\mu}_2^\top)\boldsymbol{\Sigma}_2(\mathbf{x} - \boldsymbol{\mu}_2) - \ln\left(\frac{|\boldsymbol{\Sigma}_1|}{|\boldsymbol{\Sigma}_2|}\right) > 0 \\ 0 \quad \text{otherwise.} \end{cases}$$

The training samples are used to estimate the different parameters ($\boldsymbol{\mu}_1$, $\boldsymbol{\Sigma}_1$, $\boldsymbol{\mu}_2$ and $\boldsymbol{\Sigma}_2$). Such approaches, which assume that $p(\mathbf{x}|y)$ follows a specific distribution, are called *parametric approaches*. In many remote sensing applications it is assumed that the distributions follow the form of a multivariate normal Gaussian model. In addition, many processes can be described by this model assumption and it is computationally relatively simple because the model is only described by the mean and the covariance matrix. However, when dealing with multisensor and multi-temporal data sets, the class distributions can often not be modelled by adequate multivariate statistical models. Recall our general machine learning application, i.e. the detection of buildings in a remote sensing data set. What if the background pixels do not follow a Gaussian distribution or we would like to include additional information, such as texture or shape descriptors? In these cases a new model $p(\mathbf{x}|y)$ should be defined.

Non-parametric approaches seem particularly interesting in this context because they are not constrained to prior assumptions on the distribution of input data. *Kernel methods*, for example, enable the definition of general functions that can be tuned directly during the training step. These functions typically have the following form:

$$f(\mathbf{x}) = \sum_{i=1}^{\ell} \alpha_i K(\mathbf{x}_i, \mathbf{x}) + b, \tag{1.4}$$

where k is a kernel function and $\left((\alpha_i)_{i=1}^{\ell},\ b\right)$ are the parameters of the decision function.

1.1.3 Paradigms in remote sensing

Machine learning in remote sensing comprises several different paradigms such as classification, regression, clustering, feature extraction, dimensionality reduction and density estimation. These aspects are often interdependent, e.g., before performing a classification one might extract some additional texture features and also reduce the dimensionality of the data set with feature selection techniques (Figure 1.2). Perhaps the most commonly undertaken applications in remote sensing are feature reduction, clustering and classification. In the following, a brief formal description of these machine learning concepts is given.

Feature extraction and selection

Terms such as feature extraction and feature selection are closely related to dimensionality reduction, which refers to the mapping of the data from the original feature space into a space of a lower dimension, without discarding any meaningful information. Furthermore, *meaningful*

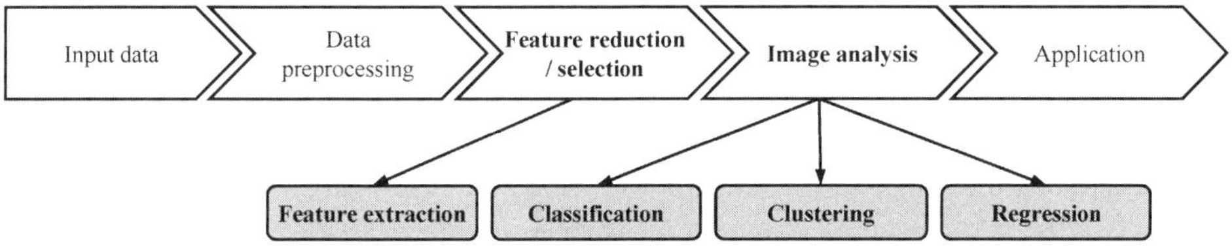

Figure 1.2 Overview on a computational remote sensing application. Bold lettering indicates where machine learning approaches are most effective.

information is defined according to the final application. Hyperspectral imaging, for example, is characterized by high spectral resolution with up to hundreds of data channels with narrow bands, ranging from the visible to the short-wave infrared region of the electromagnetic spectrum. Although such data provide detailed spectral information, theoretical and practical problems arise with increasing dimensionality of the data in the spectral domain. The idea of the *dimension* is intuitive, driven by experiments in one-, two- or three-dimensional space, and geometric concepts that are self-evident in these spaces do not necessarily apply in higher-dimensional space (Kendall 1961; Landgrebe 2003). For example, normally-distributed data have a tendency to concentrate in the tails, which seems to contradict its bell-shaped density function. For the purpose of classification, these problems are related to the *curse of dimensionality*. In particular, Hughes showed that with a limited training set, beyond a certain limit, the classification accuracy decreases as the number of features increases (Hughes 1968). This is paradoxical, since with a higher spectral resolution one can discriminate more classes and have a finer description of each class, but the data complexity leads to poorer classification.

To mitigate this phenomenon, a dimensionality reduction is performed by *feature selection* and *feature extraction*. A feature reduction algorithms can be supervised or unsupervised. The objective is to find f:

$$\begin{aligned} f : \mathbb{R}^n &\rightarrow \mathbb{R}^d \\ \mathbf{x} &\mapsto \tilde{\mathbf{x}} \end{aligned} \tag{1.5}$$

with $d \ll n$.

Feature selection refers to a selection of a subset of relevant features, whereas feature extraction combines and transforms the original features–both in order to obtain a relevant representation of the data in a lower-dimensional space. Such strategies were initially designed in accordance with both the specific characteristics of the remote sensors and the objectives (e.g., crop monitoring). Well-known extraction methods are vegetation indices and transformations such as the *Tasseled Cap Transformation* (Kauth and Thomas 1976). The Tassled Cap was proposed to underline significant spectral characteristics in crop development in a multi-spectral Landsat image, to improve the differentiation between crops and other vegetation types. However, such transformations are sensor-dependent and the physical analysis associated with each transformation can be intractable for hyperspectral data.

Transformations based on statistical analysis have already proved to be useful for classification, detection, identification or visualization of remote sensing data (Chang 2003; Keshava 2004; Richards and Jia 1999; Ünsalan and Boyer 2004). Two main approaches can be defined. (1) *Supervised* transformation is in general well suited to pre-processing for the task of classification, since the transformation improves class separation. However, its effectiveness is correlated with how well the training set represents the whole data set. Moreover, this transformation can be extremely time-consuming. *Unsupervised* methods do not focus on class discrimination, but aim for another representation of the data in a lower-dimensional space, satisfying some given criterion. For Principal Component Analysis (PCA), the data are projected into a subspace that minimizes the reconstruction error in the mean square sense. Note that both the unsupervised and supervised cases can be also divided into *linear* and *nonlinear* algorithms.

PCA plays an important role in the processing of remote sensing images. Even though its theoretical limitations for hyperspectral data analysis have been pointed out (Landgrebe 2003; Lennon 2002), in a practical situation the results obtained using the PCA are still competitive for the purpose of classification (Journaux *et al.* 2006; Lennon *et al.* 2001). The advantages of the PCA are its low complexity and the absence of parameters. However, the PCA only considers the second-order statistic, which can limit the effectiveness of this method. Part IV of this book presents a nonlinear version of the PCA, namely *Kernel Principal Component Analysis* (KPCA), which considers higher-order statistics.

Clustering algorithms

In contrast to supervised approaches, which construct the decision boundaries from training data, clustering algorithms are based on a set of unlabelled data. In general, cluster algorithms aim to indentify data (unknown) structure, such as natural groups or clusters within the multi-dimensional feature space by measuring similarities between different pixels. The pixels within a cluster or group are more similar to each other than those pixels belonging to other clusters (Jain *et al.* 2000).

As for classification, the goal is to learn a function f

$$\begin{aligned} f : \mathbb{R}^n &\rightarrow \mathbb{N} \\ \mathbf{x} &\mapsto \hat{c}, \end{aligned} \tag{1.6}$$

where no labelled data are available and the training relies only on the observations $\mathbf{x}$.

Thanks to kernel functions, the similarity can be computed easily (Hofmann *et al.* 2008). Moreover, it can be adapted to the data by using an appropriate kernel. Nevertheless, the most common similarity measurement is the Euclidean distance. However, when dealing with data that contains clusters of different shapes and sizes the definition of clusters and the selection of an adequate similarity measure are critical and thus the clustering process can be difficult (Jain *et al.* 2000). After measuring the similarity, the cluster algorithm merges the individual pixels into groups. Of course, the same data set can result in several potential combinations of clusters and some kind of accuracy assessment is required to evaluate the quality of the generated clusters. A common quality measurement is the squared error measure that is also used as the cluster criterion. Clustering techniques are also used for unsupervised classifications. After generating groups of pixels with similar properties, the user usually assign a class label to each cluster. Clustering approaches, such as ISODATA and k-means are well known in remote sensing. They usually assign each pixel to only one cluster (*hard techniques*), whereas *fuzzy approaches* as the fuzzy c-mean results in a degree of membership for all clusters.

Probably the widest used is the ISODATA (iterative self-organizing data analysis). In contrast to the k-means clustering, the approach enables an optimization of the individual clusters (e.g., splitting and merging) during the clustering process. The basic algorithm might be described as follows.

1. *Initialization*: The number of clusters is defined (by the user) or estimated. For each cluster a pixel is (often randomly) selected as cluster centre.
2. *Clustering process*: Each pixel is assigned the nearest cluster, using the distance measurement (e.g. *Euclidean distance*) between the pixel and the cluster centre.
3. *Test*: Compute the new cluster centres and compare them with the previous centres. If a centre has changed return to step 2, otherwise the procedure is terminated.
4. *Optimization*: Optimize clusters (at the end and any suitable processing step) by merging clusters that are close together or contain only a few pixels and splitting incompact clusters.

Other approaches as the *agglomerative hierarchical clustering* techniques do not rely on specific cluster centres. In the beginning each pixel is handled as a separate cluster. During the process neighbouring pixels (or clusters) are sequentially merged, depending on a distance measurement. The process of clustering can be represented in a tree diagram or so-called *dendogram*. This figure shows how the pixels are grouped and at what distance measurement individual pixels and clusters are merged. In addition to agglomerative algorithms that sequentially reduce the number of clusters, by merging similar groups (bottom-up approach), divisive techniques procedures start with all the pixels belonging to one cluster and split it iteratively into smaller clusters (bottom-down approach). An alternative clustering approach is based on *mixture models*. This approach is based on the assumption that each natural group is described by a different probability distribution. Whereas the form of the distribution is similar for all groups, the parameter values are different. In addition, a mixture can contain different models and more complex density functions can be modelled (Jain *et al.* 2000).

Supervised classification algorithms

The major part of machine learning algortihms for remote sensing image analysis is aiming perhaps for a supervised classification of the data. The machine learning problem is to learn the function f, from a set of labelled examples, i.e. our training samples whose class membership is known:

$$\begin{aligned} f : \mathbb{R}^n &\to \mathbb{N} \\ \mathbf{x} &\mapsto \hat{c}, \end{aligned} \tag{1.7}$$

which assigns each sample to a particular class $\hat{c}$. In general such learning problems are ill-posed and it is necessary to restrict the space of possible functions f. Using a Bayesian framework, this is done by choosing an appropriate prior density function. In Section 1.2 different supervised classifiers are introduced.

In machine learning, a trade-off is made between how well the function fits the data and how complicated the function is. Several measures can be used to assess the complexity of a function. One that has played an important role in the last decades is the so-called

VC-dimension (Vapnik 1998). The trade-off leads to solving the following general minimization problem:

$$\min_f \frac{1}{\ell} \sum_{i=1}^{\ell} V(f(\mathbf{x}_i), c_i) + \lambda \|f\|^2, \tag{1.8}$$

where the norm $\| \cdot \|$ is a measure of complexity (low norm corresponding to a simple function) and V is a loss function. For instance, $V(f(\mathbf{x}_i), c_i) = \max(0, 1 - f(\mathbf{x}_i)c_i)$ for the SVM. Note that the resolution of (1.8) can involve supervised or semi-supervised algorithms. Different supervised classifiers are introduced in Section 1.2, semi-supervised kernel algorithms are discussed in Chapters 9 and 10.

Regression problem are very similar to classification problem. The only change occurs for the outputs of the function that is now defined on $\mathbb{R}$. Actually, many classification algorithms can be turned to regression ones by a simple change in the loss function: for the SVM, the loss function is changed to the ϵ-insensitive function (Vapnik 1998). As for classification, the regression may be supervised or semi-supervised. In Chapter 13 the use of support vector machines for regression is discussed.

1.2 Supervised classification: algorithms and applications

1.2.1 Bayesian classification strategy

Classification approaches that rely on the Bayesian framework are one of the basic concepts in (statistical) pattern recognition. The (Gaussian) maximum likelihood classifier, which is derived from the Bayes rule when classes have equal priors, is doubtlessly the most common supervised classification technique in the field of remote sensing. In contrast to geometrical and discriminant analysis-based approaches, Bayesian concepts are density based and assume specific density functions for each information class.

Following the Bayesian decision strategy a pixel $\mathbf{x}$ is classified to class y_i if $p(y_i|\mathbf{x}) > p(y_j|\mathbf{x})$, for all $i \neq j$, where $p(y_i|\mathbf{x})$ is called the *a-posteriori probabilities* and refers to the probability that $\mathbf{x}$ belongs to class $p(y_i|\mathbf{x})$. In this context it is assumed that our feature vector $\mathbf{x}$ have a specific probability density, that is dependent on the observed information class. The pixel $\mathbf{x}$ belonging to class y_i is understood as an observation drawn randomly from the *class-conditional probability density function* $p(\mathbf{x}|y_i)$. Following *Bayes' theorem*, the required a-posteriori probabilities can be computed by the *class-conditional probability density function* or *likelihoods* and the *a-priori probabilities*, which describe the probability that a specific class occurs in the data set.

Generally, the Bayesian decision rule is aiming at minimizing the conditional risk and expected loss, respectively, that is associated with each potential classification. A so-called loss function enables a weighting of the different classifications, which can be important if the costs of misclassification vary with different types of classes. However, often a so-called 0/1 loss function is used, which weights all misclassifications equally.

Thus, the final decision rule is based only on the a-posteriori probabilities. This classification rule is also referred to as the maximum a posteriori rule (MAP). However, the conditional probability density function $p(\mathbf{x}|y_i)$ are generally not known and must be estimated from the training data. Often (particularly in remote sensing) it is assumed that the data follow the form of a multivariate normal (Gaussian) model, which is an adequate assumption for many remote sensing data sets. Moreover the normal model is defined by only two parameters, the mean and the covariance matrix, which are derived from the training data. However, the use of a Bayesian classifier might be critical if an adequate multivariate statistical model is not available and the number of training samples is limited (with respect to the dimensionality of our data set). In such cases, the performance of the classifier can be affected, resulting in a lower classification accuracy. However, if the form of the density function is unknown, it is possible to operate in a non-parametric mode: Approaches such as histogram methods, k-nearest neighbour and kernels methods enable an estimation of the class-conditional densities and, thus, a classification in the context of the Bayesian framework.

1.2.2 Neural networks

In the past there has been great research focus on neural networks (NN) in the context of pattern recognition and remote sensing (Atkinson and Tatnall 1997; Benediktsson *et al.* 1990). In several studies the performance of NN was compared with widely-used statistical methods, as for the classification of multi-spectral (Bischof *et al.* 1992; Paola and Schowengerdt 1995), SAR (Bruzzone *et al.* 2004; Del Frate *et al.* 2003), hyperspectral (Benediktsson *et al.* 1995) and multi-source data sets (Benediktsson *et al.* 1990). The results in several of these studies demonstrate that NN approaches perform better or at least equally as well as conventional approaches. In the context of remote sensing image analysis, several different NN models have been proposed (Benediktsson *et al.* 1997, 1990; Serpico *et al.* 1996). A general introduction to neural networks in the context of pattern recognition is given for example in Bishop (1995), an overview on remote sensing applications can be found in Atkinson and Tatnall (1997).

Like other machine learning techniques, neural networks have an advantage over traditional statistical methods in that they are distribution-free, i.e. they do not rely on any underlying statistical distribution of the data. In the context of classification, a neural network can be considered as a black box model that receives a set of input vectors and produces responses from its output neurons, where the number of neurons depends on the number of information classes (i.e. land cover classes). A neural network is an interconnection of neurons, where a neuron receives input signals that represent the activity at the input or the momentary frequency of neural impulses delivered by another neuron to this input. The output value or the frequency of the neuron is often represented by a function, which is controlled, among others factors, by the so-called synaptic efficacies or weights. The weights are modified by an adaptive training process in which a set of training samples is presented at the input. The network gives an output response for each sample. The actual output response is compared with the desired response for the sample and the error difference between the desired output and the actual output is used to modify the weights in the neural network. The training process ends when the error is reduced to a pre-specified threshold or cannot be minimized any further (i.e, the network structure is stabilized and the weights do not change or are less that a specific threshold).

Neural network approaches enable the determination of the weights for each data source during the classification procedure, which makes them particularly valuable for multi-source applications (e.g., multi-temporal and multisensor land cover classifications). In this context an adequate multivariate statistical model is often not known, and traditional statistical approaches can be limited. However, the NN approach can be computationally complex and requires a large number of training samples. Moreover, compared with statistical methods, the approach has more problems, classifying unknown pixels that are not identical to training samples.

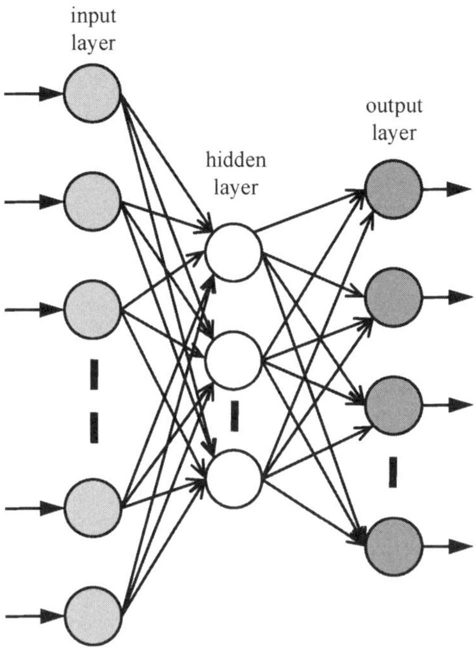

Figure 1.3 Schematic diagram of a fully connected feed forward neural network.

Consequently the performance of a neural network strongly depends on representative training samples, whereas statistical models require an appropriate model for each class (Benediktsson *et al.* 1990). Whereas the performance of networks with a single layer of adaptive weights is limited, the use of several layers enables a more general application. It has been shown that with two hidden layers a neural network can approximate any function in the mean square sense. However, in remote sensing applications, one hidden layer is most often sufficient. A schematic overview of a fully connected feed forward neural network is given in Figure 1.3.

Several different neural network algorithms have been proposed. A widely used approach, for instance, is a multilayer perceptron network (MLP) with an error backpropagation algorithm. It is a multilayer feedforward algorithm that can be used to discriminate data that are not linearly separable. During the training process, the error from the outputs of the network is propagated backwards and the weights are changed in order to reduce this error. However this method has some limitations, e.g., the slow convergence of the error backpropagation, the potential convergence to a local minimum, and the incapacity to detect that an input sample has fallen in a region of the feature space without training data (Bruzzone *et al.* 2004; Serpico *et al.* 1996). This is a serious weakness, especially when the dimensionality of the data is very high. Therefore, neural networks require an efficient feature extraction method when dealing with very high dimensional data sets.

A radial basis function (RBF) network can solve some of the problems mentioned above and the training can be much faster (Bishop 1995; Bruzzone and Prieto 1999). The RBF approach is based only on a single hidden layer and is considered as a design for a curve-fitting problem in a high-dimensional space. In regard to this, learning is comparable to defining a surface in a multi-dimensional space that provides a best fit to the training samples, with the criterion for best fit being measured in some statistical sense. In the RBF neural network, the hidden units provide a set of functions that constitute an arbitrary basis for the input patterns when they are expanded into the hidden unit space (radial basis functions). The transformation from the input space to the hidden-unit space is nonlinear, whereas the transformation from the hidden-unit space to the output space is linear.

Another development is the family of ART (Adaptive Resonance Theory) and ARTMAP networks, proposed by Carpenter and Grossberg (Carpenter *et al.* 1997, 1991, 1992), which is different from other networks, such as the backpropagation algorithms. The original ART framework is an unsupervised approach that groups pixels into stable clusters. The number of clusters depends on the so-called vigilance parameter. Using a distance metric vigilance defines how outspread individual groups might be distributed in the feature space. The ARTMAP approach, an supervised extension of ART, is iteratively modifying the vigilance parameter in case of a misclassification (Muchoney and Williamson 2001).

The last decade has seen the emergence of more efficient supervised machine learning algorithms: *Regularization networks* (RN) and *support vector machines* (SVM) (Girosi *et al.* 1995; Vapnik 1998). While classical neural networks and radial basis classifiers learn the classification problem by minimizing the *Empirical Risk* (i.e. the number of errors made on the training set), the RN and the SVM implement Structuring Risk Minimization (SRM): the optimal solution is found by selecting the function that minimizes both the Empirical Risk and a measure of the complexity of the function.

The RN and SVM can be described using the framework of VC theory (Evgeniou *et al.* 1999): for instance the use of kernel functions in Reproducing Kernel Hilbert Space (RKHS) as regularizer and SRM principle. However, SVM and RN do not provide equal solution since they do not use the same loss functions to compute the training errors (Evgeniou *et al.* 1999). Classically, the RN uses the square errors while SVM uses the hinge loss (Chapelle 2007). It implies that SVM finds solutions that emphasize the separating surface and provide very good results in real world problem.

SVM have been intensively studied in remote sensing. The following section reviews some remote sensing applications of SVM.

1.2.3 Support Vector Machines (SVM)

Support Vector Machines (SVM) were introduced to context pattern recognition and machine learning and are a relatively recent development in the context of remote sensing and still exhibit further improvement and modification, e.g., in the context of semi-supervised approaches (Bruzzone and Carlin 2006; Chi and Bruzzone 2007) and the introduction of new kernel concepts (Camps-Valls *et al.* 2008, 2006). SVM were originally introduced as binary classifiers. The concept is based on an optimal linear separating hyperplane that is fitted to the training patterns of two classes within a multi-dimensional feature space. The optimization problem that has to be solved relies on structural risk minimization and is aiming at a maximization of the margins between the hyperplane and the closest training samples. For linearly not separable samples the input data are mapped into a high dimensional space. In doing so a linear hyperplane can be fitted, which is nonlinear in the original input space. The computationally extensive mapping process is handled by using a positive definite kernel function. This kernel-trick enables us to work within the newly transformed feature space, without explicitly knowing the map but only the kernel function.

In the context of remote sensing, binary classification problems generally do not exist, thus the binary nature of SVM requires a method to solve multi-class problems. The most frequently used approaches are the one-against-one rule, which separates each pair-wise class-combination, and the one-against-all, which is based on the separation of each class from the rest. For a detailed introduction to SVM the reader is referred to the following chapters.

SVMs have been used successfully in several remote sensing studies (Foody and Mathur 2004; Huang *et al.* 2002; Pal and Mather 2005). In many studies they performed more accurately than other classifiers or performed at least equally well. Because the approach considers only samples close to the class boundary, it works well with small training sets, even when high dimensional data sets are classified (Melgani and Bruzzone 2004; Pal and Mather 2006). In van der Linden *et al.* (2007) the use of SVM on segmenting HyMap data urban areas was discussed. SVM achieved high overall accuracies using a purely pixel-based classification, underlining the high efficiency of SVM to describe complex class distributions. The approach differentiated broad thematic classes without a preliminary definition of spectrally homogeneous sub-classes or separate treatment of dark areas. In another study a framework for the classification of multi-spectral and hyperspectral data was presented (Farag *et al.* 2005), which is based on the maximum a posteriori (MAP) concept. The MAP approach aims on maximizing the posterior probabilities that are represented as the product of the class prior probability and the class conditional probability (CCP). In the proposed approach, SVM were used to estimate the required CCP.

In other experiments, spectral and spatial image information were combined to improve the classification accuracy (Camps-Valls *et al.* 2006; Fauvel *et al.* 2006). An alternative approach for classifying very high resolution urban imagery is based on the use of different segmentation results at various scales (Bruzzone and Carlin 2006). The result demonstrates that different levels provide different types of information, which seems particularly interesting regarding high- resolution images.

Although SVM achieved promising results in several studies, SVM was rarely applied to multi-source classification problems. Whereas some of these application are based on the use of the original SVM and common kernel functions (Koetz *et al.* 2008; Song *et al.* 2005), in other experiments the SVM approach was extended and modified for classifying multi-source data sets. In Halldorsson *et al.* (2003) a common kernel function was extended for classifying a multi-source data set containing Landsat MSS data and topographical information. In Camps-Valls *et al.* (2006) so-called composite kernels were introduced, for combing contextual and spectral information from hyperspectral imagery.

Another approach for combining spatial and spectral information was proposed by Fauvel *et al.* (2008). It is based on the use of both the spatial and spectral information for classification. The spatial information is modelled by the spatial response to morphological filters while the spectral information is the spectral components themselves. The approach was successfully applied on very high spatial resolution hyperspectral images.

In another study, SVM were used for combining a set of multi-temporal SAR data and multi-spectral imagery (Waske and Benediktsson 2007). The overall accuracy was further improved by training individual SVM classifiers on the two data sources and fusing the preliminary outputs (i.e. distances to the hyperplane) by a third SVM. This concept was further extended by a multilevel component (Waske and van der Linden 2008). In doing so they include different segmentation scales from each source, i.e. the SAR data and multi-spectral imagery. The main reason for the success of these strategies could be the different nature of the used data types, i.e. the multi-spectral imagery and the SAR data. The multi-temporal

SAR data and the multi-spectral images provide different information and may not be equally reliable. Furthermore, different aggregation levels contribute unequally to the classification of the various classes and hence the sources do not appear equally reliable. The application of a single SVM for the whole heterogeneous data sets requires the definition of one single kernel function. However, it seems more adequate to define the kernel functions for each data source separately and fuse the derived outputs

In the context of multi-temporal and multi-source classification by SVM, new kernel functions were proposed in (Camps-Valls *et al.* 2008). In that study a general framework based on composite kernels for remote sensing data was presented that combines spectral, spatial, multi-temporal and multisensor information. The classification results underline the good performance of these concepts, which outperform a traditional stacked-vector approach in terms of accuracy. Other developments in the context of SVM and remote sensing are the introduction of semi-supervised techniques (Bovolo *et al.* 2008; Bruzzone *et al.* 2006; Chi and Bruzzone 2007).

Toy classification example: kernel method versus parametric method

To illustrate the ability of kernel approaches to deal with hyperspectral remote sensing data, a toy classification example is presented. We compare three techniques: one parametric, the quadratic Bayesian classifier and two machine learning algorithms, one naive, the minimum distance to the mean classifier and the most powerful SVM. The decision rule are:

1. *Minimum distance to the mean classifier*: Samples are assigned to whichever class has the smallest Euclidean distance to its mean. The decision boundary is

$$\left\{\mathbf{x}|2\mathbf{x}^{\top}(\boldsymbol{\mu}_{-}-\boldsymbol{\mu}_{+})+(\boldsymbol{\mu}_{+}^{\top}\boldsymbol{\mu}_{+}-\boldsymbol{\mu}_{-}^{\top}\boldsymbol{\mu}_{-})=0\right\}. \tag{1.9}$$

2. *The linear SVM classifier*: We use the soft margin formulation, where the decision boundary is

$$\left\{\mathbf{x}\left|\sum_{i=1}^{\ell}\alpha_i y_i\langle\mathbf{x}^i,\mathbf{x}\rangle+b=0\right.\right\}. \tag{1.10}$$

3. *The quadratic Bayesian classifier*: Under the Gaussian assumption and with the same covariance matrix, the decision boundary is

$$\left\{\mathbf{x}|\left((\boldsymbol{\mu}_{+}-\boldsymbol{\mu}_{-})^{\top}\boldsymbol{\Sigma}^{-1}\right)\mathbf{x}-\frac{1}{2}(\boldsymbol{\mu}_{+}^{\top}\boldsymbol{\Sigma}^{-1}\boldsymbol{\mu}_{+}-\boldsymbol{\mu}_{-}^{\top}\boldsymbol{\Sigma}^{-1}\boldsymbol{\mu}_{-})=0\right\}. \tag{1.11}$$

To test the behaviour of SVM in the 'high dimensional space / small training set' situation, a typical setting in remote sensing, two Multivariate Gaussian clusters were generated with increasing dimension, and classification was performed using the three classifiers, but with a fixed training set of 40 samples per class. For each dimension, 100 experiments with 1000 test samples per class were performed and the mean results are plotted in Figure 1.4. The minimum distance classifier performs the worst and, as in the previous experiment, the SVM and Gaussian classifiers perform equally well for low and moderate dimensions. But, above a certain dimension (≈ 28), classification accuracy decreases for the Gaussian classifier. The

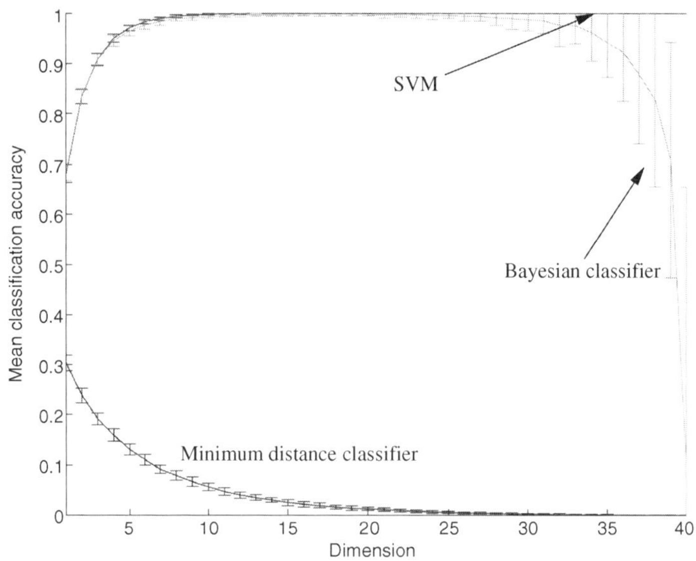

Figure 1.4 Toy example: two Gaussian clusters in high-dimensional space. The line shows the mean classification accuracy and the bar is the standard deviation over 100 experiments.

problem is related to the estimated covariance matrix, which becomes badly conditioned and hence noninvertible. Unlike the other classifiers, SVM does not suffer from the dimensionality and performs perfect classification.

These experiments reveal some properties of SVM that render it suitable for remote-sensing applications. However, it is well known that classes of interest in remote sensing are partially overlapped (Richards and Jia 1999). Hence the choice of a linear function may not be optimal. Hopefully, the use of kernel methods will make it possible *to have both* the effective linear training model *and* the powerful discriminant ability of a nonlinear model.

To illustrate the effectiveness of nonlinear SVM, we performed another experiment. We generated data with two clusters: one Gaussian cluster with zero mean and one cluster with *ring* distribution with zero mean, see Figure 1.5. Linear classifiers cannot handle this sort of data set. For the experiments, we use a polynomial kernel with $p = 2$ and $q = 0$ with the SVM. The minimum distance classifier can also be *kernelized*, since it can be expressed in terms of inner product. For the Bayesian classifier, we made the assumption of two Gaussian clusters, with identical means but different covariance. This leads to the following decision rules:

1. *Minimum distance to the mean classifier*:

$$\left\{\mathbf{x} \middle| \frac{2}{\ell}\sum_{i,k=1}^{\ell/2}\langle\mathbf{x}^i,\mathbf{x}^k\rangle^2 - \frac{2}{\ell}\sum_{j,l=1}^{\ell/2}\langle\mathbf{x}^j,\mathbf{x}^l\rangle^2 - \frac{2}{\ell}\sum_{m=1}^{\ell} y_m\langle\mathbf{x}^m,\mathbf{x}\rangle^2 = 0\right\}. \tag{1.12}$$

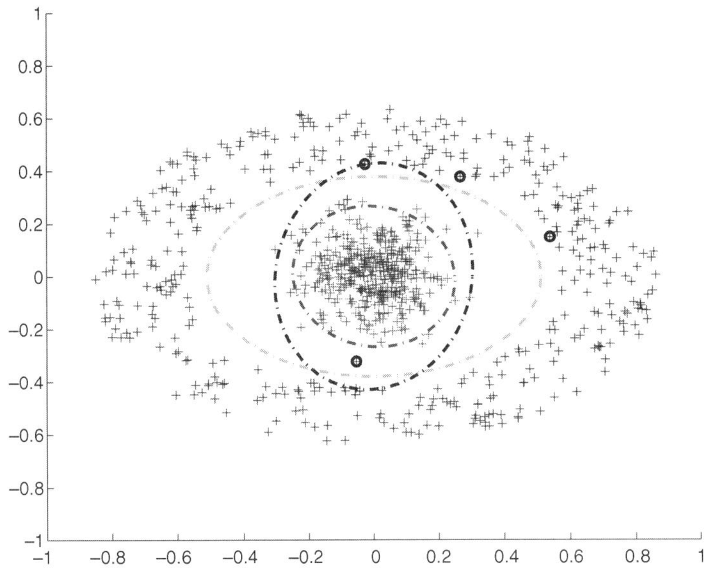

Figure 1.5 *Toy example: two nonlinearly separable clusters. Black circles: support vectors. Grey line: decision boundary minimum distance classifier, light grey line: SVM, black line: Bayesian classifier.*

2. *The SVM classifier*:

$$\left\{\mathbf{x}\Big|\sum_{i=1}^{\ell}\alpha_i y_i\langle\mathbf{x}^i,\mathbf{x}\rangle^2+b=0\right\}. \tag{1.13}$$

3. *The quadratic Bayesian classifier*:

$$\left\{\mathbf{x}|\mathbf{x}^\top\left(\Sigma_+^{-1}-\Sigma_-^{-1}\right)\mathbf{x}+\log\left(\frac{\det(\Sigma_+)}{\det(\Sigma_-)}\right)=0\right\}. \tag{1.14}$$

The decision functions are plotted in Figure 1.5. The Bayesian classifier is unable to classify the data correctly. In this situation, the Gaussian assumption is not verified and thus the classification is not optimal. The SVM classifier fits the ellipsoidal geometry of the data well, which is not true for the minimum distance classifier. It is important to note that if the SVM performs better than the Bayesian classifier, it is because the Gaussian assumption does not hold and the data are linearly separable in the feature space induced by the polynomial kernel.

1.2.4 Use of multiple classifiers

Classifier ensembles, also known as multiple classifier systems, have been used successfully in recent years and are particularly interesting for multi-source and high-dimensional data sets. Several applications in the context of remote sensing and pattern recognition have shown that the classification accuracy can be increased by combining different independent classifiers (e.g. Briem *et al.* 2002; Gislason *et al.* 2006; Ham *et al.* 2005; Waske and van der Linden 2008). These concepts rely on the combination of different classifier methods or the combination of variants of the same classifier. In the latter approach, a set of independent classifier is generated by training the base-classifier on modified input data (e.g., a subset of training samples or input features). The final results are most often determined by a voting scheme, and often a simple majority vote is used. Because the individual classifiers are using only a fraction of the input data, each single classifier performs worst in context of classification accuracy (i.e. also known as weak classifiers). However, the concept is based on the assumption that independent classifiers produce individual errors that are not produced by the majority of the other classifiers. Consequently, the errors are eliminated by combining the outputs. Although classifier ensembles can be generated with different classifier algorithms, e.g., neural networks (Hansen and Salamon 1990) and SVM (Kim *et al.* 2003) often tree-like classifiers are used. Numerous concepts for generating classifier ensembles have been introduced, whereas bagging and boosting are the main approaches, which are discusssed below.

Boosting, bagging and variants

Boosting is a concept to improve the performance of any classifier. The AdaBoost.M1 approach widely used the boosting concept in remote sensing (Freund and Schapire 1996). In the beginning of the iterative boosting procedure, all training samples are equally weighted and the classifier $\mathcal{C}_B$ is the same base classifier. The weights of the training samples are modified after each step and the next classifier within the ensemble uses the newly distributed reweighted samples. Whereas misclassified samples are assigned a stronger weight, the weights of correctly classified samples go down. Thus the classifier is concentrating on the difficult samples and the overall accuracy is generally improved. AdaBoost is appropriate for classification problems that include training samples that are not equally difficult to classify. It tends to exhibit virtually no overfitting when the data is noiseless. Moreover boosting can reduce both the variance and the bias of the classification. Nevertheless, in contrast to other ensemble techniques that can be performed simultaneously (e.g., bagging), boosting is based on a sequentially process. Consequently, the computation time is relatively long. The sensitivity to noisy training data is another disadvantage of boosting (Bauer and Kohavi 1999). The AdaBoost.M1 approach can be described as follows:

Input: A training set $T = \{(\mathbf{x}_j, y_j)\}_{j=1}^{\ell}$, base classifier $\mathcal{C}_B$ and number of classifiers I.

1. $T_1 = S$ and weight$(\mathbf{x}_j) = 1$ for $j = 1 \ldots l$ $(\mathbf{x} \in S_1)$
2. FOR $i = 1$ to $I\{$
3. $\quad C_i = \mathcal{C}_B(S_i)$
4. $\quad$ calculate error rate ϵ_i
5. $\quad$ if $\epsilon_i > 0.5$, terminate procedure
6. $\quad$ calculate weight $\beta_i = \epsilon_i/(1 - \epsilon_i)$

7. for each $\mathbf{x}_j \in T_i\{$ if $C_i(\mathbf{x}_j) \neq y_j$ then
   weight($\mathbf{x}_j$) = weight($\mathbf{x}_j$) $\cdot \beta_i\}$.
8. normalize weights that the total sum of weights is $1\}$.
9. END
10. $C^*(\mathbf{x}) = \arg\max \sum_{C_i(x)=y} \log(1/\beta_i)$

A new framework on boosting was discussed in Rätsch *et al.* (2002), introducing a boosting like one class classifier algorithm. In this study it was shown that the SVM approach can be theoretically transferred into an equivalent boosting like algorithm and vice versa. Although the general similarities between the two concepts have already been discussed in different papers, the authors in Rätsch *et al.* (2002) have recalled these statements to make it clearer. The boosting like approach was derived by using equivalence between the boosting and SVM on the mathematical concepts underlying these algorithms. The authors finally concluded that the presented concept shows new research prospects for boosting: unsupervised learning.

Bagging is the abbreviation of bootstrap aggregating and was originally introduced by Breiman (1996). It is based on combining independent classifier outputs, which were generated by training the base classifier using different training sample sets (also known as bootstrapped aggregates or bags). Although both bagging and boosting rely on the generation of various classifiers and their combination by a vote, the two approaches are dissimilar: Bagging always uses resampling instead of reweighting, thus it does not change the distribution of the samples and all classes are equally weighted. In general a random and uniform selection is performed, selecting ℓ samples from a training set of same size ℓ. This process is performed *with replacement*, i.e. a training sample can be selected several times in the same sample set and perhaps another sample is not considered in this particularly bag. Each individual training set is used to generate a separate classifier, resulting in various independent outputs. The final classification map is created by combining the individual outputs, and often a simple majority vote is performed. The approach can be described as follows:

A training set $T=\{(\mathbf{x}_j, y_j)\}_{j=1}^{\ell}$, the base classifier $\mathcal{C}_B$ and number of randomly generated training sets I.

1. FOR $i = 1$ to $I\{$
2. T_i = bootstrapped bag from S
3. $C_i = \mathcal{C}(T_i)\}$
4. END
5. the class with the maximum number of votes is chosen

If the base classifier is unstable, i.e., a small change in training samples can result in a large change in classification accuracy, then bagging can improve the classification accuracy significantly. On the other hand, the use of a stable base classifier, such as, for example, a k-NN classifier, results in reduced classification accuracy, because each classifier receives less information (Briem *et al.* 2002). In contrast to the aforementioned modification of the training patterns by bagging and boosting, other concepts are based on resampling of input features (Bryll *et al.* 2003; Ho 1998), also referred to as *random feature selection*. For each classifier a subset of features is created. These techniques can outperform other concepts in terms of accuracy (Bryll *et al.* 2003; Ho 1998) and also seems adequate for classifying remote sensing data (Waske *et al.* 2006). Breiman (2001) introduced *random forests* (RF), a tree-based ensemble, which was applied in several remote sensing studies (Gislason *et al.* 2006; Ham *et al.* 2005; Waske and van der Linden 2008). RF further improves the classification performances and overcome the shortcomings of the previous approaches of bagging and boosting. It combines the conventional bootstrap aggregating and a random feature selection method: in training, the algorithm creates multiple decision trees, each trained on a bootstrapped sample of the original training data. Moreover the decision rule at each split node is determined using only a randomly selected feature subset of the input data. A simple majority vote is used to create the final classification result. By reducing the number of features at each split, the computationally complexity of the individual DT classifier is simplified, resulting in a computationally lighter method than bagging and boosting. This enables an RF to handle high-dimensional data sets. In addition the correlation between the classifiers is decreased, which generally improves the performance of the ensemble.

Combination of different classifiers

In contrast to the methods above, which are based on the same classifier algorithm, other approaches combine different classifiers (Benediktsson and Kanellopoulos 1999; Fauvel *et al.* 2006; Steele 2000; Waske and van der Linden 2008). Consequently the advantages of different methods can be combined. Perhaps each classifier was developed in a diverse context and enables a different description of the same classification problem. Furthermore, each method may have its own region in the feature space where it outperforms other classifiers in terms of accuracy (Jain *et al.* 2000).

In many remote sensing studies the different outputs were combined by decision fusion. Decision fusion can be described as combining information from different data sources, after each individual data set has been classified previously. It has often been based on the consensus theory, which employs single probability functions to summarize estimates from various data sources using consensus rules (Benediktsson and Swain 1992). However, these techniques are relatively costly and computationally less demanding voting concepts like majority voting and complete agreement have been proposed (Benediktsson and Kanellopoulos 1999).

1.3 Conclusion

In recent years, machine learning algorithms have been designed to address various applications in remote sensing. Besides the development in the field of (supervised) classification machine learning concepts provide a significant contribution to other related fields in the context of remote sensing, such as feature extraction and the retrieval of surface parameters by regression models. Nevertheless, most of the machine learning applications deal with supervised classification and this chapter briefly discussed the use of different algorithms such as SVM, neural networks and ensemble techniques. All these algorithms provide very promising results in the context of different data sets. Regarding the recent developments of Earth observation platforms with increased spatial resolutions and higher revisit times as well as the availability of multisensory data sets, future trends will rise from the following subjects: (1) integration of time-series, which seems particularly interesting over regions that are characterized by typical temporal variability, e.g., agricultural regions. The research might focus on the handling of the temporal variability within the image time series on the one hand and the determination of adequate acquisition dates on the other. (2) Spatial high resolution data sets are often integrated into urban applications. In this context the combination of spectral and spatial features as well as the extraction of multiple features is particularly interesting. For instance, the spectral reflectance, which characterizes the physical nature of different surface materials, is complementary to the spatial information (describing the shape and geometry of the different natural objects, such as buildings, trees etc.). Thus the development of advanced feature extraction methods is useful to derive spatial information. Moreover, adequate classifier concepts are needed to take advantage of these complementary sources. (3) The combination of different data sets and classifiers, e.g. by decision fusion, will be the important focus for multisensor imagery. In this context the reliability of each source is a critical issue when generating a fusion strategy, requiring adequate methods to assess the reliability of different data sources and individual classifiers. (4) Finally, land cover classifications provide an important input to several environmental monitoring and decision support systems. In the context of operational applications, user interaction is often minimized in these applications. Consequently, future developments in will aim towards automatic or semiautomatic land cover classification methods.

Acknowledgments

This work is partially funded by the Research Fund of the University of Iceland and a PostDoc grant by the Icelandic Research Fund.

References

Atkinson, P.M. and Tatnall, A.R. (1997) Introduction: neural networks in remote sensing. *International Journal of Remote Sensing*, **18**(4), 699–709.

Bauer, E. and Kohavi, R. (1999) An empirical comparison of voting classification algorithms: bagging, boosting, and variants. *Machine Learning*, **36**(1), 105–139.

Benediktsson, J.A. and Kanellopoulos, I. (1999) Classification of multi-source and hyperspectral data based on decision fusion. *IEEE Transactions on Geoscience and Remote Sensing*, **37**(3), 1367–1377.

Benediktsson, J.A. and Swain, P.H. (1992) Consensus theoretic classification methods. *IEEE Transactions on Systems Man and Cybernetics*, **22**(4), 688–704.

Benediktsson, J.A., Sveinsson, J.R. and Amason, K. (1995) Classification and feature extraction of AVIRIS data. *IEEE Transactions on Geoscience and Remote Sensing*, **33**(5), 1194–1205.

Benediktsson, J.A., Sveinsson, J.R., Ersoy, O.K. and Swain, P.H. (1997) Parallel consensual neural networks. *IEEE Transactions on Neural Networks*, **8**(1), 54–64.

Benediktsson, J.A., Swain, P.H. and Ersoy, O.K. (1990) Neural network approaches versus statistical-methods in classification of multi-source remote-sensing data. *IEEE Transactions on Geoscience and Remote Sensing*, **28**(4), 540–552.

Bischof, H., Schneider, W. and Pinz, A.J. (1992) Multi-spectral classification of Landsat-images using neural networks. *IEEE Transactions on Geoscience and Remote Sensing*, **30**(3), 482–490.

Bishop, C. (1995) *Neural Networks for Pattern Recognition*. Oxford University Press.

Bovolo, F., Bruzzone, L. and Marconcini, M. (2008) A novel approach to unsupervised change detection based on a semi-supervised SVM and a similarity measure. *IEEE Transactions on Geoscience and Remote Sensing*, **46**(7), 2070–2082.

Breiman, L. (1996) Bagging predictors. *Machine Learning*, **24**, 123–140.

Breiman, L. (2001) Random forests. *Machine Learning*, **45**(1), 5–32.

Briem, G.J., Benediktsson, J.A. and Sveinsson, J.R. (2002) Multiple classifiers applied to multi-source remote sensing data. *IEEE Transactions on Geoscience and Remote Sensing*, **40**(10), 2291–2299.

Bruzzone, L. and Carlin, L. (2006) A multilevel context-based system for classification of very high spatial resolution images. *IEEE Transactions on Geoscience and Remote Sensing*, **44**(9), 2587–2600.

Bruzzone, L. and Prieto, D. (1999) A technique for the selection of kernel-function parameters in RBF neural networks for classification of remote-sensing images. *IEEE Transactions on Geoscience and Remote Sensing*, **37**(2), 1179–1184.

Bruzzone, L., Chi, M.M. and Marconcini, M. (2006) A novel transductive SVM for semi-supervised classification of remote-sensing images. *IEEE Transactions on Geoscience and Remote Sensing*, **44**(11), 3363–3373.

Bruzzone, L., Marconcini, M., Wegmuller, U. and Wiesmann, A. (2004) An advanced system for the automatic classification of multi-temporal SAR images. *IEEE Transactions on Geoscience and Remote Sensing*, **42**(6), 1321–1334.

Bryll, L., Gutiérrez-Osuna, R. and Quek, F. (2003) Attribute bagging: improving accuracy of classifier ensembles by using random feature subsets. *Pattern Recognition*, **36**(6), 1291–1302.

Camps-Valls, G., Gómez-Chova, L., Muñoz-Marí, J., Rojo-Álvarez, J.L. and Martínez-Ramón, M. (2008) Kernel-based framework for multi-temporal and multi-source remote sensing data classification and change detection. *IEEE Transactions on Geoscience and Remote Sensing*, **46**(6), 1822–1835.

Camps-Valls, G., Gómez-Chova, L., Muñoz-Marí, J., Vila-Francés, J. and Calpe-Maravilla, J. (2006) Composite kernels for hyperspectral image classification. *IEEE Geoscience and Remote Sensing Letters*, **3**(1), 93–97.

Carpenter, G.A., Gjaja, M.N., Gopal, S. and Woodcock, C.E. (1997) ART neural networks for remote sensing: vegetation classification from Landsat TM and terrain data. *Geoscience and Remote Sensing, IEEE Transactions on*, **35**(2), 308–325.

Carpenter, G.A., Grossberg, S. and Reynolds, J.H. (1991) ARTMAP: Supervised real-time learning and classification of non-stationary data by a self-organizing neural network. *Neural Networks*, **4**(5), 565–588.

Carpenter, G.A., Grossberg, S., Markuzon, N., Reynolds, J.H. and Rosen, D.B. (1992) Fuzzy ARTMAP: A neural network architecture for incremental supervised learning of analog multi-dimensional maps. *Neural Networks, IEEE Transactions on*, **3**(5), 698–713.

Chang, C. (2003) *Hyperspectral imaging. Techniques for spectral detection and classification*. Kluwer Academic.

Chapelle, O. (2007) *Training a Support Vector Machine in the Primal Neural Information Processing*, pp. 29–50. MA: MIT Press Cambridge.

Chi, M.M. and Bruzzone, L. (2007) Semi-supervised classification of hyperspectral images by SVMs optimized in the primal. *IEEE Transactions on Geoscience and Remote Sensing*, **45**(6), 1870–1880.

Del Frate, F., Schiavon, G., Solimini, D., Borgeaud, M., Hoekman, D.H. and Vissers, M.A.M. (2003) Crop classification using multiconfiguration C-band SAR data. *IEEE Transactions on Geoscience and Remote Sensing*, **41**(7), 1611–1619.

Duda, R.O., Hart, P.E. and Stork, D.G. (2001) *Pattern Classification*. 2nd edn. John Wiley & Sons.

Evgeniou, T., Pontil, M. and Poggio, T. (2000) Regularization networks and support vector machines. *Advances on Computational Mathematics*, **13**(1), 1–50.

Farag, A.A., Mohamed, P.M. and El-Baz, A. (2005) A unified framework for MAP estimation in remote sensing image segmentation. *Geoscience and Remote Sensing, IEEE Transactions on*, **43**(7), 1617–1634.

Fauvel, M., Benediktsson, J.A., Chanussot, J. and Sveinsson, J.R. (2008) Hyperspectral data using SVMs and morphological profiles. *IEEE Trans. Geoscience and Remote Sensing*, **46**(11), 3804–3814.

Fauvel, M., Chanussot, J. and Benediktsson, J. (2006) A combined support vector machines classification based on decision fusion *Proc. IEEE International Conference on Geoscience and Remote Sensing Symposium IGARSS 2006*, pp. 2494–2497.

Foody, G.M. and Mathur, A. (2004) A relative evaluation of multiclass image classification by support vector machines. *IEEE Transactions on Geoscience and Remote Sensing*, **42**(6), 1335–1343.

Freund, Y. and Schapire, R.E. (1996) Experiments with a new boosting algorithm *13th International Conference of Machine Learning*, Bari, Italy.

Girosi, F., Jones, M. and Poggio, T. (1995) Regularization theory and neural network architectures. *Neural Computation* **7**, 219:269.

Gislason, P.O., Benediktsson, J.A. and Sveinsson, J.R. (2006) Random forests for land cover classification. *Pattern Recognition Letters*, **27**(4), 294–300.

Halldorsson, G.H., Benediktsson, J.A. and Sveinsson, J.R. (2003) Support vector machines in multisource classification. *Proc. IEEE International Conference on Geoscience and Remote Sensing IGARSS 2003*, pp. 2054–2056.

Ham, J., Chen, Y.C., Crawford, M.M. and Ghosh, J. (2005) Investigation of the random forest framework for classification of hyperspectral data. *IEEE Transactions on Geoscience and Remote Sensing*, **43**(3), 492–501.

Hansen, L.K. and Salamon, P. (1990) Neural network ensembles. *IEEE Transactions on Pattern Analysis and Machine Intelligence*, **12**(10), 993–1001.

Ho, T.K. (1998) The random subspace method for constructing decision forests. *IEEE Transactions on Pattern Analysis and Machine Intelligence*, **20**(8), 832–844.

Hofmann, T., Schölkpof, B. and Smola, A.J. (2008) Kernel methods in machine learning. *The Annals of Statistics*, **36**(3), 1171–1220.

Huang, C., Davis, L.S. and Townshend, J.R.G. (2002) An assessment of support vector machines for land cover classification. *International Journal of Remote Sensing*, **23**(4), 725–749.

Hughes, G.F. (1968) On the mean accuracy of statistical pattern recognizers. *IEEE Transactions on Information Theory*, **IT-14**, 55–63.

Jain, A.K., Duin, R.P.W. and Mao, J.C. (2000) Statistical pattern recognition: a review. *IEEE Transactions on Pattern Analysis and Machine Intelligence*, **22**(1), 4–37.

Journaux, L., Xavier, X., Foucherot, I. and Gouton, P. (2006) Dimensionality reduction techniques: an operational comparison on multi-spectral satellite images using unsupervised clustering *IEEE 7th Nordic Signal Processing Symposium*.

Kauth, R.J. and Thomas, G.S. (1976) The tasseled cap–a graphic description of the spectral temporal development of agricultural crops as seen by Landsat. *Proceedings of the Symposium on Machine Processing of Remotely Sensed Data*, 4–B41–4–B51.

Kendall, M.G. (1961) *A course in the geometry of n-dimensions*. New York: Dover Publication.

Keshava, N. (2004) Distance metrics and band selection in hyperspectral processing with application to material identification and spectral libraries. *IEEE Transactions on Geoscience and Remote Sensing*, **42**, 1552–1565.

Kim, H.C., Pang, S., Je, H.M., Kim, D. and Yang Bang, S. (2003) Constructing support vector machine ensemble. *Pattern Recognition*, **36**(12), 2757–2767.

Koetz, B., Morsdorf, F., van der, Linden, S., Curt, T. and Allgöwer, B. (2008) Multi-source land cover classification for forest fire management based on imaging spectrometry and LiDAR data. *Forest Ecology and Management*, **256**(3), 263–271.

Landgrebe, D.A. (2003) *Signal Theory Methods in Multi-spectral Remote Sensing*. New Jersey: John Wiley & Sons.

Lennon, M. (2002) *Méthodes d'analyse d'images hyperspectrales. Exploitation du capteur aéroporté CASI pour des applications de cartographie agro-environnementale en Bretagne*. PhD thesis Université de Rennes 1.

Lennon, M., Mercier, G., Mouchot, M. and Hubert-Moy, L. (2001) Curvilinear component analysis for nonlinear dimensionality reduction of hyperspectral images. *Proceedings SPIE*, 4541, 157–168.

Melgani, F. and Bruzzone, L. (2004) Classification of hyperspectral remote sensing images with support vector machines. *IEEE Transactions on Geoscience and Remote Sensing*, **42**(8), 1778–1790.

Muchoney, D. and Williamson, J. (2001) A Gaussian adaptive resonance theory neural network classification algorithm applied to supervised land cover mapping using multi-temporal vegetation index data. *IEEE Transactions on Geoscience and Remote Sensing*, **39**(9), 1969–1977.

Pal, M. and Mather, P. (2006) Some issues in the classification of DAIS hyperspectral data. *International Journal of Remote Sensing*, **27**(14), 2895–2916.

Pal, M. and Mather, P.M. (2005) Support vector machines for classification in remote sensing. *International Journal of Remote Sensing*, **26**(5), 1007–1011.

Paola, J.D. and Schowengerdt, R.A. (1995) A detailed comparison of backpropagation neural network and maximum-likelihood classifiers for urban land use classification. *IEEE Transactions on Geoscience and Remote Sensing*, **33**(4), 981–996.

Richards, J.A. (2005) Analysis of remotely sensed data: The formative decades and the future. *IEEE Transactions on Geoscience and Remote Sensing*, **43**(3), 422–432.

Richards, J.A. and Jia, X. (1999) *Remote Sensing Digital Image Analysis: An Introduction*. Springer.

Rätsch, G., Mika, S. and Müller, K.R. (2002) Constructing boosting algorithms from SVMs: an application to one-class classification. *IEEE Transactions on Pattern Analysis and Machine Intelligence*, **24**(9), 1184–1199.

Serpico, S., Bruzzone, L. and Roli, F. (1996) An experimental comparison of neural and statistical non-parametric algorithms for supervised classification of remote-sensing images. *Pattern Recognition Letters*, **17**(13), 1331–1341.

Song, X.M., Fan, G.L. and Rao, M. (2005) Automatic CRP mapping using nonparametric machine learning approaches. *IEEE Transactions on Geoscience and Remote Sensing*, **43**(4), 888–897.

Steele, B.M. (2000) Combining multiple classifiers: an application using spatial and remotely sensed information for land cover type mapping. *Remote Sensing of Environment*, **74**(3), 545–556.

Ünsalan, C. and Boyer, K.L. (2004) Linearized vegetation indices based on a formal statistical framework. *IEEE Transactions on Geoscience and Remote Sensing*, **42**(7), 1575–1585.

van der, Linden, S., Janz, A., Waske, B., Eiden, M. and Hostert, P. (2007) Classifying segmented hyperspectral data from a heterogeneous urban environment using support vector machines. *Journal of Applied Remote Sensing*, 1(013543).

Vapnik, V. (1998) *Statistical Learning Theory*. New York: John Wiley & Sons.

Waske, B. and Benediktsson, J.A. (2007) Fusion of support vector machines for classification of multi-sensor data. *IEEE Transactions on Geoscience and Remote Sensing*, **45**(12), 3858–3866.

Waske, B. and van der, Linden, S. (2008) Classifying multilevel imagery from SAR and optical sensors by decision fusion. *IEEE Transactions on Geoscience and Remote Sensing*, **46**(5), 1457–1466.

Waske, B., Schiefer, S. and Braun, M. (2006) Random feature selection for decision tree classification of multi-temporal SAR data *Proc. IEEE International Conference on Geoscience and Remote Sensing Symposium IGARSS 2006*, pp. 168–171.

2

An introduction to kernel learning algorithms

Peter V. Gehler and Bernhard Schölkopf
Max Planck Institute for Biological Cybernetics

Kernel learning algorithms are currently becoming a standard tool in the area of machine learning and pattern recognition. In this chapter we review the fundamental theory of kernel learning. As the basic building block we introduce the kernel function, which provides an elegant and general way to compare possibly very complex objects. We then review the concept of a reproducing kernel Hilbert space and state the representer theorem. Finally we give an overview of the most prominent algorithms, which are support vector classification and regression, Gaussian Processes and kernel principal analysis. With multiple kernel learning and structured output prediction we also introduce some more recent advancements in the field.

2.1 Introduction

Algorithms making use of positive definite kernels have considerably influenced the field of machine learning, pattern recognition and related fields over the last decade. For example the prominent Support Vector Machine has been applied with much success to a variety of tasks and belongs nowadays to the standard toolbox of every practitioner. In addition to their empirical success kernel methods have a solid theoretical foundation and have also been studied in the mathematics and statistics communities. In this chapter we will review the basic mathematical concepts of kernel learning and introduce some prominent algorithms. In contrast to the majority of work on Support Vector Learning we will avoid using duality theory but rather use the regularized risk formulation as the underlying basis for the derivation.

Kernel Methods for Remote Sensing Data Analysis Edited by Gustavo Camps-Valls and Lorenzo Bruzzone

Image processing and in particular remote sensing applications are very challenging because they confront us with a variety of different problems. To begin with, image data, and representations thereof, are typically of very high dimension while, on the other hand, one has access to only very few labelled examples. Algorithms have to account for noise in the observations in a robust way. In particular the acquisition of hyperspectral image data is prone to different noise sources such as instrumental and observational noise. Furthermore, image data may stem from different sources demanding ways of combining information. Kernel algorithms are suited to tackle such problems. With the design of a kernel function it is possible to combine different feature entities, different feature dimensionalities and account for high dimensional data. With the framework of regularized risk minimization, kernel methods are efficient even with little training data and offer ways to incorporate unlabelled examples through semi-supervised models. Such algorithms have already proven to be a valuable tool for image processing applications such as image coding, image de-noising, image segmentation and image classification (Camps-Valls *et al.* 2007; Kim *et al.* 2005). In particular for classification of hyperspectral images, kernel classifiers have found to yield state-of-the-art results, as also reported in Part 2 of this book.

The selection of topics for this chapter is by no means comprehensive and aims to obtain a self-contained exposition of the basic concepts of kernel learning and the most used algorithms. For a more detailed introduction to the field we refer to the articles by Hofmann *et al.* (2008), Shawe-Taylor and Cristianini (2004) and Schölkopf and Smola (2002), who also cover the statistical background of kernel algorithms.

This chapter is divided into three sections. We start in Section 2.2 with a basic introduction to the notion of kernels and introduce the reproducing kernel Hilbert space. In Section 2.3 we state the representer theorem, which serves as the foundation of all the algorithms that are presented in Section 2.4.

2.2 Kernels

2.2.1 Measuring similarity with kernels

Suppose that we are given empirical data

$$(\mathbf{x}_1, y_1), \ldots, (\mathbf{x}_n, y_n) \in \mathcal{X} \times \mathcal{Y}.$$

We will call $\mathbf{x}_i$ the *inputs* that are taken from the nonempty set $\mathcal{X}$ and $y_i \in \mathcal{Y}$ the *targets*. The problem of learning is to use this data in order to make statements about previously unseen elements $\mathbf{x} \in \mathcal{X}$. For example in binary classification where the training data stems from two classes with labels $\mathcal{Y} = \{-1, +1\}$ one aims to construct a function $f : \mathcal{X} \to \mathcal{Y}$ that assigns to each element of $\mathcal{X}$ a class label. The function one is interested in should not be arbitrary but one that *generalizes* well, that is, loosely speaking, making few errors on unseen data from the same problem. In the classification example this corresponds to making as few mistakes as possible when inferring the class labels. In order to enable generalization we need to exploit the structure of the training examples and in order to impose a structure we will define a similarity between pairs of data points. The most general setting would be to define such a similarity between pairs of inputs $(\mathbf{x}, y)$ including the targets. For now we will restrict ourselves and define similarities only between inputs $\mathbf{x} \in \mathcal{X}$ and refer to Section 2.4.5 for a generalization.

There was no other assumption about $\mathcal{X}$ other than it being a set and, in particular, nothing has been said about inputs being similar to each other. Therefore we will first map the data into a space where we have a notion of similarity, namely a dot product space $\mathcal{H}$, using the mapping

$$\phi : \mathcal{X} \rightarrow \mathcal{H}, \mathbf{x} \mapsto \phi(\mathbf{x}).$$

The similarity between the elements in $\mathcal{H}$ can now be measured using its associated dot product $\langle \cdot, \cdot \rangle_{\mathcal{H}}$. For convenience we introduce the following function that does exactly that

$$K : \mathcal{X} \times \mathcal{X} \rightarrow \mathbb{R}, (\mathbf{x}, \mathbf{x}') \mapsto K(\mathbf{x}, \mathbf{x}'),$$

which we require to satisfy for all $\mathbf{x}, \mathbf{x}' \in \mathcal{X}$

$$K(\mathbf{x}, \mathbf{x}') = \langle \phi(\mathbf{x}), \phi(\mathbf{x}') \rangle_{\mathcal{H}}. \tag{2.1}$$

This function is called a *kernel*. The mapping ϕ is referred to as its *feature map* and the space $\mathcal{H}$ as its *feature space*.

Although this construction seems a bit inconspicuous we will see that it has far reaching consequences. Sometimes we will drop the subscript specifying the origin of the dot product in those cases where it should be clear from the context.

2.2.2 Positive definite kernels

The construction of the similarity measure as the dot product in some space $\mathcal{H}$ is rather general. Different measures of similarity can be obtained by simply varying the feature map ϕ. A particular simple case is when $\mathcal{X}$ is itself a dot product space in which case one may choose ϕ to be the identity. We will now show that the class of kernels that can be written in the form of (2.1) coincides with the class of positive definite kernels. This yields a very comfortable situation due to the following observation. Algorithms that operate on the data only in terms of dot product can be used with any positive definite kernel by simply replacing $\langle \phi(\mathbf{x}), \phi(\mathbf{x}') \rangle_{\mathcal{H}}$ with kernel evaluations $K(\mathbf{x}, \mathbf{x}')$, a technique also known as the *kernel trick* (Schölkopf *et al.* 1998). Another direct consequence is that for a positive definite kernel one does not need to know the explicit form of the feature map since it is implicitly defined through the kernel. We will even encounter examples where $\mathcal{H}$ is infinite dimensional and thus the replacement of the dot product is crucial in order to evaluate the dot product at all.

We need some definitions before we can state the equivalence between $K(\mathbf{x}, \mathbf{x}')$ and $\langle \phi(\mathbf{x}), \phi(\mathbf{x}') \rangle_{\mathcal{H}}$.

Definition 2.2.1 (Gram matrix) *Given a kernel K: $\mathcal{X} \times \mathcal{X} \rightarrow \mathbb{R}$ and inputs $\mathbf{x}_1, \ldots, \mathbf{x}_n \in \mathcal{X}$. We call the $n \times n$ matrix $\mathbf{K}$ with entries*

$$\mathbf{K}_{ij} = K(\mathbf{x}_i, \mathbf{x}_j) \tag{2.2}$$

the Gram matrix *or the* kernel matrix *of K with respect to $\mathbf{x}_1, \ldots, \mathbf{x}_n$.*

Definition 2.2.2 (Positive definite matrix) *A real symmetric $n \times n$ matrix $\mathbf{K}$ is called* positive definite *if for all $c_1, \ldots, c_n \in \mathbb{R}$*

$$\sum_{i,j=1}^{n} c_i c_j \mathbf{K}_{ij} \geq 0. \tag{2.3}$$

If equality in (2.3) occurs only for $c_1 = \ldots = c_n = 0$ then the matrix is called strictly positive definite.

A positive definite kernel is one that always produces a positive definite Gram matrix for elements in $\mathcal{X}$. More precisely:

Definition 2.2.3 (Positive definite kernel) *If for all $n \in \mathbb{N}$ and for all $\mathbf{x}_1, \ldots, \mathbf{x}_n \in \mathcal{X}$ the Gram matrix $\mathbf{K}_{ij} = K(\mathbf{x}_i, \mathbf{x}_j)$ is positive definite we call the kernel a* positive definite kernel. *If furthermore for all $n \in \mathbb{N}$ and distinct $\mathbf{x}_1, \ldots, \mathbf{x}_n \in \mathcal{X}$ the kernel K gives rise to a strictly positive definite Gram matrix we will call it a* strictly positive definite kernel.

Now we are ready to state one of the most important observations for kernel methods. To this end we need the concept of a Hilbert space. Recall that a Hilbert space $\mathcal{H}$ is a real (or complex valued) inner product space that is complete by the inner product $\langle \cdot, \cdot \rangle_{\mathcal{H}}$. Some simple examples of Hilbert spaces are $\mathbb{R}^d$ and $\mathbb{C}^d$.

Proposition 2.2.4 *A function $K : \mathcal{X} \times \mathcal{X} \to \mathbb{R}$ is a positive definite kernel* if and only if *there exists a Hilbert space $\mathcal{H}$ and a feature map $\phi : \mathcal{X} \to \mathcal{H}$ such that for all $\mathbf{x}, \mathbf{x}' \in \mathcal{X}$ we have $K(\mathbf{x}, \mathbf{x}') = \langle \phi(\mathbf{x}), \phi(\mathbf{x}') \rangle_{\mathcal{H}}$.*

Proof: '$\Leftarrow$' Assume the kernel can be written in the form (2.1). It being positive definite is a simple consequence of the bilinearity of the dot product $\langle \cdot, \cdot \rangle_{\mathcal{H}}$

$$\sum_{i,j=1}^{n} c_i c_j \langle \phi(\mathbf{x}_i), \phi(\mathbf{x}_j) \rangle_{\mathcal{H}} = \left\langle \sum_{i=1}^{n} c_i \phi(\mathbf{x}_i), \sum_{j=1}^{n} c_j \phi(\mathbf{x}_j) \right\rangle_{\mathcal{H}} = \left\| \sum_{i=1}^{n} c_i \phi(\mathbf{x}_i) \right\|_{\mathcal{H}}^2 \geq 0.$$

'$\Rightarrow$' In the next section (2.2.3) we will present how to construct, given a positive definite kernel, a Hilbert space along with a feature map ϕ with the desired properties. This will conclude the proof. □

Owing to this equivalence we will sometimes refer to a positive definite kernel simply as a *kernel*. Although kernels compute dot products in some space $\mathcal{H}$, they should not be mistaken to be themselves dot products in the input space. For example they are in general not bilinear. However they share important properties such as the Cauchy–Schwarz inequality.

Proposition 2.2.5 (Cauchy–Schwarz) *If K is a positive definite kernel, and $\mathbf{x}_1, \mathbf{x}_2 \in \mathcal{X}$, then*

$$K(\mathbf{x}_1, \mathbf{x}_2)^2 \leq K(\mathbf{x}_1, \mathbf{x}_1) K(\mathbf{x}_2, \mathbf{x}_2).$$

Proof: Since K is positive definite, so is the 2×2 Gram matrix $\mathbf{K}_{ij} = K(\mathbf{x}_i, \mathbf{x}_j)$. Therefore the eigenvalues of $\mathbf{K}$ are non-negative and so is its determinant. Writing out the determinant completes the proof

$$0 \leq \det(\mathbf{K}) = K(\mathbf{x}_1, \mathbf{x}_1)K(\mathbf{x}_2, \mathbf{x}_2) - K(\mathbf{x}_1, \mathbf{x}_2)^2.$$

□

2.2.3 Constructing the reproducing kernel Hilbert space

Using the positive definite kernels as building blocks we will now go one step further and consider functions that can be expressed as linear combinations of kernels. This leads us to the concept of a reproducing kernel Hilbert space (RKHS) . In the following we will present a construction scheme for a fixed kernel K, which will also conclude the proof of Proposition 2.2.4. The main idea is to construct a Hilbert space whose elements are functions. For a given kernel K we define the following set

$$F = \left\{ f(\cdot) = \sum_{i=1}^{n} \alpha_i K(\cdot, \mathbf{x}_i);\ n \in \mathbb{N}, \alpha_i \in \mathbb{R}, \mathbf{x}_i \in \mathcal{X} \right\} \subseteq \mathbb{R}^{\mathcal{X}},$$

where the element $K(\cdot, \mathbf{x}) \colon \mathcal{X} \to \mathbb{R}$ denotes a function. Obviously $K(\cdot, \mathbf{x})$ itself is an element in F. It is easy to see that this set can be turned into a vector space if we endow it with the two operations addition $(f + g)(\mathbf{x}) = f(\mathbf{x}) + g(\mathbf{x})$ and multiplication by a scalar $(\lambda f)(\mathbf{x}) = \lambda f(\mathbf{x})$, $\lambda \in \mathbb{R}$. Now we define an inner product between two elements of this space

$$f(\cdot) = \sum_{i=1}^{n} \alpha_i K(\cdot, \mathbf{x}_i) \quad \text{and} \quad g(\cdot) = \sum_{j=1}^{n'} \beta_j K(\cdot, \mathbf{x}'_j)$$

with $n, n' \in \mathbb{N}$, $\alpha_i, \beta_j \in \mathbb{R}$, $\mathbf{x}_i, \mathbf{x}'_j \in \mathcal{X}$, simply as

$$\langle f, g \rangle_F := \sum_{i=1}^{n} \sum_{j=1}^{n'} \alpha_i \beta_j K(\mathbf{x}_i, \mathbf{x}'_j).$$

This is a well defined construction and does not depend on the choice of the expansion coefficients of f or g. To see this note that we can also write

$$\sum_{j=1}^{n'} \beta_j f(\mathbf{x}'_j) = \langle f, g \rangle_F = \sum_{i=1}^{n} \alpha_i g(\mathbf{x}_i),$$

where the left term does not depend on the expansion of f and the right term does not depend on the expansion of g. This also shows that $\langle \cdot, \cdot \rangle_F$ is bilinear. Furthermore it is symmetric and positive definite, which follows directly from the positive definiteness of the kernel K, since

$$\langle f, f \rangle_F = \sum_{i,j=1}^{n} \alpha_i \alpha_j K(\mathbf{x}_i, \mathbf{x}_j) \geq 0$$

implies that for any functions $f_1, \ldots, f_p \in F$ and any coefficients $c_1, \ldots, c_p \in \mathbb{R}$ we have

$$\sum_{i,j=1}^{p} c_i c_j \langle f_i, f_j \rangle_F = \left\langle \sum_{i=1}^{p} c_i f_i, \sum_{j=1}^{p} c_j f_j \right\rangle_F \geq 0.$$

From the last equation we see that $\langle \cdot, \cdot \rangle_F$ is a positive definite kernel defined on a vector space of functions.

If we choose $g(\cdot) = K(\cdot, \mathbf{x})$ it follows directly from the definition of the inner product that

$$\langle f, K(\cdot, \mathbf{x}) \rangle_F = \sum_{i=1}^{n} \alpha_i K(\mathbf{x}_i, \mathbf{x}) = f(\mathbf{x}), \tag{2.4}$$

and in particular

$$\langle K(\cdot, \mathbf{x}), K(\cdot, \mathbf{x}') \rangle_F = K(\mathbf{x}, \mathbf{x}').$$

This property is known as the *reproducing property* of the kernel. A function f can thus be represented as a linear function defined by an inner product in the vector space F.

We still need to show definiteness of the inner product. Applying (2.4) and the Cauchy–Schwarz inequality we obtain

$$|f(\mathbf{x})|^2 = |\langle K(\cdot, \mathbf{x}), f \rangle_F|^2 \leq K(\mathbf{x}, \mathbf{x}) \cdot \langle f, f \rangle_F,$$

which proves that $\langle f, f \rangle_F = 0 \Leftrightarrow f = 0$. The space we constructed can be completed by adding all limit points of sequences that are convergent in the norm $\|f\|_F = \sqrt{\langle f, f \rangle}$, which yields the Hilbert space $\mathcal{H}$, see e.g. Aronszajn (1950) for details.

Owing to the property (2.4) this space is called a *reproducing kernel Hilbert space* (RKHS) for K. The RKHS uniquely determines K and vice versa. This is the statement of the following theorem.

Theorem 2.2.6 (Moore–Aronszajn, see Aronszajn (1950)) *To every positive definite kernel K there exists a unique reproducing kernel Hilbert space $\mathcal{H}$ whose kernel is K and vice versa.*

Proof: It remains to show uniqueness. Let $\mathcal{H}'$ be another Hilbert space for which K is the reproducing kernel. Then for all $\mathbf{x}, \mathbf{x}' \in \mathcal{H}$

$$\langle K(\cdot, \mathbf{x}), K(\cdot, \mathbf{x}') \rangle_{\mathcal{H}} = K(\mathbf{x}, \mathbf{x}') = \langle K(\cdot, \mathbf{x}), K(\cdot, \mathbf{x}') \rangle_{\mathcal{H}'}.$$

Owing to the linearity of the dot product and the uniqueness of the completion it must hold $\mathcal{H} = \mathcal{H}'$. Now assume $K, K' \in \mathcal{H}$, $K \neq K'$ are two different reproducing kernels of $\mathcal{H}$. Then there exists a $\mathbf{x} \in \mathcal{X}$ for which

$$\begin{aligned} 0 &< \|K(\cdot, \mathbf{x}) - K'(\cdot, \mathbf{x})\|_{\mathcal{H}}^2 = \left\langle K(\cdot, \mathbf{x}) - K'(\cdot, \mathbf{x}), K(\cdot, \mathbf{x}) - K'(\cdot, \mathbf{x})\right\rangle_{\mathcal{H}} \\ &= \langle K(\cdot, \mathbf{x}), K(\cdot, \mathbf{x}) - K'(\cdot, \mathbf{x})\rangle_{\mathcal{H}} - \langle K'(\cdot, \mathbf{x}), K(\cdot, \mathbf{x}) - K'(\cdot, \mathbf{x})\rangle_{\mathcal{H}} = 0, \end{aligned}$$

which is a contradiction. □

We have constructed a Hilbert space that can act as the feature space for our kernel. The corresponding feature map for this space is the so-called *Aronszajn map*

$$\phi : \mathcal{X} \to \mathbb{R}^{\mathcal{X}}, \mathbf{x} \mapsto K(\cdot, \mathbf{x}).$$

For this ϕ it is easy to see that the kernel K is indeed of the form (2.1). We want to point out that although there exists a unique RKHS to each kernel K, it might well be that the kernel computes the inner product of different spaces as well. If the two feature maps ϕ_1, ϕ_2 map into the feature spaces $\mathcal{H}_1$, resp. $\mathcal{H}_2$ then it might be the case that

$$K(\mathbf{x}, \mathbf{x}') = \langle \phi_1(\mathbf{x}), \phi_1(\mathbf{x}')\rangle_{\mathcal{H}_1} = \langle \phi_2(\mathbf{x}), \phi_2(\mathbf{x}')\rangle_{\mathcal{H}_2},$$

for all $\mathbf{x}, \mathbf{x}' \in \mathcal{X}$, but that does not necessarily imply that $\phi_1 = \phi_2$. For our purposes however we can consider the two spaces identical since we are only interested in the kernel evaluation as the dot product and this remains identical.

We will see examples of kernels for which the corresponding RKHS is infinite dimensional. To prevent complications we generally assume that the RKHS are separable, i.e. there exists a countable complete orthonormal system. For this special case we have the following theorem, which states that there in fact is *only one* such space.

Theorem 2.2.7 *All separable infinite dimensional Hilbert spaces are isometrically isomorphic.*

2.2.4 Operations in RKHS

We will turn our attention to some basic operations in the reproducing kernel Hilbert space and show how they can be computed in terms of kernel function evaluations. Although the space $\mathcal{H}$ can be very high dimensional or even infinite dimensional, in some cases basic operations can still be computed. Essentially this is the case whenever we can express its elements in terms of kernel evaluations.

Translation

A translation in feature space can be written as the modified feature map $\tilde{\phi}(\mathbf{x}) = \phi(\mathbf{x}) + \Gamma$ with $\Gamma \in \mathcal{H}$. We expand the dot product for $\langle \tilde{\phi}(\mathbf{x}), \tilde{\phi}(\mathbf{x}')\rangle_{\mathcal{H}}$ to write

$$\langle \phi(\mathbf{x}) + \Gamma, \phi(\mathbf{x}') + \Gamma\rangle = \langle \phi(\mathbf{x}), \phi(\mathbf{x}')\rangle + \langle \phi(\mathbf{x}), \Gamma\rangle + \langle \Gamma, \phi(\mathbf{x}')\rangle + \langle \Gamma, \Gamma\rangle.$$

In general only the first term can be evaluated via the kernel function. But if we restrict Γ to lie in the span of the functions $\phi(\mathbf{x}_1), \ldots, \phi(\mathbf{x}_n) \in \mathcal{H}$ we can calculate the translated dot product

$$\langle\tilde{\phi}(\mathbf{x}), \tilde{\phi}(\mathbf{x}')\rangle = K(\mathbf{x}, \mathbf{x}') + \sum_{i=1}^{n} \alpha_i K(\mathbf{x}, \mathbf{x}_i) + \sum_{i=1}^{n} \alpha_i K(\mathbf{x}', \mathbf{x}_i) + \sum_{i,j=1}^{n} \alpha_i \alpha_j K(\mathbf{x}_i, \mathbf{x}_j). \quad (2.5)$$

Centring

The translation operations allows us to centre data $\mathbf{x}_1, \ldots, \mathbf{x}_n \in \mathcal{X}$ *in the feature space*. The mean of the data is $\phi_\mu = \frac{1}{n}\sum_{i=1}^{n} \phi(\mathbf{x}_i)$ and thus fulfills the requirements for the translation element Γ. Applying (2.5) with $\Gamma = -\phi_\mu$ thus yields a feature map for which $0 = \frac{1}{n}\sum_{i=1}^{n} \tilde{\phi}(\mathbf{x}_i)$.

Computing distances

With the kernel corresponding to a dot product in a Hilbert Space $\mathcal{H}$ and thus inducing a norm it is natural to ask if one can also compute the distances of the images of elements in $\mathcal{X}$. Such a distance can be evaluated entirely in terms of kernel evaluations as is evident from

$$d(\mathbf{x}_1, \mathbf{x}_2) = \|\phi(\mathbf{x}_1) - \phi(\mathbf{x}_2)\|_{\mathcal{H}} = \sqrt{K(\mathbf{x}_1, \mathbf{x}_1) + K(\mathbf{x}_2, \mathbf{x}_2) - 2K(\mathbf{x}_1, \mathbf{x}_2)}$$

for $\mathbf{x}_1, \mathbf{x}_2 \in \mathcal{X}$ and ϕ being the feature map for K. This is a very elegant way to measure the similarity between arbitrary objects, for example between two graphs or two sentences.

Subspace projections

For two points Ψ, Γ in the RKHS the projection of Ψ onto the subspace spanned by Γ is

$$\Psi' = \frac{\langle\Gamma, \Psi\rangle_{\mathcal{H}}}{\|\Gamma\|_{\mathcal{H}}^2}\Gamma.$$

We immediately see that if we have a kernel expansion of the points Ψ, Γ we can compute the projection Ψ' and express it also in terms of kernel evaluations.

2.2.5 Kernel construction

The following proposition states some operations that preserve the positive definiteness of kernels. These can be used to create new, possibly complicated, kernels from existing ones.

Proposition 2.2.8 *Let $K_1, K_2, \ldots$ be arbitrary positive definite kernels on $\mathcal{X} \times \mathcal{X}$, where $\mathcal{X}$ is a nonempty set. Then*

(i) $\alpha_1 K_1 + \alpha_2 K_2$ is positive definite for $\alpha_1, \alpha_2 \geq 0$.

(ii) If $K(\mathbf{x}, \mathbf{x}') := \lim_{n\to\infty} K_n(\mathbf{x}, \mathbf{x}')$ exists for all $\mathbf{x}, \mathbf{x}'$, then K is positive definite.

(iii) The pointwise product $K_1 K_2$ is positive definite.

(iv) *Assume for $i = 1, 2$, K_i is a positive definite kernel on $\mathcal{X}_i \times \mathcal{X}_i$, where $\mathcal{X}_i$ is nonempty. The tensor product $K_1 \otimes K_2$ and the direct sum $K_1 \oplus K_2$ are positive definite kernels on $(\mathcal{X}_1 \times \mathcal{X}_2) \times (\mathcal{X}_1 \times \mathcal{X}_2)$.*

(v) *The function $K(\mathbf{x}, \mathbf{x}') := f(\mathbf{x})f(\mathbf{x}')$ is a positive definite kernel for any function $f: \mathcal{X} \to \mathbb{R}$.*

Proof: For proofs see Berg et al. (1984). □

The first two statements of the last proposition state that the set of positive definite kernels is a closed convex cone. Loosely speaking the operations of (i)–(iv) are the only simple operations that preserve positive definiteness. Without stating the result we mention that it is possible to fully characterize the class of functions that preserve positive definiteness (FitzGerald *et al.* 1995; Hofmann *et al.* 2008).

2.2.6 Examples of kernels

With the closure properties of the last result we finally turn our view to some concrete examples of kernel functions. We concentrate on the most prominent ones and also introduce two kernels that have been used for several image processing tasks. For a more general overview including examples for other data structures such as graphs, trees, strings, etc. we refer to Bakir *et al.* (2007), Hofmann *et al.* (2008), Shawe-Taylor and Cristianini (2004) and Schölkopf and Smola (2002). We will first present three simple kernels that which operate on real vector spaces. The later kernels are more suited for the task of image processing, that is whenever the elements that need to be compared are themselves images.

Linear kernel

The most simple kernel is arguably the ordinary dot product in $\mathbb{R}^d$. Functions that are build upon this kernel are of the form

$$f(\mathbf{x}) = \sum_{i=1}^{n} \alpha_i \langle \mathbf{x}, \mathbf{x}_i \rangle = \langle \mathbf{x}, \sum_{i=1}^{n} \alpha_i \mathbf{x}_i \rangle = \langle \mathbf{x}, \mathbf{w} \rangle, \tag{2.6}$$

where we defined $\mathbf{w} = \sum_{i=1}^{n} \alpha_i \mathbf{x}_i$. Thus the RKHS of a linear kernel consists of all hyperplanes in $\mathbb{R}^d$.

Polynomial kernel

The *homogeneous polynomial* kernel $K(\mathbf{x}, \mathbf{x}') = \langle \mathbf{x}, \mathbf{x}' \rangle^p$ is positive definite for all $p \in \mathbb{N}$ and $\mathbf{x}, \mathbf{x}' \in \mathbb{R}^d$ as a direct consequence of Proposition 2.2.8(iii). This again is an example where the input space $\mathcal{X}$ is itself a dot-product space. One can write down the corresponding feature space; it is of finite dimension and consists of all monomials of degree p in the input coordinates. Another prominent polynomial kernel is the *inhomogeneous polynomial kernel* that computes the inner product of all monomials *up to* degree p: $K(\mathbf{x}, \mathbf{x}') = \left(\langle \mathbf{x}, \mathbf{x}' \rangle + c\right)^p$.

Gaussian kernel

From the Taylor series expansion of the exponential function $e^z = \sum_{i=0}^{\infty} \frac{1}{i!} z^i$ and Proposition 2.2.8(ii) we see that

$$K(\mathbf{x}, \mathbf{x}') = e^{\gamma \langle \mathbf{x}, \mathbf{x}' \rangle}$$

is a positive definite kernel for all $\mathbf{x}, \mathbf{x}' \in \mathbb{R}^d$, $\gamma \in \mathbb{R}$. It follows immediately that the Gaussian function

$$e^{-\gamma \|\mathbf{x}-\mathbf{x}'\|^2} = e^{-\gamma \langle \mathbf{x}, \mathbf{x} \rangle} e^{2\gamma \langle \mathbf{x}, \mathbf{x}' \rangle} e^{-\gamma \langle \mathbf{x}', \mathbf{x}' \rangle} \tag{2.7}$$

is a valid positive definite kernel. Furthermore we observe from the Taylor series expansion that it is a polynomial kernel with infinite degree. The corresponding Hilbert space is infinite dimensional, in fact it corresponds to a mapping into $\mathcal{C}^\infty$ the space of smooth functions.

Set kernels

So far we have presented kernels for rather simple input spaces only but the power of kernel methods also stems from the fact that we can measure similarity between possibly complex objects. Assume an input space to be the power set of some other set, e.g. $\mathcal{X} = 2^{\mathcal{K}}$ for some finite dictionary $\mathcal{K}$. A similarity on this space could for example be defined by just counting the number of equal elements in the sets. The corresponding kernel for $X, X' \in \mathcal{X}$ reads

$$K(X, X') = \sum_{\mathbf{x} \in X} \sum_{\mathbf{x}' \in X} \delta_{\mathbf{x}, \mathbf{x}'},$$

where we used upper case letters for the elements of $\mathcal{X}$ to make explicit that those elements are in fact sets. This kernel is widely used in text classification and is also referred to as the *sparse vector kernel*, see for example Joachims (1998). Consider a text document as being represented as an unordered collection of words, the so-called *bag-of-words* representation. This kernel is measuring the similarity between two texts by just counting the number of common words.

A more general kernel can be defined on the same input set with the use of a base kernel K_0. It sums up the similarities between all elements in the two sets

$$K(X, X') = \sum_{\mathbf{x} \in X} \sum_{\mathbf{x}' \in X} K_0(\mathbf{x}, \mathbf{x}').$$

In Haussler (1999) it was shown that this is a kernel if and only if K_0 itself is a kernel.

Histogram kernels

Suppose an image is represented as a collection of image features $\mathbf{f}_i$, e.g. small image patches or some detected keypoints. Two images that are to be compared might have a different number of such features. Quantization into a histogram is a convenient way to transform such data into a feature vector of fixed length.

In order to obtain a histogram one needs a codebook of finite size $\mathcal{K}$; this is usually obtained by clustering in the feature space, e.g. k-means. Using this codebook the features are quantized to yield a histogram $\mathbf{h} \in \mathbb{N}^{\mathcal{K}}$. Each histogram bin $\mathbf{h}_j \in \mathbb{N}$ stores the number of features $\mathbf{f}_i$ which were quantized to the jth codebook entry. In the context of image data the elements of such a codebook are also referred to as 'visual words' and the representation itself as a 'bag-of-visual-words'. Similar to the bag-of-words representation for text data, all spatial structure of the features is disregarded.

This pipeline of transforming features into histograms is of course very general and applicable to many tasks. Normalized histograms $\mathbf{h}_i / \sum_{j=1}^{d} \mathbf{h}_j$ can also be thought of as probability distributions and the following two histogram kernels are in fact special cases of the larger class of kernels between probability distributions (Hein and Bousquet 2005). For two histograms $\mathbf{h}, \mathbf{h}'$ we define the *histogram intersection kernel* as

$$K(\mathbf{h}, \mathbf{h}') = \sum_{i=1}^{d} \min\left(\mathbf{h}_i, \mathbf{h}'_i\right).$$

It can be understood as computing the overlap between the two histograms. Another prominent choice is the the χ^2 *kernel* for histograms

$$K(\mathbf{h}, \mathbf{h}') = \exp\left(-\gamma \sum_{i=1}^{d} \frac{\left(\mathbf{h}_i - \mathbf{h}'_i\right)^2}{\mathbf{h}_i + \mathbf{h}'_i}\right),$$

where we used the convention $0/0 := 0$.

Spatial kernels

A spatial kernel for image processing is an instructive example of how special structure in the data can be incorporated into the kernel function. The idea is very simple and dates back to the local receptive fields of neural networks. A kernel value is computed using not all features of an image but only those that fall into some subwindow, e.g. only those of the upper half of the image. This kernel was first used in Schölkopf (1997) for an application to digit classification. Recently it gained some attention in the field of visual object classification. The *spatial pyramid match kernel* was proposed in Lazebnik *et al.* (2006) as the spatial variant of the so called *pyramid match kernel* (Grauman and Darrell 2005).

Different spatial layouts are conceivable but we will only present the pyramidal representation that turned out to give good performance for visual object classification. The image is subdivided into a grid of $1 \times 1, 2 \times 2, \ldots$ equally spaced subwindows. Each such subdivision is referred to as a *level* of the pyramid. Starting with level 0 as the entire image, the lth level has 4^l non-overlapping subwindows. For each level l an histogram $\mathbf{h}_l$ is computed by concatenating the histograms of all subwindows within the level. Two images $\mathbf{x}, \mathbf{x}'$ are compared by combining the similarity of the individual levels

$$K(\mathbf{x}, \mathbf{x}') = \sum_{l=0}^{L-1} d_l K(\mathbf{h}_l, \mathbf{h}'_l),$$

where $d_l \in \mathbb{R}_+$ is an extra weighting parameter for each level. In both Grauman and Darrell (2005) and Lazebnik *et al.* (2006) it is proposed to set $d_l = 2^{L-l}$ in order to give more weight to a finer griding of the image, but in general these parameters can also be optimized over (cf. Section 2.4.4). There are several choices for the kernel K, using the χ^2 kernel described above good results on visual object classification datasets have been achieved (Bosch *et al.* 2007).

2.3 The representer theorem

In the last chapter we saw that a kernel centred at a point $\mathbf{x} \in \mathcal{X}$ is an element in an RKHS $\mathcal{H}$ associated with the kernel K. Other functions $f \in \mathcal{H}$ can be represented as linear combinations of kernel expansions but have a possibly infinite number of expansion coefficients. The representer theorem (Cox and O'Sullivan 1990; Kimeldorf and Wahba 1971) states that the solutions of a large class of optimization problems are expressible by only a finite number of kernel functions. We present a slightly more general version of the theorem from Schölkopf *et al.* (2001).

Theorem 2.3.1 (Representer theorem) *Let $\Omega : [0, \infty) \to \mathbb{R}$ be a strictly monotonic increasing function and $V : \left(\mathcal{X} \times \mathbb{R}^2\right)^n \to \mathbb{R} \cup \{\infty\}$ be an arbitrary loss function. Furthermore let $\mathcal{H}$ be a RKHS with reproducing kernel K. Each minimizer $f \in \mathcal{H}$ of the regularized functional*

$$V\left((\mathbf{x}_1, y_1, f(\mathbf{x}_1)), \ldots, (\mathbf{x}_n, y_n, f(\mathbf{x}_n))\right) + \Omega\left(\|f\|^2_{\mathcal{H}}\right) \tag{2.8}$$

admits a representation of the form

$$f(\mathbf{x}) = \sum_{i=1}^{n} \alpha_i K(\mathbf{x}, \mathbf{x}_i), \tag{2.9}$$

with $\alpha_i \in \mathbb{R}$.

Proof: *We decompose an element $f \in \mathcal{H}$ into the part $f^{\parallel}$ that is inside the span of kernel functions $K(\cdot, \mathbf{x}_1), \ldots, K(\cdot, \mathbf{x}_n)$ and its orthogonal complement $f^{\perp}$ and show that the latter is always zero. We write*

$$f(\mathbf{x}) = f^{\parallel}(\mathbf{x}) + f^{\perp}(\mathbf{x}) = \sum_{i=1}^{n} \alpha_i K(\mathbf{x}, \mathbf{x}_i) + f^{\perp}(\mathbf{x})$$

with $\alpha_i \in \mathbb{R}$ and $\langle f^{\perp}, K(\cdot, \mathbf{x}_i)\rangle_{\mathcal{H}} = 0, \forall i \in \{1, \ldots, n\}$. Using the reproducing property of $\mathcal{H}$ we can write

$$f(\mathbf{x}) = \langle f, K(\cdot, \mathbf{x})\rangle_{\mathcal{H}} = \sum_{i=1}^{n} \alpha_i K(\mathbf{x}, \mathbf{x}_i) + \langle f^{\perp}, K(\cdot, \mathbf{x})\rangle_{\mathcal{H}} = \sum_{i=1}^{n} \alpha_i K(\mathbf{x}, \mathbf{x}_i).$$

We thus see that the term $f^{\perp}$ is irrelevant for the value of c in (2.8). Making use of the fact that Ω is monotonically decreasing we get the following inequality

$$\Omega\left(\|f\|_{\mathcal{H}}^2\right) = \Omega\left(\left\|\sum_{i=1}^{n}\alpha_i K(\cdot,\mathbf{x}_i)\right\|_{\mathcal{H}}^2 + \|f^{\perp}\|_{\mathcal{H}}^2\right) \geq \Omega\left(\left\|\sum_{i=1}^{n}\alpha_i K(\cdot,\mathbf{x}_i)\right\|_{\mathcal{H}}^2\right).$$

Thus for any fixed $\alpha_i \in \mathbb{R}$ the function Ω in (2.8) is minimized for $f^{\perp} = 0$. □

This theorem has far reaching consequences. It tells us that whenever we can formulate an objective function in the form of (2.8) we can rest assured that the solution can be expressed in terms of finitely many kernel evaluations. Even if the function space is infinite dimensional, we only need to search for the n expansion coefficients.

Monotonicity of Ω does not ensure a unique minimizer of (2.8), we have to require convexity to prevent the possibility of several solutions and indeed many algorithms make use of convex loss functions. The strictness of the monotonicity of Ω can be discarded but then there might be minimizers of (2.8) that do not admit the form (2.9); it still follows that there is at least one other solution that is as good and that does admit the expansion form.

The minimizer of the regularized risk formulation should on the one hand minimize the training error, as measured by the cost term c and, on the other hand, have a low norm. Since the function spaces are usually extremely rich, for most problems there will be functions that incur no cost at all, for example by just memorizing the examples. However such solutions will be arbitrarily complex and therefore will not generalize well. The regularizer Ω can be understood as resolving this issue. Loosely speaking it can be seen as favouring smooth functions, where smoothness measured by the RKHS norm $\|\cdot\|_{\mathcal{H}}$ (see Schölkopf and Smola (2002) for a detailed review of its regularization properties).

2.4 Learning with kernels

With the ingredients of the last two sections we can now introduce some kernel based algorithms. Given a training set of observations $(\mathbf{x}_1, y_1), \ldots, (\mathbf{x}_n, y_n) \in \mathcal{X} \times \mathcal{Y}$ we aim to find a function $f: \mathcal{X} \to \mathcal{Y}$ that minimizes the empirical risk on this dataset. For binary classification problems where $\mathcal{Y} = \{-1, +1\}$, this can be posed as the search of a function $f: \mathcal{X} \to \mathbb{R}$ that maximizes the agreement between $\operatorname{sgn} f(\mathbf{x})$ and the label of the pattern $y(\mathbf{x})$. The space of functions we are searching over is implicitly defined by the kernel that is used as a similarity measure between the data points. From the representer theorem we know which parameters we have to search over, namely the coefficients of the kernel expansion. Note that one can think of the functions as being linear functions in a high, or even infinite dimensional, space. The functions can thus also be written as $f(\mathbf{x}) = \langle \mathbf{w}, \phi(\mathbf{x})\rangle_{\mathcal{H}}$ or an affine function $f(\mathbf{x}) = \langle \mathbf{w}, \phi(\mathbf{x})\rangle_{\mathcal{H}} + b$, with $\mathbf{w} \in \mathcal{H}$ and $b \in \mathbb{R}$.

Minimizing the empirical risk with respect to the parameters $(\mathbf{w}, b)$ confronts us with several problems. To begin with, the minimization is a NP hard problem already shown by Minsky and Papert (1969). Even approximately minimizing the empirical risk is NP hard not only for linear functions but also for other simple geometrical objects such as spheres (Ben-David *et al.* 2000). Furthermore the true target function, the indicator function

$\delta_{f(\mathbf{x}) \neq y}$ is discontinuous and small changes in f may lead to large changes in both empirical and expected risk.

In order to overcome the difficulties arising from the exact minimization of the empirical risk we will use upper bounds and minimize those instead. This is not only computationally effective but has also the benefit of yielding consistent estimators (Hofmann *et al.* 2008).

2.4.1 Support vector classification

Consider binary classification with input data $\{(\mathbf{x}_1, y_n), \ldots, (\mathbf{x}_n, y_n)\} \subset \mathcal{X} \times \{-1, +1\}$. We seek a function assigning to each point $\mathbf{x} \in \mathcal{X}$ its label sign. The search for $f : \mathcal{X} \to \mathbb{R}$ over the function space $\mathcal{H}$ is implemented as the following regularized risk functional (Cortes and Vapnik 1995; Vapnik 1995)

$$\min_{f \in \mathcal{H}} \frac{1}{2} \|f\|_{\mathcal{H}}^2 + C \sum_{i=1}^{n} V(y_i, f(\mathbf{x}_i)). \tag{2.10}$$

Typical choices for the loss function are $V(y, t) = \max\{0, 1 - yt\}^p$ with $p = 1, 2$. For $p = 1$ the loss is usually referred to as the *Hinge loss*, for $p = 2$ as the *quadratic loss*. Both choices are upper bounds on the indicator function $\delta_{y \neq f(\mathbf{x})}$ and both functions are convex in the second argument. It is due to this convexity that the minimizer of (2.10) is unique. We have also introduced the regularization parameter $C \in \mathbb{R} \cup \infty$ to the optimization problem to control the trade-off between the smoothness of the function measured by $\|f\|_{\mathcal{H}}$ and the ability of the function to explain the data correctly. We may as well set $C = \infty$ in which case we enforce the solution f to incur no loss at all (Vapnik and Lerner 1963). For this choice it might be possible that the problem becomes infeasible because the function class $\mathcal{H}$ may not contain a function perfectly separating the data. The regularization constant C is usually selected using a cross-validation scheme as it is done for selecting the appropriate kernel for the problem at hand.

The representer theorem states that the solution of (2.10) can be written in terms of kernel expansions around the training points. This fact can also be seen from taking the derivative of (2.10) with respect to f, assuming a differentiable loss. At the optimal solution f^* the gradient vanishes

$$f^* + C \sum_{i=1}^{n} \frac{\partial V}{\partial t}(y_i, f^*(\mathbf{x}_i)) K(\cdot, \mathbf{x}_i) = 0,$$

where we made use of the reproducing property of the kernel and denote by $\partial V / \partial t$ the partial derivative of V w.r.t. its second argument. This allows us to reformulate the optimization problem and obtain the equivalent problem to (2.10)

$$\min_{\alpha \in \mathbb{R}^n} \sum_{i,j=1}^{n} \alpha_i \alpha_j K(\mathbf{x}_i, \mathbf{x}_j) + C \sum_{i=1}^{n} V\left(y_i, \sum_{j=1}^{n} \alpha_j K(\mathbf{x}_i, \mathbf{x}_j)\right). \tag{2.11}$$

The SVM formulation (2.10) is a quadratic program and can thus be solved efficiently (Boyd and Vandenberghe 2004). Several algorithmic strategies have been proposed for this particular problem, see for example the SMO algorithm (Platt 1999). But of course for a differentiable loss it is conceptually easier to resort to simple gradient descent techniques. A detailed analysis of a Newton optimization scheme can be found in Chapelle (2007).

2.4.2 Support vector regression

As the next example of a kernel based learning algorithm we consider the task of regression with target space $\mathcal{Y} \subseteq \mathbb{R}$. Again we use a regularized risk functional and write

$$\min_{f \in \mathcal{H}} \frac{1}{2} \|f\|^2_{\mathcal{H}} + C \sum_{i=1}^{n} V(y_i, f(\mathbf{x}_i)).$$

Several loss functions can be used, for example the *ϵ-insensitive loss* $V_\epsilon(y, t) = \max\{0, |y - t| - \epsilon\}$ as proposed in Drucker *et al.* (1997), Vapnik (1995) and Vapnik *et al.* (1997) or the simple quadratic loss $V(y, t) = (y - t)^2$ yielding penalized least squares regression (Hoerl and Kennard 1970; Morozov 1984; Tikhonov 1963; Wahba 1990). The representer theorem always ensures a finite representation of the optimal function. Plugging this into the problem and using the quadratic loss we obtain the following closed form solution for the optimal parameters α^*, where we assume $C > 0$

$$\begin{aligned}\alpha^* &= \arg\min_{\alpha \in \mathbb{R}^n} \frac{1}{2} \alpha^\top \mathbf{K}\alpha + C\|\mathbf{K}\alpha - \mathbf{y}\|^2 \\ &= \left(\mathbf{K} + \frac{1}{2C}\mathbf{I}\right)^{-1} \mathbf{y},\end{aligned}$$

with $\mathbf{y} = (y_1, \ldots, y_n)^\top$. Due to the structure of the solution this is also referred to as *kernel ridge regression*, the 'ridge' $1/2C$ added to the kernel matrix is a consequence of the regularizer. Using the shorthand $\mathbf{K}_\mathbf{x}(\cdot) = (K(\cdot, \mathbf{x}_1), \ldots, K(\cdot, \mathbf{x}_n))^\top$ the prediction function becomes

$$f(\mathbf{x}) = \mathbf{K}_\mathbf{x}(\mathbf{x})^\top \left(\mathbf{K} + \frac{1}{2C}\mathbf{I}\right)^{-1} \mathbf{y}.$$

2.4.3 Gaussian processes

Another way of looking at the regression problem is from the viewpoint of Gaussian Processes (GPs). Gaussian processes provide a probabilistic approach for kernel learning and are not limited to regression. For a more general overview of the GP framework we refer to the work of MacKay (1998) and Rasmussen and Williams (2006).

A GP defines a distribution over functions $f : \mathcal{X} \to \mathbb{R}$ and is fully described by a mean $m : \mathcal{X} \to \mathbb{R}$ and covariance function $K : \mathcal{X} \times \mathcal{X} \to \mathbb{R}$

$$m(\mathbf{x}) = \mathbb{E}[f(\mathbf{x})], \qquad K(\mathbf{x}, \mathbf{x}') = \mathbb{E}[(f(\mathbf{x}) - m(\mathbf{x}))(f(\mathbf{x}') - m(\mathbf{x}'))].$$

For notational simplicity we will set m to be the zero function. Given a finite collection of data $\mathbf{x}_1, \ldots, \mathbf{x}_n$ we first compute its covariance matrix $\mathbf{K}_{\mathbf{xx}}$ in the same way that we did for the Gram matrix (2.2). The covariance matrix defines a distribution over the vector of output values $f_{\mathbf{x}} = (f(\mathbf{x}_1), \ldots, f(\mathbf{x}_n))^\top$

$$f_{\mathbf{x}} \sim \mathcal{N}(\mathbf{0}, \mathbf{K}_{\mathbf{xx}}), \tag{2.12}$$

which is a multivariate Gaussian distribution. Therefore the specification of the covariance function implies the form of the distribution over the functions. The role of the covariance for GPs is the same as the role of kernels we used so far, both specify the notion of similarity. This is also the reason why we choose K to denote both quantities.

Let us consider again the task of real-valued regression. Given training data pairs of inputs $(\mathbf{x}_1, y_1), \ldots, (\mathbf{x}_n, y_n)$ the goal is to predict the output value y^* for a new test data point $\mathbf{x}^*$. We will assume that the output values we have access to are only noisy observations of the true underlying function $y = f(\mathbf{x}) + \epsilon$. Furthermore we assume the noise to be additive independently identically Gaussian distributed with zero mean and variance σ. For notational convenience we define the following vectors, using bold symbols for vectorial variables: the stacked output values $\mathbf{y} = (y_1, \ldots, y_n)^\top$, the covariance terms of the test point $\mathbf{K}_* = (K(\mathbf{x}^*, \mathbf{x}_1), \ldots, K(\mathbf{x}^*, \mathbf{x}_n))^\top$ as well as $\mathbf{K}_{**} = K(\mathbf{x}^*, \mathbf{x}^*)$. From the model assumption (2.12) we know that the output values are distributed according to

$$\begin{bmatrix} \mathbf{y} \\ f(\mathbf{x}^*) \end{bmatrix} \sim \mathcal{N}\left(\mathbf{0}, \begin{pmatrix} (\mathbf{K} + \sigma^2\mathbf{I}) & \mathbf{K}_* \\ \mathbf{K}_*^\top & \mathbf{K}_{**} \end{pmatrix}\right).$$

The predictive equations for the Gaussian process we are interested in are then obtained by computing the conditional distribution

$$f(\mathbf{x}^*)|\mathbf{y}, \{\mathbf{x}_1, \ldots, \mathbf{x}_n\}, \mathbf{x}^* \sim \mathcal{N}\left(\mathbf{K}_*^\top(\mathbf{K} + \sigma^2\mathbf{I})^{-1}\mathbf{y}, \mathbf{K}_{**} - \mathbf{K}_*^\top(\mathbf{K} + \sigma^2\mathbf{I})^{-1}\mathbf{K}_*\right).$$

Comparing this with (2.4.2) we observe that the GP mean predictor is exactly the same solution that we have obtained for Kernel Ridge Regression. The noise variance term σ^2 appeared as a regularization constant in the kernel ridge regression case. Also note that we have witnessed yet another manifestation of the representer theorem, which we not used explicitly. The difference from Kernel Ridge Regression is that not only a mean prediction is defined but we obtained a full distribution over the output values including an *uncertainty* of the prediction.

2.4.4 Multiple kernel learning

A new development in the field of kernel learning that goes in the direction of automatically selecting an appropriate kernel for a given task is the framework of Multiple Kernel Learning (Bach *et al.* 2004; Lanckriet *et al.* 2004; Rakotomamonjy *et al.* 2008; Sonnenburg *et al.* 2006). Instead of using only one kernel K, or equivalently only one feature map ϕ one uses a set of M different kernels $K_1, \ldots, K_M$ and optimizes for a linear combination of those. This corresponds to the search of a function $f(\mathbf{x}) = \sum_{m=1}^{M} f_m(\mathbf{x})$ where each function $f_m \in \mathcal{H}_m$ is an element of the RKHS $\mathcal{H}_m$ of the kernel K_m. The associated optimization problem can be formulated as a joint convex problem in all parameters. We use the convention that $0/0 := 0$

and $x/0 = \infty$, $x \neq 0$ to write the MKL formulation as

$$\min_{f_m \in \mathcal{H}_m, d_m} \frac{1}{2} \sum_{m=1}^{M} \frac{1}{d_m} \| f_m \|_{\mathcal{H}_m}^2 + C \sum_{i=1}^{n} V \left(y_i, \sum_{m=1}^{M} f_m(\mathbf{x}_i) \right) \tag{2.13}$$

$$\text{subject to} \sum_{m=1}^{M} d_m = 1, d_m \geq 0 \quad m = 1, \ldots, M.$$

Skipping some details we directly mention that the final kernel will be of the form

$$K(\mathbf{x}, \mathbf{x}') = \sum_{m=1}^{M} d_m K_m(\mathbf{x}, \mathbf{x}')$$

with $d_1, \ldots, d_M \in [0, 1]$ and thus the final function can be written as

$$f(\mathbf{x}) = \sum_{i=1}^{n} \alpha_i \sum_{m=1}^{M} d_m K_m(\mathbf{x}, \mathbf{x}_i).$$

The same formulation can also be applied with an infinite number of kernels $K_1, K_2, \ldots$, e.g. all Gaussian kernels with positive definite covariance matrix. The optimum is still expressible, that is there is only a finite number of nonzero weights $d_m > 0$ (Argyriou *et al.* 2006; Gehler and Nowozin 2008; Özöğür-Akyüz and Weber 2008).

Comparing this with the SVM optimization problem (2.11) we see that the only difference is the new parameters d_m, which are also minimized over and that the final kernel is now a convex combination of kernels. The weight d_m controls the influence of the norm of the function f_m on the overall objective value. Whenever $d_m = 0$ the corresponding function f_m must be the zero function in order to yield a finite objective value. The simplex constraint of the d_m ensures that the resulting kernel is always positive definite. One could in principle allow for negative weights, in which case it is not assured that the resulting kernel is positive definite. This would have to be enforced as a constraint, turning the problem into a semi-definite programming problem (SDP) for which the best known algorithm scales $O(n^6)$ for an $n \times n$ constraint matrix (Lanckriet *et al.* 2004).

With the new parameters $d_1, \ldots, d_M$ one has the additional freedom to search also over a suitable feature space from which the final function is taken. Therefore the optimization problem (2.13) is over both the feature space and a function therein. Disregarding some details, for a fixed choice of the parameters d_m the associated feature space is the vector product of all participating feature spaces $\mathcal{H}_1, \ldots, \mathcal{H}_M$ for which $d_m > 0$. A detailed derivation can be found in Rakotomamonjy *et al.* (2008).

Several different algorithms have been proposed to solve this problem, e.g. Rakotomamonjy *et al.* (2007) and Sonnenburg *et al.* (2006). For example, Chapelle *et al.* (2002) and Rakotomamonjy *et al.* (2007) derive a simple gradient descent algorithm that alternates between updating the parameters d_m and the SVM parameters. For fixed weights d_m one can compute the joint Gram matrix and optimize over the SVM parameter using any standard SVM algorithm.

With MKL one can optimize over a linear combination of kernels instead of selecting between them via cross-validation. One has to bear in mind that the latter might be a better estimator for the empirical risk, which is what we are ultimately interested in. In visual object classification MKL has been used with much success to combine different image representations (Bosch *et al.* 2007). For example, one kernel could be defined solely on colour information while another is defined on shape or appearance information. For the spatial kernel one can select between different spatial layouts associating to each subwindow a different kernel and learning the combination via MKL. Sonnenburg *et al.* (2006) have used the feature selection correspondence of MKL to build *interpretable* SVMs for splice data from computational biology.

2.4.5 Structured prediction using kernels

So far we have considered only very simple target spaces $\mathcal{Y}$, for example $\mathcal{Y} = \{-1, +1\}$ and $\mathcal{Y} = \mathbb{R}$. This however is a very limited scenario as in many tasks the objects of interest are more complex than being only binary class membership. For example ranking a set of web pages according to their relevance for a given query is a task that is not easily expressible in the previously used framework. Making predictions about graph layouts, entire image patches or multi-label problems are a few more of examples that call for a more general framework.

We want to apply kernel methods to all these kinds of problems and the following simple modification of the kernel function allows us to reuse the results we have obtained so far (Altun *et al.* 2004; Cai and Hofmann 2004; Tsochantaridis *et al.* 2005). We extend the class of functions to be of the form

$$f : \mathcal{X} \times \mathcal{Y} \to \mathbb{R}, \quad (\mathbf{x}, y) \mapsto f(\mathbf{x}, y),$$

i.e. they depend on elements of *both* input and target space. Since the predictions we are interested in live in the space $\mathcal{Y}$ we will use the following prediction rule

$$y^*(\mathbf{x}) = \arg\max_{y \in \mathcal{Y}} f(\mathbf{x}, y). \tag{2.14}$$

Note that besides this new prediction function little has changed from the setup we have developed so far. We just extended the input space and restricted the output space to always be $\mathbb{R}$. The feature map is of the form $\phi : \mathcal{X} \times \mathcal{Y} \to \mathcal{H}$, the corresponding kernel is the dot product in the RKHS $\mathcal{H}$

$$K(\mathbf{x}, y, \mathbf{x}', y') = \langle \phi(\mathbf{x}, y), \phi(\mathbf{x}', y') \rangle_{\mathcal{H}}$$

and the representer theorem ensures us that the solutions of regularized risk functionals can be expressed in terms of expansions around training data points. The dependency of f on the target space opens up the possibility of taking into account its structure.

The loss function can also encode the structure in the target set. In binary classification there is no such structure beyond two elements being equal (belonging to the same class) or different (belonging to separate classes). Now consider the task of retrieving a ranked list of websites given some text query. Missing the most relevant website should incur a higher cost than missing the tenth most relevant website. This is encoded using a loss function of the type $\Delta : \mathcal{Y} \times \mathcal{Y} \to \mathbb{R}_+$. We will think of $\Delta(y, y')$ as the cost of predicting y' where it should have been y and therefore set $\Delta(y, y) = 0\ \forall y \in \mathcal{Y}$. If the maximum of (2.14) is taken at the correct labelling no cost is produced. Predicting a different y incurs a cost which, however, depends on the similarity of the true and the predicted output. This gives rise to the following regularized risk formulation (Tsochantaridis *et al.* 2005)

$$\min_{f\in\mathcal{H}} \frac{1}{2}\|f\|^2_{\mathcal{H}} + C\sum_{i=1}^{n} \max\left\{0, \max_{y\in\mathcal{Y}} (\Delta(y_i, y) - (f(\mathbf{x}_i, y_i) - f(\mathbf{x}_i, y)))\right\}.$$

This is a convex problem and can be efficiently solved whenever the inner maximization, that is (2.14), can be solved efficiently. Usually this is not the case and one can only hope for approximate solutions. A standard way of solving the problem is by dualizing the problem and using column-generation techniques (Bennett *et al.* 2000; Hettich and Kortanek 1993).

For the regularized risk formulation above the optimal function is of the form

$$f(\cdot, \cdot) = \sum_{i=1}^{n}\sum_{y\in\mathcal{Y}} \alpha_{iy} K((\cdot, \cdot), (\mathbf{x}_i, y)).$$

We will outline only a few applications of this model and refer to (Bakir *et al.* 2007) for a more detailed overview.

- The classical binary setup is recovered by simply setting $\phi(\mathbf{x}, y) = y\phi(\mathbf{x})$ and $\Delta(y, y') = \delta_{y=y'}$. The inner maximum reduces to $1 - 2y_i \sum_{j=1}^{n} \alpha_j y_j K(\mathbf{x}_i, \mathbf{x}_j)$, which (ignoring some offsets) yields exactly the SVM optimization problem (2.10) with the hinge loss.
- Multiclass classification (Allwein *et al.* 2000; Collins 2002; Crammer and Singer 2000; Rätsch *et al.* 2003) can be cast in this framework. Let N denote the number of classes. Then $y \in \{1, \ldots, N\}$ and the loss function is a multicategory version of the $0 - 1$ loss, namely $\Delta(y, y') = 1 - \delta_{y,y'}$. Corresponding kernels are typically chosen to be $\delta_{y,y'} K(\mathbf{x}, \mathbf{x}')$.
- In the case of multilabel estimation one is interested in predicting a set of labels $y \in 2^{\{1,\ldots,n\}}$ for each input point. This is described in Elisseeff and Weston (2001), where a ranking scheme is devised such that $f(\mathbf{x}, i) > f(\mathbf{x}, j)$ if the label $i \in y$ but $j \notin y$.

2.4.6 Kernel principal component analysis

Principal Component Analysis (PCA) is a widely used algorithm with many applications such as feature extraction, dimensionality reduction and data visualization. Its benefits are that it is easy to compute and easy to interpret. Given some data $\mathbf{x}_1, \ldots, \mathbf{x}_n$ the PCA is an orthogonal projection of these points onto their principal axis, which are those which minimize the average projection cost measured as the squared distance between the points and their projections. The PCA algorithm boils down to an eigenvalue decomposition of the empirical covariance matrix of the data $C_{\text{emp}} = \mathbf{E}_{\text{emp}}[(\mathbf{x} - \mathbf{E}_{\text{emp}}(\mathbf{x}))(\mathbf{x} - \mathbf{E}_{\text{emp}}(\mathbf{x}))^\top]$, i.e. solving the system of equations $\mathbf{C}_{\text{emp}}\mathbf{v}_k = \lambda_k \mathbf{v}_k$. For d-dimensional data $\mathbf{x}_i$ this problem can be solved in $O(d^3)$ time.

In Schölkopf *et al.* (1998) this problem is posed in the feature space by simply replacing $\mathbf{x}$ with $\phi(\mathbf{x})$. Since the empirical covariance lies in the span of $\{\phi(\mathbf{x}_1), \ldots, \phi(\mathbf{x}_n)\}$ we can compute it in terms of kernel evaluations at the data points. For notational convenience we assume that we already centred the data in the feature space such that $\sum_{i=1}^{n} \phi(\mathbf{x}_i) = 0$. Then we can write out the eigenvalue problem as

$$C_{\text{emp}} \mathbf{v}_k = \frac{1}{n} \sum_{i=1}^{n} \phi(\mathbf{x}_i) \phi(\mathbf{x}_i)^{\top} \mathbf{v}_k = \lambda_k \mathbf{v}_k \tag{2.15}$$

and thus see that the eigenvectors are of the form $\mathbf{v}_k = \sum_{i=1}^{n} \alpha_{ki} \phi(\mathbf{x}_i)$. Resubstituting this into (2.15) we find that the coefficients α are easily computed by the Eigenvalue problem

$$\mathbf{K}\alpha = \lambda \alpha$$

with $\mathbf{K}$ being the kernel matrix of the data. Having solved this eigenvalue problem we can compute the projection of any point $\mathbf{x}$ onto the kth principal component of the data as $\langle \mathbf{v}_k, \phi(\mathbf{x}) \rangle = \sum_{i=1}^{n} \alpha_{ki} K(\mathbf{x}, \mathbf{x}_i)$.

Kernel PCA can be used as a preprocessing step for algorithms which are not 'kernelizable', that is algorithms not based entirely on dot products.

2.4.7 Applications of support vector algorithms

Probably much of the success of kernel based algorithms and the SVM and SVR formulation in particular is due to the empirical success on diverse practical problems. Just to name a few we mention that SVMs were used to achieve the best classification scores on the MNIST benchmark dataset for handwritten recognition (DeCoste and Schölkopf 2002) and visual object recognition (Blanz *et al.* 1996; Bosch *et al.* 2007; Chapelle *et al.* 1999). Other applications include Object Detection (Romdhani *et al.* 2004), microarray processing tasks (Brown *et al.* 2000), text categorization (Dumais 1998), ranking (Herbrich *et al.* 2000), novelty detection (Hayton *et al.* 2001) and many more. More recently interdependent label problems have been attacked using SVMs (McCallum *et al.* 2005; Tsochantaridis *et al.* 2005).

Several authors applied kernel learning algorithms to image processing applications. While this book provides an overview of the special area of remote sensing we refer the reader to Camps-Valls *et al.* (2007) for applications on the classical problems of image processing, namely image coding, image de-noising and image segmentation. In Kim *et al.* (2005) image models based on Kernel PCA are proposed and it is shown that they perform well on image de-noising and super-resolution tasks.

2.4.8 Available software

There are several software packages using kernel algorithms available. For example for SVM and SVR optimization the most prominent to date are LibSVM (Chang and Lin 2001), SVM-Struct (Tsochantaridis *et al.* 2005), SVMlight (Joachims 1999) but also plain stochastic gradient descent (Bottou 2008). The website www.mloss.org, accompanied by a special track in the *Journal of Machine Learning Research* on open source software, makes an effort to support the dissemination of machine learning software. It provides a good source to obtain implementations for many kernel learning algorithms including those reviewed in this chapter.

2.5 Conclusion

In this chapter we introduced the most basic concepts of positive definite kernels and presented some algorithms that build on those. The main idea is that positive definite kernels provide a measure of similarity between possibly complex objects. With the regularized risk framework one can implement the search over rich classes of functions and still obtain functions that can be expressed in finitely many terms of kernel evaluations. Another benefit is that this search can be made convex and thus yield problems that can be solved efficiently but on the other hand guarantee global optimality.

References

Allwein, E.L., Schapire, R.E. and Singer, Y. (2000) Reducing multiclass to binary: a unifying approach for margin classifiers. In P. Langley (ed.) *Proc. Intl. Conf. Machine Learning*, pp. 9–16. San Francisco, California: Morgan Kaufmann Publishers.

Altun, Y., Smola, A.J. and Hofmann, T. (2004) Exponential families for conditional random fields. *Uncertainty in Artificial Intelligence (UAI)*, pp. 2–9. AUAI Press, Arlington, Virginia.

Argyriou, A., Hauser, R., Micchelli, C.A. and Pontil, M. (2006) A dc-programming algorithm for kernel selection. *ICML '06*, pp. 41–48. ACM, New York, NY, USA.

Aronszajn, N. (1950) Theory of reproducing kernels. *Trans. Amer. Math. Soc.* **68**, 337–404.

Bach, F.R., Lanckriet, G.R.G. and Jordan, M.I. (2004) Multiple kernel learning, conic duality, and the SMO algorithm. *ICML '04: Proceedings of the Twenty-First International Conference on Machine Learning*, p. 6. ACM, New York, NY, USA.

Bakir, G., Hofmann, T., Schölkopf, B., Smola, A., Taskar, B. and Vishwanathan, S.V.N. (2007) *Predicting Structured Data*. MIT Press, Cambridge, Massachusetts.

Ben-David, S., Eiron, N. and Long, P.M. (2000) On the difficulty of approximately maximizing agreements. *Proc. 13th Annual Conference on Computer Learning Theory*, pp. 266–274. Morgan Kaufmann, San Francisco.

Bennett, K.P., Demiriz, A. and Shawe-Taylor, J. (2000) A column generation algorithm for boosting. In P. Langley (ed.) *Proc. Intl. Conf. Machine Learning*, pp. 65–72. San Francisco, Morgan Kaufmann Publishers.

Berg, C., Christensen, J.P.R. and Ressel, P. (1984) *Harmonic Analysis on Semigroups*. Springer, New York.

Blanz, V., Schölkopf, B., Bülthoff, H., Burges, C., Vapnik, V. and Vetter, T. (1996) Comparison of view-based object recognition algorithms using realistic 3D models. In C. von der Malsburg, W. von Seelen, J.C. Vorbrüggen and B. Sendhoff (eds). *Artificial Neural Networks ICANN'96*, vol. 1112 of *Lecture Notes in Comput. Science*, pp. 251–256. Springer-Verlag, Berlin.

Bosch, A., Zisserman, A. and Munoz, X. (2007) Representing shape with a spatial pyramid kernel. *CIVR '07: Proceedings of the 6th ACM International Conference on Image and Video Retrieval*, pp. 401–408. ACM, New York, NY, USA.

Bottou, L. (2008) Software available at http://leon.bottou.org/projects/sgd.

Boyd, S. and Vandenberghe, L. (2004) *Convex Optimization*. Cambridge University Press, Cambridge, UK.

Brown, M., Grundy, W., Lin, D., Cristianini, N., Sugnet, C., Furey, T., Ares, M. and Haussler, D. (2000) Knowledge-based analysis of microarray gene expression data by using support vector machines. *Proc. Natl. Acad. Sci.* **97**, 262–267.

Cai, L. and Hofmann, T. (2004) Hierarchical document categorization with support vector machines. *Proceedings of the Thirteenth ACM Conference on Information and Knowledge Management*, pp. 78–87. ACM Press, New York, NY, USA.

Camps-Valls, G., Rojo-Álvarez, J.L. and Martínez-Ramón, M. (2007) *Kernel Methods in Bioengineering, Signal and Image Processing*. Idea Group.

Chang, C.C. and Lin, C.J. (2001) *LIBSVM: A Library for Support Vector Machines*. Software available at http://www.csie.ntu.edu.tw/~cjlin/libsvm.

Chapelle, O. (2007) Training a support vector machine in the primal. *Neural Computation*, **19**(5), 1155–1178.

Chapelle, O., Haffner, P. and Vapnik, V. (1999) SVMs for histogram-based image classification. *IEEE Transactions on Neural Networks*, **10**(5), 1055–1064.

Chapelle, O., Vapnik, V., Bousquet, O. and Mukherjee, S. (2002) Choosing multiple parameters for support vector machines. *Machine Learning*, **46**(1–3), 131–159.

Collins, M. (2002) Discriminative training methods for hidden Markov models. *Proceedings of the Conference on Empirical Methods in Natural Language Processing*, vol. 10, pp. 1–8. Association for Computational Linguistics, Morristown, NJ, USA.

Cortes, C. and Vapnik, V. (1995) Support vector networks. *Machine Learning*, **20**(3), 273–297.

Cox, D.D. and O'Sullivan, F. (1990) Asymptotic analysis of penalized likelihood and related estimators. *The Annals of Statistics*, **18**(4), 1676–1695.

Crammer, K. and Singer, Y. (2000) On the learnability and design of output codes for multiclass problems. In N. Cesa-Bianchi and S. Goldman (eds), *Proc. Annual Conf. Computational Learning Theory*, pp. 35–46. Morgan Kaufmann Publishers, San Francisco, CA.

DeCoste, D. and Schölkopf, B. (2002) Training invariant support vector machines. *Machine Learning*, **46**, 161–190. Also: Technical Report JPL-MLTR-00-1, Jet Propulsion Laboratory, Pasadena, CA, 2000.

Drucker, H., Burges, C.J.C., Kaufman, L., Smola, A.J. and Vapnik, V. (1997) Support vector regression machines. In M.C. Mozer, M.I. Jordan and T. Petsche (eds), *Advances in Neural Information Processing Systems 9*, pp. 155–161. MIT Press, Cambridge, MA.

Dumais, S. (1998) Using SVMs for text categorization. *IEEE Intelligent Systems*, **13**(4), 18–28. In M.A. Hearst, B. Schölkopf, S. Dumais, E. Osuna and J. Platt (eds), *Trends and Controversies–Support Vector Machines*.

Elisseeff, A. and Weston, J. (2001) A kernel method for multi-labeled classification. *Advances in Neural Information Processing Systems 14*, pp. 681–687. MIT Press, Cambridge, MA.

FitzGerald, C.H., Micchelli, C.A. and Pinkus, A. (1995) Functions that preserve families of positive semidefinite matrices. *Linear Algebra Appl.* **221**, 83–102.

Gehler, P.V. and Nowozin, S. (2008) Infinite kernel learning. Technical Report 178, Max Planck Institute for Biological Cybernetics.

Grauman, K. and Darrell, T. (2005) The pyramid match kernel: discriminative classification with sets of image features. *ICCV '05: Proceedings of the Tenth IEEE International Conference on Computer Vision*, pp. 1458–1465. IEEE Computer Society, Washington, DC, USA.

Haussler, D. (1999) Convolutional kernels on discrete structures. Technical Report UCSC-CRL-99-10, Computer Science Department, UC Santa Cruz.

Hayton, P., Schölkopf, B., Tarassenko, L. and Anuzis, P. (2001) Support vector novelty detection applied to jet engine vibration spectra. In T.K. Leen, T.G. Dietterich and V. Tresp (eds), *Advances in Neural Information Processing Systems 13*, pp. 946–952. MIT Press, Cambridge, MA.

Hein, M. and Bousquet, O. (2005) Hilbertian metrics and positive definite kernels on probability measures. In Z. Ghahramani and R. Cowell (eds), *Proc. of AI & Statistics*, vol. 10.

Herbrich, R., Graepel, T. and Obermayer, K. (2000) Large margin rank boundaries for ordinal regression. In A.J. Smola, P.L. Bartlett, B. Schölkopf and D. Schuurmans (eds), *Advances in Large Margin Classifiers*, pp. 115–132. MIT Press, Cambridge, MA.

Hettich, R. and Kortanek, K.O. (1993) Semi-infinite programming: theory, methods, and applications. *SIAM Rev.* **35**(3), 380–429.

Hoerl, A.E. and Kennard, R.W. (1970) Ridge regression: biased estimation for nonorthogonal problems. *Technometrics*, **12**, 55–67.

Hofmann, T., Schölkopf, B. and Smola, A.J. (2008) Kernel methods in machine learning. *Annals of Statistics*, **36**, 1171–1220.

Joachims, T. (1998) Text categorization with support vector machines: learning with many relevant features. *Proceedings of the European Conference on Machine Learning*, pp. 137–142. Springer, Berlin.

Joachims, T. (1999) Making large-scale SVM learning practical. In B. Schölkopf, C.J.C. Burges and A.J. Smola (eds), *Advances in Kernel Methods – Support Vector Learning*, pp. 169–184. MIT Press, Cambridge, MA.

Kim, K.I., Franz, M.O. and Schölkopf, B. (2005) Iterative kernel principal component analysis for image modelling. *IEEE Transactions on Pattern Analysis and Machine Intelligence*, **27**(9), 1351–1366.

Kimeldorf, G.S. and Wahba, G. (1971) Some results on Tchebycheffian spline functions. *J. Math. Anal. Appl.* **33**, 82–95.

Lanckriet, G.R.G., Cristianini, N., Bartlett, P., Ghaoui, L.E. and Jordan, M.I. (2004) Learning the kernel matrix with semidefinite programming. *Journal of Machine Learning Research*, **5**, 27–72.

Lazebnik, S., Schmid, C. and Ponce, J. (2006) Beyond bags of features: spatial pyramid matching for recognizing natural scene categories. *CVPR '06: Proceedings of the 2006 IEEE Computer Society Conference on Computer Vision and Pattern Recognition*, pp. 2169–2178. IEEE Computer Society, Washington, DC, USA.

MacKay, D.J.C. (1998) Introduction to Gaussian processes. In C.M. Bishop (ed.) *Neural Networks and Machine Learning*, Springer Berlin pp. 133–165.

McCallum, A., Bellare, K. and Pereira, F. (2005) A conditional random field for discriminatively-trained finite-state string edit distance. *Conference on Uncertainty in AI (UAI)*, p. 388. AUAI Press, Arlington, Virginia.

Minsky, M. and Papert, S. (1969) *Perceptrons: An Introduction To Computational Geometry*. MIT Press, Cambridge, MA.

Morozov, V.A. (1984) *Methods for Solving Incorrectly Posed Problems*. Springer-Verlag, New York.

Özöğür-Akyüz, S. and Weber, G.W. (2008) Learning with infinitely many kernels via semi-infinite programming. *Proceedings of Euro mini conference on 'Continuous Optimization and Knowledge Based Technologies'*.

Platt, J. (1999) Fast training of support vector machines using sequential minimal optimization. In B. Schölkopf, C.J.C. Burges and A.J. Smola (eds), *Advances in Kernel Methods – Support Vector Learning*, pp. 185–208. MIT Press, Cambridge, MA.

Rakotomamonjy, A., Bach, F., Canu, S. and Grandvalet, Y. (2007) More efficiency in multiple kernel learning. *ICML '07, Proceedings of the 24th International Conference on Machine Learning*, pp. 775–782. ACM Press, New York, NY, USA.

Rakotomamonjy, A., Bach, F., Grandvalet, Y. and Canu, S. (2008) SimpleMKL. *Journal of Machine Learning Research*, **9**, 2491–2521.

Rasmussen, C.E. and Williams, C.K.I. (2006) *Gaussian Processes for Machine Learning*. MIT Press, Cambridge, MA.

Rätsch, G., Mika, S. and Smola, A.J. (2003) Adapting codes and embeddings for polychotomies. In S. Becker, S. Thrun and K. Obermayer (eds), *Advances in Neural Information Processing Systems 15*. MIT Press, Cambridge, MA, pp. 513–520.

Romdhani, S., Torr, P.H.S., Schölkopf, B. and Blake, A. (2004) Efficient face detection by a cascaded support vector machine expansion. *Proc. R. Soc. Lond. Ser. A Math. Phys. Eng. Sci.* **460**, 3283–3297.

Schölkopf, B. (1997) *Support Vector Learning*. R. Oldenbourg Verlag, Munich. Download: http://www.kernel-machines.org.

Schölkopf, B. and Smola, A. (2002) *Learning with Kernels*. MIT Press, Cambridge, MA.

Schölkopf, B., Herbrich, R. and Smola, A.J. (2001) A generalized representer theorem. In D.P. Helmbold and B. Williamson (eds), *Proc. Annual Conf. Computational Learning Theory*, pp. 416–426 number 2111 in *Lecture Notes in Comput. Sci.* Springer-Verlag, London, UK.

Schölkopf, B., Smola, A.J. and Müller, K.R. (1998) Nonlinear component analysis as a kernel eigenvalue problem. *Neural Comput.* **10**, 1299–1319.

Shawe-Taylor, J. and Cristianini, N. (2004) *Kernel Methods for Pattern Analysis*. Cambridge University Press, Cambridge, UK.

Sonnenburg, S., Rätsch, G., Schäfer, C. and Schölkopf, B. (2006) Large scale multiple kernel learning. *J. Mach. Learn. Res.* **7**, 1531–1565.

Tikhonov, A.N. (1963) Solution of incorrectly formulated problems and the regularization method. *Soviet Math. Dokl.* **4**, 1035–1038.

Tsochantaridis, I., Joachims, T., Hofmann, T. and Altun, Y. (2005) Large margin methods for structured and interdependent output variables. *J. Mach. Learn. Res.* **6**, 1453–1484.

Vapnik, V. (1995) *The Nature of Statistical Learning Theory*. Springer, New York.

Vapnik, V. and Lerner, A. (1963) Pattern recognition using generalized portrait method. *Autom. Remote Control*, **24**, 774–780.

Vapnik, V., Golowich, S. and Smola, A.J. (1997) Support vector method for function approximation, regression estimation, and signal processing. In M.C. Mozer, M.I. Jordan and T. Petsche (eds), *Advances in Neural Information Processing Systems 9*, pp. 281–287. MIT Press, Cambridge, MA.

Wahba, G. (1990) *Spline Models for Observational Data*, vol. 59 of *CBMS-NSF Regional Conference Series in Applied Mathematics*. SIAM, Philadelphia.

II
Supervised image classification

3

The Support Vector Machine (SVM) algorithm for supervised classification of hyperspectral remote sensing data

J. Anthony Gualtieri

NASA/GSFC, CISTO & Global Science and Technology

The Support Vector Machine (SVM) algorithm provides an effective way to perform supervised classification of hyperspectral remote sensing data. The problem is to learn from a training set of examples – hyperspectral data with class labels attached – and then generalize to find the class labels of hyperspectral data outside the training set. The *high dimensionality* of hyperspectral data, due to the many spectral channels a sensor simultaneously measures, causes problems for other supervised classification algorithms, both parametric and nonparametric. A key feature of the nonparametric SVM supervised classification algorithm is its ability to use high-dimensional data *without* the usual recourse to a feature selection step to reduce the dimensionality of the data, as is required by many other algorithms in order for them to work. Additionally, SVM can utilize the Kernel method of projecting the data into a different space to improve performance. We give an introduction to hyperspectral data and its acquisition and for most of the sequel focus on using only the *spectral* information available in the data for performing the classification. We then present an overview of the mathematical foundations of statistical learning theory, show how the *large margin* SVM, appropriate to supervised

This chapter has been adapted from *Recent Advances in Hyperspectral Signal and Image Processing*, pp. 351–397. Edited by Chein-I Chang, Transworld Research Network, 2006.

Kernel Methods for Remote Sensing Data Analysis Edited by Gustavo Camps-Valls and Lorenzo Bruzzone

classification, can be derived in the context of these very general results, show its realization as a quadratic optimization problem, and indicate the Kernel method, which extends the efficacy of the SVM by using nonlinear transformation of the training data. These results are then applied to several benchmark hyperspectral data sets, and the SVM results are compared with other supervised classification methods. Then we indicate how using the *spatial* content of the data can further improve the classification results. Finally we close with an exploration of the reasons why the SVM offers improved performance over other algorithms and summarize with a brief conclusion.

3.1 Introduction

Remote sensing with imaging spectrometers that provide hundreds of channels over contiguous wavelength bands, possibly with some gaps, has immense promise as a tool to provide data for scientific and commercial applications. Some diverse examples include environmental monitoring for the U.S. Environmental Protection Agency remediation of old mine tailings (Swayze *et al.* 1997), mapping of invasive species (Ustin *et al.* 2005), disaster management in New York City after the Sept. 11, 2001 attack (Clark *et al.* 2001), precision farming (Garegnani *et al.* 2000; Goel *et al.* 2003), archeology (Bianchi *et al.* 1997) and depth measurement and shallow water benthic mapping of coral (Goodman and Ustin 2002; Gualtieri and Howard 2003; Lee *et al.* 2001). A valuable collection of work concerning development of hyperspectral remote sensing and many applications can be found at the NASA/JPL AVIRIS website, where workshop publications dating from 1987–2004 are available online (Green 2004), and where a large hyperspectral data archive can also be found. AVIRIS is an acronym for the NASA Airborne Visible InfraRed Imaging Spectrometer and is a premier instrument in the realm of Earth remote sensing from aircraft platforms. The AVIRIS programme has fostered the development and application of instrumentation, algorithms, and applications demonstrating the unique capabilities of hyperspectral remote sensing for geoscience.

To put the promise and challenge of hyperspectral remote sensing into context, at a recent AVIRIS workshop Joe Boardman of Analytical Imaging and Geophysics (Boardman 2005) put it thus: 'Our algorithms and processing systems must grow exponentially in capability to match the information content in the current and future (we're really just getting started, still).' In this chapter we will address an important hyperspectral remote sensing application, performing supervised classification on hyperspectral data, which has provided a new approach to addressing Boardman's challenge. The approach, introduced in the machine learning community in 1992, is the Support Vector Machine (SVM). The term Machine in Support Vector Machine is only a name and does not imply a hardware device. The SVM algorithm was first applied to hyperspectral data in 1998 (Gualtieri and Cromp 1998; Gualtieri *et al.* 1999), and has produced a quantitative improvement in classification accuracy of hyperspectral data, and provided a new set of tools for geoscientists to render hyperspectral data into Earth science products.

In the sequel we will summarize some aspects of hyperspectral data acquisition so that algorithm practitioners are aware of what has been done to the data that they will use in their algorithms. We will introduce the Support Vector Machine and place it in the context of the more general problem of supervised classification. We will discuss an earlier expectation that the large number of measurements associated with a pixel should have improved previous supervised classification algorithm accuracies, but have in fact led to reduced accuracies (Hughes 1968). This led to pre-processing techniques such as hand selecting channels to use, or more principled approaches, such as Principle Components Analysis (PCA), (Duda *et al.* 2001, p. 115). However, we will show for the SVM this 'curse of dimensionality' (Bellman 1961) is, in principle and in practice, not a limitation, thereby making SVM a valuable tool for supervised classification for hyperspectral data. In what follows we will relate some very general results from statistical learning theory, and from them lay out in a principled manner how this leads to a SVM called the *large margin* SVM. We go on to indicate a quadratic version that implements this idea. Then we describe its implementation in software, and show its application to a toy problem to give insight into how SVM works. We go on to apply SVM to a small, but well known benchmark data set, describe the dependencies of the classification accuracy on adjustable parameters in the algorithm, and give comparative results for its classification accuracy versus other supervised techniques. Following this we will describe some more recent developments that further improve the classification accuracy, and give some answers as to why SVMs perform so well in comparison with other methods.

3.2 Aspects of hyperspectral data and its acquisition

In hyperspectral remote sensing the data are derived from a sensor, an imaging spectrometer on an aircraft or satellite, that measures the at-sensor upwelling radiance from the ground at many contiguous spectral wavelengths. The basic data object is a hyperspectral cube, which can be thought of as a registered stack of spectral images of a scene on the ground. Thus a data element is a spectral measurement of the upwelling at sensor radiance in an instantaneous field of view from a region (pixel) on the ground in a narrow wavelength channel centred at a particular spectral wavelength λ. The cube is indexed by two ground spatial coordinates, and a spectral wavelength. More details of the actual acquisition and production that creates the cube will be described below.

Typical sensors are constructed with one or more CCD arrays (depending on the desired wavelength coverage), and image at each time step a single cross-track of ground pixels that are then dispersed across the two-dimensional CCD surface to provide an array of measurements with spectral wavelength in one direction and cross-track spatial location in the other direction. The forward motion of the sensor over time then creates the perpendicular along-track other spatial dimension. This approach is called *push broom* imaging. For push broom sensors, the wavelength ranges, depending on CCD technology, can cover 400–900 nm, with of the order of a hundred or more contiguous channels with 3–10 nm widths, or can cover from the visible out to the the near infrared with bands each of 10 nm width using multiple CCDs.

By comparison, NASA's AVIRIS sensor is unique in that it uses four spectrometers, each covering a part of the full range of 400–2500 nm, in 224 contiguous channels, with each acquiring only a single pixel at a time. A mirror sweeps perpendicular to the forward direction of the aircraft to create the cross-track. This approach is called *whisk boom* imaging. This has the advantage that no correction for variation in the individual regions of the CCD is required. However AVIRIS is mechanically more complex and expensive to build.

The set of radiances measured at one pixel can be thought of as spectra and described as a vector. The collection of these spectra comprising a scene is then assembled into a three-dimensional array indexed by the two spatial coordinate of the pixel and one spectral wavelength index, and is called a hyperspectral cube.

Typically this at-sensor radiance is post-processed to geo-register and geo-rectify the data. Geo-registration means registering the data in the scene with an Earth based coordinate system such as latitude longitude coordinates. This can be accomplished using either known control points on the ground, or data derived from on-board global positioning data to obtain platform spatial coordinates and inertial measurement data to obtain platform angular coordinates of orientation. From these measurements, geometrical computation yields the ground coordinates for each pixel. Geo-rectifiying the data means converting the position and size data of pixels taken in the frame of reference of the sensor platform, to a ground based reference system of the scene, and restructuring them so that the pixels form a rectangular grid. Owing to the motion of the sensor platform, especially on low flying aircraft, which are subject to turbulence and other motion in addition to forward translation, this implies resampling and possible interpolation will occur during geo-rectification. Thus while the data before geo-rectification is a rectangular structure, its coverage of the ground after geo-rectification may be a substantially distorted rectangle due to the effects of the platform movement. Geo-registration can be performed on the data before or after geo-rectification. That these processing steps can have dramatic consequences is seen in Figure 3.1, (Boardman 1999), which shows the dramatic distortions and corrections that can occur.

Practitioners have pointed out that it would be better to perform classification on scenes *before* geo-rectification, thereby removing errors and biases due to the re-sampling and interpolation, and then subsequently apply geo-rectification. These biases and interpolations mean that classification after geo-rectification is performed on changed spectral data, due to combining spectral information from nearby pixels in the resampling process. By comparison satellite platforms are extremely stable, and this distortion is not present, though off-nadir pixels will suffer some distortion due to the imaging geometry.

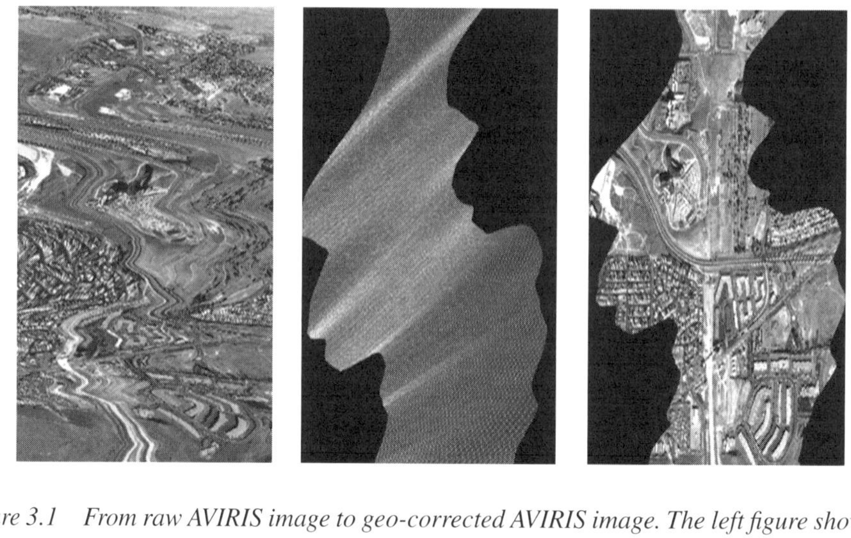

Figure 3.1 From raw AVIRIS image to geo-corrected AVIRIS image. The left figure shows an image constructed from a few bands in a hyperspectral cube acquired at low altitude. The data are rectangular, but the ground image is dramatically distorted due to aircraft motion as the sensor acquires cross-tracks sequentially from top to bottom along the flight path. The middle image shows how the cross-track actually registers to ground coordinates, The AVIRIS whisk broom acquisition is shown in the left to right lines on the diagonals. The right image shows the corrected image. Now the shapes on the ground are as expected, but the image borders are far from straight. Images courtesy NASA/JPL-Caltech.

Often the data have also been post-processed to perform atmospheric correction. This transforms the measured at-sensor based upwelling spectral radiance to a ground based spectral reflectance, a projected intrinsic property of the surface being imaged. Spectral reflectance is a dimensionless quantity with values in the range [0 1], derived from the ratio of upwelling radiance scattered from the pixel to the solar down welling radiance illuminating the pixel at a given wavelength in a small wavelength range. Atmospheric correction, (Gao and Davis 1997; Gao *et al.* 1993; Montes *et al.* 2001; Tanré *et al.* 1983, 1979), is used to remove the effects of the intervening atmosphere between the sensor and the ground on the hyperspectral data. These effects include absorption and scattering by atmospheric gases, and scattering by aerosols in the path that the radiation travels between the Sun and the ground and back to the sensor. Water vapour can vary rapidly in position and time, and thus atmospheric correction is best done on a pixel by pixel basis. Modern software packages such as ENVI, produced by ITT, provide a range of tools to handle these prepossessing steps, as well as many tools to process hyperspectral cubes into products.

We note there are other potential corrections that could be applied, such as topographic correction that recognizes that the surface scattering the light may lie at some orientation relative to level surface, or that the surface elevation has an effect on the atmospheric correction, due to varying optical path lengths from the Sun to the surface and back to the sensor.

At this point the data has been rendered by processing into a hyperspectral cube, where each element of the cube contains either a measurement of the upwelling radiance or a reflectance from a small patch of surface area described by two spatial coordinates and centred at a particular spectral wavelength covering a small spectral range. We take as our stating point – the basic data element – a vector of spectral radiances or spectral reflectances and call it a feature vector associated to a spatial pixel. It is conventional to call the number of channels the *dimensionality* of the data, though for hyperspectral data this is not the number of *independent* coordinates that the feature vector possesses. The use of dimension to describe the number of channels in hyperspectral data does not imply that the spectral radiances in the various channels are all independent of each other. Indeed, depending on the complexity of the scene and the Signal-to-Noise Ratio (SNR) of the sensor, one can estimate the number of independent pieces of information available in a scene. Various definitions of the SNR for hyperspectral sensors are possible. A simple scene dependent definition is the ratio of the mean power in the signal of an image at a given wavelength to the root mean square of the signal in the image at that wavelength. The SNR is wavelength dependent. A study across 510 scenes measured by the AVIRIS sensor taken in 1999 over varied terrain including open water, deserts, vegetation, and cities showed that there were at least 60 independent dimension in AVIRIS's 224 channels that were significantly above the noise floor of the sensor (Boardman and Green 2000).

Our problem in supervised classification is then, given a training set of feature vectors, each with a class label that attaches meaning to that vector, to use them to train a learning machine, a SVM, to provide class labels to the feature vectors without labels, called the testing set. For hyperspectral data the training set labels are based on ground truth that can be obtained by field work where ground locations are typically found by using a global positioning device or by using data from other sensors or maps. Labels can be found simply by the observation of what is on the ground, to hyperspectral measurements made with portable devices that directly measure a spectrum at a single location. At best, the ground truth is gathered coincident in time with the overflight of the sensor. However the gathering of ground truth is expensive in term of peoples' time, and typically only covers a small fraction of the scene being studied. The training set is often taken from the hyperspectral cube at hand, but can come from another source.

In summary, we wish to use these labelled examples to construct an algorithm that will subsequently classify feature vectors from the pixels in the hyperspectral cube without class labels, called the test set, according to the list of labels from the training set.

3.3 Hyperspectral remote sensing and supervised classification

In order to understand more fully why the SVM provides good classification accuracy as a supervised classifier on high dimensional data, we present an overview of some foundations. To set the stage, we quote from Burges's helpful tutorial on Support Vector Machines (Burges 1998):

> The problem which drove the initial development of SVM's occurs in several guises–the bias variance trade off (Geman *et al.* 1992), capacity control (Guyon *et al.* 1992), over fitting (Montgomery and Peck 1992), complexity regularization trade-off (Girosi *et al.* 1995)–but the basic idea is the same. Roughly speaking, for a given learning task, with a given finite amount of training data, the best generalization classification accuracy will be achieved if the right balance is struck between the accuracy attained on that particular training set, and the capacity of the machine, that is, the ability of the machine to learn any training set without error. A machine with too much capacity is like a botanist with a photographic memory who, when presented with a new tree, concludes that it is not a tree because it has a different number of leaves from anything she has seen before; a machine with too little capacity is like the botanist's lazy brother, who declares that if it is green, it is a tree. Neither can generalize well. The exploration and formalization of these concepts has resulted in one of the shining peaks of the theory of statistical learning (Vapnik 1998).

We also wish to give understanding to the issue of dimensionality of the data. Traditionally, many supervised classifiers use the training set data to build models of the underlying density in the feature space of the data for each of the various classes in the training set. Thus each class can be thought of as a distribution of points in a space of the dimension of the feature space, the number of spectral channels. Thus we can imagine a local density of such points in the feature space, and thereby model training data by local densities in the feature space.

We then seek a set of separating surfaces which partition the feature space so that each cluster of points described by its local density is separated from the other clusters. However this density estimation in high-dimensional spaces suffers from the Hughes effect (Hughes 1968; Landgrebe 1999): for a fixed amount of training data, the classification accuracy as a function of the dimension of the data (the number of hyperspectral bands) reaches a maximum and then declines with increasing dimension, because there is a limited amount of training data to estimate the larger and larger number of parameters needed to describe the model of the densities of each class. To deal with this, usually a feature selection step is first performed on the high-dimensional data to reduce its dimensionality.

In fact, first doing density estimation as a step toward supervised classification is unnecessary. As pointed out by Vapnik (1995, pp. 28, 169):

When solving a given problem, try to avoid solving a more general problem as an intermediate step.

This cuts to the heart of the supervised classification problem. We should seek to directly find the separating surface, which by definition is characterized by the dimension of the feature space and not by the dimensions of the structures that describe the density distributions. Those structures, if thought of as skewed ellipsoids, can have of the order of the square of the dimension of the feature space number of parameters to describe them. This is part of the explanation of the Hughes effect in which the classification accuracy first increases with the dimensionality of the data, but then goes down as the dimensionality increases further. Other approaches make assumptions of the distributions of the data in the training set, called parametric supervised classification, but this introduces biases.

In what follows we will restrict our attention to using only the *spectral* information associated to each pixel (carrying the spatial information along as ancillary information). We will not utilize the spatial information available from the spatial correlation among neighbouring pixels to help perform classification. Indeed the results we obtain would be the same were we to randomly permute the pixel location in the image. However later, in Section 3.8, we will allude to how spatial information can be integrated into classification methods. This is an active area of research.

The task at hand is how to perform supervised classification on such data using all the available channels – that is, how to attach class labels to pixels that describe objects with similar characteristics.

3.4 Mathematical foundations of supervised classification

To formalize the problem of supervised classification, we adopt the following abstraction and notation. For classification, a known set of examples is given, called the training set, taken from a *world* of examples, where each example consists of a class label and a feature vector. You desire to find a classifier function that gives correct answers on these examples, and has low generalization error, meaning it gives good results for the class labels when applied to feature vector inputs that it has not seen before, called the test set. For the moment we will allow for only two class labels. The case of more than two classes will be taken up later. Thus we are given l training pairs, $\{(y_i, \mathbf{x}_i)\ i = 1, \ldots l\}$, each consisting of a class label, $y_i \in \{1, -1\}$, and an N-dimensional feature vectors, $\mathbf{x}_i = (x_{i1}, \ldots, x_{iN}) \in \mathbb{R}^N$.

We wish to find a function, $f(\cdot;\alpha) : \mathbf{x} \mapsto y$, that represents the classifier $y = f(\mathbf{x};\alpha)$, where $\alpha \in \Lambda$, represents the parameters of a particular classifier, and Λ is the space of all possible parameters values for that classifier. Figure 3.2 shows a block diagram of the process associated with the abstract classifier model. We introduce this model as it will illustrate the notions involved in classification. For a given problem there are data vectors characterized by a cumulative probability distribution, $p(\mathbf{x})$, describing the world of examples that is unknown. A supervisor obtains samples, $\mathbf{x}$ of the world, such as a hyperspectral measurement from a sensor, and ascertains a label y for that measurement from ground truth with a conditional probability $p(y|\mathbf{x})$.

The classifier generates an output y^* when applied to $\mathbf{x}$. We wish to examine the error the classifier makes called the *Expected Risk*, $R(\alpha)$,

$$R(\alpha) = \int dp(y, \mathbf{x}) Q(y, f(\mathbf{x}, \alpha)),$$

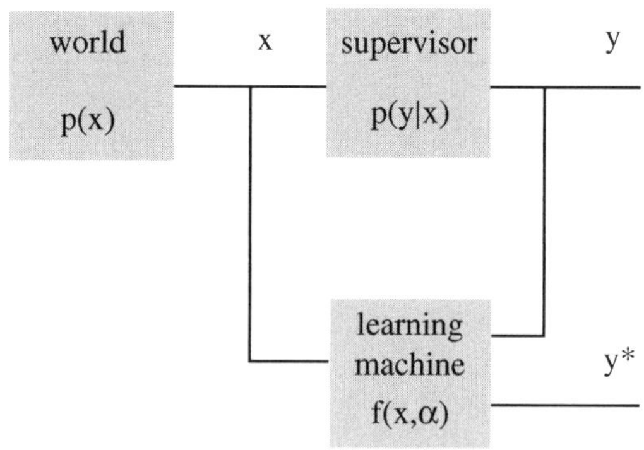

Figure 3.2 Abstract model of a learning machine being trained by a supervisor.

where

$$Q(y, f(\mathbf{x}, \alpha)) = \begin{cases} 0 & y = y^* \\ 1 & y \neq y^* \end{cases}$$

is an indicator function for counting correct answers, the case where $y^* = y$.

The best classifier will be the one that minimizes the expected risk. However, in general we can not obtain the Expected Risk, $R(\alpha)$, since $p(\mathbf{x})$ is unknown, and thus we cannot perform the minimization. But we can measure the error on the training set, called the Empirical Risk, $R_{emp}(\alpha_l)$, given by

$$R_{emp}(\alpha_l) = \frac{1}{l} \sum_{i=1}^{l} Q(y_i, f(\mathbf{x}_i, \alpha_l)).$$

3.4.1 Empirical risk minimization

A standard way to find the best classifier is to use Empirical Risk Minimization (ERM), that is, to search the space of classifiers $\alpha \in \Lambda$ to find the particular classifier, $\hat{\alpha}_l$, that minimizes $R_{emp}(\alpha)$:

$$\hat{\alpha}_l = \min_{\alpha_l \in \Lambda} \frac{1}{l} \sum_{i=1}^{l} Q(y_i, f(\mathbf{x}_i, \alpha_l))$$

The hope is then that this classifier will also work well on unlabelled data, that is, the classifier will be able to generalize. This was the more restrictive view of classification before the developments of Statistical Learning Theory.

3.4.2 General bounds for a new risk minimization principle

Here we will report theoretical results on mathematical bounds found by Vapnik and coworkers that form the underpinning of the specific implementation of the SVM we will use, and show that a *nonparametric* supervised classifier can be constructed in which the dimensionality of the data does not appear. Though the bounds found are mathematically precise, they provide only a guide as to how to obtain good classification results. But, we need to do more than finding the $\hat{\alpha}_l$. To go beyond ERM we must also show that the classifier has good generalization capability. Thus we seek an approach that provides for consistency–conditions under which the expected risk and empirical risk converge to the same limit as $\lim_{l\to\infty}$, and to characterize that rate of convergence. It is here that Vapnik (1982, 1995, 1998) and Vapnik and Chervonenkis (1971, 1974a, 1991) laid the groundwork for a general answer to this problem. They developed a set of bounds independent of the unknown underlying probability distribution of the data $p(\mathbf{x})$. A paper by Vapnik summarizes succinctly this theory (Vapnik 1999).

Vapnik and Chervonenkis showed for the pattern recognition problem that, independent of the distribution function, $p(\mathbf{x})$, the empirical risk, $R_{emp}(\alpha)$ and the actual risk, $R(\alpha)$, converge in probability, to the same value as as the training set grows, and does so exponentially fast. And, that there are general ways to characterize the difference between the expected risk and the empirical risk. This characterization is given in terms of a quantity, called the growth function, $G^\wedge(l)$, to be defined below, and this is the quantity that describes the capacity of the learning machine to learn in terms of the number of training vectors (Vapnik and Chervonenkis 1991; Vapnik 1995, p. 54; Vapnik 1998, pp. 117–120). Convergence in probability of value $R(\alpha_l) \xrightarrow[l\to\infty]{P} R(\alpha_0)$ means that for any $\varepsilon > 0$ and for any $\eta > 0$, there exists a number $l_0 = l_0(\varepsilon, \eta)$, such that for any $l > l_0$ with probability at least $1 - \eta$, $R(\alpha_l) - R(\alpha_0) < \varepsilon$. In equations (Vapnik 1995, p. 71, Equation 3.10, Equation 3.11; p.72, Equation 3.15, p. 73, Equation 3.16; Vapnik 1998, p. 148, Equation 4.46, Equation 4.47, Equation 4.48):

$$\begin{aligned}
&\text{IF} && \lim_{l\to\infty} G^\wedge(l)/l = 0 \\
&\text{THEN} && P\left\{\sup_{\alpha\in\Lambda} \left|R(\alpha) - R_{emp}(\alpha)\right| > \varepsilon\right\} \le 4\,\exp\left\{\left(\frac{G^\wedge(2l)}{l} - \varepsilon^2\right) l\right\}, \\
& && P\left\{\sup_{\alpha\in\Lambda} (R(\alpha) - R_{emp}(\alpha))/\sqrt{R(\alpha)}) > \varepsilon\right\} \le 4\,\exp\left\{\left(\frac{G^\wedge(2l)}{l} - \varepsilon^2/4\right) l\right\},
\end{aligned} \tag{3.1}$$

where $\varepsilon > 0$. An alternative form of the first bound above[1], assuming $\lim_{l\to\infty} G^\wedge(l)/l = 0$), is

$$P\left\{R(\alpha) - R_{emp}(\alpha) \le \sqrt{\frac{G^\wedge(2l) - \ln(\frac{\eta}{4})}{l}}\right\} \ge 1 - \eta, \tag{3.2}$$

where $1 > \eta > 0$. Here $P\{u(l) > \varepsilon\}$ in Equation (3.1) means the probability of u being larger than ε is bounded by $\exp(-c\varepsilon^2 l)$ (c is a positive constant), and means exponentially rapid convergence. The term that bounds the difference of the expected risk and the empirical risk

[1] To see this define $\eta = 4\,\exp\left\{\left(\frac{G^\wedge(2l)}{l} - \varepsilon^2/4\right) l\right\}$ and solve for ε in terms of η to get $\varepsilon = \sqrt{\frac{G^\wedge(2l) - \ln(\frac{\eta}{4})}{l}}$. Now because total probability over all events adds to 1, $P\left\{\sup_{\alpha\in\Lambda}|R(\alpha) - R_{emp}(\alpha)| \le \varepsilon\right\} = 1 - P\left\{\sup_{\alpha\in\Lambda}|R(\alpha) - R_{emp}(\alpha)| > \varepsilon\right\} \ge 1 - \eta$. From the second bound we see $R(\alpha) > R_{emp}(\alpha)$ and this is always less than the sup bound, thus we may remove the $\sup_{\alpha\in\Lambda}|\quad|$ and substituting for ε gives the result.

is called the risk bound or the confidence term. What this means is that if the Growth function grows with l slower than linearly, then the worst case difference between the actual risk and the empirical risk converges exponentially fast to 0 as the number of examples goes to infinity, for all possible classifiers and for any underlying probability distribution of data vectors. And the the actual risk is always larger then the empirical risk. In the past it was assumed that the empirical risk as $\lim_{l \to \infty}$ would always converge to the actual risk. Vapnik and Chervonenkis showed under what conditions this would be true for any distribution for the pattern recognition problem, and where it would *not* be true.

The growth function, $G^\wedge(l)$, characterizes the capacity of the learning machine to learn (generalize), and how that generalization scales with the number of training examples, l. To explain the growth function, think of the l training examples as points in an N-dimensional space and then define $\mathcal{N}^\wedge(l)$ as the largest number of ways that *any* arrangement of those points can be divided into two classes. This is a characterization of the capacity of the learning machine to learn. In particular $G^\wedge(l)$ is defined by

$$G^\wedge(l) = \ln\left(\sup_{\mathbf{x_1},\ldots,\mathbf{x_l}} \mathcal{N}^\wedge(\mathbf{x_1},\ldots,\mathbf{x_l})\right),$$

where $\sup_{\mathbf{x_1},\ldots,\mathbf{x_l}}$ means find the largest value of $\mathcal{N}^\wedge(\mathbf{x}_1,\ldots,\mathbf{x}_l)$ over all possible arrangements of points $(\mathbf{x_1},\ldots,\mathbf{x_l})$. If the classifier can always separate any arrangement of l points for any l into two classes then $\mathcal{N}^\wedge(l) = 2^l$ and $G^\wedge(l) = l\ln 2$, which in the theorem above means we *cannot* make any claims of fast convergence of the empirical risk to the actual risk as the risk bound (the square root quantity in Equation (3.2)) does not go to 0 as $l \to \infty$. But what Vapnik and Chervonenkis were able to show is that any growth function either satisfies the equality

$$G^\wedge(l) = l\ln 2$$

or is bounded by the inequality

$$G^\wedge(l) \le h(\ln\frac{l}{h} + 1),$$

where h is an integer called the VC-dimension, such that when $l = h$

$$G^\wedge(h) = h(\ln 2)$$
$$G^\wedge(h+1) < (h+1)\ln 2.$$

Figure 3.3 shows two possible growth functions, and a growth function that cannot occur: for the lower curve the actual risk and the empirical risk converge to each other; for the upper curve they do not converge, and that the middle curve is not possible.

We can then see that if a finite VC dimension exists, then the risk bound will go to 0 exponentially fast as $\lim_{l \to \infty}$.

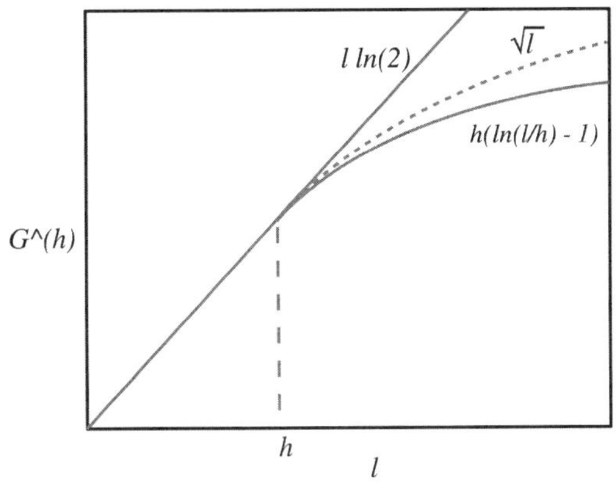

Figure 3.3 The two forms of the growth function are shown in solid lines. A growth function scaling as $\sqrt{l}$ (dotted line) cannot occur according to the theorem proved by Vapnik and Chervonenkis.

To gain some intuition as to the value of the VC dimension consider a linear classifier for data in a space $\mathbb{R}^2$. Here the classifier is represented by a straight line in the plane separating the two classes. A line can always separate any arrangement of three points into any of the $2^3 = 8$ arrangements of two classes, but a line cannot always separate four points into all possible arrangements $2^4 = 16$. In fact it can be shown that a hyperplane in N dimensions can separate at most $N + 1$ points into two classes (Burges 1998, p. 160).

The VC-dimension is then a dimensionless number that characterizes the capacity of the classifier through its role in the growth function as the scaling exponent of the number of examples that can be separated. For a set of indicator functions $Q(y, f(\mathbf{x}, \alpha)), \alpha \in \Lambda$, the VC dimension, h, is equal to the largest number of vectors $\mathbf{x}_1, \ldots \mathbf{x}_l$ that can be separated into two different classes in all the 2^h possible ways using this set of functions. For points in $\mathbb{R}^2$, $h = 3$.

Inserting the bound for the growth function $G^\wedge(l) \leq h(\ln \frac{l}{h} + 1)$ into Equation (3.2) gives

$$P\left\{R(\alpha) \leq R_{emp}(\alpha) + \sqrt{\frac{h\left(\ln \frac{2l}{h} + 1\right) - \ln(\frac{\eta}{4})}{l}}\right\} \geq 1 - \eta. \tag{3.3}$$

We can use this bound in a constructive way to formulate a new minimization principle for finding the best classifier, with the caveat that we are only using Equation (3.3) as a guide, because this gives (with some chosen probability) an upper bound on the actual risk. This does not prevent a particular machine with the same value for empirical risk, and whose function set has higher VC dimension, from having better classification accuracy.

3.4.3 Structural risk minimization

Now for minimization, we must minimize over both the empirical risk and the term describing the capacity of the machine. This is called Structural Risk Minimization (SRM). For a fixed

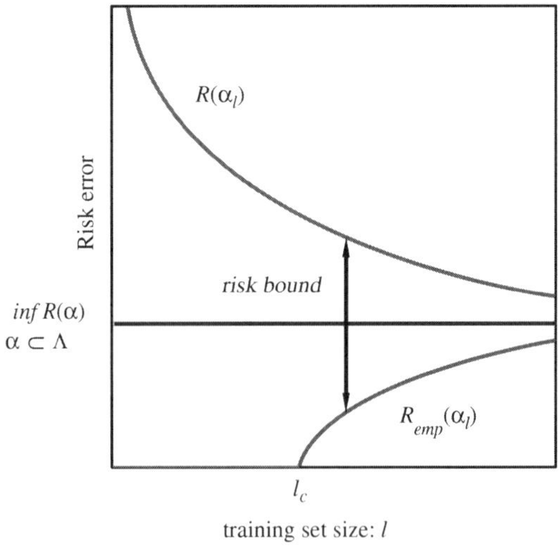

Figure 3.4 Learning curve.

number of training vectors, l, the two terms inside {} to the right of the $\leq$ in Equation (3.3), represent the complexity regularization trade-off. As the class of functions parametrized by α gets larger (more complex) $R_{emp}(\alpha)$ will get smaller as there is more flexibility in $f(\mathbf{x}, \alpha)$ to fit the data, but the VC dimension will get larger. Conversely if the class of functions parametrized by α gets smaller (less complex), then $R_{emp}(\alpha)$ will get larger, but the VC dimension will get smaller and accordingly the second term will get smaller. This gives the intuition that the uniform deviation between expected risk and empirical risk decreases with number of data points l, but increases with the VC-dimension, and therefore with the complexity of the classifier functions $f(\mathbf{x}, \alpha)$, as shown in Figure 3.4.

Thus we are now minimizing the empirical risk and the risk bound. Now the risk bound is defined for a class of functions, so in order to minimize this term we must impose a structure on the class of all functions. Following Vapnik we do this using nested subsets:

$$\Lambda_1 \subset \Lambda_2 \subset \ldots \subset \Lambda$$

for which the VC-dimension for the classes of functions satisfy

$$h_{\Lambda_1} \leq h_{\Lambda_2} \leq \ldots \leq h_{\Lambda}.$$

With this in place, the principle of Structural Risk Minimization can be cast as: Minimize $R(\alpha)$ over α in each subset class of functions, Λ_i, and over the sequence of nested subsets of classes of functions $\Lambda_i \in \Lambda$ in Equation (3.3) for a fixed η.

$$\hat{\alpha} = \underset{\substack{\alpha \in \Lambda_i \\ \Lambda_i \subset \Lambda}}{\arg\min} \left(R_{emp}(\alpha) + \sqrt{\frac{h_{\Lambda_i}\left(\ln \frac{2l}{h_{\Lambda_i}} + 1\right) - \ln(\frac{\eta}{4})}{l}} \right).$$

3.5 From structural risk minimization to a support vector machine algorithm

Here we summarize the steps from SRM to an algorithm we can solve for supervised classification (Niyogi *et al.* 1999).

3.5.1 SRM for hyperplane binary classifiers

We begin with supervised binary classification and later generalize to multiple class classification, which can be handled by aggregating results from binary classification. A simple binary classifier functions is defined by a hyperplane that attempts to separate the data into classes. Thus define a parametrization of the classifier function Λ from SRM to be

$$\Lambda = \{\alpha : \mathbb{R}^N \rightarrow \{-1, 1\} \mid f(\mathbf{x}, \alpha) = \text{sgn}(\mathbf{w} \cdot \mathbf{x} + b)\}, \quad \alpha = (\mathbf{w}, b),$$

where $\mathbf{w}$ is a vector perpendicular to the hyperplane and b is the closest distance of the hyperplane to the origin, and $\mathbf{x} \in \mathbb{R}^N$. Now we can extend the ordering relationship imposed by increasing values of the VC dimension, h, for SRM by finding a relationship between h and the classifier parameters $\alpha = (\mathbf{w}, b)$. For this we use theorem of (Vapnik 1995, Theorem 5.1, p. 128; Vapnik 1998, Theorem 10.3, p. 413; Burges 1998, Theorem 6, p. 30) that bounds the VC dimension for a subset of canonical hyperplanes by a bound on the hyperplane properties. Canonical means a particular normalization that defines a length scale in the problem. Given the space in which the training vectors lie is $\mathbf{x}_i \in \mathbb{R}^N$, then

IF $B_{\mathbf{x}_1 \ldots \mathbf{x}_l} = \{\mathbf{x} \in \mathbb{R}^N \mid \|\mathbf{x} - \mathbf{a}\| \leq D, \ \mathbf{a} \in \mathbb{R}^N, \ \|\mathbf{x}_i - \mathbf{a}\| \leq D, \ i = 1, \ldots l\}$
is the smallest ball, with radius D, centred at $\mathbf{a}$, containing the training set, and
$\Lambda_A = \{\alpha = \{\mathbf{w}, b\} \mid f(\mathbf{x}, \alpha) = \text{sgn}(\mathbf{w} \cdot \mathbf{x} + b), \ \|\mathbf{w}\| \leq A\}$
is a subclass of hyperplanes in canonical form,
THEN $h \leq min(\lceil D^2 A^2 \rceil, l) + 1.$ (3.4)

Thus the natural structure for SRM on hyperplanes classifiers is the monotonic dependence of h on the parameter A, which for a fixed set of training data are an upper bound on the norm of $\mathbf{w}$. We write the ordering relationship of the SRM as

$$\begin{aligned} &h_{\Lambda_1} \leq h_{\Lambda_2} \leq \ldots \leq h_{\Lambda} \\ &A_1 \leq A_2 \leq \ldots \leq A \\ &\Lambda_{A_1} \subset \Lambda_{A_2} \subset \ldots \subset \Lambda_A \end{aligned}$$

and the SRM principle becomes

$$\hat{\Lambda} = \underset{\substack{\alpha \in \Lambda_{A_i} \\ \Lambda_{A_i} \in \Lambda,}}{\text{arg min}} \ (R_{emp}(\mathbf{w}, b) + \gamma \mathbf{w} \cdot \mathbf{w}) \quad \text{or} \quad (\hat{\mathbf{w}}, \hat{b}) = \text{arg min}_{\mathbf{w}, b}(R_{emp}(\mathbf{w}, b) + \gamma \mathbf{w} \cdot \mathbf{w}), \tag{3.5}$$

where γ is a parameter that trades off the fit to the training data, $R_{emp}(\mathbf{w}, b)$ with a measure of the model complexity or capacity of the classifier to learn, $\mathbf{w} \cdot \mathbf{w}$. Note this is a related result to that given in Chapters 1 and 2 for the case of quadratic loss.

Though we have parametrized the classifier, there is no recourse to a parametrization of the distribution of the training data. Also no consideration of the dimensionality of the data has been made. Thus at least at this point the issue of the *curse of dimensionality* is not explicit. There is continuing work to refine these bounds, and derive new theoretical results for a variety of cases, and we point the interested reader to the following papers: Ikeda (2003), Malzahn and Opper (2005), Mukherjee *et al.* (2006), Schaback and Werner (2006), Zhou (2003), Zhou and Jetter (2006).

3.5.2 SVM algorithm

To proceed further, consider the two cases for training a hyperplane binary classifier: either the training data can or cannot be separated by a hyperplane. Figure 3.5 shows these two case for two dimensional data. For separable training data $R_{emp}(\mathbf{w}, b)$ is zero. The condition of separability is that all training vectors of class +1 lie on one side of the hyperplane and all training vectors of class -1 lie on the other side. The positive distance d_i of any training vector from the hyperplane is given by $d_i = y_i \frac{(\mathbf{w} \cdot \mathbf{x}_i + b)}{\|\mathbf{w}\|} > 0$, where multiplying by y_i (recall $y_i = \pm$) assures all such distance are positive. Because an arbitrary change in scale of $\mathbf{w} \rightarrow c\mathbf{w}$, $b \rightarrow \alpha c$, $d_i \rightarrow cd_i$, where c is a constant, does not affect the minimization equations, we will remove this arbitrariness by choosing the length scale in the problem to be the minimum

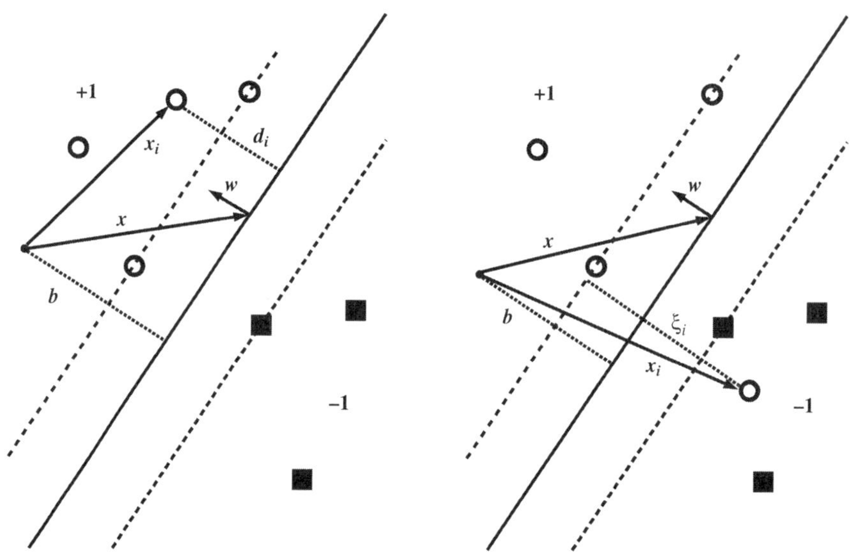

Figure 3.5 Left: Schematic of separable data in $\mathbb{R}^2$. The circles are feature vectors in class +1 and the squares are feature vectors in class −1. The placement of the hyperplane shown is optimal. The dotted lines are the margin boundaries. Right: Schematic of non-separable data in $\mathbb{R}^2$. The circles are feature vectors in class +1 and the squares are feature vectors in class −1. There is one feature vector that is not separable.

of the distances d_i. This will make the distance from the separating hyperplane to the nearest training vectors $1/|\mathbf{w}|$. This is called the canonical form. With this scaling we then have

$$y_i(\mathbf{w} \cdot \mathbf{x}_i + b) \geq 1, \tag{3.6}$$

and the statement of our minimization problem for finding the separable hyperplane is

$$\begin{aligned} &\min_{\mathbf{w},b} \quad \tfrac{1}{2}\|\mathbf{w}\|^2 \\ &y_i(\mathbf{w} \cdot \mathbf{x}_i + b) - 1 \geq 0 \quad i = 1, \ldots, l. \end{aligned} \tag{3.7}$$

By minimizing $\frac{1}{2}\|\mathbf{w}\|^2$, we are making $1/\|\mathbf{w}\|$ the largest, and from the discussion above, that is maximizing the distance of the hyperplane from the nearest training vectors of the two classes. That distance is called the margin and lends the name maximum margin SVM to this type of classifier. The boundaries beyond which all the training vectors of each class lie are called the margin boundaries. One can see that to solve the minimization problem we are solving a quadratic optimization problem that is always guaranteed to have a unique solution. The optimum solution giving the best hyperplane is where the hyperplane is placed so that the distance of the closest feature vectors in the two classes are the furthest they can be from the hyperplane. Thus the optimization problem will find those closest vectors and locate the hyperplane to lie between them so that perpendicular distance from the hyperplane to the those feature vectors are $1/\|\mathbf{w}\|$. Clearly the more distant feature vectors from the hyperplane do not affect the solution. These special closest feature vectors are called the *support vectors*, and they then determine the position and orientation of the separating hyperplane. Because of the linearity of the problem we can write the hyperplane parameters $\mathbf{w}$, b as linear combinations of the training vectors. Write

$$\mathbf{w} = \sum_{i=1}^{l} \lambda_i y_i \mathbf{x}_i, \tag{3.8}$$

where the linear coefficients λ_i are the indicator values for the support vectors when $\lambda_i > 0$, and where we have explicitly included y_i so that $\lambda_i \geq 0$. Using any one of the constraint equations from above we can also express the other hyperplane parameter, b, as

$$b = 1 - \sum_{i=1}^{l} \lambda_i y_i \mathbf{x}_i \cdot \mathbf{x}_j. \tag{3.9}$$

Thus the outcome of the minimization is to find the λ_i. Note that the dimensionality of the feature vector does not explicitly enter our minimization optimization problem.

The standard method used to actually solve the minimization's equations is to recast this problem into a simpler form by using Lagrange undetermined multipliers – the λ_i we have introduced above – in order to transform the problem into the *dual* representation, (Gualtieri and Cromp 1998),

$$\begin{aligned} &\max_{\lambda_1 \ldots \lambda_l} \quad \left[-\tfrac{1}{2} \textstyle\sum_{i=1}^{l} \sum_{j=1}^{l} \lambda_i y_i (\mathbf{x}_i \cdot \mathbf{x}_j) y_j \lambda_j + \sum_{i=1}^{l} \lambda_i \right] \\ &\lambda_i \geq 0 \quad i = 1, \ldots, l \\ &\textstyle\sum_{i=1}^{l} \lambda_i y_i = 0. \end{aligned} \tag{3.10}$$

For the non-separable problem, Figure 3.5 (right), R_{emp} is no longer 0. We can construct a version of R_{emp} for this case by introducing a variable ξ_i, associated with each training vector, which measures the distance of that training vector beyond the margin boundary if it is a non-separable, and $\xi_i = 0$ if it is a separable training vector. Then we write $R_{emp} = C\sum_{i=1}^{l} \xi_i$, where C is a constant, to be set, that measures the strength of the non-separable contribution relative to the capacity of the classifier to learn in the term $\frac{1}{2}\|\mathbf{w}\|^2$. We also must change Equation (3.6) so as to relax the requirement that all training vectors of the same class lie on the same side of the hyperplane. We now have

$$\begin{aligned} &y_i(\mathbf{w} \cdot \mathbf{x}_i + b) \geq 1 - \xi_i \\ &\xi_i \geq 0. \end{aligned} \tag{3.11}$$

This can also be cast in a *dual* form using Lagrange undetermined multipliers, (Gualtieri and Cromp 1998), and gives

$$\begin{aligned} &\max_{\lambda_1 \ldots \lambda_l} \quad \left[-\frac{1}{2}\sum_{i=1}^{l}\sum_{j=1}^{l} \lambda_i y_i (\mathbf{x}_i \cdot \mathbf{x}_j) y_j \lambda_j + \sum_{i=1}^{l} \lambda_i \right] \\ &C \geq \lambda_i \geq 0 \quad i = 1, \ldots, l \\ &\sum_{i=1}^{l} \lambda_i y_i = 0 \quad i = 1, \ldots, l. \end{aligned} \tag{3.12}$$

Note, the only difference from the dual of the separable case, Equation (3.10), is that the λ_i are bounded above by C, reflecting the fact that the original inequality constraint, Equation (3.6), holds only while $\xi_i = 0$ and then becomes *soft* when $\xi_i > 0$. Knowing the solutions λ_i, we can again find $\mathbf{w}$ and b for the hyperplane.

3.5.3 Kernel method

Up to this point we have dealt only with classification as a linear function of the training data – the decision surface is a hyperplane defined by linear equations on the training data. However, it can be the case that no hyperplane exists to separate the data. The non-separable method provides one way to deal with this. As an alternative we would like a way to build a *nonlinear* decision surface. An extremely useful generalization that can give nonlinear decision surfaces and improved separation of the training data are possible using the following idea, first introduced by Aizerman, Braverman and Rozoner (Aizerman *et al.* 1964), and incorporated into machine learning as part of the Support Vector Machine by Boser, Guyon, and Vapnik (Boser *et al.* 1992). We refer the reader to Chapter 2 for details of this method, where the SVM optimization problem Equation (2.10) is transformed by means of a Kernel transformation, K, to Equation (2.11).

For our purposes we give a short synopsis of how to make a nonlinear transformation that projects the input space to a a higher dimensional Euclidean space, by means of a nonlinear vector function: $\boldsymbol{\phi} : \mathbb{R}^N \mapsto \mathcal{H}$, and in this space pose the problem of finding a hyperplane that best separates the projected training vector data. Then we may again pose the optimal margin problem in the space $\mathcal{H}$ by replacing $\mathbf{x}_i \cdot \mathbf{x}_j$, by $\boldsymbol{\phi}(\mathbf{x}_i) \cdot \boldsymbol{\phi}(\mathbf{x}_j)$. Then, as before, solve the optimization problem for the λ_i. This finds the support vectors among the transformed vectors,

$\boldsymbol{\phi}(\mathbf{x}_i)$, by association with the $\lambda_i > 0$. We then use these to build the classifier function:

$$f(\mathbf{x}, \lambda_1, \ldots, \lambda_l) = \operatorname{sgn}\left(\sum_{i=1}^{l} \lambda_i y_i \boldsymbol{\phi}(\mathbf{x}_i) \cdot \boldsymbol{\phi}(\mathbf{x}) + b\right). \tag{3.13}$$

The Kernel method is to note that there are classes of functions $\boldsymbol{\phi}$ that satisfy the special property of a *Kernel function K* where

$$K(\mathbf{x}_i, \mathbf{x}_j) = \boldsymbol{\phi}(\mathbf{x}_i) \cdot \boldsymbol{\phi}(\mathbf{x}_j). \tag{3.14}$$

Then everywhere that $\mathbf{x}_i \cdot \mathbf{x}_j$ occurs, we replace it with $K(\mathbf{x}_i, \mathbf{x}_j)$. A Kernel function is a function defined on two variables, $\mathbf{u} \in \mathbb{R}^k$ and $\mathbf{y} \in \mathbb{R}^k$, such that $K : \mathbb{R}^k \times \mathbb{R}^k \to \mathbb{R}$, where the form of K is given by $K(\mathbf{u}, \mathbf{v}) = \boldsymbol{\phi}(\mathbf{u}) \cdot \boldsymbol{\phi}(\mathbf{v})$, and where $\boldsymbol{\phi}()$ is a vector valued function on a vector argument.

We need not explicitly compute $\boldsymbol{\phi}(\mathbf{x})$, which could be computationally expensive, but only need compute the Kernel functions. In fact we need not have an explicit representation of $\boldsymbol{\phi}$ at all, but only K. The restrictions on what functions can qualify as Kernel functions is that they must satisfy Mercer's condition, which is the case for Kernels of positive integral powers of the dot product, such as, $K(\mathbf{x}_i, \mathbf{x}_j) = ((\mathbf{x}_i \cdot \mathbf{x}_j) + 1)^p$. See Burges (Burges 1998, Section 4.1) for further discussion and Chapter 2. What is gained is that we have moved the data into a larger space where the training data may be spread further apart and a larger margin may be found for the optimal hyperplane. In the cases where we can explicitly find $\boldsymbol{\phi}$, then we can use the inverse of $\boldsymbol{\phi}$ to construct the nonlinear separator in the original space. Clearly there is a lot of freedom in choosing the Kernel function and work has gone into the study of this idea both for SVM and for other problems (Smola *et al.* 1998). In addition, due to Mercer's conditions on Kernels, unlike in other machine learning techniques based on Neural Networks, the positive semi-definite Kernel ensures that the objective function is convex and hence there are no local minima, only a global minimum.

With respect to the curse of dimensionality, we never explicitly work in the higher dimensional space, so we are never confronted with computing the large number of vector components in that space. And there is no significant computational burden of working in the high dimensional space since the dot products are formed in the original space, and that dot product scalar is all that is transformed to the higher dimensional space.

Two popular Kernel transformations are the polynomial Kernel and the radial basis function Kernel. The polynomial Kernel function is given by

$$K(\mathbf{x}, \mathbf{y}) = (\mathbf{x} \cdot \mathbf{y} + 1)^d, \tag{3.15}$$

and the radial basis function Kernel (RBF) is given by

$$K(\mathbf{x}, \mathbf{y}) = \exp\left(-\frac{\|\mathbf{x} - \mathbf{y}\|^2}{\sigma^2}\right). \tag{3.16}$$

Further work on Kernel functions can be found in (Schölkopf *et al.* 1998, pp. 89–102) and in Chapter 2 of this book.

3.5.4 Hyperparameters

In the basic SRM equation, Equation (3.5), the parameters to be found that characterize the classifier are $\mathbf{w}$, b. In addition there are the parameters, called hyperparameters, that are used to handle the case of non-separable data, C, and with the introduction of the Kernel method there are the choices of what particular Kernel function is to be used. Thus for the polynomial Kernel function, there is d, the dimensionality of the polynomial, while for the radial basis function Kernel there is σ. The usual approach once having chosen a Kernel function is to do *k fold validation*. The training data are split into k parts of size l/k. A discrete range of possible values of the hyperparameter is chosen and for each hyperparameter value a SVM classifier is trained using $k-1$ parts and tested on the remaining 1 part from which a classification accuracy is measured since we know the labels of the data. This is repeated one by one through all k such splits of the training data into $k-1$ folds for training with testing on the remaining fold, and an average classification accuracy is obtained. This is repeated for each of the hyperparameter values and the value with the highest classification accuracy is chosen. If $k = l - 1$, then this approach, also called *leave one out*, will give an unbiased estimate of the hyperparameter. Clearly even for a moderate value of k this can be an expensive computation. Work has been done to find computationally inexpensive ways to estimate or bound the hyperparameters (Duan *et al.* 2003).

3.5.5 A toy example

As an illustration of the Kernel Method, consider its application in $N = 2$. Using the package LIBSVM (Chang and Lin 2001), we show for two dimensions in Figure 3.6, the shape of the separating surface for several values of d. Note that only when $d = 7$ is reached does a separating surface exist that completely separates the two classes.

3.5.6 Multi-class classifiers

When we have $K > 2$ classes, labelled $1, 2, \ldots K$, we choose to create all the possible pair classifiers, $K(K-1)/2$ in number, which are trained using training set data from class, 1 vs. 2, 1 vs. 3, ..., 1 vs. K, 2 vs. 3, ..., K-1 vs. K. To decide a class label in the testing phase, apply the $K(K-1)/2$ pair classifiers to the test vector and record which label each of the pair classifiers gives. From each such pair classifier place a vote into one of K bins labelled $1, 2, \ldots, K$, and when all of the pair classifiers have voted, record the bin with the most votes, as the classification label for that test vector. For ties randomly select one of the competing bins.

We have chosen this approach, called in the literature *one-against-one* (OAO) over creating K classifiers that are trained from training sets of 1 vs. $\{$ 2 3 4 ... K $\}$, 2 vs. $\{$1 3 ... $K\}$, ..., K vs. $\{$1 2 ... K-1 $\}$ called *one-against-all* (OAA), because it appear to be more accurate (Hsu and Lin 2002). We note that for training the time complexity is less for OAO vs. OAA. This can be seen in Table 3.1, which gives the time complexities $\mathcal{O}(K^2m^2)$, $\mathcal{O}(K^3m^2)$ for OAO and OAA respectively. Here we assume that the training sets are each of size $m/2$ and that the time complexity of a single pair classifier training on m total training vectors is $\mathcal{O}(m^2)$, because we are solving a quadratic optimization problem for training. Other work (Melgani and Bruzzone 2004) comparing OAO, OAA, and two hierarchical methods indicate that OAO and OAA have similar classification accuracy and are superior to hierarchical methods, but that OAO is faster for training and slower for testing than OAA.

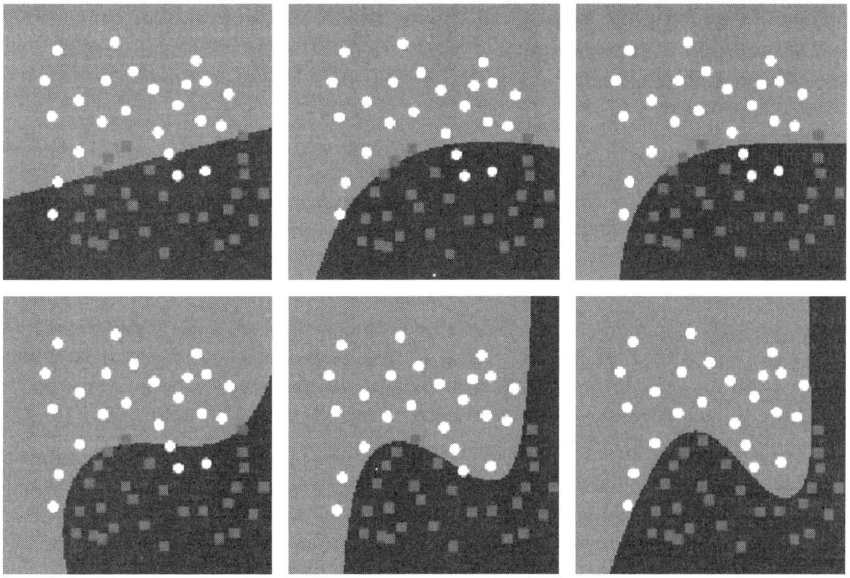

Figure 3.6 The white dots and the grey squares in each of the six panels represent the same training data, while the light grey and black backgrounds show the division of the space into two contiguous regions with various separating curves between them. The polynomial Kernel function $K(\mathbf{x}, \mathbf{y}) = (\mathbf{x} \cdot \mathbf{y} + 1)^d$ is used with SVM for d = 1, 2, 3, 4, 5, 7 (left to right top to bottom), with $C = 1000$. For $d = 7$ the separating curve completely separates the two classes of points. The software package LIBSVM (Chang and Lin 2001) was used to produce this example.

3.5.7 Data centring

An important preprocessing step for all applications of SVM is to mean centre and scale the data according to

$$\mathbf{x}'(\mathbf{r}, \lambda) = \frac{\mathbf{x}_i(\mathbf{r}, \lambda) - \sum_{\mathbf{r}} \mathbf{x}_i(\mathbf{r}, \lambda) / \sum_{\mathbf{r}} 1}{S},$$

where $\mathbf{x}(\mathbf{r}, \lambda)$ is the original feature vector (spectra) at pixel position $\mathbf{r}$ and at band λ and S is a constant scale factor. The reason for doing this is that hyperspectral data are always all positive

Table 3.1 Time complexity on a single processor for performing SVM training for a classifier of K classes, each class containing $m/2$ exemplars

Pairwise classification	One-versus-the-rest
$\mathcal{O}(K^2m^2)$	$\mathcal{O}(K^3m^2)$
$K(K-1)/2 * \mathcal{O}(m^2)$	$K * \mathcal{O}(((K-1)\frac{m}{2} + \frac{m}{2})^2)$

in value and thus the original data lies only in the positive 'quadrant' occupying $1/2^N$ of the available space of $\mathbb{R}^N$, whereas the mean centred data fills the entire space of $\mathbb{R}^N$, thereby dramatically spreading out the data, which improves the SVM classification accuracy.

3.6 Benchmark hyperspectral data sets

A particular hyperspectral data set that has become something of a benchmark for testing hyperspectral supervised classification algorithms is the AVIRIS 1992 Indian Pines data set AVIRIS, first used by David Landgrebe and hist students (Jackson and Landgrebe 2001; Jiménez and Landgrebe 1996; Kuo, and Landgrebe 2001; Tadjudin and Landgrebe 1998a). These data were acquired on June 12, 1992 in the northern part of the state of Indiana, in the U.S. The data set and ground truth are available at the website (Landgrebe 1992). The data consists of 145 × 145 pixels (each pixel is 18 m × 18 m in size) by 220 bands in band interleaved format (BIL) format of at-sensor spectral radiance data as 2 byte unsigned integers digital numbers (DN) that have been scaled and offset from the at-sensor spectral radiance, *Rad*, in units W [cm^2 nm sr], to give digital numbers (DN) according to DN = 500 * em Rad + 1000. No atmospheric correction has been performed. However. because of atmospheric water absorption the bands [104 : 108, 150 : 163] are not meaningful, as the sensor only reports noise in these bands. Thus they are removed as they will only serve to decrease the accuracy of the SVM classifier. Also band 220 is very noisy. Thus for analysis these bands should be removed leaving a 200 band data set. Ground truth gathered by Landgrebe and students is also available and delineates 16 classes. The scene consists of about two-thirds agriculture, and one-third forest or other natural perennial vegetation. There are two major dual lane highways, a rail line, as well as some low density housing, other built structures, and smaller roads. Since the scene is taken in June some of the crops present, corn, soybeans, are in early stages of growth with less than 5% coverage. The ground truth available is designated into sixteen classes and is not all mutually exclusive. This data set has been used as the 16 class full scene, as the 9 class full scene, and in a 4 class subset scene.

3.6.1 The 4 class subset scene

The subset scene consists of pixels [27 − 94] × [31 − 116] for a size of 68 × 86. (Upper left in the original scene is at (1, 1).) There is ground truth for over 75% of this scene and it comprises the three row crops, Corn-notill, Soybean-notill, Soybean-mintill, and Grass-Trees, where notill and mintill describe tillage practices when planting these crop types. Table 3.2

Table 3.2 Data description of the Indian Pines 4 class subset scene

Class name	Number of ground truth vectors	Number of training vectors	Number of test vectors
Corn-notil	1008	201	807
Soybean-notill	727	145	582
Soybean-mintill	1926	385	1541
Grass-Trees	732	146	586

Table 3.3 Data description of the 16 class Indian Pines Full scene

Class name	Number of ground truth vectors	Number of training vectors	Number of test vectors
Alfalfa	54	10	44
Corn-notil	1434	286	1148
Corn-min	834	166	668
Corn	234	46	188
Grass-Pasture	497	99	398
Grass-Trees	747	149	598
Grass-Pasture-mowed	26	5	21
Hay-windrowed	489	97	392
Oats	20	4	16
Soybean-notill	968	193	775
Soybean-mintill	2468	493	1975
Soybean-clean	614	122	492
Wheat	212	42	170
Woods	1294	258	1036
Bldgs-Grass-Trees-Drives	380	76	304
Stone-steel-towers	95	19	76

gives further details. We choose a random selection of pixels with ground truth to yield a training/testing split of 20%/80%.

3.6.2 The 16 class scene

For the full 16 class scene we also used a random selection of 20% of the ground truth data and tested on the remaining 80%. Table 3.3 describes this data.

A difference with the data and results reported by (Tadjudin and Landgrebe 1998a,b) is that they studied the scene using 17 classes whereas we only used 16. The difference being that they further resolved the class Soybeans-notill into two subclasses of Soybeans-notill based on fields that were in different locations in the full scene.

3.6.3 The 9 class scene

Papers by Camps-Valls and Bruzzone (2005) and Melgani and Bruzzone (2004) have studied the same Indian Pines 1992 data set using SVM, Kernel Fisher Discriminants, AdaBoost and Regularized RBF neural networks. From the 16 different land-cover classes available in the original ground truth, 7 were discarded as these authors deemed there were insufficient number of training samples available. The remaining nine classes were used to generate a set of 4757 training samples (for the learning phase of the classifiers) and 4588 test samples (for validating their classification accuracy). See Table 3.4 (Camps-Valls and Bruzzone 2005, Table II, p. 1357), which summarizes their use of the data. Their training data were about 2.5 times larger than the 16 class data set above, indicating they were training on roughly 50% of the available data as compared with our 20%.

Table 3.4 Indian Pines 9 class scene. Number of Training and Test Samples used in the the experiments of Camps-Valls and Bruzzone

Class	Training	Test
C1 Corn-notil	742	692
C2 Corn-mintil	442	392
C3 Grass/Pasture	260	237
C4 Grass/Trees	380	358
C5 Hay-windrow	236	253
C6 Soybean-notil	487	481
C7 Soybean-mintil	1245	1223
C8 Soybean-cleantil	305	309
C9 Woods	651	643
Total	4757	4588

3.7 Results

3.7.1 SVM implementation

SVM has been implemented using two open source codes: SVM^{light} by T. Joachim (Joachims 1998a) together with quadratic optimization code written by A. Smola (Smola 1998) and LIBSVM (Chang and Lin 2001). These codes consist of a learning module that finds the support vectors, given two sets of training vectors, and a specification of the non-separable parameter C, and the Kernel function (in this work a polynomial with parameter d), and a classification module that classifies any test vector into one of the pair of classes. To handle multiple classes we have embedded theses code in calling routines that implement the OAO classifier built on $K(K-1)/2$ pair classifiers [2]. Central to the learning module is the quadratic optimization as formulated above in Equations (3.12). Direct application of quadratic optimization to large numbers of training vectors can be computationally slow. For SVM^{light} Joachim (Joachims 1998b) has shown a way to reformulate the problem as a series of smaller optimization problems. The solution of these smaller optimizations is accomplished using a quadratic optimization code written by A. Smola (Smola 1998). For LIBSVM the optimization builds on Platt's Sequential Minimal Optimization (Platt 1999).

3.7.2 Effect of hyperparameter *d*

To test the effect of changing the hyperparameter parameter d in the kernel function, Equation (3.15), in the 4 class subset scene, five different random choices of training testing vectors were used to obtain classification accuracy for $d = 1, 2, \ldots, 15$ and $C = 1000$. The results are shown in Figure 3.7. A significant increase in classification accuracy occurs by going to a nonlinear classifier with $d \geq 2$. The five lines are for each of the different random choices of training vectors respectively. Also note that the classification accuracies of the five different cases all lie within a $\pm 1\%$ envelope, suggesting that differences in classification results greater than 1% is significant. We did not vary the parameter C. Later work to be reported

[2] LIBSVM has options built in for multiclass classification, but it was not suitable for our application.

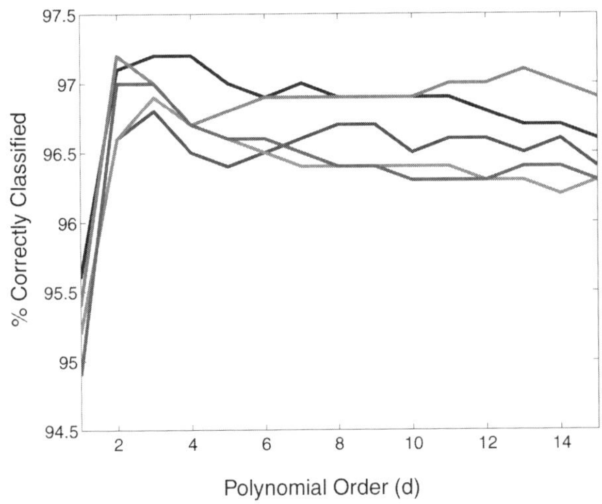

Figure 3.7 Effect on classifier classification accuracy by varying the kernel parameter d, in $K(\mathbf{x}, \mathbf{y}) = (\mathbf{x} \cdot \mathbf{y} + 1)^d$. The five lines are the results for each of the different random choices of training vectors respectively. In all cases the parameter $C = 1000$.

below investigated the effect of varying this hyperparameter for the case of the radial basis function kernel (RBF).

3.7.3 Measure of accuracy of results

To gain a measure of the accuracy of the classification results five different trials for the 4 class subset scent were conducted with different random selection of the 20%/80% training testing splits with $d = 7$ and $C = 1000$. For a trial, the overall classification accuracy is the sum of the number of samples correctly labelled for each class in the test set divided by the total number of samples in the test set. Table 3.5 summarizes these results and shows that the results are consistent within one percent and the average classification accuracy was 96%.

Table 3.5 Summary of trials on SVM classifier for the 4 class subset scene

Trial	Overall	Class correct(%)			
	Correct(%)	Corn-notill	Grass-Trees	Soybean-notill	Soybean-mintill
1	96.3	94.3	100.0	96.1	95.7
2	95.8	92.8	99.8	95.7	96.0
3	96.1	95.2	99.8	95.7	94.7
4	95.5	94.7	100.0	95.1	93.5
5	95.6	95.7	99.8	94.8	93.3
Average	95.9	94.5	99.9	95.5	94.6

Table 3.6 A comparison of results for the Indian Pines 4 class subset scene (68 × 86 pixels) and the 16 class full scene (145 × 145 pixels). The results labelled bLOOC+DAFE+ECHO and Euclidean are taken from their works (Tadjudin and Landgrebe 1998a, 1999), and represent the best classifier results reported for this scene in that work. All training is based on 20% of the ground truth and testing on the remaining 80%

Method	Performance	
	4 class subset scene	16 class full scene
Support Vector Machine	95.9%	87.3%
bLOOC+DAFE[+ECHO]	93.5%	82.9%
Euclidean	66.7%	48.2%

3.7.4 Classifier results for the 4 class subset scene and the 16 class full scene

SVM Results for the 4 class subset scene and the 16 class full scene are given in Table 3.6 and compared with the results of the earlier work and was first studied by D. Landgrebe and students (Tadjudin and Landgrebe 1998a, 1999) who developed results for the Indian Pines 4 class subset scene with a classifier called bLOOC+DAFE, and for the 16 class scene, with a classifier called bLOOC+DAFE+ECHO. Here bLOOC stands for Bayesian Leave One Out Covariance Estimator (Tadjudin and Landgrebe 1999), DAFE stands for Discriminant Analysis Feature Extraction (Fukunaga 1990), and ECHO stands for Extraction and Classification of Homogeneous Objects (Kettig and Landgrebe 1976). Additionally they reported results for a Euclidean classifier, which we include below. The Euclidean classifier is not expected to give good accuracy for hyperspectral data, as it simply finds the mean spectral feature vectors of each of the training classes, and then classifies testing data by using a Euclidean distance of those means to the test spectra, selecting as the class of that spectra the one with the smallest such distance.

Results for the full scene were produced with a polynomial kernel with hyperparameters $d = 7$ and $C = 1000$. No preprocessing of the data was used, besides mean centering of the data.

For both the 4 class subset scene and the 16 class full scene, we see that the SVM outperforms bLOOC+DAFE+ECHO by values outside the variation due to random selection of the training/testing split, and both outperform the Euclidean classifier as expected.

3.7.5 Results for the 9 class scene and comparison of SVM with other classifiers

Camps-Valls and Bruzzone (2005) give SVM results for the Gaussian kernel (RBF) and the polynomial kernel. The hyperparameters to be found for the RBF SVM are the Gaussian width σ, and the regularization parameter C. They tuned over the ranges $\sigma = (1, \ldots, 50)$ (same for all RBFs) and $C = \{10^{-2}, \ldots, 10^{6}\}$. For the polynomial kernel they tuned d, and the the regularization parameter C with values in the range $d = \{1, \ldots, 15\}$ and $C = \{10^{-2}, \ldots, 10^{6}\}$. Also in their results they adopted the OAA approach for handling classification with more than two classes. In Melgani and Bruzzone (2004) they report training/testing times (in seconds) for OAA 2361 s/341 s on a single processor machine, and for OAO 212 s. / 554 s giving rough agreement with Table 3.1.

Now consider Table 3.7, which shows a classification accuracy comparison of the linear discriminant analysis (LDA), regularized RBF neural network (Reg-RBFNN), SVMs with RBF kernel (SVM-RBF) and with polynomial kernel (SVM-Poly), kernel Fisher discriminant (KFD) analysis (with RBF kernel) and regularized AdaBoost (Reg-AB), for the 200 band data in the upper part of the table and for the 220 band (noisier due to inclusion of noisy channels) data set. In both cases, SVM-Poly has the best classification accuracy, closely by followed by SVM-RBF and then Reg-AB. In a later section we offer some insight as to the nature of these results.

3.7.6 Effect of training set size

Also in Camps-Valls and Bruzzone (2005) was an investigation of the effect of training set size as shown in Figure 3.8. Here SVMs and Reg-AB have a clear advantage of 3% to 8 % over KFD and Reg-RBFNN. Note that LDA show a threshold.

3.7.7 Effect of simulated noisy data

Camps-Valls and Bruzzone (2005) also studied the effect of adding random Gaussian and uniform distributed noise and impulsive noise spikes only to the the test set, but using the classifiers trained as above using training sets without additive noise. What was found is that the SVM classifiers, both polynomial kernel and RBF kernel are more robust with respect to Gaussian and uniform distributed noise than linear discriminant analysis (LDA), regularized

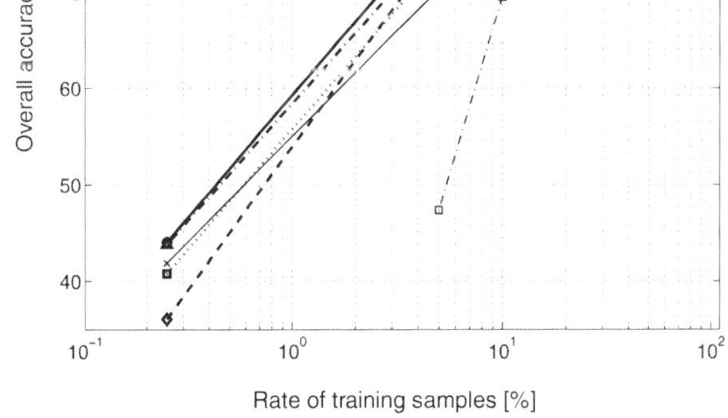

Figure 3.8 The effect of size of the training set on classifier accuracy. Taken with permission from Camps-Valls and Bruzzone (2005, Fig. 6). ©2005 IEEE.

Table 3.7 The 9 class scene. Results after a coarse feature selection (200 input bands). Bottom: results in the original (noisy) data set (220 input bands). Several accuracy measures are included: Users, Producers, Overall Accuracy (OA[%]), and Kappa statistic (κ) (Congalton 1991) in the test set for different kernel classifiers: linear discriminant analysis (LDA), regularized RBF neural network (Reg-RBFNN), SVMs with RBF kernel (SVM-RBF) and with polynomial kernel (SVM-Poly), kernel Fisher discriminant (KFD) analysis (with RBF kernel), and regularized AdaBoost (Reg-AB). The column features gives some information about the final models. The best scores for each class are highlighted in bold. The OA[%] being statistically different (at 95% confidence level) from the best model are underlined, as tested through paired Wilcox rank sum test are reproduced from their work and show their results. Taken with permission from Camps-Valls and Bruzzone (2005,Table III). ©2005 IEEE

Method	Features	Producers/Users									OA[%]	κ
		C1	**C2**	**C3**	**C4**	**C5**	**C6**	**C7**	**C8**	**C9**		
LDA	–	78.27	68.05	88.53	88.28	**100.00**	65.30	84.78	73.24	100.00	<u>82.08</u>	0.79
		84.83	75.51	81.43	98.88	**100.00**	82.95	66.07	80.58	97.51		
Reg-RBFNN	nodes: 48,	87.43	80.61	94.94	**99.72**	**100.00**	83.78	90.92	91.91	99.22	91.39	0.90
	λ=1e-7	91.11	92.67	96.57	95.20	99.61	84.13	86.88	88.47	**99.53**		
SVM-RBF	γ=379.27,	**92.82**	94.95	**95.78**	96.74	**100.00**	88.80	92.67	92.74	**99.53**	94.31	0.93
	C=46.42	91.47	91.07	95.78	99.44	99.60	90.64	92.97	**95.15**	98.76		
SVM-Poly	d=7,	**92.82**	94.95	95.78	96.84	**100.00**	**88.89**	**92.76**	91.35	99.43	**94.44**	**0.93**
	C=63.10	91.47	91.16	95.88	**99.55**	99.60	**90.74**	**93.06**	94.88	97.93		
KFD	$\gamma = 144$,	89.78	**95.50**	95.73	96.22	99.60	84.91	86.34	91.40	99.37	91.54	0.90
	$C = 15$	88.87	81.12	94.51	99.44	99.60	81.91	91.50	92.88	98.76		
Reg-AB	λ=1e-6, T=10	91.47	86.98	94.09	99.16	**100.00**	88.56	90.67	**94.17**	99.22	93.50	0.92
	nodes: 8, $\phi = 1/2$	**92.00**	**94.19**	**96.95**	97.79	**100.00**	85.71	89.87	90.93	**99.53**		
LDA	–	77.92	69.58	88.24	89.20	**100.00**	65.89	84.97	72.30	100.00	<u>82.32</u>	0.79
		85.69	75.26	82.28	99.16	**100.00**	82.74	66.56	80.26	97.36		
Reg-RBFNN	nodes: 33,	83.53	73.98	93.67	**98.88**	**100.00**	75.47	86.26	80.58	98.60	88.36	0.86
	λ=2.15e-8	85.38	89.51	94.87	95.16	99.22	74.23	81.97	80.06	99.22		
SVM-RBF	γ=379.27,	85.67	87.47	92.92	94.65	99.21	85.59	87.19	86.62	99.37	91.34	0.90
	C=46.42	86.42	85.46	94.09	**98.88**	99.60	81.50	89.04	83.82	97.82		
SVM-Poly	d=5,	85.38	**88.47**	93.67	94.65	**100.00**	85.88	86.88	86.58	**99.68**	**91.74**	**0.90**
	C=3000	86.42	85.51	94.87	**98.88**	99.60	81.97	89.04	83.22	97.88		
KFD	$\gamma = 100$,	87.57	88.36	**94.17**	96.43	**100.00**	**86.15**	85.37	**93.73**	99.53	90.69	0.89
	$C = 100$	**88.58**	85.20	95.36	98.04	**100.00**	76.30	**89.70**	**91.91**	98.91		
Reg-AB	λ=1e-6,T=10	**88.29**	83.42	94.09	98.32	**100.00**	80.67	**89.04**	90.61	98.91	90.65	0.89
	nodes: 10, $\phi = 1/2$	87.91	**91.34**	**95.71**	95.91	**100.00**	**83.44**	85.88	90.32	**99.53**		

RBF neural network (Reg-RBFNN), kernel Fisher discriminant (KFD) analysis (with RBF kernel) and regularized AdaBoost (Reg-AB).

3.8 Using spatial coherence

Clearly there is useful information in the local spatial coherence of the spectral data. By this we mean that we expect the spectral shape in nearby pixels to be correlated, and vary slowly except near edges between classes. In Section 3.3 we noted that we would restrict our attention to only using the spectral information associated with each pixel (carrying the spatial information along as ancillary information). Approaches have been developed in *unsupervised classification* (Gualtieri and Tilton 2002; Plaza *et al.* 2002) that demonstrate methods that use both the spectral and spatial information of hyperspectral data without dimension reduction, however with substantial computational cost. These results could subsequently be used with *supervised classification* to yield improved results.

Here we mention results of how the spatial data can be incorporated into the SVM approach to improve subsequent classification. Mercier *et al.* (2002) use a preprocessing step that rearranges the spectral data by using the Minimum Noise Fraction transform, Green *et al.* (1988) and Lee *et al.* (1990) followed by vector anisotropic diffusion (Lennon *et al.* 2002). The idea is that since the spatial correlation of the signal is strong (except across edges between classes), as compared with the spatial correlation of the noise, then a noise covariance matrix can be estimated for each band by computing the mean of the spatial differences (in the x and y direction). With this estimate of the noise covariance matrix, the hyperspectral image cube is then transformed in the wavelength regime to yield a new image cube where the strength of the signal-to-noise ratio (SNR) is ordered by increasing band number in the transformed cube. This image cube is then subjected to anisotropic diffusion, a nonlinear sequential filtering operation that smooths regions with spectral similarity (regions of the same class) and sharpens the differences where there are rapid changes in spectral similarity (the boundary edges between classes). Because the MNF transform is reversible the nonlinearly processed cube is back transformed to the original wavelength domain, where now the image cube has noise removed, and regions that are spectrally similar are more uniform and the boundaries between classes are sharpened.

When a supervised classification method is applied to this transformed image using 10, and then 20 training vectors for each of ten classes, the overall classification accuracy of SVM using the OAO approach for multiple classes with a polynomial kernel transformation, as compared with classical techniques such as Spectral Angle Mapper (SAM), and Gaussian Maximum Likelihood (GML), the classification accuracy improves by 5% and 8% respectively. But when the data were preprocessed with the MNF/anisotropic diffusion, the SVM saw a 12% and 10–14% improvement in classification accuracy over SAM and GML results for 10 and 20 training vectors respectively. Compared with SVM on non-preprocessed data, SVM with preprocessed data for both 10 and 20 training sample data classification accuracy was improved by 10% to 82% and 92% respectively. Thus SVM can provide superior results when used in concert with another method for very small training samples for multiple class problems.

In another development of incorporating spatial information with spectral information Camps-Valls *et al.* (2005) first create from neighbourhood textural spatial information indices associated with each pixel for each wavelength. Then they created kernel transformations for spatial and spectral information separately and formed both direct products of the two kernels as well as *cross information* kernels. When compared with the earlier results found above for the two Indian Pines scenes, subset and full, in Table 3.6, Camps-Valls *et al.* demonstrated 2–3% improvement and 8–9% improvement for overall classification accuracy respectively, depending on the particulars of the spatial-spectral kernel used.

3.9 Why do SVMs perform better than other methods?

We have already mentioned in Section 3.3 Vapnik's dictum about solving the appropriate problem, and the problem of the Hughes effect and why density estimation as a first step in solving for the separating surface for supervised classification is inappropriate. For supervised classification with two classes, we only need to find a separating single surface. The Support Vector Machine directly seeks that separating surface by finding the exemplars that form the *boundaries* of the classes, the *support vectors*. This is significant because it is usually the case that there are a small subset of all the training data that are involved in defining the separating surface, i.e., those training vectors that are closest to the separating surface. However, we quickly see that in the original space finding such *linear* separating surface will not in general work as shown in Figure 3.5(right). However, adding a penalty term for finding training vectors that lie on the wrong side of the separating surface, and using the Kernel transformation we can generalize the kinds of separating surfaces that can effectively be used. The nonlinear transform of the data to a higher dimensional space effectively spreads the data out making a linear separation more feasible, and reduces the need for the penalty term. For cases of the Kernel transformation where the high-dimensional linear separating surface can be back transformed to the original space, this yields *nonlinear* separating surfaces as seen in the toy example of Figure 3.6. Thus the limited training data are being used to find a small number of high-dimensional objects, the support vectors, and this computation is not directly affected by the data dimensionality, as it is for the density estimation approach. This aspect of SVM is described by saying that it uses a sparse representation, meaning with a few objects, each with substantial complexity, the SVM is capable of modelling the object sought – the separating surface.

To further examine why SVMs work better than feed forward neural networks we consider an observation from Vapnik (Vapnik 1995, Section 5.64) which elucidates a comparison of SVMs and neural networks. If the Kernel function for the SVM is chosen as $K(\mathbf{x}, \mathbf{x}') = S(\mathbf{x} \cdot \mathbf{x}')$, where $S(u) = tanh(au + b)$ (a, b are constants) is the sigmoid function, then for a restricted set of values of a, b that satisfy the Mercer condition for valid Kernels, the architecture of the SVM is that of a two layer neural network where the number of support vectors are the number of hidden layer nodes, and the weight vectors for neural network are the coefficients λ_i, the Lagrange multipliers. Thus the SVM automatically optimally computes the number of hidden layer nodes and their strengths. However while the quantity that a Neural Network is minimizing is the empirical risk,

$$\sum_i (y_i - f(\mathbf{x}_i, \alpha))^2, \tag{3.17}$$

the quantity that the SVM version of a NN above is minimizing is the upper bound of the generalization error as derived from the Structural Minimization Principle, which we have argued in Section 3.4.3 is more robust and has its derivation based on general principles applicable to these types of problems. Camps-Valls and Bruzzone (2005) also elucidate the effectiveness of SVMs in comparison with regularized RBF neural networks (Reg-RBFNN) and Kernel Fisher Discriminant (KFD). SVMs working from the Structural minimization Principle (SRM) minimize both the Empirical Risk and the Risk Bound, which is a measure of the classifier model's complexity or learning capacity, whereas KFD and Reg-RBFNN minimize only the Empirical Risk. Since empirical risk is defined for KFD and Reg-RBFNN is defined by a sum of squared deviations, Equation (3.17), where $(y_i, \mathbf{x}_i)$ are the training data, and $f(\mathbf{x}_i, \alpha)$ is the desired classifier function, all the data are included equally, whereas the SVM approach explicitly finds the most important training data (those nearest the separating surface) and only uses those in the object to be minimized.

3.10 Conclusions

We have described an approach to building a supervised learning machine called the Support Vector Machine and applied it to classify hyperspectral remote sensing data. We have given some insight into the issues of how hyperspectral data are acquired and processed before any classification is performed to alert users to issues in the data.

The inherent high dimensionality of this data is challenging for traditional classifiers due to the Hughes effect and the curse of dimensionality. We have indicated some of the theoretical foundations of SVMs that point to why the dimensionality of the data is not a handicap as it has been for traditional classifiers, and how the SVMs give improved results, and thus are suitable for use with hyperspectral data. The results we have obtained show SVMs to be competitive with other classifiers for hyperspectral data for hyperspectral scenes with as many as 16 classes. And more recent developments show how to incorporate local spatial information and show further improvement is possible.

References

Aizerman, M., Braverman, E. and Rozoner, L.I. (1964) Theoretical foundations of the potential function method in pattern recognition. *Automation and Remote Control,* **25**, 821–837.

Bellman, R.E. (1961) *Adaptive Control Processes: A Guided Tour*. Princeton, New Jersey: Princeton University Press.

Bianchi, R., Cavalli, R., Marino, C., Pignatti, S., Colosi, F. and Poscolliri, M. (1997) Airborne hyperspectral MIVIS data over Selinunte ancient town area (Sicily, Italy) as a support classical archaeological investigation *Proceedings of the Third International Airborne Remote Sensing Conference and Exhibitio*, pp. 761–768.

Precision geocoding of low altitude AVIRIS data: Lessons learned in 1998. *In AVIRIS Airborne Geoscience Workshop* (ed. Green R) NASA/JPL.

Boardman, J. (2005) There's more to it than that: exploring the complex, dimensionality, and information content of AVIRIS data. In *AVIRIS Airborne Geoscience Workshop* (ed. Green R) NASA/JPL.

Boardman, J.W. and Green, R.O. (2000) Exploring the spectral variability of the Earth as measured by AVIRIS in 1999. In *AVIRIS 2000 Workshop Proceedings* (ed. Green RO). NASA/JPL. http://popo.jpl.nasa.gov/docs/workshops/00_docs/toc.html.

Boser, B.E., Guyon, I.M. and Vapnik, V.N. (1992) A training algorithm for optimal margin classifiers *Fifth Annual Workshop on Computational Learning Theory*, pp. 144–152. ACM.

Burges, C.J.C. (1998) A tutorial on support vector machines for pattern recognition. *Data Mining and Knowledge Discovery* **2**(2), 121–167.

Camps-Valls, G. and Bruzzone, L. (2005) Kernel-based methods for hyperspectral image classification. *IEEE Transactions on Geoscience and Remote Sensing* **43**(6), 1351–1362.

Camps-Valls, G., Gómez-Chova, L., Muñoz-Marí, J., Vila-Francés, J. and Calpe-Maravilla, J. (2005) Composite kernels for hyperspectral image classification. *IEEE Geoscience and Remote Sensing Letters*. **3**(1), 93–97

Chang, C.C. and Lin, C.J. 2001 *LIBSVM: a library for support vector machines*. Software available at http://www.csie.ntu.edu.tw/~cjlin/libsvm.

Clark, R.N., Green, R.O., Swayze, G.A., Meeker, G., Sutley, S., Hoefen, T.M., Livo, K.E., Plumlee, G., Pavri, B., Sarture, C., Wilson, S., Hageman, P., Lamothe, P., Vance, J.S., Brownfield, J.B.I., Gent, C., Morath, L.C., Taggart, J., Theodorakos, P.M. and Adams, M. (2001) Environmental studies of the world trade centre area after the September 11, 2001 attack. Technical report, U.S. Geological Survey. http://speclab.cr.usgs.gov/wtc/.

Congalton, R. (1991) A review of assessing the accuracy of classifications of remotely sensed data. *Remote sensing of environment* **37**(1), 35–46.

Cortes, C. (1995) *Prediction of Generalization Ability in Learning Machines* PhD thesis, University of Rochester.

Cortes, C. and Vapnik, V.N. (1995) Support vector networks. *Machine Learning* **20**, 1–25.

Duan, K., Keerthi, S.S. and Poo, A.N. (2003) Evaluation of simple performance measures for tuning SVM hyperparameters. *Neurocomputing* **51**, 41–59.

Duda, R.O., Hart, P.E. and Stork, D.G. (2001) *Pattern Classification* '2nd edn. Wiley-Interscience.

Fukunaga, K. (1990) *Introduction to Statistical Pattern Recognition*. Academic Press.

Gao, B.C. and Davis, C.O. (1997) Develpment of a line-by-line-based atmosphere removal algorithm for airborne and spaceborne imaging spectrometers *Imaging Spectrometry III, Vol. 3118*, pp. 132–141 SPIE.

Gao, B.C., Heidebrecht, K.B. and Goetz, A.F.H. (1993) Derivation of scaled surface reflectance's from AVIRIS data. *Remote Sensing of Environment* **44**, 165–178.

Garegnani, J., Gualtieri, J.A., Chettri, S., Robinson, J., Hunt, J.P., Bechdol, M. and Vermeulen, A. (2000) Preliminary tests of the utility of hyperspectral image data to precision farming *International Geoscience and Remote Sensing Symposium*, IEEE.

Geman, S., Bienenstock, E. and Doursat, R. (1992) Neural networks and the bias/variance dilemma. *Neural Computation* **4**(1), 1–58.

Girosi, F., Jones, M. and Poggio, T. (1995) Regularization theory and neural networks architectures. *Neural Computation* **7**(2), 219–269.

Goel, P.K., Prasher, S.O., Landry, J.A., Patel, R.M., Viau, A.A. and Miller, J.R. (2003) Estimation of crop biophysical parameters through airborne and field hyperspectral remote sensing. *Transactions of the ASAE* **46**, 1235–1246.

Goodman, J.A. and Ustin, S.L. (2002) Hyperspectral image calibration in a coral reef environment: An empirical approach. *Int. Geoscience and Remote Sensing Symposium*, IEEE.

Green, A.A., Berman, M., Switzer, P. and Craig, M.D. (1988) A tranformation for ordering multi-spectral data in terms of immage quality with implication for noise removal. *IEEE Trans. Geoscience and Remote Sensing* **26**(1), 65–74.

Green, R. (2004) AVIRIS airborne workshops. Technical report, JPL. Series covers 1987-2004. See http://aviris.jpl.nasa.gov/html/aviris.documents.html.

Gropp, W., Lusk, E., and Skjellum, A. (1996) *Using MPI: Portable Parallel Programming with the Message Passing Interface*. MIT Press.

Gualtieri, J.A. (2004) A parallel processing algorithm for remote sensing classication *AVIRIS Airborne Geoscience Workshop* NASA/JPL. ftp://popo.jpl.nasa.gov/pub/docs/workshops/04_docs/gualtieri_aviris_200%4.pdf.

Gualtieri, J.A. and Cromp, R.F. (1998) Support vector machines for hyperspectral remote sensing classification In *27th AIPR Workshop, Advances in Computer Assisted Recognition, Proceeding of the SPIE, Vol. 3584* (ed. Merisko RJ), pp. 221–232. SPIE.

Gualtieri, J.A. and Howard, E. (2003) Hyperspectral sensing opportunities for coastal zones and sea grasses *Proceedings of the 30th International Symposium on the Remote Sensing of the Environment.* International Society for Photogrammetry and Remote Sensing, Hawaii.

Gualtieri, J.A. and Tilton, J.C. (2002) Hierarchical segmentation of hyerspectral data *Proc. 11th JPL Airborne Earth Science Workshop* NASA/JPL. JPL Pub. 03-4. http://popo.jpl. nasa.gov/docs/workshops/02_docs/toc.html.

Gualtieri, J.A., Chettri, S.R., Cromp, R.F. and Johnson, L.F. (1999) Support vector machine classifiers as applied to AVIRIS data. In *Summaries of the Eighth JPL Airborne Earth Science Workshop:JPL Publication 99-17* (ed. Green R), pp. 217–227. NASA/JPL. http://aviris.jpl.nasa.gov/docs/workshops/99_docs/toc.htm.

Guyon, I., Vapnik, V., Boser, B., Bottou, L. and Solla, S.A. (1992) Structural risk minimization for character-recognition. *Advances In Neural Information Processing Systems* **4**, 471–479.

Hsu, C.W. and Lin, C.J. (2002) A comparison of methods for multiclass support vector machines. *IEEE Trans. Neural Nets* **13**, 415–425.

Hughes, G.F. (1968) On the mean accuracy of statistical pattern recognizers. *IEEE Transactions on Information Theory* **14**(1), 55–63.

Ikeda, K. (2003) Generalization error analysis for polynomial kernel methods: algebraic geometrical approach. *Artificial neural networks and neural information processing*.

Jackson, Q.Z. and Landgrebe, D. (2001) *Design of an Adaptive Classification Procedure for the Analysis of High-dimensional Data With Limited Training Samples.* PhD thesis School of Electrical Computer Engineering, Purdue University. http://dynamo.ecn.purdue.edu/~landgreb/JacksonTR.pdf.

Jiménez, L.O. and Landgrebe, D. (1996) *High dimensional feature reduction via projection pursuit.* PhD thesis School of Electrical Computer Engineering, Purdue University. http://dynamo.ecn.purdue.edu/~landgreb/JimenezTR.pdf.

Joachims, T. (1998a). The SVM^{light} package is written in gcc and distributed for free for scientific use from http://www-ai.cs.uni-dortmund.de/FORSCHUNG/VERFAHREN_ LIGHT/svm_light.eng.htm%l It can use one of several quadratic optimizers. For our application we used the A. Smola's PR_LOQO package (Smola 1998).

Joachims, T. (1998b) Making large-scale SVM learning practical. In *Advances in Kernel Methods–Support Vector Learning* (ed. Schölkopf B, Burges C and Smola A) MIT Press pp. 169–184. http://www-ai.cs.uni-dortmund.de/DOKUMENTE/Joachims_98b.ps.gz.

Kettig, R.L. and Landgrebe, D.A. (1976) Classification of Multi-spectral Image Data by Extraction and Classification of Homogeneous Objects. *Transactions on Geoscience Electronics* **14**(1), 19–26.

Kuo, B.C., and Landgrebe, D. (2001) *Improved Statistics Estimation And Feature Extraction For Hyperspectral Data Classification.* PhD thesis School of Electrical Computer Engineering, Purdue University. http://dynamo.ecn.purdue.edu/~landgreb/KuoTR.pdf.

Landgrebe, D. (1992) Indian pines AVIRIS hyperspectral radiance data:92av3c. A 145 × 145 pixel and 220 band subset in BIL format of radiance data that has been scaled and offset to digital numbers, DN, from radiance, rad in units $W/(cm^2 * nm * sr)$ according to DN = 500 * rad + 1000 as 2 bit signed integers. It is available at ftp://ftp.ecn.purdue.edu/biehl/MultiSpec/92AV3C.lan. Ground truth is also available at ftp://ftp.ecn.purdue.edu/pub/biehl/PC_MultiSpec/ThyFiles.zip. The full data set is called Indian Pines 1 f920612t01p02_r02 as listed in the AVIRIS JPL repository (http://aviris.jpl.nasa.gov/ql/list92.html). The flight line is Lat_Start: 40:36:39 Long_Start: -87:02:21 Lat_End: 40:10:17 Long_End: -87:02:21 Start_Time:19:42:43GMT End_Time:19:46:2GMT.

Landgrebe, D. (1999) Information extraction principles and methods for multispectral and hyperspectral image data. In *Information Processing for Remote Sensing* (ed. Chen H) World Scientific Publishing, Chapter 1. http://dynamo.ecn.purdue.edu/~landgreb/publications.html.

Lee, J.B., Woodyatt, S. and Berman, M. (1990) Enhancement of high spectral resolution remote sensing data by a noise adjusted principle components transform. *IEEE Trans. Geoscience and Remote Sensing* **28**(3), 295–304.

Lee, Z., Carder, K.L., Chen, R.F. and Peacock, T.G. (2001) Properties of the water column and bottom derived from airborne visible infrared imaging spectrometer (AVIRIS) data. *J. G. R.* **106**(C6), 11639–11651.

Lennon, M., Mercier, G. and Hubert-Moy, L. (2002) Nonlinear filtering of hyperspectral images with anisotropic diffusion *International Geoscience and Remote Sensing Symposium*, IEEE.

Malzahn, D. and Opper, M. (2005) A statistical physics approach for the analysis of machine learning algorithms on real data. *Journal of Statistical Mechanics: Theory and Experiment*, P11001.

Melgani, F. and Bruzzone, L. (2004) Classification of hyperspectral remote-sensing images with support vector machine. *IEEE Transactions on Geoscience and Remote Sensing* **42**(8), 1778–1790.

Mercier, G., Lennon, M. and Hubert-Moy, L. (2002) Classification of hyperspectral images with nonlinear filtering and support vector machines *International Geoscience and Remote Sensing Symposium*. IEEE.

Montes, J., Gao, B.C. and Davis, C.O. (2001) A new algorithm for atmospheric correction of hyperspectral remote sensing data. *Geo-Spatial Image and Data Exploitation II, Proceedings of the SPIE Vol. 4383*, pp. 23–30, SPIE.

Montgomery, D. and Peck, E. (1992) *Introduction to Linear Regression Analysis*. John Wiley & Sons, Inc.

Mukherjee, S., Niyogi, P., Poggio, T. and Rifkin, R. (2006) Learning theory: stability is sufficient for generalization and necessary and sufficient for consistency of empirical risk minimization. *Advances in Computational Mathematics* **25**(1), 161–193.

Niyogi, P., Burges, C. and Ramesh, P. (1999) Distinctive feature detection using support vector machines. *Int. Conf. Acoustics, Speech, and Signal Processing*. http://research.microsoft.com/~cburges/papers/icassp98.ps.gz.

Platt, J. (1999) Fast training of support vector machines using sequential minimal optimization. In *Advances in Kernel Methods–Support Vector Learning* (ed. Schölkopf B, Burges C and Smola A) MIT Press pp. 185–208. http://research.microsoft.com/ ~jplatt/abstracts/SMO.html.

Plaza, A., Martinez, P., Gualtieri, J. and Perez, R. (2002) Automated identification of endmembers from hyperspectral data using mathematical morphology. *Proc. SPIE–Int. Soc. Opt. Eng. (USA)* **4541**, 278–287.

Schaback, R. and Werner, J. (2006) Linearly constrained reconstruction of functions by kernels with applications to machine learning. *Advances in Computational Mathematics* **25**(1), 237–258.

Schölkopf, B., Burges, C.J. and Smola, A.J. (1999) *Advances in Kernel Methods: Support Vector Learning*. MIT Press.

Smola, A. (1998). The PR_LOQO quadratic optimization package is distributed for research purposes by A. Smola at http://svm.first.gmd.edu.de/software/loqosurvey.html.

Smola, A.J., Schölkopf, B. and Müller, R.J. (1998) The connection between regularization operators and support vector kernels. *Neural Networks* **11**(4), 637–649.

Swayze, G.A., Clark, R.N., Pearson, R.M. and Livo, K.E. (1997) Mapping acid-generating minerals at the California gulch Superfund site in Leadville, Colorado using imaging spectroscopy.

Summaries of the 6th Annual JPL Airborne Earth Science Workshop NASA, JPL. http://speclab.cr.usgs.gov/PAPERS.Leadville95/leadville1.html.

Tadjudin, S. and Landgrebe, D. (1998a) *Classification of High Dimensional Data with Limited Training Samples.* PhD thesis School of Electrical Engineering and Computer Science, Purdue University. available as TR-ECE-98-9 from http://dynamo.ecn.purdue.edu/~landgreb/Saldju_TR.pdf.

Tadjudin, S. and Landgrebe, D. (1998b) Covariance estimation for limited training samples. *Int. Geoscience and Remote Sensing Symposium*, vol. 5, pp. 2688–2690. IEEE, Seattle, WA. http://dynamo.ecn.purdue.edu/+~+landgreb/SaldjuCovarEst.pdf.

Tadjudin, S. and Landgrebe, D. (1999) Covariance estimation with limited training samples. *IEEE Transactions on Geoscience and Remote Sensing* **37**(4), 2113–2118.

Tanré, D., Herman, M. and Deschamps, P.Y. (1983) Influence of the atmosphere on space measurements of bidirectional properties. *Applied Optics* **22**, 733–741.

Tanré, D., Herman, M., Deschamps, P.Y. and de Leffe, A. (1979) Atmospheric modelling for space measurements of ground reflectances including bidirectional properties. *Applied Optics* **18**, 3587–3594.

Ustin, S., Underwood, E., Mulitsch, M.J., Greenberg, J.A., Kefauver, S.C., Whiting, M.L, Rueda, CA, Ramirez, C.M., Scheer, G.J. and Ross, L.F. (2005) Mapping invasive plant species in the Sacramento–San Joaquin delta region using hyperspectral imagery. Technical report, Centre for Spatial Technologies and Remote Sensing (CSTARS) and California Space Institute Centre of Excellence (CalSpace) University of California, Davis.

Vapnik, V.N. (1982) *Estimation of Dependencies Based on Empirical Data.* Springer. First published in Russian: Nauka, Moscow, 1979.

Vapnik, V.N. (1995) *The Nature of Statistical Learning Theory* 1st, edn., Springer.

Vapnik, V.N. (1998) *Statistical Learning Theory* 1st. edn., John Wiley and Sons, Inc.

Vapnik, V.N. (1999) An overview of statistical learning theory. *IEEE-Trans. on Neural Networks* **10**(5), 988–998.

Vapnik, V.N. and Chervonenkis, A.J. (1971) On the uniform comvergemce of relative frequencies of events to their probabilities. *Theory of Probability and its Applications* **16**(3), 264–280. English translation of 1968 work in Russian.

Vapnik, V.N. and Chervonenkis, A.J. (1974a) *Theory of Pattern Recognition.* Nauka. In Russian.

Vapnik, V.N. and Chervonenkis, A.J. (1974b) *Theory of Pattern Recognition.* Nauka. In Russian.

Vapnik, V.N. and Chervonenkis, A.J. (1991) The necessary and sufficient conditions for the uniform convergence of the method of empirical risk minimization. *Pattern Recognition and Image Analysis* **1**(3), 284–305. English translation of 1989 work in Russian.

Zanghirati, G. and Zanni, L. (2003) A parallel solver for large quadratic programs in training support vector machines. *Parallel Computing* **29**(4), 535–551.

Zhou, D. (2003) Capacity of reproducing kernel spaces in learning theory. *Information Theory, IEEE Transactions on* **49**(7), 1743–1752.

Zhou, D. and Jetter, K. (2006) Approximation with polynomial kernels and SVM classifiers. *Advances in Computational Mathematics* **25**(1), 323–344.

4

On training and evaluation of SVM for remote sensing applications

Giles M. Foody
School of Geography, University of Nottingham, UK

The design of the training and testing stages of a supervised classification is, to differing degrees, classifier-dependent. This chapter provides an overview of the some of the key issues in the design of training and testing stages for image classification, with particular regard to classification by SVM. A key issue stressed is that only effective support vectors are required in training an SVM. This feature enables SVM to derive accurate classifications from small training sets. The accuracy of an SVM may be assessed using a variety of approaches but some may not be practical for all types of SVM classification.

4.1 Introduction

Remote sensing has been used in a wide variety of application areas. In particular, satellite remote sensing has been widely used as a source of thematic information such as land cover. This is valuable for many reasons, not least because land cover and land cover change are major components of global environmental change. Indeed, land cover and its dynamics greatly impact on issues of environmental, social and economic significance and are recognized as grand

Kernel Methods for Remote Sensing Data Analysis Edited by Gustavo Camps-Valls and Lorenzo Bruzzone

challenges for research (Aspinall, 2008). As a result of the importance of land cover as an environmental variable, accurate and up-to-date land-cover maps are required for many applications.

Remote sensing can be an efficient tool for the provision of up-datable information on land cover information at frequent intervals and for large regions. Despite remote sensing's considerable potential as a source of land cover information many problems are encountered. In particular, the accuracy of land cover maps derived from remote sensing has often be viewed as being insufficient by many in the user community (Wilkinson 1996; Foody 2002, 2008; Gallego 2004). A wide range of factors may be responsible for this situation including the nature of the classes being studied, the properties of sensing system used to acquire the imagery and the techniques used to extract thematic information from the imagery (Foody 2002; Pal and Mather 2003). Although aspects of accuracy assessment in remote sensing are pessimistically biased (Foody 2008) it appears, therefore, that many challenges have to be addressed in mapping land cover accurately from remotely sensed data. Attention in this chapter is focused on some of the issues connected with the extraction of land cover information via a SVM-based classification.

4.2 Classification for thematic mapping

Information on land cover and land cover change is commonly derived from remotely sensed data with the aid of a supervised image classification analysis. A variety of classification approaches are available and these can differ greatly in detail. Some analyses may, for example, use only the remotely sensed data on a per-pixel basis while others may make use of ancillary data and be undertaken on a per-parcel basis. The basic nature of the classification process is, however, the same for most scenarios. For simplicity, this chapter will assume that a per-pixel based classification using only spectral information (e.g. reflectance) is used but the discussion could be generalized for other approaches.

Assuming that the remotely sensed data have been fully and rigorously pre-processed, a supervised classification analysis may be considered to comprise three stages: training, allocation and testing (Figure 4.1). In the first, training, stage, regions of known ground identity (class membership) are typically identified on the image. The spectral response of these training sites or areas may be used to generate descriptive statistics for the land cover classes to inform the second, class allocation, stage of the classification analysis. The class allocation stage essentially acts to convert the remotely sensed data set into a thematic map depicting the spatial distribution of the classes of interest in the region represented by the image. In the third, testing, stage, the accuracy of the derived thematic map is evaluated. This evaluation is typically based on the class allocations made for the cases in the testing set, a sample of pixels that ideally were not used in training the classifier. The assessment of classification accuracy is typically based on a confusion matrix, which is a cross-tabulation of the actual and classifier predicted class labels for the cases contained in the testing set. The accuracy statement derived in the testing stage is an important summary of the quality of the classification and its suitability for use. The end product of the supervised classification analysis is, therefore, a thematic map of known accuracy.

For many applications, the value of the thematic map derived with a supervised classification analysis is a function of its accuracy. The latter is a function of many variables, including

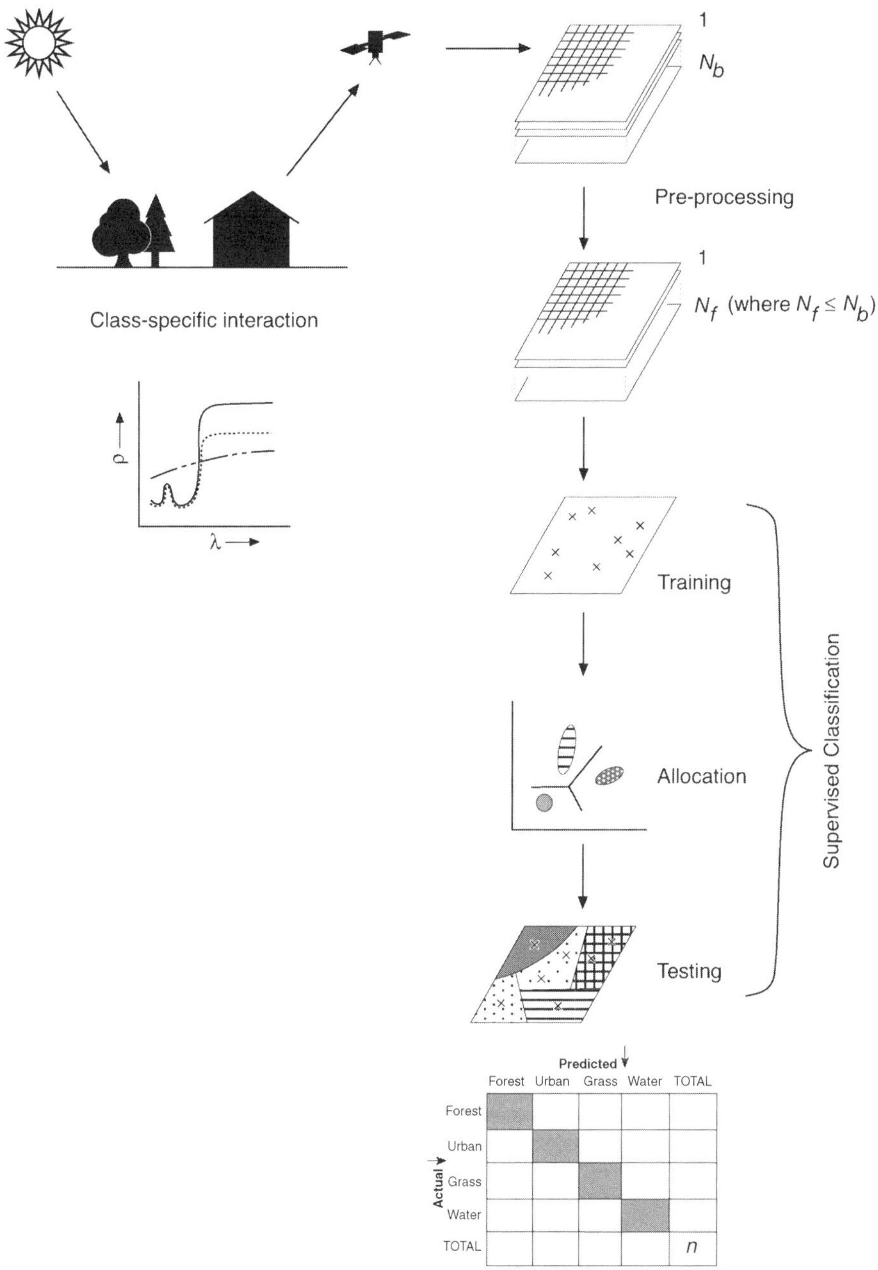

Figure 4.1 Overview of thematic mapping from remotely sensed data with a supervised classification. Note one popular pre-processing analysis is feature reduction, in which the original data set of dimensionality N_b is reduced to a smaller size, N_f, without substantial loss of information. The end of the process is a thematic map or classification of known accuracy (the latter typically estimated from a confusion matrix).

a suite of issues connected with the initial two stages of the classification analysis over which the analyst has considerable influence. Since classification accuracies have often been viewed negatively, much research effort has focused on the training and class allocation stages with an overall aim of increasing the accuracy of classification. Considerable work has addressed the potential of a variety of different classifiers with much recent interest focused on kernel based techniques such as the SVM. The SVM has become popular as it has often been observed in many studies to classify data sets with an accuracy equivalent to or higher than that derived from the application of a range of alternative classifiers (Huang *et al.* 2002; Foody and Mathur 2004a; Melgani and Bruzzone 2004; Dixon and Candade 2008; Oommen *et al.* 2008). In addition, there is scope to increase the accuracy of SVM classification further, as, for example, is evident in studies such as those directed at including prior knowledge into the analysis (e.g. Lauer and Bloch 2008).

This chapter focuses on the first (training) and third (testing) stages of supervised classification by SVM; details on the nature of SVM-based classifiers is given elsewhere in this book (see especially Chapters 2–4). For completeness and to ensure that this discussion is self-contained, however, a very brief overview of some of the salient features of classification by SVM is given before addressing the issues connected with training a SVM classification and evaluating its accuracy. The discussion relating to both the training and testing stages will be placed within the context of the general approaches used in remote sensing before considering concerns connected more specifically with SVM-based classifiers. A key focus in this discussion of the testing stage will be on issues connected with the comparison of classification accuracy statements as this is often the basis for evaluations of the relative performance of image classifiers in remote sensing.

4.3 Overview of classification by a SVM

Research aimed at increasing the accuracy with which thematic maps may be derived from remote sensing has witnessed the evaluation of many classifiers. Recent history has seen a progression from conventional statistical classifiers such as the maximum likelihood classifier through artificial neural networks to kernel based classifiers. The discussion here is focused mainly on issues connected with one particularly widely used kernel-based approach, the SVM. SVM were originally designed for binary classification problems. Although binary applications are sometimes encountered (e.g. Sánchez-Hernández *et al.* 2007) most studies involve multiple classes. Fortunately, the basic binary SVM approach to classification can be extended for the common multi-class classification problem. Frequently, this type of classification is based on a series of multiple binary analyses following either the one-against-all or one-against-one strategies (Huang *et al.* 2002; Gualtieri and Cromp 1998). With these strategies, a multi-class problem is sub-divided into to a set of binary problems, allowing the basic binary approach of SVM to be utilized to yield a multi-class classification. These approaches can involve considerable time in training all of the component classifiers, especially if there are many classes or if the data set is large. Various approaches may be used to speed up the analysis including methods that aim to reduce the size of the input data (e.g. Cervantes *et al.* 2008) or reduce the number of binary analyses required through the exclusion of redundant sub-classifiers (Ye and Teng 2007). An alternative and computationally efficient strategy for multi-class classification is the adoption of a one-shot multi-class SVM classifier (Hsu and Lin 2002). Irrespective of the precise method used, the fundamental feature of classification by a SVM is the nature in which the optimal separating hyperplane (OSH) is fitted between classes. The OSH is oriented in space such that it is placed at maximum distance between the classes, maximizing the margin between them. This latter property should allow a SVM to generalize more accurately on unseen cases as compared with classifiers such as the multilayer perceptron neural network that aim to minimize the training error.

A detailed mathematical explanation of SVM can be found in Vapnik (1995) with examples in a remote sensing context including the discussions provided by Huang *et al.* (2002), Melgani and Bruzzone (2004), Pal and Mather (2005) and Watanachaturaporu *et al.* (2008), as well as in other chapters in this book. Here, only some of the main features are discussed to aid the discussion on the training stage below. The key feature is to note that the OSH is formulated from some of the training samples that lie at the edge of the class distribution in feature space (Figure 4.2). This, often small, sub-set of a standard training set are the support vectors that are fundamental to classification by a SVM. The classification decision function may often be considered to be

$$f(\mathbf{x}) = \sum_{i}^{l} \alpha_i y_i K(\mathbf{x}, \mathbf{x}_i), \tag{4.1}$$

where, for each of the l training cases, a vector, $\mathbf{x}_i$, represents the spectral response of the case together with a definition of class membership, y_i, α_i, $i = 1, \ldots, l$ are Lagrange multipliers and $K(\mathbf{x}, \mathbf{x}_i)$ is a kernel function. The magnitude of α_i is determined by the parameter C and lies on a scale of 0–C (Belousov *et al.* 2002). More detailed discussions are provided in Chapters 3 and 4.

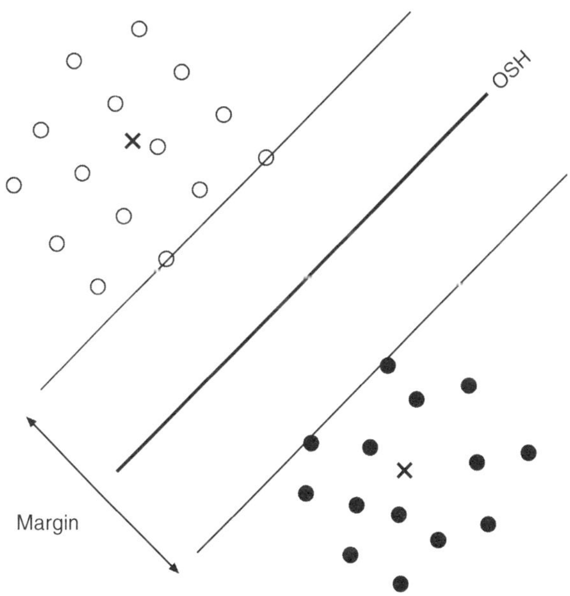

Figure 4.2 The fitting of the OSH between 2 classes (open and closed circles).

The accuracy of an SVM classification varies as a function of the magnitude of the parameters C and γ. For example, if γ and/or C are set at large values then there is a tendency for the SVM to over-fit to the training data, which can negatively impact on the ability to generalize and accurately label previously unseen pixels. Since a high degree of generalizability is typically desirable in thematic mapping from remotely sensed data, the values for the parameters C and γ must be determined carefully for the specific task in hand. A variety of approaches may be used for this application. For example, the training data set could be sub-divided into two independent sets: training and validation. A variety of parameter value scenarios could then be used with those that provide the most accurate classification of the validation set adopted. This approach, the hold out method, may not be attractive if training data are scarce or difficult to acquire (Bishop 1995). A popular alternative is to adopt a cross-validation analysis. With this approach, the training set is typically broken down into S independent subsets. An SVM may then be trained with $S - 1$ subsets and its accuracy evaluated on the remaining subset. By repeating this process S times, ensuring that each subset is used only once for validation purposes, and calculating the average validation accuracy an overall measure of the generalizability, a particular SVM classifier, is obtained that may be used to help select parameter settings. This approach does allow the use of a large fraction of training data for the purpose of training, as opposed to validation, but does require the training process to be repeated S times (Bishop 1995). A related alternative is to adopt bootstrapping, which is based on subsamples rather than subsets of the training data set. From the various methods available to evaluate the generalizability of a SVM the use of a cross-validation analysis is particularly popular (e.g. Belousov *et al.* 2002).

The key issue to note in this chapter is that the SVM classification is based on just the training samples that lie at the edge of the class distributions, the support vectors. The latter are the training samples for which $\alpha_i > 0$. All of the other training cases ($\alpha_i = 0$) do not contribute to the formulation of the classifier (Equation (4.1)) and, are, therefore, irrelevant and unnecessary. Such training cases may be removed from a training set without negative impact on the accuracy of the classification. Perhaps more constructively, effort and resources need not be spent in the acquisition of such training cases as they are not used in fitting the OSH. With the classification dependent on just the support vectors, it is possible to derive an accurate classification from a SVM trained with only a small training set (Foody and Mathur 2004b; Foody *et al.* 2006). A similar situation occurs for related classifiers such as the one-class SVDD (Muñoz-Marí *et al.* 2007; Sánchez-Hernández *et al.* 2007) and RVM, which is a probabilistic counterpart to the SVM (Bowd *et al.* 2005); although the relevance vectors for a RVM are generally of a more anti-boundary nature than the support vectors of a SVM analysis (Tipping 2001).

Having provided a brief overview of the nature of class allocation by a SVM, attention now turns to the two other stages of the supervised classification. Again it is worth stressing that the discussion is based around the use of a conventional per-pixel based classification but should provide general material of relevance to a wide variety of other classification scenarios.

4.4 Training stage

The basic aim of the training stage is, essentially, to provide descriptive statistics for each class in the image that may be used to inform the class labelling process undertaken by the classifier.

Unless using a resource such as a spectral library, the typical approach adopted involves the acquisition of a sample of pixels of known class membership from the image to characterize the classes spectrally. These selected pixels represent the training set that describes the classes upon which the remainder of the supervised classification analysis is based.

Typically, the descriptive statistics derived from the training pixels are used to characterize the classes statistically and ultimately convey the information needed to partition feature space so that class membership may be determined for all image pixels in the allocation stage of the classification. In the latter, each pixel of unknown class membership is allocated to the class with which it appears to have the greatest spectral similarity. Obtaining an accurate description of each class is therefore often seen as fundamental to the derivation of an accurate classification (Kuo and Landgrebe 2002; Mather 2004).

Since the training stage provides the class descriptors upon which all allocations are based the quality of the training data set used is of fundamental importance to a supervised classification and a major determinant of classification accuracy. Many studies have shown that the accuracy of a classification varies as a function of a range of training set properties (Zhuang *et al.* 1994; Foody *et al.* 1995; Foody and Arora 1997; Staufer and Fischer 1997; Fardanesh and Ersoy 1998; Foody 1999; Tsai and Philpot 2002). Indeed the nature of the training stage can have a larger impact on classification accuracy than the type of classifier used (Campbell 2002). This situation has prompted research on the design of the training stage. The latter has addressed issues such as those connected with the sampling design used to select training sites (Campbell 2002; Chen and Stow 2002), size of the training set (Congalton 1991; Foody and Arora 1997; Foody and Mathur 2004a), composition of the training set (Foody *et al.* 1995) as well as issues such as the spacing of training samples, the time of sampling with respect to that of image acquisition and the potential to automate aspects of the process (Huang *et al.* 2008). However, most attention has focused on the size of training set, the number of training cases acquired. This issue has frequently attracted attention because of the costs, especially in terms of time and finance, involved in the acquisition of a training set (Buchheim and Lillesand 1989; Jackson and Landgrebe 2001) and its known importance for many classifiers including SVM (see Chapter 3).

4.4.1 General recommendations on sample size

The remote sensing literature contains a range of recommendations on the size of the training set required for a supervised classification. Much of this literature is based on a classical statistical view of the classification process and an assumption that a complete description of each class in feature space is required. One common recommendation in the literature is that the size of the training set be defined as a function of the number of discriminatory variables used. For the common situation of a classification using only the data acquired in a set of spectral wavebands, it is often suggested that a sample comprises at least 10–$30N_b$ cases, where N_b defines the number of spectral wavebands used (Piper 1992; Mather 2004; van Niel *et al.* 2005). The sample size estimated by this method can be large, especially if a hyperspectral data set is to be classified. Alternatively, as the derivation of an accurate and unbiased description of each class is often seen as an aim in the training stage, conventional statistical theory may be used to define the required sample size. To derive the representative and unbiased description of the classes that is commonly perceived as a requirement of training, basic sampling theory suggests the use of some variant of random sampling to define the location of training sites. With simple random sampling, the sample size required to characterize the spectral response of a class, assuming it follows a normal distribution, could be estimated using

$$n = \frac{\sigma^2 z^2}{W^2}, \tag{4.2}$$

where W is a specified half-width of the confidence interval, σ the planning or estimated value for the population standard deviation and z is the value of the z score at a specified level of confidence.

Equation (4.2) provides an estimate of the sample size required to estimate the mean value of a distribution with a specified degree of precision. The theoretical basis of this equation can also used to determine the sample size required in a testing set for accuracy assessment, where the aim is typically to determine the sample size required to estimate a proportion. Equation (4.2), like the $10\text{–}30N_b$ heuristic, defines the minimum sample required and a 'bigger is better' attitude is often held by researchers. Note for example that from Equation (4.2) it is evident that the precision of the estimate is linked to sample size; the more precise the estimate of the mean needed, and so the smaller W, the larger the sample size that is required.

Conventional guidance is, therefore, to collect as large a sample of training cases as possible. This can have advantages in helping to control for unspecified variables that may have an effect on the analysis. The acquisition of a large sample by random sampling will, for example, help to accommodate for geographical variation in spectral response (e.g. variations in crop reflectance arising as a consequence of regional variations in variables such as soil background reflectance and growth stage or condition). Consequently, analysts are encouraged to embark on an expensive programme of training data acquisition driven by a desire to have a large number of training samples from locations spread over the entire image area capture and represent the full spectral variability of the classes. Although this can be costly in time and resources the literature does, however, show that, for a variety of classifiers and environments, classification accuracy is positively related to training set size (Pal and Mather 2003; Zhuang *et al.* 1994; Foody *et al.* 1995; Arora and Foody 1997; Foody and Mathur 2004a) and so that the effort expended in aquiring the training data set would seem to be worthwhile. It should be noted, however, that the standard recommendations on training set size estimation are of a very general nature and are often made without any regard to the specific study area, the complexity of the classes to be mapped (e.g. the Anderson level) or the classifier to be used and the aim of the analysis (Foody *et al.* 2006). The adoption of a standard approach to training will require the acquisition of a large training set, which may be undesirable as training data acquisition is often expensive (Chi and Bruzzone 2005; Mantero *et al.* 2005) and may be unnecessary as some of the training data may be of little or no value. The latter, as discussed below, is important in the context of classification by an SVM.

The size of training set is not the only variable to consider in the design of the training stage. The acquisition of and post-processing of training data often includes actions that may have an important impact on the training data and so ultimately the classification and its accuracy. For example, it is desirable that only pixels that actually represent an area of the class being described are used in its description. Thus, the pixels selected for training purposes should ideally be pure members of the relevant classes. To achieve this researchers often deliberately mask out or exclude boundary regions where the mixing of class spectral responses may occur (e.g. Arikan 2004). Additionally, some analysts apply post-acquisition refinement operations to the training data that remove outliers or down-weight the contribution of cases perceived to be atypical of the class being characterized (Buttner *et al.* 1989; Aria 1992; Ediriwickrema and Khorram 1997; Mather 2004). Alternatively, some researchers use seed functions in the acquisition of training data (e.g. Sun *et al.* 2003), which do not allow the inclusion of pixels with values greatly different from the seed. The use of such post-acquisition refinement operations and seed functions will generally act to shift focus towards the purest exemplars of the classes. This situation may be useful if interest is focused on the characterization of the class centroid but, as discussed below, may be unsuitable if using a SVM.

In summary, therefore, the aim of the training stage has conventionally been to acquire a large sample of pure pixels to describe the classes. Some approaches for training set definition or refinement adopted may, however, act to bias the description by placing emphasis on cases typical of the class centroid. This may be suitable for some classification analyses but need not be generally desirable. Indeed, the basis of the conventional approach to training set design may not actually be focused on the provision of the critical information needed for an accurate image classification. It may be preferable instead to tailor the approach to training data acquisition to the particular problem in hand (Foody *et al.* 2006; Sánchez-Hernández *et al.* 2007), with particular regard to the selected classifier.

Classifiers may differ greatly in how they use the training data in order to partition feature space for class allocation. It is, therefore, unsurprising that different classifiers applied to an image often produce dissimilar allocations even if using the same training set (Huang *et al.* 2002; Foody and Mathur 2004a). As classifiers differ in the way they partition feature space into classes, the value of individual training cases and the nature of an ideal training set may vary considerably from one application to another. For example, the training data for a parallelepiped classification should ideally describe accurately the extremities of the classes in feature space, while it is more important to estimate accurately the class centroids for a minimum distance to means classifier. Classifiers also differ in how the training data set is used. The popular maximum likelihood classifier, for instance, uses parameters such as the mean and covariance matrix that summarize the spectral response of each class while a feedforward neural network such as the multi-layer perceptron uses each training case directly. Classifiers can, therefore, be expected to differ greatly in terms of the required training information for an accurate classification and individual training cases may vary greatly in value. It is possible that a training set which could be used to derive a highly accurate classification from one classifier may yield a considerably lower accuracy if used with another classifier (Foody 1999). For classification by a multi-layer perceptron neural network or a SVM, the training cases that lie both at the border of class distributions and between class centroids in data space are the most important in terms of helping to derive an accurate classifier (Foody 1999; Foody and Mathur 2004b). These training data would, however, poorly describe the typical spectral response of the classes and so be expected to be inappropriate for use with, for instance, a minimum distance to means classifier.

As individual training cases vary in their value to a classifier and training sets vary in suitability for use with different classification algorithms, the nature of the classifier selected for a particular application should inform the design of a training data collection programme. The typically classifier-independent design recommendations in the literature should be viewed critically but may still be used to inform the development of a design that is suited to the classifier selected. Critically, for a SVM, the variation in the value of the individual training samples may be used constructively as it points to a potential to limit total training set size and cost of training through a focus on the informative training cases. Indeed the way that some classifiers operates makes it possible to tailor the training stage in a way that could require a relatively small training sample (Foody and Mathur 2004b; Foody *et al.* 2006). This is the case for the SVM where the potential for accurate classification from small training sets stems from the fundamentals of SVM-based classification, namely that only the training cases that are support vectors are needed for accurate discrimination with all others being redundant. In recognition of the variation in the importance of the individual training cases some studies have sought to identify and use only the most important training samples (Foody and Mathur 2004b) or to weight training cases by importance (Yang *et al.* 2007). Some of the major issues are discussed in the following section.

4.4.2 Training a SVM

The training cases in a SVM-based classification vary greatly in importance, with those lying near the hyperplanes being most informative (Foody and Mathur 2004b). Thus with a SVM, the desire need not be to obtain as large a training sample as possible but one that contains the most useful training cases. These are the training samples that lie at the edge of the class distributions but between the class centroids in feature space (the support vectors) and only these are needed in the establishment of the decision surface; the other training cases can effectively be discarded without impacting negatively on the accuracy of the classification as they do not contribute to the fitting of the OSH (Brown *et al.* 2000; Belousov *et al.* 2002; Wang *et al.* 2005). Thus, the accuracy of a SVM classification depends not so much on the size of the training set but more on the location of training data cases in the feature space. Moreover, since the computation of the decision surface is not dependent on the dimensionality of the data, SVM can accurately classify data in high dimensional space with a limited number of training data. SVM may, therefore, possibly overcome, or at least be less sensitive to, the curse of dimensionality or the Hughes phenomenon which may negatively impact on other classifiers (Pal and Mather 2004; Oommen *et al.* 2008). Consequently, it may also sometimes be unnecessary to undertake pre-processing operations such as feature-reduction analysis (Melgani and Bruzzone 2004), although this can sometimes remain a useful part of a classification project (Neumann *et al.* 2005).

The potential to use small training sets that contain appropriate support vectors for the derivation of an accurate classification could allow considerable savings in training data acquisition to be achieved relative to the use of conventional practices. However, the realization of this potential requires and ability to identify the most useful training cases (Foody and Mathur 2004b). A variety of approaches have been suggested for the identification of informative training sites. For example, one approach to identify potential support vectors is to identify the extremities of the class distributions in feature space with the aid of knowledge on the variables controlling the spectral response of the classes. The principles of this knowledge-based or intelligent approach to training have been identified (Foody and Mathur 2004b) and may be illustrated by a simple hypothetical example.

The basis of the intelligent training approach is to use knowledge of the factors that influence the spectral response of the classes to help identify informative training sites. Consider a project that aims to derive a thematic map of a stereotypical tropical island. Aside from the large sandy beaches that ring the island, the island is covered in forest. This simple scenario involves three classes: vegetation, beach and water. Each class has a distinctive interaction with the electromagnetic radiation used commonly in remote sensing but there is a degree of intra-class variation which is summarized in Figure 4.3. As a result of the intra-class

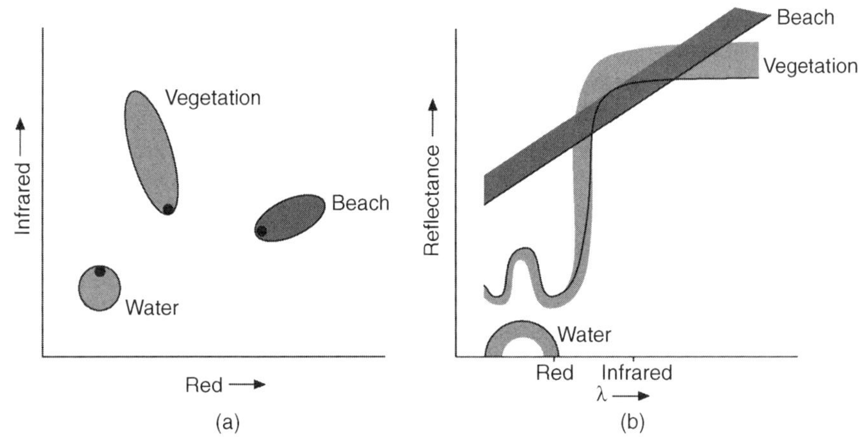

Figure 4.3 The spectral properties of the three classes in the hypothetical example. (a) Classes in feature space defined by the infrared and red wavebands, circles highlight the possible location of candidate support vectors. (b) The spectral response curves of the classes, the solid line in each class indicates the response of a candidate location highlighted in (a).

variation, the spectral response observed from a location of a particular class may lie within an envelope of possible responses for the class (Figure 4.3(b)). Furthermore, the intra-class variation may be evident in the area of feature space that the classes occupy (Figure 4.3(a)). The training of a supervised classifier requires descriptive information on the location of the classes in feature space. Rather than characterize the spectral response of each class fully, a SVM only needs some of the relatively extreme cases of each class to be used in training. Potential candidate training cases are highlighted in feature space by circles in Figure 4.3(a). The spectral response associated with each location is then highlighted as a line plotted within the spectral response envelope for each class in Figure 4.3(b). To make the intelligent training approach operational there is, therefore, a need to be able to locate training sites that may have the spectral response similar to the highlighted cases. Fortunately, this is often possible. For example, the most informative training sites for the water class have a relatively high reflectance in each of the wavebands defining the feature space. Basic knowledge of the interactions of radiation with water bodies (e.g. Curran 1985) should allow the analyst to deduce that regions of deep clear water would be inappropriate training sites as they are typically associated with low reflectance but regions of shallow water overlying a bright toned substrate or of turbid water would be expected to have a high reflectance. Again, a simple deduction is that candidate locations for useful training sites for water would lie close to the shoreline, where the water is shallow and likely to have relatively high turbidity. Similarly, for the beach class it is known that a major determinant of the reflectance of bare sand its moisture content (Curran 1985). Dry sand sites would be expected to have a relatively high reflectance and so be expected to be relatively poor candidate locations for training site acquisition. However, wet sand would be expected to have a low reflectance and so training sites acquired from such a region would be expected to have spectral properties near the highlighted part of the spectral response in Figure 4.3(a). Note also that these would be expected to lie near the shoreline and so close to where the training sites for the water class would be located, providing a basis for an efficient fieldwork programme to collect the ground data set. Finally, in this example, with the vegetation class, the aim would be to locate vegetated regions that had a relatively low infra-red reflectance and high red reflectance. From knowledge of the impacts of variables such as vegetation amount (e.g. leaf area index) and its condition (e.g. chlorophyll content) on the remotely sensed response (Curran 1985) the analyst might deduce that locations likely to make suitable candidate sites for training a SVM are regions of relatively low vegetation amount, while locations of more luxuriant coverage should be avoided. Thus to map the land covers of the region, quite basic knowledge may be used to locate sites that may be expected to be of particular informative value to a SVM classification. Clearly the example used is very basic and simplified (e.g. impacts of variable such as shadow or surface roughness have been ignored for simplicity) but its potential has been illustrated in practice. For example, Mathur and Foody (2008a) used easy to acquire information on variables such as crop maturity status, proximity to water bodies and soil type to direct fieldwork to define informative training sites for a SVM classification. The use of this approach yielded a classification of comparable accuracy to that obtained when a larger training set, acquired following conventional practice, was used and reduced costs by approximately a quarter (Mathur and Foody 2008a). It is also worth noting that a focus on extreme training cases has been found to be valuable in related analyses such as support vector regression analysis (Guo and Zhang 2007).

An alternative approach to defining useful training data is to focus on areas of spectral mixing. This is because the most useful training cases are those that lie close to where the hyperplane is to be fitted and this lies between the classes and so in the region in which mixtures of the class spectral responses occur. Consequently, mixed spectral responses and mixed pixels may sometimes represent useful and informative training cases for a classification (Foody and Mathur 2006). Such locations are normally deliberately avoided in contemporary approaches to the training stage that focus on homogeneous regions (pure pixels). This approach would not, however, be expected to work well if the spectral response observed from a mixture of two classes resembled that of another class in the image.

The nature of the thematic mapping programme may also be amenable to use of a very small training set. This is particularly apparent when attention is focused on one or just a small sub-set of the classes present in the area imaged. This is the case in a wide range of applications including studies focused on mapping alien species, particular crops or specific habitats for conservation (e.g. Foody *et al.* 2006; Goodwin *et al.* 2005;. Ramsey *et al.* 2002; Underwood *et al.* 2003; Laba *et al.* 2005; Hill *et al.* 1980; Pinter *et al.* 2003; Boyd *et al.* 2006). While conventional approaches to image classification require a large training sample as the set of classes must be exhaustively defined and each class characterized fully, the use of SVM-based classifiers offers the potential to derive the desired information from a small training set in which effort has been focused on the class(es) of interest rather than what may be considered as background. Here, the binary nature of the SVM is attractive as it may allow the analyst to focus on separating the class of interest from all others (Boyd *et al.* 2006). Furthermore, when attention is focused entirely on a single class, mis-classification amongst the classes that are not of interest is completely unimportant and can be ignored in the testing stage of the classification. Finally, alternative scenarios for classification by SVM exist. Of particular relevance to this section of the book is the potential for semi-supervised classification, which makes use of unlabelled cases in training (Bruzzone *et al.* 2006; Camps-Valls *et al.* 2007).

4.4.3 Summary on training

Conventional recommendations on the design of the training stage appear to be focused on describing the classes. This approach can also be biased towards the purest cases (and to the class centroid), which is far from ideal for SVM classification as the most informative training cases, which make good support vectors, are located towards the edge of the class distribution in feature space and not the centre. Thus, the standard approach to training may not actually be based on meeting the fundamental aim of training. The latter is not to describe the classes accurately but to provide information on the classes that will aid the fitting of classification decision boundaries or hyperplanes to separate them accurately. Approaches to training should therefore be viewed as being classifier-dependent. This is especially important for a SVM as the targetting of extreme cases could be seen as running directly in opposition to the conventional guidance, especially as a small, intelligently selected training sample drawn deliberately from sites that could be viewed as relatively atypical, even impure members, of the classes is stressed as useful for a SVM classification. This may be a successful approach because only the support vectors are needed and hence the highly unrepresentative sample of cases acquired with a judgemental sample design may provide the required information. A variety of approaches may be used depending on the specific nature of the mapping task in-hand (Foody *et al.* 2006). Critically, it is possible to suggest the spectral properties of candidate support vectors and so target training data acquisition activities to locations likely to have the desired spectral characteristics. This presents opportunities for substantial savings in time and effort. For example, Mathur and Foody (2008a) report ~26% reduction in cost of ground data collection activities without any significant loss in the ability to discriminate classes through the use of the intelligent approach to training site selection.

Although the general basis of SVM operation applies to the various strategies of multi-class classification there are some important differences that are worthy of discussion. Note, for example, that the size of the training set needed may vary with different strategies to SVM classification. For example, a one shot multi-class based SVM, which requires the solution to a single optimization problem, may require fewer support vectors than either of the two widely used strategies for multi-class classification based on a series of binary analyses, the one-against-one and one-against-all approaches (Mathur and Foody 2008b). Even smaller training sets may be used in some situations. For example, if interest is focused on a single class using a one-class classifier such as the SVDD, which is based on the principles of the SVM, a very small training set may provide all the necessary information for an accurate classification (Tax and Duin 2004; Sánchez-Hernández *et al.* 2007). In all situations, the key feature is that the training data are not required to describe the classes precisely but to enable them to be separated and with a SVM this only requires effective support vectors.

4.5 Testing stage

The main aim of the testing stage of a supervised classification analysis is to convey information on the quality of the class allocations made by the classifier. The quality of a classification may be expressed and measured in various ways but attention is most commonly focused on the accuracy of the class allocations. Accuracy is a measure of bias (or more strictly unbias) and precision (Atkinson and Foody 2002). In terms of a standard image classification analysis, an allocation is correct if the label predicted by the classifier matches that contained in the ground reference data set. Classification accuracy may be evaluated from various perspectives, with particular attention focused on measures of overall accuracy (i.e. that of the entire multi-class classification as a whole) as well as per-class accuracy (e.g. when interest is focused on just one of the classes). Measures of classification accuracy provide a basis on which potential users of the classification may evaluate its fitness for a proposed application.

Although the provision of a classification accuracy statement to describe the quality of a classification is a major aim it is not the sole issue to consider in the testing stage of a supervised classification. Classification accuracy assessment is, for example, central to studies that have sought to evaluate different classifiers with the superiority of one classifier over another commonly based on observed differences in classification accuracy. In the design of the testing stage this comparative aspect may need explicit consideration as it can have important implications, not least on the required size of the testing set (Foody 2009a). Indeed, while classification accuracy assessment and accuracy comparison may seem relatively simple issues there are many challenges to be faced in their undertaking (Foody 2002, 2004, 2008, 2009a). With many challenges to address, the subject is one still open for research and development. Here, the discussion is focused on some fundamental issues in accuracy assessment and comparison with particular regard to SVM-based classification.

4.5.1 General issues in testing

Classification accuracy assessment has developed considerably in recent years and many methods may be used to estimate and convey accuracy information (e.g. Congalton 1991, 1994; Congalton and Green 1999; Pontius 2000, 2002; Foody 2002; Pontius and Cheuk 2006; Lu and Weng 2007; Liu *et al.* 2007). Most commonly, classification accuracy is assessed through the use of site-specific techniques based on the analysis of the entries in a confusion or error matrix (Congalton and Green 1999; Foody 2002). This matrix is a simple cross-tabulation of the predicted and actual class labels observed for the sample of cases contained in the testing set (Figure 4.4). A properly constructed confusion matrix should provide a simple summary of classification accuracy and highlight the two types of misclassification error that may occur: omission (cases of a class incorrectly allocated to another class, and so omitted from the class of interest) and commission (cases of another class incorrectly allocated to the class of interest, and so commissioned by the class of interest). There are, of course, many problems associated with the derivation of a confusion matrix and measures of accuracy from it. This includes, amongst other things, issues connected with the sample design used to acquire the testing set and the quality of the ground data (Foody 2002, 2009b). Thus some of the issues encountered in the training stage, such as determination of required sample size, also feature in the testing stage. For simplicity, it will be assumed that the analyst has been able to generate a confusion matrix following a simple random sampling design. Discussion on other sample designs and their implications as well as general background to the testing stage is given in the literature (e.g. Stehman 1999, 2000; Wickham *et al.* 2004). Attention will also focus only on the standard situation in which a raw or unprocessed confusion matrix is used as some, albeit popular, approaches such as matrix normalisation are often inappropriate, especially if interest is focused on an individual class (Stehman 2004; Liu *et al.* 2007).

The confusion matrix may be used to derive a variety of measures to express the accuracy of a classification (Trodd 1995; Stehman 1997; Liu *et al.* 2007). The two most widely used measures of overall classification accuracy are the proportion of correctly allocated cases and

(a)

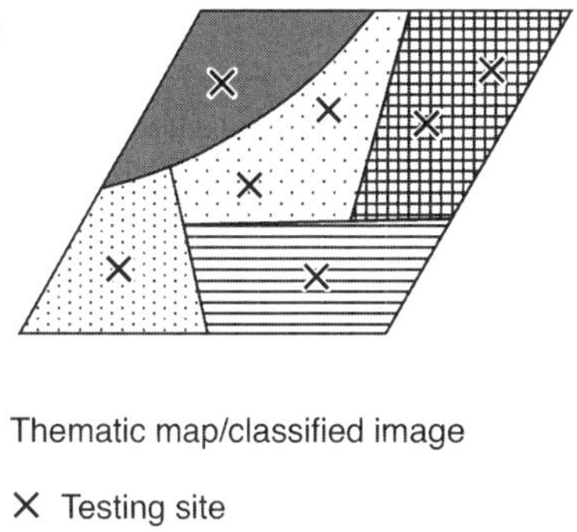

Case	Actual	Predicted
1	Forest	Forest
2	Urban	Water
3	Water	Urban
4	Grass	Urban
5	Grass	Urban
6	Grass	Grass
..	..	..
..	..	..
n	..	..

(b)

Actual ↓ / Predicted →	Forest	Urban	Grass	Water	TOTAL
Forest					
Urban					
Grass					
Water					
TOTAL					*n*

Figure 4.4 The assessment of image classification accuracy. (a) The actual and classifier predicted class label of each of the n testing cases is defined. (b) The cross-tabulation of the actual and predicted labels yields a confusion matrix from which various measures of accuracy may be derived. The total in each row or column defines the matrix marginal values. Note there is no accepted style of matrix presentation and so the row and column labels can be swapped but their meaning must be remembered when producing per-class estimates of accuracy.

the kappa coefficient of agreement (Trodd 1995). The proportion of correctly allocated cases is simply the sum of cases contained in the main diagonal of the confusion matrix divided by the total number of cases used to define the confusion matrix. This value is often multiplied by 100 to yield the percentage of correctly allocated cases, which provides a basic measure of the overall accuracy of a classification. If interest is focused on the accuracy with which a specific class has been classified rather than on the overall classification accuracy, the key concern is the proportion of cases of that class that have been correctly classified. In this situation, the measure of per-class accuracy is based on the ratio of cases contained in the relevant element of the matrix's main diagonal to the relevant marginal value. Note that the latter may be derived from two perspectives depending on whether the matrix is viewed along the rows or columns. As there is no accepted standard way of presenting the matrix, with rows, for example, sometimes used for the ground data and at other times for the predictions of the classifier it may be appropriate to consider the two perspectives in terms of the mis-classification errors to which they relate. If attention is focused on omission errors, accuracy may be expressed by what is commonly referred to as producer's accuracy. The latter is the number of correctly allocated cases of the class divided by the total number of cases of that class; the latter would be represented by the row marginal of the matrix depicted in Figure 4.4(b). Conversely, if attention was focused on commission error, accuracy could be expressed by what is commonly referred to as the user's accuracy. User's accuracy is the number of correctly allocated cases of the class divided by the total number of cases predicted to belong to that class (i.e. the column marginal of the matrix depicted in Figure 4.4(b).

The kappa coefficient of agreement has been widely used in remote sensing as a measure of classification quality. The kappa coefficient is a re-scaled measure of the proportion of correctly allocated cases that accounts for the effects of chance agreement (Congalton *et al.* 1983; Congalton and Green 1999; Smits *et al.* 1999; Wilkinson 2005) and the general formula for its calculation is

$$\hat{\kappa} = \frac{m_o - m_c}{1 - m_c}, \tag{4.3}$$

where m_o is the proportion of cases in agreement (i.e., the proportion of correctly allocated cases in the testing set) and m_c is the proportion of agreement that is expected by chance. The resulting value provides a measure of overall classification quality. A per-class based measure, conditional kappa, may also be derived if interest is focused on the accuracy with which a particular class has been classified (Congalton and Green 1999).

The kappa coefficient has often been promoted on the basis that its calculation includes a correction for chance agreement and that a variance term can be calculated for it which facilitates statistical comparisons (e.g. Congalton *et al.* 1983; Monserud and Leemans 1992; Janssen and van der Wel 1994; Smits *et al.* 1999; Wheeler and Alan 2002). The latter issue is especially important as there is often a desire to compare classifications in order to aid the evaluation of two or more classifiers. For example, one might wish to determine if a SVM provided a more accurate classification than some other classifier. The variance term is critical here as the classification accuracy statements derived typically are estimates of accuracy and so the simple summary measures of accuracy derived from a confusion matrix should not be compared directly without regard to their estimated variances. Instead a statistically rigorous approach that compares the derived summary measures with regard to the variances should be followed to determine the statistical significance of the differences.

The kappa coefficient has been widely promoted in the remote sensing literature as a measure for accuracy assessment because of the ability to compare kappa coefficients rigorously (Congalton and Mead 1983; Congalton *et al.* 1983; Janssen and van der Wel 1994; Smits *et al.* 1999). From a classification accuracy comparison perspective, the aim is to determine if the difference in the derived estimates can be inferred to indicate a difference in the associated population parameters of map accuracy. The statistical significance of the difference in accuracy between two classifications with independent kappa coefficients, represented by $\hat{\kappa}_1$ and

$\hat{\kappa}_2$, may be evaluated with the normal curve deviate,

$$z = \frac{\hat{\kappa}_1 - \hat{\kappa}_2}{\sqrt{\hat{\sigma}^2_{\kappa_1} - \hat{\sigma}^2_{\kappa_2}}}, \tag{4.4}$$

where $\hat{\sigma}^2_{\kappa_1}$ and $\hat{\sigma}^2_{\kappa_2}$ represent the estimated variances of the derived coefficients.

The significance of the difference between the two kappa coefficients and so, by inference, the difference in classifier performance may assessed by comparing the value of z calculated from Equation (4.4) against tabulated values. At the widely used 0.05 significance level, a value of $z > 1.96$ indicates that two coefficients are significantly different (Congalton *et al.* 1983; Congalton and Green 1999). Although this approach has been widely used in classifier comparisons its use may often be inappropriate. This is because the kappa coefficient itself is of debatable value as an index of classification accuracy and the assumptions underlying the comparison of coefficients are often invalid in remote sensing applications. Indeed the meaning and value of the kappa coefficient in classification accuracy assessment has often been questioned (Stehman 1997; Turk 2002; Jung 2003; Foody 2008) with each of the commonly argued reasons for using the kappa coefficient as a measure of map accuracy open to criticism (Foody 2008). In relation to comparative studies, a key issue is that the approach outlined assumes that the samples used to estimate the kappa coefficients are different and independent. This is often not the case in remote sensing studies. In the latter, it is common to use the same testing set of cases in the evaluation of each classification. The samples are, therefore, related and this feature needs to be recognized in the comparison (McKenzie *et al.* 1996; Donner *et al.* 2000). Methods to compare related kappa coefficients could be used instead, which account for the covariance between kappa statistics due to the use of a related sample (e.g. Donner *et al.* 2000), but the limitations of kappa coefficients as a measure of classification accuracy remain. Fortunately, however, there are other approaches that could be used to measure classification accuracy and to compare accuracies. One simple approach is to use the proportion of correctly allocated cases as a measure of accuracy. A variance term can be derived for this, and other measures of accuracy, which also facilitates accuracy comparison (Foody 2004). For example, the statistical significance of differences in the proportion of correctly allocated cases observed from two classifications, m_1 and m_2, may be estimated instead from

$$z = \frac{m_1 - m_2}{\sqrt{\bar{m}(1 - \bar{m})[\frac{1}{n_1} + \frac{1}{n_2}]}}, \tag{4.5}$$

where $\bar{m} = (o_1 + o_2)/(n_1 + n_2)$, in which o_1 and o_2 are the number of correctly allocated cases in two independent samples of size n_1 and n_2 respectively. Equation (4.5) may be viewed as the alternative to (4.4) for independent samples. If the samples are related the statistical significance of the difference between two proportions may be evaluated using McNemar's test. The latter is based focused on the discordant cases, those that were correctly classified by one classifier but mis-classified by the other. This test has been used in the comparison of classifications derived from remotely sensed data (Foody 2004; Ouyang *et al.* 2006; De Leeuw *et al.* 2006) and is based on a matrix of the type shown in

Case	Classifier 1	Classifier 2
1	Correct	Incorrect
2.	Correct	Incorrect
3	Incorrect	Incorrect
4	Incorrect	Correct
5	Correct	Correct
6	Correct	Correct
7	Correct	Correct
..	..	..
..	..	..
..	..	..
n		

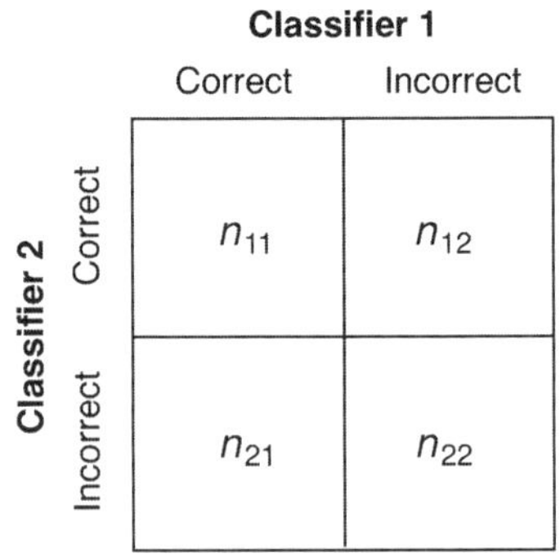

Figure 4.5 The basis of the McNemar test for comparing classifiers. Each of the n cases is allocated to the relevant element of the 2 × 2 matrix. For example, cases 1and 2 would be contributors to the subset of cases that form n_{21} while case 3 would contribute to n_{22}. The test focuses on the discordant subset of cases (contained in n_{12} and n_{21}).

Figure 4.5. Using the notation defined in Figure 4.5, the McNemar test is typically based on

$$z = \frac{n_{12} - n_{21}}{\sqrt{n_{12} + n_{21}}}. \tag{4.6}$$

Irrespective of whether the samples are independent or related, the size of the testing set needs to be large enough to detect a meaningful difference in accuracy (Foody 2009a). Determining the required sample size requires specification of a minimum important difference in accuracy (effect size) together with the desired significance level and power (Foody 2009a). For example, the sample size required for a typical remote sensing scenario using the McNemar test may be estimated from

$$n = \frac{|z_\alpha \Psi^{1/2} + z_\beta(\Psi - \delta^2)^{1/2}|^2}{\delta^2}, \tag{4.7}$$

where z_α and z_β are appropriate values of z for the selected significance level (α) and power $(1 - \beta)$ of analysis respectively, Ψ is the proportion of mismatched cases and δ the effect size or minimum detectable difference (Connor 1987). Further details in a remote sensing context are given in Foody (2009a).

Although popular, hypothesis testing based approaches to accuracy comparison, such as those outlined above, are not problem-free. One major limitation is that the hypothesis text based approach only gives a basic dichotomous outcome, in which the hypothesis (e.g. that a difference in accuracy exists) is 'accepted' or rejected but conveys no information on the magnitude of the difference. An alternative to hypothesis test based approaches for classifier comparison is to base evaluations of differences on the confidence interval fitted to derived estimates (Di Stefano 2004). The latter can provide a richer basis on which to evaluate differences in accuracy. In addition, the approach may also be easily used for a range of scenarios commonly encountered in remote sensing studies, including testing for a difference, equivalence or non-inferiority. A discussion of this approach for these three scenarios is given in Foody (2009c).

4.5.2 Specific issues for SVM classification

Although classification accuracy assessment is often thought to be independent of the classifier this is not strictly the case. In particular, with a SVM one problem is that it may sometimes be impossible to generate a full confusion matrix to summarize the class allocations made. This is evident with multi-class classification following the one-against-all strategy. With this approach it is possible for a case to be of uncertain class membership (Mathur and Foody 2008b) and so make it impossible to calculate measures such as the kappa coefficient. While this is not a major problem given the limitations of the kappa coefficient as a measure of accuracy it can still be undesirable for some applications. For example, some users may wish to use the confusion matrix for post-classification analyses such as rescaling estimates of class areal extent on the basis of the pattern of misclassification depicted in the matrix (e.g. Prisley and Smith 1987). The inability to define the matrix fully means that the full nature of misclassification is unknown and this may limit some post-classification activities. Thus, the classifier to be used in a study should be considered as a variable impacting on the design of the testing stage.

4.6 Conclusion

The training and testing stages of a supervised image classification require careful planning and design in order to optimize a classification analysis. Both stages are, to differing degrees, classifier-dependent. For classification by a SVM, the nature of the SVM operation can have important implications in both the training and testing stages.

In training the classifier, the dependence of the SVM on only the training cases that lie at the edge of the class distributions between class centroids (the support vectors) provides an opportunity to derive an accurate classification from a small training set. The realization of this potential requires a means of predicting the most useful training areas and fortunately this may sometimes be possible. Knowledge of the variables that influence the spectral response of the classes to be mapped may, for example, allow the nature of the most informative training cases to be defined and this information used to steer the training data acquisition programme. By focusing attention upon only these training cases it is possible to derive a classification that is as accurate as one derived from a much larger training set derived following standard procedures. Critically, rather than acquire a large and representative sample to describe the classes it may be more appropriate with a SVM to acquire a small, unrepresentative sample from deliberately selected locations. This enables the use of a small, potentially inexpensive, training set for the derivation of an accurate image classification; the accuracy should be comparable to that derived from a classification based on a much larger training set acquired following conventional practices.

Although the testing stage is less classifier-dependent than the training stage there are some critical issues of relevance to studies based on SVM. First, it must be stressed that with the one-against-all strategy to multi-class classification it may not always be possible to formulate a complete confusion matrix. Although the partial matrix derived may be perfectly adequate for the derivation of some accuracy measures such as the proportion of correctly allocated cases it may not, for example, be possible to calculate some measures of overall (e.g. the kappa coefficient) or some per-class indices of accuracy. Second, in relative comparisons of classification accuracy comparing SVM against other classifiers it is common to use the same test set. The statistical significance in the difference between two classifications of the same test set may be evaluated with the use of a test such as the McNemar test or through comparison of confidence intervals fitted to the derived estimates of accuracy. In comparative analyses, however, it is important to consider issues of sample size, significance level and power if the analysis is to have the potential to yield useful results.

Acknowledgments

This chapter builds on work undertaken over several years that often benefited from input from colleagues, notably Ajay Mathur, Doreen Boyd and Carolina Sánchez-Hernández, and this is gratefully acknowledged. I am also grateful for the review comments received on the original version of this chapter.

References

Airkan, M. (2004) Parcel-based crop mapping through multi-temporal masking classification of Landsat 7 images in Karacabey, Turkey. *International Archives of Photogrammetry, Remote Sensing and Spatial Information Science*, 34. *Proceedings of the ISPRS Symposium*, Istanbul.

Aria, K. (1992) A supervised Thematic Mapper classification with a purification of training samples. *International Journal of Remote Sensing*, **13**, 2039–2049.

Arora, M.K. and Foody, G.M. (1997) Log-linear modelling for the evaluation of the variables affecting the accuracy of probabilistic, fuzzy and neural network classifications. *International Journal of Remote Sensing*, **18**, 785–798.

Aspinall, R.J. (2008) Basic and applied land use science. In R.J. Aspinal and M.J. Hill (eds), *Land Use Change: Science, Policy and Management*, Boca Raton: CRC Press.

Atkinson, P.M. and Foody, G.M. (2002) Uncertainty in remote sensing and GIS: Fundamentals. In G.M. Foody and P.M. Atkinson (eds), *Uncertainty in Remote Sensing and GIS*, John Wiley & Sons, Chichester, 1–18.

Belousov, A.I., Verzakov, S.A. and von Frese, J. (2002) A flexible classification approach with optimal generalisation performance: support vector machines. *Chemometrics and Intelligent Laboratory Systems*, **64**, 15–25.

Bishop, C.M. (1995) Neural Networks for Pattern Recognition, Oxford University Press, Oxford.

Boyd, D.S., Sánchez-Hernández, C. and Foody, G.M. (2006) Mapping a specific class for priority habitats monitoring from satellite sensor data. *International Journal of Remote Sensing*, **27**, 2631–2644.

Bowd, C., Medeiros, F.A., Zhang, Z., Zangwill, L.M., Hao, J., Lee, T.-W., Sejnowski, T.J., Weinreb, R.N. and Goldbaum, M.H. (2005) Relevance vector machine and support vector machine classifier analysis of scanning laser polarimetry retinal nerve fiber layer measurements. *Investigative Opththalmology and Visual Science*, **46**, 1322–1329.

Brown, M., Lewis, H.G. and Gunn, S.R. (2000) Linear spectral mixture models and support vector machines for remote sensing. *IEEE Transactions on Geoscience and Remote Sensing*, **38**, 2346–2360.

Bruzzone, L., Chi, M.M. and Marconcini, M. (2006) A novel transductive SVM for semi-supervised classification of remote-sensing images. *IEEE Transactions on Geoscience and Remote Sensing*, **44**, 3363–3373.

Buchheim, M.P. and Lillesand, T.M. (1989) Semi-automated training field extraction and analysis for efficient digital image classification. *Photogrammetric Engineering and Remote Sensing*, **55**, 1347–1355.

Buttner, G., Hajos, T. and Korandi, M. (1989) Improvements to the effectiveness of supervised training procedures. *International Journal of Remote Sensing*, **10**, 1005–1013.

Campbell, J.B. (2002) *Introduction to Remote Sensing*. 3rd edn, London: Taylor and Francis.

Camps-Valls, G., Marheva, T.V.B. and Zhou, D.Y. (2007) Semi-supervised graph-based hyperspectral image classification. *IEEE Transactions on Geoscience and Remote Sensing*, **45**, 3044–3054.

Cervantes, J., Li., X., Yu, W. and Li, K. (2008) Support vector machine classification for large data sets via minimum enclosing ball clustering. *Neurocomputing*, **71**, 611–619.

Chen, D.M. and Stow, D. (2002) The effect of training strategies on supervised classification at different spatial resolutions. *Photogrammetric Engineering and Remote Sensing*, **68**, 1155–1161.

Chi, M. and Bruzzone, L. (2005) A semi-labelled-sample-driven bagging technique for ill-posed classification problems. *IEEE Geoscience and Remote Sensing Letters*, **2**, 69–73.

Congalton, R.G. (1991) A review of assessing the accuracy of classifications of remotely sensed data. *Remote Sensing of Environment*, **37**, 35–46.

Congalton, R.G. and Green, K. (1999) *Assessing the Accuracy of Remotely Sensed Data: Principles and Practices*, Boca Raton, Lewis Publishers.

Congalton, R.G. and Mead, R.A. (1983) A quantitative method to test for consistency and correctness in photointerpretation. *Photogrammetric Engineering and Remote Sensing*, **49**, 69–74.

Congalton, R.G., Oderwald, R.G. and Mead, R.A. (1983) Assessing Landsat classification accuracy using discrete multivariate analysis statistical techniques, P*hotogrammetric Engineering and Remote Sensing*, **49**, 1671–1678.

Connor, R.J. (1987) Sample size for testing differences in proportions for the paired-sample design. *Biometrics*, **43**, 207–211.

Curran, P.J. (1985) *Principles of Remote Sensing*, Longman, Harlow.

De Leeuw, J., Jia, H., Yang, L., Liu, X., Schmidt, K. and Skidmore, A.K. (2006) Comparing accuracy assessments to infer superiority of image classification methods. *International Journal of Remote Sensing*, **27**, 223–232.

Di Stefano, J. (2004) A confidence interval approach to data analysis. *Forest Ecology and Management*, **187**, 173–183.

Dixon, B. and Candade, N. (2008) Multi-spectral landuse classification using neural networks and support vector machines: one or the other, or both?, *International Journal of Remote Sensing*, **29**, 1185–1206.

Donner, A., Shourkri, M.M., Klar, N. and Bartfay, E. (2000) Testing the equality of two dependent kappa statistics. *Statistics in Medicine*, **19**, 373–387.

Ediriwickrema, J. and Khorram, S. (1997) Hierarchical maximum-likelihood classification for improved accuracies. *IEEE Transactions on Geoscience and Remote Sensing*, **35**, 810–816.

Fardanesh, M.T. and Ersoy, O.K. (1998) Classification accuracy improvement of neural network classifiers by using unlabelled data. *IEEE Transactions on Geoscience and Remote Sensing*, **36**, 1020–1025.

Foody, G.M. (1999) The significance of border training patterns in classification by a feedforward neural network using backpropagation learning. *International Journal of Remote Sensing*, **20**, 3549–3562.

Foody, G.M. (2002) Status of land cover classification accuracy assessment. *Remote Sensing of Environment*, **80**, 185–201.

Foody, G.M. (2004) Thematic map comparison: evaluating the statistical significance of differences in classification accuracy. *Photogrammetric Engineering and Remote Sensing*, **70**, 627–633.

Foody, G.M. (2008) Harshness in image classification accuracy assessment. *International Journal of Remote Sensing*, **29**, 3137–3158.

Foody, G.M. (2009a) Sample size determination for image classification accuracy assessment and comparison. *International Journal of Remote Sensing*, (in press).

Foody, G.M. (2009b) The impact of imperfect ground reference data on the accuracy of land cover change estimation. *International Journal of Remote Sensing*, (in press).

Foody, G.M. (2009c) Classification accuracy comparison: hypothesis test and the use of confidence intervals in evaluation of difference, equivalence and non-inferiority. *Remote Sensing of Environment*, (in press).

Foody, G.M. and Arora, M.K. (1997) An evaluation of some factors affecting the accuracy of classification by an artificial neural network. *International Journal of Remote Sensing*, **18**, 799–810.

Foody, G.M. and Mathur, A. (2004a) A relative evaluation of multiclass image classification by support vector machines. *IEEE Transactions on Geoscience and Remote Sensing*, **42**, 1335–1343.

Foody, G.M. and Mathur, A. (2004b) A. Toward intelligent training of supervised image classifications: directing training data acquisition for SVM classification. *Remote Sensing of Environment*, **93**, 107–117.

Foody, G.M., McCulloch, M.B. and Yates, W.B. (1995) The effect of training set size and composition on artificial neural network classification. *International Journal of Remote Sensing*, **16**, 1707–1723.

Foody, G.M., Mathur, A., Sánchez-Hernández, C. and Boyd, D.S. (2006) Training set size requirements for the classification of a specific class. *Remote Sensing of Environment*, **104**, 1–14.

Gallego, F.J. (2004) Remote sensing and land cover area estimation. *International Journal of Remote Sensing*, **25**, 3019–3047.

Goodwin, N., Turner, R. and Merton, R. (2005) Classifying eucalyptus forests with high spatial and spectral resolution imagery: an investigation of individual species and vegetation communities. *Australian Journal of Botany*, **53**, 337–345.

Gualtieri, J.A. and Cromp, R.F. (1998) Support vector machines for hyperspectral remote sensing classification. *Proceedings SPIE*, 3584, 221–232.

Guo, G. and Zhang, J.-S. (2007) Reducing examples to accelerate support vector regression. *Pattern Recognition Letters*, **28**, 2173–2183.

Hill, J.D., Strommen, N.D., Sakamoto, C.M. and Leduc, S.K. (1980) LACIE – application of meteorology for United-States and foreign wheat assessment. *Journal of Applied Meteorology*, **19**, 22–34.

Hsu, C.-W. and Lin, C.-J. (2002) A comparison of methods for multiclass support vector machines. *IEEE Transactions on Neural Networks*, **13**, 415–425.

Huang, C., Davis, L.S. and Townshend, J.R.G. (2002) An assessment of support vector machines for land cover classification. *International Journal of Remote Sensing*, **23**, 725–749.

Huang, C., Song, K., Kim, S., Townshend, J.R.G., Davis, P., Masek, J.G. and Goward, S.N. (2008) Use of a dark object concept and support vector machines to automate forest cover change analysis. *Remote Sensing of Environment*, **112**, 970–985.

Hixson, M., Scholz, D. and Fuhs, N. (1980) Evaluation of several schemes for classification of remotely sensed data. *Photogrammetric Engineering and Remote Sensing*, **46**, 1547–1553.

Jackson, Q. and Landgrebe, D.A. (2001) An adaptive classifier design for high-dimensional data analysis with a limited training data set. *IEEE Transactions on Geoscience and Remote Sensing*, **39**, 2664–2679.

Janssen, L.L.F. and van der Wel, F.J.M. (1994) Accuracy assessment of satellite derived land-cover data: a review. *Photogrammetric Engineering and Remote Sensing*, **60**, 419–426.

Jung, H.-W. (2003) Evaluating interrater agreement in SPICE-based assessments. *Computer Standards and Interfaces*, **25**, 477–499.

Kuo, B.C. and Landgrebe, D.A. (2002) A covariance estimator for small sample size classification problems and its application to feature extraction. *IEEE Transactions on Geoscience and Remote Sensing*, **40**, 814–819.

Laba, M., Tsai, F., Ogurcak, D., Smith, S. and Richmond, M.E. (2005) Field determination of optimal dates for the discrimination of invasive wetland plant species using derivative spectral analysis. *Photogrammetric Engineering and Remote Sensing*, **71**, 603–611.

Lauer, F. and Bloch, G. (2008) Incorporating prior knowledge in support vector machines for classification: a review. *Neurocomputing*, **71**, 1578–1594.

Liu, C., Frazier, P. and Kumar, L. (2007) Comparative assessment of the measures of thematic classification accuracy. *Remote Sensing of Environment*, **107**, 606–616.

Lu, D. and Weng, Q. (2007) A survey of image classification methods and techniques for improving classification performance. *International Journal of Remote Sensing*, **28**, 823–870.

Mantero, P., Moser, G. and Serpico, S.B. (2005) Partially supervised classification of remote sensing images through SVM-based probability density estimation. *IEEE Transactions on Geoscience and Remote Sensing*, **43**, 559–570.

Mather, P.M. (2004) *Computer Processing of Remotely-Sensed Images*, 3rd edn, Chichester: John Wiley & Sons.

Mathur, A. and Foody, G.M. (2008a) Crop classification by a support vector machine with intelligently selected training data for an operational application. *International Journal of Remote Sensing*, **29**, 2227–2240.

Mathur, A and Foody, G.M. (2008b) Multi-class and binary SVM classification: implications for training and classification users. *IEEE Geoscience and Remote Sensing Letters*, **5**, 241–245.

McKenzie, D.P., Mackinnon, A.J., Peladeau, N., Onghena, P., Bruce, P.C., Clarke, D.M., Haarrigan, S. and McGorry, P.D. (1996) Comparing correlated kappas by resampling: Is one level of agreement significantly different from another?, *Journal of Psychiatric Research*, **30**, 483–492.

Melgani, F. and Bruzzone, L. (2004) Classification of hyperspectral remote sensing images with support vector machines. *IEEE Transactions on Geoscience and Remote Sensing*, **42**, 1778–1790.

Monserud, R.A. and Leemans, R. (1992) Comparing global vegetation maps with the kappa statistic. *Ecological Modelling*, **62**, 275–293.

Muñoz-Marí, J., Bruzzone, L. and Camps-Valls, G. (2007) A support vector domain description approach to supervised classification of remote sensing images. *IEEE Transactions on Geoscience and Remote Sensing*, **45**, 2683–2692.

Neumann, J., Schnorr, C. and Steidl, G. (2005) Combined SVM-based feature selection and classification. *Machine Learning*, **61**, 129–150.

Oommen, T., Misra D., Twarakavi, N.K.C., Prakash, A., Sahoo, B., and Bandopadhyay, S. (2008) An objective analysis of support vector machine based classification for remote sensing. *Mathematical Geosciences*, **40**, 409–424.

Ouyang, Y., Ma, J. and Dai, Q. (2006) Bayesian multi-net classifier for classification of remote sensing data. *International Journal of Remote Sensing*, **27**, 4943–4961.

Pal, M. and Mather, P.M. (2003) An assessment of the effectiveness of decision tree methods for land cover classification. *Remote Sensing of Environment*, **86**, 554–565.

Pal, M. and Mather, P.M. (2004) Assessment of the effectiveness of support vector machines for hyperspectral data. *Future Generation Computer Systems*, **20**, 1215–1225.

Pal, M and Mather, P.M. (2005) Support vector machines for classification in remote sensing. *International Journal of Remote Sensing*, **26**, 1007–1011.

Pinter, P.J., Ritchie, J.C., Hatfield, J.L. and Hart, G.F. (2003) The agricultural research service's remote sensing program: An example of interagency collaboration. *Photogrammetric Engineering and Remote Sensing*, **69**, 615–618.

Piper, J. (1992) Variability and bias in experimentally measured classifier error rates. *Pattern Recognition Letters*, **13**, 685–692.

Pontius, R.G. (2000) Quantification error versus location error in comparison of categorical maps. *Photogrammetric Engineering and Remote Sensing*, **66**, 1011–1016.

Pontius, R.G. (2002) Statistical methods to partition effects of quantity and location during comparison of categorical maps at multiple resolutions. *Photogrammetric Engineering and Remote Sensing*, **68**, 1041–1049.

Pontius, R.G. and Cheuk, M.L. (2006) A generalised cross-tabulation matrix to compare soft-classified maps at multiple resolutions. *International Journal of Geographical Information Science*, **20**, 1–30.

Prisley, S.P. and Smith, J.L. (1987) Using classification error matrices to improve the accuracy of weighted land-cover models. *Photogrammetric Engineering and Remote Sensing*, **53**, 1259–1263.

Ramsey, E.W., Nelson, G.A., Sapkota, S.K., Seegar, E.B. and Martella, K.D. (2002) Mapping Chinese tallow with colour-infrared photography. *Photogrammetric Engineering and Remote Sensing*, **68**, 251–255.

Sánchez-Hernández, C., Boyd, D.S. and Foody, G.M. (2007) One-class classification for monitoring a specific land cover class: SVDD classification of fenland. *IEEE Transactions on Geoscience and Remote Sensing*, **45**, 1061–1073.

Smits, P.C., Dellepiane, S.G. and Schowengerdt, R.A. (1999) Quality assessment of image classification algorithms for land-cover mapping: a review and proposal for a cost-based approach. *International Journal of Remote Sensing*, **20**, 1461–1486.

Staufer, P. and Fischer, M.M. (1997) Spectral pattern recognition by a two-layer perceptron: effects of training set size. In I. Kanellopoulos, G.G. Wilkinson, F. Roli and J. Austin (eds), *Neurocomputation in Remote Sensing Data Analysis*, (Berlin: Springer), 105–116.

Stehman, S.V. (1997) Selecting and interpreting measures of thematic classification accuracy. *Remote Sensing of Environment*, **62**, 77–89.

Stehman, S.V. (1999) Basic probability sampling designs for thematic map accuracy assessment. *International Journal of Remote Sensing*, **20**, 2423–2441.

Stehman, S.V. (2000) Practical implications of design-based sampling inference for thematic map accuracy assessment. *Remote Sensing of Environment*, **72**, 35–45.

Stehman, S.V. (2004) A critical evaluation of the normalised error matrix in map accuracy assessment, *Photogrammetric Engineering and Remote Sensing*, **70**, 743–751.

Sun, W.X., Heidt, V., Gong, P. and Xu, G. (2003) Information fusion for rural land-use classification with high-resolution satellite imagery. *IEEE Transactions on Geoscience and Remote Sensing*, **41**, 883–890.

Tax, D.M.J. and Duin, R.P.W. (2004) Support vector data description. *Machine Learning*, **54**, 45–66.

Tipping, M.E. (2001) Sparse Bayesian learning and the relevance vector machine. *Journal of Machine Learning Research*, **1**, 211–244.

Trodd, N.M. (1995) Uncertainty in land cover mapping for modelling land cover change, In *Proceedings of RSS95 – Remote Sensing in Action*, Nottingham: Remote Sensing Society.

Tsai, F. and Philpot, W.D. (2002) A derivative-aided hyperspectral image analysis system for land cover classification. *IEEE Transactions on Geoscience and Remote Sensing*, **40**, 416–425.

Turk G. (2002) Map evaluation and 'chance correction', *Photogrammetric Engineering and Remote Sensing*, **68**, 123.

Underwood, E., Ustin, S. and DiPietro, D. (2003) Mapping nonnative plants using hyperspectral imagery. *Remote Sensing of Environment*, **86**, 150–161.

Van Niel, T.G., McVicar, T.R. and Datt, B. (2005) On the relationship between training sample size and data dimensionality of broadband multi-temporal classification. *Remote Sensing of Environment*, **98**, 468–480.

Vapnik, V. (1995) *The Nature of Statistical Learning Theory*. Springer-Verlag New York.

Wang, J.G., Neskovic, P. and Cooper, L.N. (2005) Training data selection for support vector machines. *Lecture Notes in Computer Science*, 3610, 554–564.

Watanachaturaporn, P., Arora, M.K. and Varshney, P.K. (2008) Multi-source classification using support vector machines: an empirical comparison with decision tree and neural network classifiers. *Photogrammetric Engineering and Remote Sensing*, **74**, 239–246.

Wheeler, A.P. and Allen, M.S. (2002) Comparison of three statistical procedures for classifying the presence–absence of an aquatic macriphyte from microhaibitat observations. *Journal of Freshwater Ecology*, 17, 601–608.

Wickham, J.D., Stehman, S.V., Smith, J.H., Wade, T.G. and Yang, L. (2004) A priori evaluation of two-stage cluster sampling for accuracy assessment of large-area land-cover maps. *International Journal of Remote Sensing*, **25**, 1235–1252.

Wilkinson, G.G. (1996) Classification algorithms where next? In E. Binaghi, P.A. Brivio and A. Rampini (eds), *Soft Computing in Remote Sensing Data Analysis*, World Scientific, Singapore, 93–99.

Wilkinson, G.G. (2005) Results and implications of a study of fifteen years of satellite image classification experiments. *IEEE Transactions on Geoscience and Remote Sensing*, **43**, 433–440.

Yang, X., Song, Q. and Wang, Y. (2007) A weighted support vector machine for data classification. *International Journal of Pattern Recognition and Artificial Intelligence*, **21**, 961–976.

Ye, W. and Teng, H.S. (2007) Reducing the number of sub-classifiers for pairwise multi-category support vector machines. *Pattern Recognition Letters*, **28**, 2088–2093.

Zhuang, X., Engel, B.A., Lozano-García, D.F., Fernández, R.N. and Johannsen, C.J. (1994) Optimisation of training data required for neuro-classification. *International Journal of Remote Sensing*, **15**, 3271–3277.

5

Kernel Fisher's Discriminant with heterogeneous kernels

M. Murat Dundar[1] and Glenn Fung[2]

[1] *Indiana University-Purdue University, Indianapolis, USA*

[2] *Siemens Medical Solutions Inc., Malvern, USA*

In this chapter we first present a framework suitable for obtaining a nonlinear version of the Fisher's Discriminant (KFD). Then we propose an iterative classification algorithm for KFD using heterogeneous kernel models (AKFD). In contrast with the standard KFD that requires the user to predefine a kernel function, we incorporate the task of choosing an appropriate kernel into the optimization problem to be solved. The choice of kernel is defined as a linear combination of kernels belonging to a potentially large family of different positive semi-definite kernels. Experiments on a Hymap dataset demonstrate that the AKFD algorithm outperforms the linear version of the Fisher's discriminant and also significantly reduces the time required to train the KFD algorithm while maintaining similar performance.

5.1 Introduction

In hyperspectral data analysis, materials of practical interest, such as agricultural crops, forest plantations, natural vegetation, minerals, and fields of interest in urban areas exist in a variety of states and are usually observed in a number of conditions of illumination. That is, most land-cover types do not have a single spectral response. For example a crop type will show different spectral characteristics at different times of the day and year. Similarly, roof tops are usually made of a variety of different materials including concrete, tile, bricks, glass, etc. all of which have different spectral responses. The number of such examples can easily be augmented.

Kernel Methods for Remote Sensing Data Analysis Edited by Gustavo Camps-Valls and Lorenzo Bruzzone

One possible way to deal with this problem is to model each class distribution data using Finite Mixture models (McLachlan and Peel 2004). Finite Mixture Models usually leads to competitive performance when there is enough labelled data to reveal the underlying structure of the class distributions. However, in most real world settings, this may not be the case. The price one must pay for labelled data is usually prohibitively expensive, as acquiring labelled data requires a tedious and time-consuming process of human labelling.

Another alternative for modelling multi-modal class distributions is kernel machines. Kernel machines was first introduced with Support Vector Machines (Vapnik 2000) but later adapted to several other classification algorithms including Fisher's Discriminant (Mika *et al.* 1999). Kernel concept provides the flexibility required to model complex data structures that originate from a wide range of class conditional distributions. Earlier studies show that Fisher's discriminant when implemented with kernel machines yields favourable results for the analysis of hyperspectral data with multimodal class distributions and limited training data (Dundar and Landgrebe 2004a, 2004b).

In the KFD algorithm the type of the kernel function and its parameters are usually estimated from a designated set of kernel models by cross-validation. With this approach the tuning procedure becomes quite computational for training set sizes larger than a few hundred samples. We propose an iterative classification algorithm for Kernel Fisher's discriminant (KFD) using heterogeneous kernel models. In contrast to the standard KFD that requires the user to predefine a kernel function, we incorporate the task of choosing an appropriate kernel into the optimization problem to be solved. The choice of kernel is defined as a linear combination of kernels belonging to a potentially large family of different kernels.

Preliminary results with some benchmark datasets have been presented earlier (Fung *et al.* 2004). In this study additional experimental results on a hyperspectral dataset are presented to further validate the effectiveness of the proposed algorithm in *learning* the optimal combination of kernel functions. The results demonstrate that the prediction accuracy of the proposed algorithm is not significantly different from that achieved by the standard KFD in which the kernel parameters have been tuned using cross-validation, yet the training with the proposed algorithm is multiple folds faster than that of standard KFD.

This chapter is organized as follows. In the next section we will briefly review the Linear Fisher's Discriminant (LFD). Then we will present a mathematical formulation of the Fisher's Discriminant algorithm that will form the basis for the Kernel Fisher's Discriminant (KFD). Next, we will discuss the implementation of KFD with heterogeneous kernel models and present an iterative algorithm for automatically selecting the kernels. Finally we will present results to validate the applicability of the proposed approach on a real-world problem with a hyperspectral dataset.

5.2 Linear Fisher's Discriminant

It is well known that in supervised classification problems the probability of error due to a Bayes classifier is the best that can be achieved. The Bayes classifier compares the *a-posteriori* probabilities of all classes, and assigns the sample to the class with the highest probability. However for most classes of distributions, designing an optimum Bayes classifier is very difficult if not impractical. The primary problem stems from the finite size of the training set, leading to an imperfect estimate of the class probability density functions. The most common way to mitigate this problem is to assume normal distributions for all classes. Under this hypothesis standard classifiers using quadratic and linear discriminant functions can be designed.

The well-known Linear Fisher's Discriminant (LFD) (Fukunaga 1990) arises in the special case when the considered information classes have a common covariance matrix. LFD is a classification method that projects the high dimensional data onto a line and performs classification in this one-dimensional space. This projection is chosen such that the ratio of the scatter matrices (between and within classes) or the so-called *Rayleigh quotient* is maximized. Even though LFD is mainly designed for binary classification problems its extension to multiclass classification problems is also possible. For multiclass problems the ratio of between and within class scatter matrices can be maximized by solving a generalized eigenvalue problem. This leads to a projection matrix with $K-1$ eigenvectors where K is the number of classes in the dataset (Fukunaga 1990).

In the rest of this chapter we will limit our discussion to binary classification problems. More specifically, we are given a training dataset $\{(\mathbf{x}_i, y_i)\}_{i=1}^n$, where $\mathbf{x}_i \in \mathbb{R}^d$ are input variables and $y_i \in \{-1, 1\}$ are class labels. Let $X \in \mathbb{R}^{d\times n}$ be a matrix containing all the training samples and let $X_k \subseteq X \in \mathbb{R}^{d\times n_k}$ be a matrix containing the n_k training samples for class ω_k, $k \in \{\pm\}$. Then, the LFD is the projection w, which maximizes

$$J(w) = \frac{w^\top S_B w}{w^\top S_W w}, \tag{5.1}$$

where

$$S_B = (\mu_+ - \mu_-)(\mu_+ - \mu_-)^\top \tag{5.2}$$

and

$$S_W = \sum_{k\in\{\pm\}} \frac{1}{n_k}\left(X_k - \mu_k e_{n_k}^\top\right)\left(X_k - \mu_k e_{n_k}^\top\right)^\top \tag{5.3}$$

are the between and within class scatter matrices respectively and

$$\mu_k = \frac{1}{n_k} X_k e_{n_k} \tag{5.4}$$

is the mean of class ω_k and e_{n_k} is an n_k dimensional vector of ones.

The above problem can be reformulated as follows. First notice that if w is a solution to (5.1), then so is any scalar multiple of it. Therefore, to avoid multiplicity of solutions, we impose an arbitrary constraint on $w^\top S_B w = 4$, which is equivalent to $w^\top(\mu_+ - \mu_-) = 2$. Then the optimization problem of (5.1) becomes

$$\begin{aligned} \min_{w\in\mathbb{R}^d} \quad & w^\top S_W w \\ s.t. \quad & w^\top(\mu_+ - \mu_-) = 2. \end{aligned} \tag{5.5}$$

A closed-form solution w^* for (5.5) can be obtained by optimizing the Lagrange function associated with the above problem. This gives $w^* = \lambda S_W^{-1}(\mu_+ - \mu_-)$, where λ is the Lagrange multiplier obtained as $\lambda = 1/(\mu_+ - \mu_-)^\top S_W^{-1}(\mu_+ - \mu_-)$. For $d > n$, i.e. the number of dimensionality is greater than the number of samples, S_W can be singular and thus the inverse does not exist. To avoid such ill-conditioned settings it is a common practice to replace S_W by $S_{W_\nu} = S_W + \nu I$. Here ν acts as a regularizer over the classifier.

When classes are normally distributed with equal covariance, w^* is in the same direction as the discriminant in the corresponding Bayes classifier. Hence, for this special case LFD is equivalent to the Bayes optimal classifier. Although LFD relies heavily on assumptions that are not true in most real world problems, it has proved to be very powerful. In particular when the distributions are unimodal and separated by the scatter of means, LFD becomes very effective. One reason why LFD may be preferred over more complex classifiers is that as a linear classifier it is less prone to overfitting.

For most real world data, a linear discriminant is clearly not complex enough. Classical techniques tackle these problems by using more sophisticated distributions in modelling the optimal Bayes classifier; however, these often sacrifice the closed form solution and are computationally more expensive. A relatively new approach in this domain is the kernel version of Fisher's Discriminant (Mika *et al.* 1999), which is known in the literature as Kernel Fisher's Discriminant (KFD). The main characteristic of this approach is the kernel concept, which was originally applied in Support Vector Machines and allows the efficient computation of Fisher's Discriminant in the kernel space. The linear discriminant in the kernel space corresponds to a powerful nonlinear decision function in the input space. Furthermore, different kernels can be used to accommodate the wide-range of nonlinearities that may occur in the data set. In the next section we derive the kernel version of the Fisher's Discriminant.

5.3 Kernel Fisher Discriminant

5.3.1 Mathematical programming formulation

The formulation in (5.5) is a parametric formulation of Fisher's Discriminant. The discriminative approach to the same problem can be obtained as follows. First, we define $\xi_i := w^\top(\mathbf{x}_i - \mu_k), \forall i \in \omega_k, k \in \{\pm\}$ and impose $w^\top \mu_+ = 1, w^\top \mu_- = -1$. Let ξ^k be a vector containing all the ξ_i for class ω_k and y be the vector of class labels. Then the problem in (5.5) becomes

$$\begin{aligned} \min_{(w,\gamma,\xi)\in\mathbb{R}^{n+d+1}} \quad & \tfrac{1}{2}\textstyle\sum_{k\in\{\pm\}} \frac{1}{n_k \nu} \xi^{k\top} \xi^k + \tfrac{1}{2} w^\top w \\ \text{s.t.} \quad & \xi = X^\top w - y \\ & e_{n_k}^\top \xi^k = 0, k \in \{\pm\}. \end{aligned} \tag{5.6}$$

The Lagrangian of (5.6) is given by

$$L(w, \xi, \lambda_1, \lambda_2) = \frac{1}{2}(\xi^\top D \xi + w^\top w) + \lambda_1^\top(\xi - X^\top w + y) + \lambda_2^\top B^\top \xi, \tag{5.7}$$

where D is an $n \times n$ diagonal matrix with the first n_+ entries equal to $1/(n_+\nu)$ and the remaining n_- ones equal to $1/(n_-\nu)$, B is an $n \times 2$ indicator matrix with the first n_+ entries in the first column and the last n_- entries in the second column set to one with all others being zero. Here $\lambda_1 \in \mathbb{R}^n$ and $\lambda_2 \in \mathbb{R}^2$ are the Lagrange multipliers corresponding to $\xi = X^\top w - y$ and $e_{n_k}^\top \xi^k = 0$, respectively. Solving for the gradient of (5.7) equal to zero, we obtain the Karush–Kuhn–Tucker (KKT) necessary and sufficient optimality conditions (Mangasarian

1994) for the FLD problem with equality constraints given by

$$\begin{aligned} w - X\lambda_1 &= 0 \\ D\xi + \lambda_1 + B\lambda_2 &= 0 \\ B^\top \xi &= 0 \\ X^\top w - y - \xi &= 0. \end{aligned} \tag{5.8}$$

The first two Equations of (5.8) give the following expressions for the original problem variables (w, ξ) in terms of the Lagrange multipliers λ_1 and λ_2:

$$w = X\lambda_1, \quad \xi = -D^{-1}(\lambda_1 + B\lambda_2). \tag{5.9}$$

Substituting these in the last two equalities of (5.8) gives us an explicit expression for λ_1 and λ_2 in terms of X and y as follows:

$$\begin{bmatrix} X^\top X + D^{-1} & D^{-1}B \\ B^\top D^{-1} & B^\top D^{-1} B^\top \end{bmatrix} \begin{bmatrix} \lambda_1 \\ \lambda_2 \end{bmatrix} = \begin{bmatrix} y \\ B^\top y - B^\top X^\top X \end{bmatrix}. \tag{5.10}$$

Solving the linear system of Equations (5.10) gives us the solution $\begin{bmatrix} \lambda_1^* \\ \lambda_2^* \end{bmatrix}$, which in turn yields $w^* = X\lambda_1^*$ from (5.9). Here * denotes optimal solutions. In the rest of this chapter we drop the subscript from λ_1 for notational simplicity.

The 'kernelized' version of the Fisher's Discriminant can be obtained in this framework by replacing the primal variable w by its dual equivalent $w = X\lambda$ in (5.6) to obtain:

$$\begin{aligned} \min_{(\lambda,\xi)\in\mathbb{R}^{n+d}} \quad & \tfrac{1}{2}\textstyle\sum_{k\in\{\pm\}} \frac{1}{n_k v} {\xi^k}^\top \xi^k + \tfrac{1}{2}\lambda^\top\lambda \\ \text{s.t.} \quad & \xi = X^\top X\lambda - y \\ & e_{n_k}^\top \xi^k = 0, k \in \{\pm\}, \end{aligned} \tag{5.11}$$

where the objective function has also been modified to minimize weighted 2-norm sums of the dual variables. If we now replace the $X^\top X$ by a nonlinear kernel $K(X^\top, X)$, we obtain a formulation that is equivalent to the Kernel Fisher Discriminant described in (Mika *et al.* 2000):

$$\begin{aligned} \min_{(\lambda,\xi)\in\mathbb{R}^{n+d}} \quad & \tfrac{1}{2}\textstyle\sum_{k\in\{\pm\}} \frac{1}{n_k v} {\xi^k}^\top \xi^k + \tfrac{1}{2}\lambda^\top\lambda \\ \text{s.t.} \quad & \xi = K(X^\top, X)\lambda - y \\ & e_{n_k}^\top \xi^k = 0, k \in \{\pm\}. \end{aligned} \tag{5.12}$$

Recent SVM formulations with least squares loss (Suykens and Vandewalle 1999) are much the same in spirit as the problem in (5.6). Using a similar duality analysis to the one presented here, and then 'kernelizing', the authors obtain the objective function

$$\nu\frac{1}{2}\|\xi\|^2 + \frac{1}{2}\lambda^\top K(X^\top, X)\lambda. \tag{5.13}$$

The regularization term $\lambda^\top K(X^\top, X)\lambda$ determines that the model complexity is regularized in a reproducing kernel Hilbert space (RKHS) associated with the specific kernel K where the kernel function K has to satisfy Mercer's conditions and $K(X^\top, X)$ has to be positive semi-definite.

By comparing the objective function (5.13) with problem (5.12), we can see that problem (5.12) does not regularize in terms of RKHS. Instead, the columns in a kernel matrix are simply regarded as new features $K(X^\top, X)$ of the classification task in addition to the original features X. We can then construct classifiers based on the features introduced by a kernel in the same way we build classifiers using original features X.

5.4 Kernel Fisher's Discriminant with heterogeneous kernels

The kernelized version of the Fisher's Discriminant (KFD) gives us the flexibility required to model complex data structures that originate from a wide range of class conditional distributions. Like its linear space counterpart, the statistics estimation is performed at full dimensionality (no feature extraction is needed) allowing us to exploit all the separability that the data provides without having to deal with severe numerical issues.

Like most other kernel-based approaches KFD also suffers from the computational complexity of working with kernel functions to a greater extent. The computational complexity of the algorithm is on the order of $O(n^3)$, making its use impractical for large datasets. A common way around this problem is to expand the kernel matrix in terms of a random subset of the training samples. However, the problem of selecting the best kernel function type and parameters still remains. Usually, cross-validation is used to optimize the algorithm over a large number of kernel parameters and the parameter set that maximizes the cross-validation performance is chosen as the optimum set.

Cross-validation is a method for estimating predictive error of the classifier with the training data. It splits the training dataset into k equal-sized pieces called folds. At each stage one fold is left out as testing data and the classifier is trained with the remaining $k - 1$ folds. This process is repeated until all k folds are tested and the aggregate test error is recorded as the k-fold cross-validation performance.

In this section we propose a methodology for selecting the optimum kernel function as a weighted summation of several other kernel functions where the weights of the kernel functions are learned automatically through an alternating optimization technique. To be more specific, let us suppose that instead of the kernel K being defined by a single kernel mapping (i.e., Gaussian, polynomial, etc.), the kernel K is composed of a linear combination of kernel functions K_j, $j = 1, \ldots, k$, as below

$$K(X^\top, X) = \sum_{j=1}^{k} a_j K_j(X^\top, X), \tag{5.14}$$

where $a_j \geq 0$.

As pointed out in Lanckriet *et al.* (2003), the set $\{K_1(A, A'), \ldots, K_k(X^\top, X)\}$ can be seen as a predefined set of initial 'guesses' of the kernel matrix and it could contain very different kernel matrix models, (e.g., linear, Gaussian, polynomial) with different parameter values. In this formulation parameters specific to each kernel are fixed *a priori*. Instead of fine tuning the kernel parameters for a predetermined kernel via cross-validation, we can optimize the set of values $a_i \geq 0$ in order to obtain a positive semi-definite linear combination $K(X^\top, X) = \sum_{j=1}^{k} a_j K_j(X^\top, X)$ suitable for the specific classification problem. Substituting Equation (5.14) into Equation (5.11), we obtain the KFD formulation with heterogeneous linear combinations of kernels as follows

$$\begin{aligned} \min_{(\lambda,\xi,a)\in\mathbb{R}^{n+d+k}} \quad & \tfrac{1}{2}\textstyle\sum_{k\in\{\pm\}} \frac{1}{n_k \nu} {\xi^k}^\top \xi^k + \tfrac{1}{2}\lambda^\top\lambda \\ \text{s.t.} \quad & \xi = \textstyle\sum_{j=1}^{k} a_j K_j \lambda - y \\ & e_{n_k}^\top \xi^k = 0, \; k \in \{\pm\} \\ & a_j \geq 0, \; j \in \{1, \ldots, k\}, \end{aligned} \tag{5.15}$$

where $K_j = K_j(X^\top, X)$. When considering linear combinations of kernels, the hypothesis space may become larger, making the issue of capacity control an important one. It is known that if two classifiers have similar training error, a smaller capacity may lead to better generalization on future unseen data (Cherkassky and Mulier 1998; Vapnik 2000). In order to reduce the size of the hypothesis and model space and to gain strong convexity in all variables, an additional regularization term $\frac{1}{2}a'a$ is added to the objective function of problem (5.15). The problem then becomes

$$\begin{aligned} \min_{(\lambda,\xi,a)\in\mathbb{R}^{n+d+k}} \quad & \tfrac{1}{2}\textstyle\sum_{k\in\{\pm\}} \frac{1}{n_k \nu} {\xi^k}^\top \xi^k + \tfrac{1}{2}\lambda^\top\lambda + \tfrac{1}{2}a^\top a \\ \text{s.t.} \quad & \xi = \textstyle\sum_{j=1}^{k} a_j K_j \lambda - y \\ & e_{n_k}^\top \xi^k = 0, \; k \in \{\pm\} \\ & a_j \geq 0, \; j \in \{1, \ldots, k\}. \end{aligned} \tag{5.16}$$

A new sample x is then classified by the following classifier:

$$\lambda \sum_{j=1}^{k} (a_j K_j(x, X)) = \begin{cases} \geq b, & x \in \omega_+, \\ < b, & x \in \omega_-, \end{cases} \tag{5.17}$$

where b is a predefined threshold that adjusts the trade-off between incorrectly classifying a positive sample as negative, i.e. false negative, and incorrectly classifying a negative sample as positive, i.e., false positive.

Even though the objective function in (5.16) is strictly convex in terms of the problem variables ξ, λ and a, the problem itself is not convex due to the nonconvex equality constraint $\xi = \sum_{j=1}^{k} a_j K_j \lambda - y$. However, the problem in (5.16) can be treated as a biconvex programming problem first by fixing $a = a^*$ and solving (5.16) for ξ and λ^* and then fixing $\lambda = \lambda^*$

and solving for ξ and a^*. More specifically when we fix $a = a^*$ we obtain the following subproblem

$$\begin{aligned} \min_{(\lambda,\xi)\in\mathbb{R}^{n+d}} \quad & \tfrac{1}{2}\textstyle\sum_{k\in\{\pm\}} \frac{1}{n_k\nu}{\xi^k}^\top \xi^k + \tfrac{1}{2}\lambda^\top\lambda \\ \text{s.t.} \quad & \xi = \textstyle\sum_{j=1}^{k} a_j^* K_j\lambda - y \\ & e_{n_k}^\top \xi^k = 0,\ k\in\{\pm\} \end{aligned} \tag{5.18}$$

and similarly when we fix $\lambda = \lambda^*$ we obtain the subproblem

$$\begin{aligned} \min_{(\xi,a)\in\mathbb{R}^{n+k}} \quad & \tfrac{1}{2}\textstyle\sum_{k\in\{\pm\}} \frac{1}{n_k\nu}{\xi^k}^\top \xi^k + \tfrac{1}{2}a^\top a \\ \text{s.t.} \quad & \xi = \textstyle\sum_{j=1}^{k} a_j K_j\lambda^* - y \\ & e_{n_k}^\top \xi^k = 0,\ k\in\{\pm\} \\ & a_j \geq 0,\ j\in\{1,\ldots,k\}. \end{aligned} \tag{5.19}$$

Note that both problems in (5.18) and (5.19) are strongly convex with a unique optimizer, which implies that the original objective function is guaranteed to improve at each iteration. We are now ready to describe our proposed algorithm.

5.5 Automatic kernel selection KFD algorithm

Algorithm 5.5.1 Automatic kernel selection KFD Algorithm (AKFD)
Given n data points in $\mathbb{R}^d$ *represented by the* $d \times n$ *matrix X and vector y of* ± 1 *labels denoting the class of each data point (i.e., each column of X), the parameter* ν *and an initial* $a^0 \in \mathbb{R}^k$*, we generate the nonlinear classifier (5.17) as follows:*

(0) Calculate $K_1, \ldots, K_k$ *, the k kernels on the kernel family, where for each j,* $K_j = K_j(X^\top, X)$.

For each iteration i do:

(i) Given $a^{(i-1)}$ *calculate the linear combination* $K = \sum_{j=1}^{k} a_j^{(i-1)} K_j$.

(ii) Solve subproblem (5.18) to obtain $\lambda^{(i)}$.

(iii) Calculate $K_j\lambda^{(i)}$ *for* $j = 1, \ldots, k$.

(iv) Solve subproblem (5.19) to obtain a^i.

Stop when a predefined maximum number of iterations is reached or when the change in value of the objective function (5.16) evaluated in successive iterations is less than ϵ.

The most common cases arise when $k \ll n$ (i.e., the number of kernel functions considered on the kernel family is much smaller than the number of data points). In such situations, the complexity of the AKFD algorithm 5.5.1 is approximately $O(n^3)$.

Since each of the two optimization problems ((5.18) and (5.19)) that are required to be solved by the AKFD algorithm are strongly convex and thus each of them have a unique

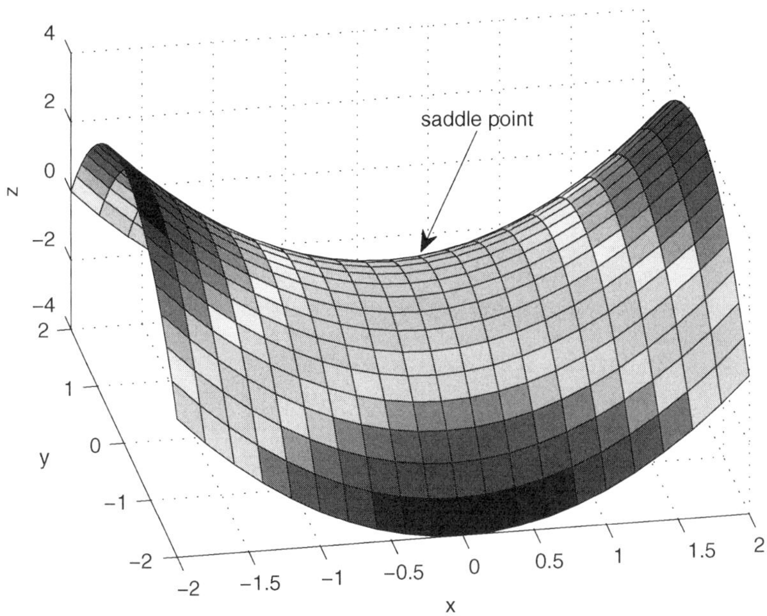

Figure 5.1 A saddle point for the function $z = x^2 - y^2$. It behaves like a local minimizer when projected along the x-axis.

minimizer, the AKFD algorithm can also be interpreted as an Alternate Optimization (AO) problem (Bezdek and Hathaway 2003). Alternate Optimization divides the entire variable space into a predefined number of subspaces and optimizes one group of variables at a time while the remaining variables are fixed. Classical instances of AO problems include fuzzy regression c-models and fuzzy c-means clustering.

The AKFD algorithm then, inherits the convergence properties and characteristics of AO problems. As stated in Bezdek and Hathaway (2002), the set of points for which Algorithm 5.5.1 can converge may include certain types of saddle points (i.e. points behaving like local minimizers only when projected along a subset of the variables, see Figure 5.1). However, it is also stated that it is extremely difficult to find examples where convergence occurs to a saddle point rather than to a local minimizer. If the initial estimate is chosen sufficiently near a solution, AO is shown to converge linearly to a local minimizer (Bezdek and Hathaway 2002). In practice, we found that Algorithm 5.5.1 typically converges in 3 or 4 iterations to a local solution of problem (5.16).

5.6 Numerical results

5.6.1 Dataset used: Purdue Campus data

This data set is a flightline over the Purdue University West Lafayette Campus. The hyperspectral data used was collected on September 30, 1999 with the airborne HYMAP system (Kruse *et al.* 2000), providing image data in 126 spectral bands in the visible and IR regions (0.4–2.4μ). The system was flown at an altitude such that the pixel size is about 5 metres.

Table 5.1 Number of labelled samples available for each class identified in the Purdue Campus Dataset

Classes	Number of samples
Roof tops	10 182
Streets	4571
Grass	1539
Trees	1743
Paths	907
Shadow	434
Cars	934
Fields	608
Total	20 918

The data set contains 358 scan lines with 390 pixels in each scan line. The list of classes and number of labelled samples for each class is given in Table 5.1. The image of the scene and the corresponding ground-truth regions of interest are shown in Figure 5.2.

5.6.2 Classifier design

We designed three different versions of Fisher's discriminant. As a baseline classifier Linear Fisher's Discriminant (LFD) is considered. To find out if the kernel version of the Fisher's Discriminant improves our baseline, we implemented the Kernel Fisher's Discriminant (KFD). Finally we design the automatic kernel selection algorithm for the Fisher's discriminant (AKFD) to see how much we save from the training time and if this is achieved while maintaining the similar performance levels achieved by KFD.

The classifier parameters namely the regularization parameter, ν, and the type of the kernel function and the corresponding parameter are optimized using a 10-fold cross-validation

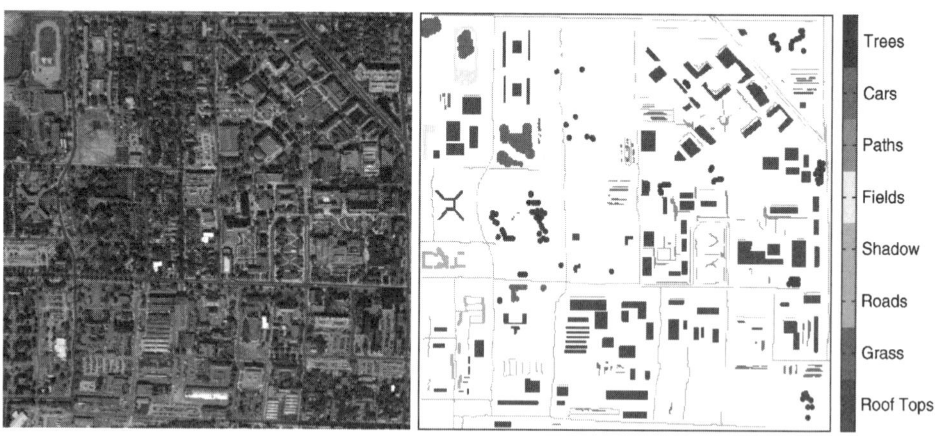

Figure 5.2 Gray-level image and Ground Truth Fields for the Purdue Campus Dataset.

Table 5.2 Percentage overall classification accuracies averaged over ten runs. Numbers in parenthesis are standard deviations

Classifiers	$n = 209$	$n = 418$	$n = 627$	$n = 836$
LFD	81.0 (5.2)	84.1 (1.0)	84.2 (0.4)	84.6 (0.8)
KFD	82.5 (3.2)	86.5 (2.9)	89.3 (1.9)	89.9 (1.5)
AKFD	80.6 (2.9)	85.8 (1.6)	87.9 (1.4)	90.1 (0.9)

approach. For the regularization parameter a discrete set of 10 values are considered, i.e. $\nu = [10^{-10}, 10^{-8}, 10^{-6}, 10^{-5}, 10^{-4}, 10^{-3}, 5 \times 10^{-2}, 10^{-2}, 5 \times 10^{-1}, 10^{-1}, 1]$. For the kernel function we considered linear and radial basis function (RBF Gaussian Functions) kernels. For the width of the RBF we considered $\sigma = [10^{-2}, 10^{-1}, 1, 10]$. The data is normalized such that each feature is between -1 and 1. One-against-all multi-class strategy is used throughout the experiments (Hsu and Lin 2002).

To be more specific for the KFD algorithm a single RBF Gaussian kernel is used. The width σ of this kernel is chosen from one of the four different values considered via cross-validation. However, for the AKFD algorithm a linear combination of five kernel functions (i.e., a linear and 4 RBF Gaussian functions one for each of the σ considered) is used.

The AKFD algorithm is initialized with a set to all ones (i.e., initially all kernels are assumed to contribute equally). The algorithm is terminated when the improvement in the objective function at any iteration over the previous iteration is less than 0.1% or a maximum number of 20 iterations is reached. We ran our experiments for four different sizes of training sets, i.e. $r = 0.01$ ($n = 209$), $r = 0.02$ ($n = 418$), $r = 0.03$ ($n = 627$), $r = 0.04$ ($n = 836$), where r denotes the ratio of the training to labelled samples. Training samples are selected randomly and each experiment is repeated ten times. The size of the expansion set S for the kernel matrices is limited to 250 randomly selected samples, i.e. kernel matrices are computed using $K(X^\top, S)$ instead of $K(X^\top, X)$ where $S \subset X$ and size of S is 250. The labelled samples that are not used for training are used for testing.

5.6.3 Analysis of the results

Table 5.2 and Table 5.3 show the percentage average classification accuracies averaged over ten runs achieved and the total time taken by each algorithm respectively. The following

Table 5.3 Total computational time (in seconds) for ten iterations of each algorithm

Classifiers	$r = 0.01$	$r = 0.02$	$r = 0.03$	$r = 0.04$
LFD	99	114	147	176
KFD	3967	26 276	35 667	74 773
AKFD	2125	5902	11 834	26 133

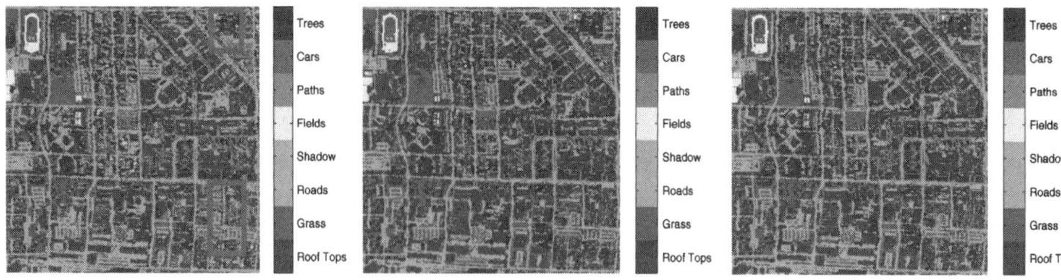

Figure 5.3 Classification maps obtained for LFD, KFD and AKFDR for $r = 0.04$, with test accuracies of 86.5%, 90.5% and 90.6%, respectively. (See plate 1)

conclusions can be drawn from these results. When the training size is small the linear version performs as well as the kernel version. As the training size increases, the linear version is no longer competent yet the prediction accuracy for the kernel version increases with increasing r. The proposed AKFD algorithm generates results comparable to the KFD algorithm yet it is multiple folds faster than the KFD algorithm. For $r = 0.04$ (roughly 800 samples) the entire training plus testing took roughly 21 hours of running time. The same task is completed in 7 hours by the AKFD algorithm. The computer used for these experiments was equipped with intel core 2 duo CPU with a 1.8 GHz clock speed.

The classification maps obtained for $r = 0.04$ (for one of the iterations) are displayed in Figure 5.3 for LFD, KFD and AKFD. As the classification maps corresponding to KFD and AKFD suggest, the difference between the predictive accuracies of KFD and AKFD is quite negligible, (i.e., test accuracy for AKFD is 90.6% and for KFD is 90.5%). The areas of the image where LFD performs poorly are annotated by the gray rectangles.

The optimized values of the kernel weights, a, obtained by the AKFD algorithm and the value of the kernel parameter, σ, selected by cross-validation for KFD are shown in Table 5.4 for each one-against-all classification task. The results suggest that AKFD favours a linear combination with mostly non-zero weights for the individual kernel functions as opposed to KFD, which uses only one kernel function selected from several others available. Despite seemingly different kernel models being used by the two algorithms, we observe almost identical predictive accuracies for the two classifiers.

Table 5.4 The optimized values of the kernel weights, a obtained by the AKFD algorithm. The last column shows the value of the σ selected by cross-validation for the KFD algorithm

Model, σ	Linear	RBF, 0.01	RBF, 0.1	RBF, 1	RBF, 10	
Roof tops	1.3	0.3	0.2	0.3	0.3	$\sigma = 0.01$
Roads	0	0.8	0	1.1	1.2	$\sigma = 0.1$
Grass	0	0.7	0	0.5	0.9	$\sigma = 0.01$
Trees	0	0.3	0	0.1	0.5	$\sigma = 0.01$
Paths	0	1.1	0	0.8	1.0	$\sigma = 0.01$
Shadow	0.9	1.0	1.2	0.9	1.0	$\sigma = 0.1$
Cars	1.2	1.1	0	0.6	0.9	$\sigma = 0.1$
Fields	1.3	0.5	0.4	0.9	1.1	$\sigma = 0.1$

5.7 Conclusion

In this chapter we first reviewed the basics of Fisher's Discriminant, then presented a mathematical programming approach for kernelizing the algorithm so as to obtain nonlinear classifiers and finally proposed an alternating optimization algorithm for automatically learning the kernel function. Unlike LFD, KFD has the potential to deal with data of complex structures such as multimodal data. However, this comes at the cost of increased computational time. The computational time for training increases, at the order of $O(n^3)$ for KFD. Moreover, to optimize the algorithm for different kernel functions and parameters, a cross-validation scheme is required. More specifically, if one is considering p different parameters for the kernel function and using a k-fold cross-validation approach, the algorithm needs to run $p \times k$ times to find the optimum kernel parameter. The proposed AKFD algorithm on the other hand eliminates the need for the cross-validation by automatically learning the weights of the different kernel functions considered. Let the number of iterations before convergence be N; then, for $N \ll p \times k$ the computational gain could be significant.

In our experimental setting the AKFD algorithm usually converged in less than five iterations. We considered five different values for the kernel width and adopted a ten-fold cross-validation framework. So instead of running the KFD algorithm $5 \times 10 = 50$ times to select the optimum kernel parameter, we just run the AKFD algorithm once. Each iteration of the AKFD algorithm takes roughly the same amount of time as KFD. Thus in the AKFD algorithm the optimum kernel is selected in around five iterations whereas in KFD this task takes fifty iterations. As for the online testing, the AKFD algorithm is slower because all of the kernel matrices needs to be computed during testing, which takes more time than computing just one kernel matrix as in KFD. As the numbers in Table 5.3 suggest, even when the testing times are included, the overall computational times clearly favours AKFD over KFD.

As the experimental results suggest, the predictive accuracy of the proposed algorithm is not significantly different from that obtained by the KFD algorithm. That is, the computational gain is achieved while maintaining similar predictive performance.

The results in Table 5.4 indicate that the kernel functions obtained by KFD and AKFD could be significantly different, yet both algorithms yield similar predictive performance. We believe further research is required to investigate how the kernel selected by the KFD algorithm correlates with that obtained by AKFD. Another area that needs attention is the initialization of the weights, a in Section 5.5. In this study we assumed that all kernel models are *a priori* likely and thus assigned equal weights for each of them. However it is worthwhile to analyse the impact of initialization on the final weights optimized and how this in turn affects the predictive performance of the algorithm.

References

Bezdek, J. and Hathaway, R. (2002) Some notes on alternating optimization. *Proceedings of the 2002 AFSS International Conference on Fuzzy Systems*, pp. 288–300. Springer-Verlag.

Bezdek, J. and Hathaway, R. (2003) Convergence of alternating optimization. *Neural, Parallel Sci. Comput.*, **11**(4), 351–368.

Cherkassky, V. and Mulier, F. (1998) *Learning from Data–Concepts, Theory and Methods*. John Wiley & Sons, New York.

Dundar, M. and Landgrebe, D. (2004a) A cost-effective semi-supervised classifier approach with kernels. *IEEE Transactions on Geoscience and Remote Sensing*, **42**(1), 264–270.

Dundar, M. and Landgrebe, D. (2004b) Toward an optimal supervised classifier for the analysis of hyperspectral data. *IEEE Transactions on Geoscience and Remote Sensing*, **42**(1), 271–277.

Fukunaga, K. (1990) *Introduction to Statistical Pattern Recognition*. Academic Press, San Diego, CA.

Fung, G., Dundar, M., Bi, J. and Rao, R.B. (2004) A fast iterative algorithm for fisher discriminant using heterogeneous kernels. In *Proceedings of the Twenty-First international Conference on Machine Learning (ICML '04)*.

Hsu, C.W. and Lin, C.J. (2002) A comparison of methods for multiclass support vector machines. *IEEE Transactions on Neural Networks*, **13**(2), 415–425.

Kruse, F.A., Boardman, J.W., Lefkoff, A.B., Young, J.M., Kierein-Young, K.S., Cocks, T.D., Jensen, R. and Cocks, P.A. (2000) Hymap: An Australian hyperspectral sensor solving global problems – results from USA Hymap data acquisitions. *Proceedings of the 10th Australian Remote Sensing and Photogrammetry Conference*.

Lanckriet, G., Cristianini, N., Bartlett. P., Ghaoui, L.E., and Jordan, M. (2003) Learning the kernel matrix with semidefinite programming. *Journal of Machine Learning Research*, **5**, 27–72.

Mangasarian, O.L. (1994) *Nonlinear Programming*. SIAM, Philadelphia, PA.

McLachlan, G. and Peel, D. (2004) *Finite Mixture Models*. Wiley-Interscience.

Mika, S., Rätsch, G. and Müller, K.R. (2000) A mathematical programming approach to the kernel Fisher algorithm. *NIPS*, pp. 591–597.

Mika, S., Rätsch, G., Weston, J., Schölkopf, B. and Müller, K.R. (1999) Fisher discriminant analysis with kernels. In Y.H. Hu, J. Larsen, E. Wilson and S. Douglas (eds), *Neural Networks for Signal Processing IX*, pp. 41–48. IEEE.

Suykens, J.A.K. and Vandewalle, J. (1999) Least squares support vector machine classifiers. *Neural Processing Letters*, **9**(3), 293–300.

Vapnik, V.N. 2000 *The Nature of Statistical Learning Theory*, 2nd edn; New York: Springer.

6

Multi-temporal image classification with kernels

Jordi Muñoz-Marí[1], Luis Gómez-Chova[1], Manel Martínez-Ramón[2], José Luis Rojo-Álvarez[3], Javier Calpe-Maravilla[1] and Gustavo Camps-Valls[1]

[1] *Image Processing Laboratory (IPL) & Dept. Enginyeria Electrònica, Universitat de València, Spain*

[2] *Dept. Signal Theory and Communications, Univ. Carlos III de Madrid, Spain*

[2,3] *Dept. Signal Theory and Communications, Univ. Rey Juan Carlos, Madrid, Spain*

In generic problems of multi-temporal classification of remote sensing images, different sources of information such as temporal, contextual or multi-sensor, are commonly available. The combination of these heterogenous sources of information is still an active research area. In this chapter, we present a classification framework based on kernel methods for multi-temporal classification of remote sensing images. The proposed kernel classifiers not only process multi-temporal images simultaneously, and with different levels of sophistication, but also allow one to properly combine different data sources, such as contextual information and multi-sensor images. In this chapter, we also present two nonlinear kernel classifiers for well-known change detection methods formulating them in an adequate high dimensional kernel-induced feature space. The developed kernels are used in two core classification machines: the binary Support Vector Machine classifier (SVM), and the one-class Support Vector Domain Description (SVDD) classifier.

Kernel Methods for Remote Sensing Data Analysis Edited by Gustavo Camps-Valls and Lorenzo Bruzzone

6.1 Introduction

The problems of classifying and detecting changes in images from the same scene taken at different instants of time are highly relevant in many application domains, including video surveillance (Collins *et al.* 2000), medical diagnosis and treatment (Bosc *et al.* 2003; Lemieux *et al.* 1998), driver assistance (Fang *et al.* 2003), and remote sensing (Bruzzone and Cossu 2002; Bruzzone and Serpico 1997; Collins and Woodcock 1996), among others (Radke *et al.* 2005). In Earth observation, the formal definition of *change detection* involves the use of multi-temporal data to discriminate areas of the land cover that changed between dates (Lillesand *et al.* 2004). Therefore, *multi-temporal classification* can be seen as a more general task, and it includes the change detection problem as a particular case. Multi-temporal classification and change detection techniques have become more useful in the last decade with the increasing multi-temporal and multi-source data available from remote sensing platforms. Excellent surveys of the remote sensing change detection literature can be found in Coppin and Bauer (1996) and Singh (2003), and the *MultiTemp* workshop (http://www.ing.unitn.it/~multi/) gathers the related researchers and disciplines periodically. As shown in these works, efficient exploitation and fusion of this unprecedented wealth of data is a critical issue and, consequently, many automatic methods have been proposed in the literature for this purpose. The procedure of identifying changed pixels in an image is of special relevance in updating digital remote-sensing databases, follow multi-seasonal crop covers phenology, or for the automatic detection of growing urbanization (Coppin and Bauer 1996; Singh 2003).

Two main approaches to the change detection problem are made in the literature: (1) *post-classification* comparison, and (2) *multi-temporal classification*, also known as *pre-classification enhancement* (Lillesand *et al.* 2004; Singh 2003). In post-classification applications, images from different dates are independently classified and co-registered, and an algorithm is used to identify those pixels whose labels change between dates. In multi-temporal classification applications, a single classification is performed on the combined image dataset for the two dates. As pointed out in Lillesand *et al.* (2004), the post-classification approach can fail as it relies on the accuracy of each classifier, while the multi-temporal approach could produce poor results if the classifier is sensitive to high input dimensionality, low number of labelled training samples, or colinearity. Therefore, in any of these approaches, we need the classifiers not only to be accurate, but also to be robust to high dimension input spaces and low number of labelled samples. Also, these algorithms should be able to combine the temporal information efficiently.

6.1.1 Multi-temporal classification methods

Given two (or more) images at times t_1 and t_2, ($t_1 < t_2$), the problem of multi-temporal classification consists in classifying pixels at time t_2 by learning the changing mapping between t_1 and t_2. If the images are *labelled*, meaning that we know the class of some of the pixels in the images (usually by a previous manual classification), *supervised* or *semi-supervised* methods can be used to build an automatic classifier, which will not only classify the image with an improved performance compared with *unsupervised* methods, but it will also give a deeper insight into the problem by further inspection of the classifier decision function. Other advantages over the unsupervised approach are the capability to explicitly detect land cover transitions, robustness to different atmospheric and light conditions at the two acquisition times, and their demonstrated ability to process multi-sensor/multi-source images (Bruzzone and Serpico 1997). To this end, many supervised methods have been developed during the last years, such as evidence reasoning (Wang 1993), generalized least squares (Morisette and Khorram 1997), neural networks (Civco 1993; Gopal and Woodcock 1996; Kushardono *et al.* 1995) and SVMs (Li and Narayanan 2003; Liu *et al.* 2005). In spite of being an excellent classification framework, many of the presented supervised methods suffer from high false alarm detection rates when the contextual information of the change is not considered. This is an important issue in multi-temporal image classification and change detection, as in practice the user is ultimately interested in detecting very precisely both the position and the spatial extent of the *class of interest* (*change*). Moreover, classifiers are often sensitive to the high dimensionality of the input space generated by putting together multi-sensor features at different time instants, increasing the well-known *curse of dimensionality* (Hughes 1968). Finally, most methods do not consider the (potentially nonlinear) cross-information between pixels at different time instants. Several strategies have been presented to tackle these problems. Multi-temporal and multi-band synthetic aperture radar (SAR) classification of urban areas using spatial analysis has been addressed with both statistical and neural approaches (Pellizzeri *et al.* 2003). A dynamic approach to link Gaussian Markov Random Fields (GMRF) at different dates was used in Melgani and Serpico (2003). In Melgani (2004), the scene classification is attained using a fuzzy fusion of the spatial, spectral, and temporal information, whereas temporal information is handled using transition probabilities. In Gamba *et al.* (2006), feature-based and pixel-based information from multiple SAR images was successfully used. Finally, it is worth noting that in most cases the examples are limited to change detection between only two dates, and thus the performance of the algorithms in long-term operational studies is unclear. In Boucher *et al.* (2006), a methodology that encompasses the use of both temporal and contextual information is presented for the classification of long time series of satellite data. The method is based on krigging-integrated variograms and Gaussian Maximum Likelihood (GML) classification, and shows very good results. It should be remarked that working with a low number of possibly high dimensional training pixels is a very challenging problem for classical methods such as GML or neural networks, something that can be alleviated using kernel methods.

6.1.2 Change detection methods

Changes in images have been analysed using *unsupervised* techniques such as multi-date principal component analysis, temporal image substraction or ratioing, change vector analysis, clustering and cross-correlation analysis (Lillesand *et al.* 2004). These techniques aim to visualize, analyse or compute the differences between sample distribution for the two dates in a low-dimensional subspace (such as two principal components or several bands). The nature of changes detected in a representative-enough space can be analysed by inspecting the spectral signatures involved in it. These techniques are unsupervised as they do not require a labelled image at time t_1 from which to learn, and then extrapolate to the subsequent image at time t_2, as supervised or semi-supervised methods do. Early approaches, considering simple threshold-based image differencing or ratioing operators were soon proved to be inefficient, whereas researchers have paid more attention to suitable thresholds using Bayesian criteria (Bruzzone and Cossu 2002; Bruzzone and Serpico 1997). Other representative examples are the Kittler–Illingworth thresholding algorithm for unsupervised SAR change detection (Moser and Serpico 2006), the fuzzy hidden Markov chains model (combined with the ratio approach) (Carincotte *et al.* 2006), the full methodology for change detection based on the analysis of the difference vectors in the polar domain (Bovolo and Bruzzone 2006), and the study of local statistics evolution with Kullback–Leibler (KL) divergence (Inglada and Mercier 2007). Nonlinear kernel-based methods have been recently pointed out as being suitable for change detection, since they allow large-margin classifications, and they intrinsically match the well-known nonlinear nature of the change (Carlotto 1997). A semi-supervised oil-slick detection is proposed in Mercier and Girard-Ardhuin (2006), using a SVDD in the wavelet decomposition of SAR images, and a SVM for abrupt change detection was presented in Potin *et al.* (2006) for detecting buried landmines from ground-penetrating radar data. Nevertheless, none of these preceding proposals has been specially designed either to consider cross-relations between time instants, or to include contextual and multi-source data in the classifier.

Summarizing, in multi-temporal image classification, one tries to classify pixels of an image at the *observation time* by using all available (instantaneous and/or previous) information whereas the aim in change detection is to identify only those pixels that have changed, according to a pre-specified criterion. They are very similar problems, but the second one will usually require less effort and information (Radke *et al.* 2005). As we have seen, unsupervised or partially supervised approaches can be used for both approaches (Bruzzone and Cossu 2002; Bruzzone and Fernandez-Prieto 2002; Bruzzone and Serpico 1997). In this chapter, however, we focus on *supervised* approaches in the sense that at least a few labelled pixels from the observation time are available.

6.1.3 The proposed kernel-based framework

Taking into account all these needs of real remote sensing applications, we present a family of powerful nonlinear classification methods for multi-temporal, contextual, and multi-source classification and change detection. These methods are developed under the *kernel methods* framework (Schölkopf and Smola 2002; Shawe-Taylor and Cristianini 2004). The conventional kernel methodology has demonstrated good results in remote sensing image classification due to their ability to work with few labelled training samples and (potentially) high dimensional spaces (Camps-Valls *et al.* 2004; Camps-Valls and Bruzzone 2005). Certainly, these are important characteriztics of kernel methods due to the intrinsically high dimension of hyperspectral pixels, which can be increased when multi-temporal features are stacked. The good classification performance shown by kernel methods using the spectral signature as input features has been recently improved by including contextual (or textural) information in the classifier by means of composite kernels (Camps-Valls *et al.* 2006b). In this chapter, we explicitly formulate a full family of kernel-based classifiers that can simultaneously take into account spectral, spatial, and local cross-information in remote sensing data for multi-temporal classification. We also give specific formulations for the peculiarities of the change detection problem, and then we propose two novel kernel developments on the basis of both the *difference* and the *ratioing* of images in the kernel space, thus constituting nonlinear difference and ratio change detection new methods. This methodology also allows efficiently integrating different information sources, such as optical and SAR data. Several scenarios are considered where labelled information at the prediction time may or may not be available. The developed methods are then tested using different kernel-based classification methods: (1) binary Support Vector Machine classifiers (SVM) (Schölkopf and Smola 2002), and (2) one-class Support Vector Domain Description (SVDD) classifiers (Tax and Duin 1999), in which only samples of a *class of interest* are used for training.

The next sections are organized as follows. In Section 6.2 we develop a full family of kernel classifiers for multi-temporal classification and change detection that considers the temporal relations between pixels. Section 6.3 further develops the methodology to integrate the contextual, textural and multi-source information in the kernel matrix. Section 6.4 shows experimental results. Finally, Section 6.5 draws some concluding remarks and provides some guidelines for future applications.

6.2 Multi-temporal classification and change detection with kernels

This section first introduces a set of formulations for the problem of multi-temporal classification. The methodology exploits some principles of functional analysis and linear algebra to construct a family of kernels of growing sophistication, in which static or dynamic multi-temporal classification is performed. Then, we formulate *kernel* versions of the well-known difference and ratio methods for change detection.

All the presented formulations are valid for any kernel method, including both for binary SVM and the one-class SVDD. The former builds the kernel among samples belonging to all labelled classes Ω, whereas the later only considers samples belonging to the class of interest. A multi-class strategy can be subsequently used, such as one-against-one, one-against-all, or error correcting codes. Finally, note that not only is the quadratic programming problem the same size as in the conventional algorithm for all the proposed composite kernels, but also the collinearity between features, and the dimensionality increase of the training samples due to stacking features, are alleviated by constructing dedicated kernels to process each information source.

6.2.1 Problem statement and notation

In multi-temporal classification, one tries to classify pixels of an image at the *observation time* t_T by using all available (instantaneous and/or previous) information, $t \leq t_T$, whereas the aim in change detection is to identify only those pixels that have changed, according to a pre-specified criterion. We follow a *cascade* strategy for classification, which means that only the previously acquired information is used to classify a given image. Another possibility is to use the information from all acquired images, as presented in Bruzzone and Cossu (2002) and Bruzzone and Serpico (1997).

We present a family of kernels to address the more general problem of *supervised multi-temporal classification* by using kernels, but we will also show that the problem of *change detection* is a special case of *multi-temporal classification* under the composite kernels framework. Two situations can take place in the *supervised learning* framework, according to the availability of previous information (see Figure 6.1):

1. *Labelled information is available only for* $t < t_T$, which is the most usual scenario. This situation discourages the use of pure supervised classifiers (as SVM) trained on the available samples, $t < t_T$, in order to classify samples at $t = t_T$, as far as poor results are usually obtained. These classifiers are trained and tested with data from different distributions, due to differences in the atmospheric and light conditions at the image acquisition dates, sensor drifts, and others.

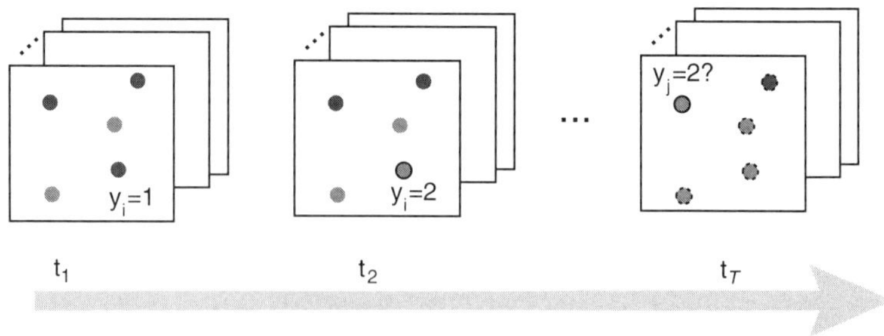

Figure 6.1 *Scheme for multi-temporal classification and change detection. The first problem consists of classifying a given pixel j at time t_T, $\mathbf{x}_j^{t_T}$ using information from $t \leq t_T$, whereas the second one aims to detect those pixels whose class label has changed.*

2. *Labelled information is available for $t \leq t_T$.* This represents a softer problem, which enables the use of supervised classifiers, such as binary and one-class schemes.

In both scenarios, we will compare the performance of SVM and SVDD classifiers according to the availability of full information on the class labels or only of the class(es) of interest, respectively.

Notationally, let $\{\mathbf{x}_i^t\} \in \mathbb{R}^N$ be a multi-temporal set of labelled training samples (pixels) at time t, and $\{y_i^t\} \in \mathbb{N}$ be their corresponding output labels, where $i = 1, \ldots, n$, and $t = 1, \ldots, t_T - 1$ or $t = 1, \ldots, t_T$, depending on the available data. In the following, we will assume that images at subsequent dates are co-registered. This implies (from a machine learning point of view) that pixels $\{\mathbf{x}_i^t\}$ are different (temporal) samples of the same object or *pixel entity* $\mathbf{x}_i$. Also, we will assume that the spatial distribution of such classes changes, but their number does not, and hence, $\Omega = \{\omega_1, \ldots, \omega_{N_C}\}$ is the set of N_C classes that characterize the geographical area at any time. This can be assumed in standard situations because the number of classes of interest is usually prespecified by the user.

6.2.2 Mercer's kernels properties

Before starting, let us review some Mercer's kernel properties that are relevant for this work. More details are provided in Chapter 2. Let K_1 and K_2 be Mercer's kernels over $\mathcal{X} \times \mathcal{X}$ with $\mathbf{x}, \mathbf{z} \in \mathcal{X} \subseteq \mathbb{R}^N$. Let us define $\mathbf{A}$ as a symmetric positive semi-definite $n \times n$ matrix, and $\mu > 0$. Then, the following are valid Mercer's kernels:

$$K(\mathbf{x}, \mathbf{z}) = K_1(\mathbf{x}, \mathbf{z}) + K_2(\mathbf{x}, \mathbf{z}) \tag{6.1}$$

$$K(\mathbf{x}, \mathbf{z}) = K_1(\mathbf{x}, \mathbf{z}) \cdot K_2(\mathbf{x}, \mathbf{z}) \tag{6.2}$$

$$K(\mathbf{x}, \mathbf{z}) = \mu K_1(\mathbf{x}, \mathbf{z}) \tag{6.3}$$

$$K(\mathbf{x}, \mathbf{z}) = \mathbf{x}^\top \mathbf{A}\mathbf{z}. \tag{6.4}$$

These properties allow one to easily combine positive definite kernel matrices.

6.2.3 Composite kernels for multi-temporal classification

The stacked input vectors kernel

The most used approach to exploit the multi-temporal information is to stack vectors at different time instants in order to predict the label of sample at t_T, such that the new input vector has $N \cdot T$ elements, $\mathbf{x}_i \equiv \{\mathbf{x}_i^1, \mathbf{x}_i^2, \ldots, \mathbf{x}_i^{t_T}\}$, and then build a generic mapping $\boldsymbol{\phi}(\cdot)$ with these new samples, which induces a kernel matrix whose terms are defined as:

$$K(\mathbf{x}_i, \mathbf{x}_j) = \langle \boldsymbol{\phi}(\mathbf{x}_i), \boldsymbol{\phi}(\mathbf{x}_j) \rangle. \tag{6.5}$$

This straightforward approach to data merging can yield a good performance compared with previously proposed methods. However, it does not consider explicit cross-relations between samples at different time instants.

The direct summation kernel

A simple composite kernel combining the *static* available information can be obtained by concatenating nonlinear transformations for each $\mathbf{x}_i^t$. Let us assume a nonlinear transformation $\varphi(\cdot)$ into corresponding Hilbert spaces $\mathcal{H}$, and $\mathbf{A}_t$ linear transformations from $\mathcal{H}_t$ to $\mathcal{H}$, respectively. Then, the following mapping to $\mathcal{H}$ can be used:

$$\boldsymbol{\phi}(\mathbf{x}_i) = \left\{ \mathbf{A}_1 \varphi\left(\mathbf{x}_i^1\right), \mathbf{A}_2 \varphi\left(\mathbf{x}_i^2\right), \ldots, \mathbf{A}_T \varphi\left(\mathbf{x}_i^{t_T}\right) \right\} \tag{6.6}$$

and the corresponding dot product can be easily computed as follows:

$$\begin{aligned}
K(\mathbf{x}_i, \mathbf{x}_j) &= \langle \boldsymbol{\phi}(\mathbf{x}_i), \boldsymbol{\phi}(\mathbf{x}_j) \rangle \\
&= \left\langle \left\{ \mathbf{A}_1 \varphi\left(\mathbf{x}_i^1\right), \ldots, \mathbf{A}_T \varphi\left(\mathbf{x}_i^{t_T}\right) \right\}, \left\{ \mathbf{A}_1 \varphi\left(\mathbf{x}_j^1\right), \ldots, \mathbf{A}_T \varphi\left(\mathbf{x}_j^{t_T}\right) \right\} \right\rangle \\
&= \sum_{t=1}^{T} \varphi\left(\mathbf{x}_i^t\right)^\top \mathbf{A}_t^\top \mathbf{A}_t \varphi\left(\mathbf{x}_j^t\right) \\
&= \sum_{t=1}^{T} K_t\left(\mathbf{x}_i^t, \mathbf{x}_j^t\right),
\end{aligned} \tag{6.7}$$

where, in the last step, we have exploited property (6.4) of Mercer's kernels (see Section 6.2.2). This composite kernel is simply a sum of the individual sample's similarities at each time instant and, again, no temporal correlation between pixels in different images is taken into account by the classifier. However, the main advantage of using a summation approach comes from the fact that individual kernels are computed with lower dimensional vectors, thus alleviating overfitting problems.

The weighted summation kernel

By exploiting Property (6.3), a composite kernel that takes into account the temporal content in (6.7) can be made:

$$K(\mathbf{x}_i, \mathbf{x}_j) = \sum_{t=1}^{T} \mu_t K_t \left(\mathbf{x}_i^t, \mathbf{x}_j^t \right), \tag{6.8}$$

where μ_t gives different weights to each time-dependent kernel. This temporal weight can be either estimated from the data or fixed by the user and, in practice, a good choice is an exponential decay, $\mu_t = \lambda^{-(t_T - t)}$, $\lambda \in (0, 1]$.

The cross-information kernel

The preceding kernel classifiers can be conveniently modified to account for the cross relationship between subsequent time instants by including an additional linear transform in the definition of the mapping function. $\mathbf{A}_t$ and $\mathbf{B}$ are linear transformations from the corresponding Hilbert space to $\mathcal{H}$:

$$\boldsymbol{\phi}(\mathbf{x}_i) = \left\{ \mathbf{A}_1 \varphi\left(\mathbf{x}_i^1\right), \mathbf{A}_2 \varphi\left(\mathbf{x}_i^2\right), \ldots, \mathbf{A}_T \varphi\left(\mathbf{x}_i^{t_T}\right), \mathbf{B}\left(\varphi\left(\mathbf{x}_i^1\right) + \cdots + \varphi\left(\mathbf{x}_i^{t_T}\right)\right) \right\}. \tag{6.9}$$

Then, the corresponding dot product can be easily computed:

$$\begin{aligned}
K\left(\mathbf{x}_i, \mathbf{x}_j\right) &= \left\langle \boldsymbol{\phi}(\mathbf{x}_i), \boldsymbol{\phi}\left(\mathbf{x}_j\right) \right\rangle \\
&= \Big\langle \left\{ \mathbf{A}_1 \varphi\left(\mathbf{x}_i^1\right), \ldots, \mathbf{A}_T \varphi\left(\mathbf{x}_i^{t_T}\right), \mathbf{B}\left(\varphi\left(\mathbf{x}_i^1\right) + \cdots + \varphi\left(\mathbf{x}_i^{t_T}\right)\right) \right\}, \\
&\qquad \left\{ \mathbf{A}_1 \varphi\left(\mathbf{x}_j^1\right), \ldots, \mathbf{A}_T \varphi\left(\mathbf{x}_j^{t_T}\right), \mathbf{B}\left(\varphi\left(\mathbf{x}_j^1\right) + \cdots + \varphi\left(\mathbf{x}_j^{t_T}\right)\right) \right\} \Big\rangle \\
&= \sum_{t=1}^{T} \varphi\left(\mathbf{x}_i^t\right)^\top \left(\mathbf{A}_t^\top \mathbf{A}_t + \mathbf{B}^\top \mathbf{B} \right) \varphi\left(\mathbf{x}_j^t\right) + \sum_{t=1}^{T} \sum_{t'=1}^{T} \varphi\left(\mathbf{x}_i^t\right)^\top \mathbf{B}^\top \mathbf{B} \varphi\left(\mathbf{x}_j^{t'}\right) \\
&= \sum_{t=1}^{T} K_t\left(\mathbf{x}_i^t, \mathbf{x}_j^t\right) + \sum_{t=1}^{T} \sum_{t'=1}^{T} K_{t,t'}\left(\mathbf{x}_i^t, \mathbf{x}_j^{t'}\right).
\end{aligned} \tag{6.10}$$

This is a complex composite kernel, given that it contains the cross-information between all possible kernel matrices computed at different time instants. This general equation can be easily simplified when considering only correlation for time instants t and $t+1$, and then the composite kernel is

$$K\left(\mathbf{x}_i, \mathbf{x}_j\right) = \sum_{t=1}^{T-1} \left[K_t\left(\mathbf{x}_i^t, \mathbf{x}_j^t\right) + K_{t+1}\left(\mathbf{x}_i^{t+1}, \mathbf{x}_j^{t+1}\right) + K_{t,t+1}\left(\mathbf{x}_i^t, \mathbf{x}_j^{t+1}\right) \right], \tag{6.11}$$

where we forced subsequent time relationships just by replacing $\mathbf{B}$ with $(\delta_{t,t'}\mathbf{B})$ in (6.10), being $\delta_{t,t'} = \mathbf{1}$ if $t' = t + 1$ and zero otherwise.

It is also worth stressing here that each kernel in the summation must not necessarily have the same structural form (for instance, RBF or polynomial).

6.2.4 Composite kernels for change detection

Two kernel-based formulations are next presented. They allow us to address the change detection problem, inspired in the traditional difference and ratioing operations. Defining these operations in a high dimensional feature space allows us to deal with nonlinear relationships between samples, and more, the (few) free parameters per feature and combination can be readily learned from the data.

Image difference in feature spaces

In change detection, changes in the scene are to be detected, and hence this is just a particular case of multi-temporal classification. The traditional procedure in remote sensing followed to identify changes consists in substracting the subsequent images and then applying a threshold, this last being tuned either heuristically or with some adequate criterion (Bovolo and Bruzzone 2006; Bruzzone and Cossu 2002; Bruzzone and Serpico 1997; Moser and Serpico 2006). This criterion can easily be formulated in the *kernel feature spaces* by defining a convenient kernel mapping function. Let us define the difference in the same sample in two subsequent images as follows:

$$\boldsymbol{\phi}(\mathbf{x}_i) = \mathbf{A}_{t_T}\boldsymbol{\varphi}\left(\mathbf{x}_i^{t_T}\right) - \mathbf{A}_{t_T-1}\boldsymbol{\varphi}\left(\mathbf{x}_i^{t_T-1}\right). \tag{6.12}$$

The corresponding dot product $\langle \boldsymbol{\phi}(\mathbf{x}_i^{t_T}), \boldsymbol{\phi}(\mathbf{x}_j^{t_T})\rangle$ can be computed as:

$$\begin{aligned} K\left(\mathbf{x}_i, \mathbf{x}_j\right) = {} & K_{t_T}\left(\mathbf{x}_i^{t_T}, \mathbf{x}_j^{t_T}\right) + K_{t_T-1}\left(\mathbf{x}_i^{t_T-1}, \mathbf{x}_j^{t_T-1}\right) \\ & -K_{t_T,t_T-1}\left(\mathbf{x}_i^{t_T}, \mathbf{x}_j^{t_T-1}\right) - K_{t_T-1,t_T}\left(\mathbf{x}_i^{t_T-1}, \mathbf{x}_j^{t_T}\right). \end{aligned} \tag{6.13}$$

This difference kernel can only be used if supervised information at time t_T is available, which makes it a particular case of the cross-information kernel defined in (6.10). Note that this kernel is the nonlinear extension of simple difference in input spaces, which is a common approach in unsupervised change detection. Difference-based methods are typically applied to optical multi-spectral data change detection problems since the noise model can be reasonably assumed as additive.

Image ratioing in feature spaces

Another classical change detection method is the ratioing between images at two different dates, which accommodates change factors such as the Sun angle or the shadows. Defining the mapping:

$$\boldsymbol{\phi}(\mathbf{x}_i) = \frac{\left\{\left(\sqrt{\gamma}\mathbf{A}_{t_T}\boldsymbol{\varphi}\left(\mathbf{x}_i^{t_T}\right)\right)^\top, \left(\mathbf{A}_{t_T-1}\boldsymbol{\varphi}\left(\mathbf{x}_i^{t_T-1}\right)\right)^\top\right\}^\top}{\sqrt{\left\langle \mathbf{A}_{t_T}\boldsymbol{\varphi}\left(\mathbf{x}_i^{t_T}\right), \mathbf{A}_{t_T}\boldsymbol{\varphi}\left(\mathbf{x}_j^{t_T}\right)\right\rangle}} \tag{6.14}$$

and computing the dot product, we obtain the ratioing operation in the kernel feature space:

$$K\left(\mathbf{x}_i, \mathbf{x}_j\right) = \gamma\mathbf{I} + \frac{K_{t_T-1}\left(\mathbf{x}_i^{t_T-1}, \mathbf{x}_j^{t_T-1}\right)}{K_{t_T}\left(\mathbf{x}_i^{t_T}, \mathbf{x}_j^{t_T}\right)}, \tag{6.15}$$

where regularization parameter γ makes the kernel matrix positive semidefinite, which ensures that this is a valid Mercer's kernel. Note that this kernel is the nonlinear extension of a simple ratio between time images in input spaces, which is typically applied to SAR data, as the noise model can be reasonably assumed to be multiplicative.

6.3 Contextual and multi-source data fusion with kernels

In this section we briefly review the formulation of this family of composite kernels that incorporates the contextual or textural information in the kernel, and also we extend it in order to deal with multi-source data.

6.3.1 Composite kernels for integrating contextual information

Let the spectral content of a pixel at time t_T be denoted as $\boldsymbol{\omega}_i^{t_T} \in \mathbb{R}^{N_\omega}$, with N_ω the number of its spectral bands, and let some (local or global) spatial feature extraction on the image yielding vector be $s_i^{t_T} \in \mathbb{R}^{N_s}$, with N_s the spatial (contextual or textural) features. A kernel-based classifier that accounts for both the spectral and the spatial features consists of stacking both vectors, $\mathbf{x}_i \equiv \{\boldsymbol{\omega}_i^{t_T}, s_i^{t_T}\}$ and using them with a standard classifier. However, the previous composite tricks allow us to define several kernel classifiers (Camps-Valls *et al.* 2006b), as follows:

- *The stacked input vectors approach:* $K(\mathbf{x}_i, \mathbf{x}_j) = \langle \boldsymbol{\phi}(\mathbf{x}_i), \boldsymbol{\phi}(\mathbf{x}_j) \rangle$.
- *The direct summation kernel:* $K(\mathbf{x}_i, \mathbf{x}_j) = K_s\left(s_i^{t_T}, s_j^{t_T}\right) + K_\omega\left(\boldsymbol{\omega}_i^{t_T}, \boldsymbol{\omega}_j^{t_T}\right)$.
- *The cross-information kernel:* $K\left(\mathbf{x}_i, \mathbf{x}_j\right) = K_s\left(s_i^{t_T}, s_j^{t_T}\right) + K_\omega\left(\boldsymbol{\omega}_i^{t_T}, \boldsymbol{\omega}_j^{t_T}\right) + K_{s\omega}\left(s_i^{t_T}, \boldsymbol{\omega}_j^{t_T}\right) + K_{\omega s}\left(\boldsymbol{\omega}_i^{t_T}, s_j^{t_T}\right)$, where $s_i^{t_T}$ and $\boldsymbol{\omega}_j^{t_T}$ must have the same dimension ($N_\omega = N_s$) for this formulation to be valid. A possibility to enable its use is to extract a spatial feature per spectral band.

6.3.2 Composite kernels for dealing with multi-source data

Multi-sensor information can also be integrated in the kernel itself in a very elegant way. If we have optical and radar information available for the same co-registered pixel at time t_T, we can define the *optical feature* vector ($\mathbf{o}_i^{t_T}$), the *radar feature* vector ($\mathbf{r}_i^{t_T}$), and its concatenation ($\mathbf{x}_i \equiv \{\mathbf{o}_i^{t_T}, \mathbf{r}_i^{t_T}\}$). Several combinations of dedicated kernels can be made:

- *The stacked features approach:* $K(\mathbf{x}_i, \mathbf{x}_j) = \langle \boldsymbol{\phi}(\mathbf{x}_i), \boldsymbol{\phi}(\mathbf{x}_j) \rangle$
- *The direct summation kernel:* $K(\mathbf{x}_i, \mathbf{x}_j) = K_o(\mathbf{o}_i^{t_T}, \mathbf{o}_j^{t_T}) + K_r(\mathbf{r}_i^{t_T}, \mathbf{r}_j^{t_T})$
- *The cross-information kernel:* $K(\mathbf{x}_i, \mathbf{x}_j) = K_o(\mathbf{o}_i^{t_T}, \mathbf{o}_j^{t_T}) + K_r(\mathbf{r}_i^{t_T}, \mathbf{r}_j^{t_T}) + K_{or}(\mathbf{o}_i^{t_T}, \mathbf{r}_j^{t_T}) + K_{ro}(\mathbf{r}_i^{t_T}, \mathbf{o}_j^{t_T})$, where $\mathbf{o}_i^{t_T}$ and $\mathbf{r}_j^{t_T}$ must have the same dimension ($N_r = N_o$) for this formulation to be valid.

6.3.3 Remarks

According to the previously presented ideas, the general problem of multi-temporal classification consists of many constituents (temporal, spectral, spatial, source, and maybe others) that can be mapped into different feature spaces, and combined there by using composite Mercer's kernels. Some of the advantages of this approach are that we are working with dedicated kernels for each information source, we are combining them linearly, and we are alleviating the problem of the curse of dimensionality, as long as stacking features are no longer necessary.

6.4 Multi-temporal/-source urban monitoring

In this section, extensive comparisons are conducted for all scenarios (partial or complete labelled information at the prediction time), for multi-temporal classification and change detection, multi-source information fusion, and for many composite kernels combinations. We illustrate the performance of the classification framework in a problem of multi-temporal and multi-source image classification and change detection with two real test sites.

6.4.1 Model development and free parameter selection

We used the linear and the RBF kernel in our experiments. The linear kernel is tested to show the performance of standard linear techniques, though it represents a more sophisticated model than the usual approaches in the literature (maximum margin learned from the data). The linear kernel classifiers yield a fair comparison to the nonlinear RBF kernel by following the same composite kernels framework for including temporal, contextual, and multi-source information. Since the number of potentially useful combinations of spatial, spectral, temporal and multi-source composite kernels is very high, we will present results for those combinations showing the best performances (Camps-Valls *et al.* 2006a,b; Camps-Valls *et al.* 2007).

For the linear kernel classifiers, only C had to be tuned. For the nonlinear RBF kernel classifiers, and depending on the composite kernel used, a different σ parameter was additionally tuned for each kernel component. The kernel sum was normalized in the feature spaces, this is

$$\hat{K}(\mathbf{x}_i, \mathbf{x}_j) = \left\langle \frac{\boldsymbol{\phi}(\mathbf{x}_i)}{\|\boldsymbol{\phi}(\mathbf{x}_i)\|}, \frac{\boldsymbol{\phi}(\mathbf{x}_j)}{\|\boldsymbol{\phi}(\mathbf{x}_j)\|} \right\rangle = \frac{K(\mathbf{x}_i, \mathbf{x}_j)}{\sqrt{K(\mathbf{x}_i, \mathbf{x}_i)K(\mathbf{x}_j, \mathbf{x}_j)}}. \tag{6.16}$$

Free parameters were tuned in the ranges $\sigma = \{10^{-3}, \ldots, 10^3\}$ and $C = \{10^{-1}, \ldots, 10^3\}$, and rejection fraction ν for the SVDD was in $\nu = \{10^{-3}, \ldots, 10^0\}$. For multi-temporal weighted summation kernel, μ was in the range [0,1]. A non-exhaustive iterative search strategy (τ iterations) was used. At each iteration, a sequential search of the minimum ν-fold cross-validation estimated kappa statistic on each parameter domain was performed (parameter range split in L points). In our experiments, $\tau = 3$ and $L = 20$ exhibited good performance. A *one-against-one* multi-classification scheme and the multi-class scheme presented in Muñoz Marí *et al.* (2007) were adopted for SVM and for one-class SVDD, respectively.

6.4.2 Data collection and feature extraction

We used images collected in the Urban Expansion Monitoring (UrbEx) ESA-ESRIN DUP project (Castracane *et al.* 2003). Results from the UrbEx project were used for analysing

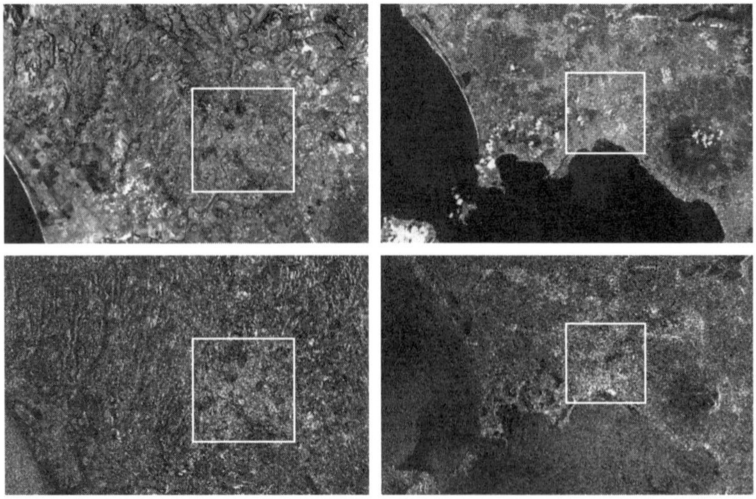

Figure 6.2 Images of the test areas of Rome (left) and Naples (right) at 1999. Both are an RGB composite from L3, L2 and L1 bands (top row) and SAR log-intensities (bottom row).

the selected test site and for validation (for further details see http://dup.esrin.esa.int/ionia/projects/summaryp30.asp). Images from ERS2 SAR and Landsat TM sensors (acquired in 1995 and 1999) of Rome and Naples (Italy) were used as test sites (see Figure 6.2). Only two time instants are available here, and the classifier complexity significantly reduces ($t_T = 2$).

An external Digital Elevation Model and a reference land cover map (Italian Institute of Statistics) were also available. The ERS2 SAR 35-day interferometric pairs were used and, in order to obtain the interferometric coherence from each complex SAR image pair, perpendicular baselines between 20 m and 150 m were selected. The available features were initially labelled as L1-L7 for Landsat bands, as In1-In2 for the SAR backscattering intensities (0–35 days), and as Co for coherence.

Given that features were acquired by different sensors, a specific processing and conditioning of optical and SAR data was previously required, and all images were co-registered. The seven bands of Landsat TM were co-registered with the ISTAT classification data, and resampled to 30×30 m (nearest-neighbour). The registration for multi-source images was at sub-pixel level, with root-mean-squared error of about 10 m, potentially enabling good urban classification ability (Gómez-Chova *et al.* 2004). For *optical* images, the seven Landsat TM spectral bands (containing three *VIS*, one *Near IR*, two *Short-Wave IR*, and one *Thermal IR* bands) were directly used, $\mathbf{o}_i = \{\text{L1}, \ldots, \text{L7}\}$. For *SAR* images, intensity and coherence were computed (Fanelli *et al.* 2000). Since *speckle* noise disturbs image interpretation, a multi-stage spatial filtering approach over coherence images was followed to increase the urban areas discrimination (Gómez-Chova *et al.* 2006), hence yielding the fourth radar input feature, Co'. Therefore, we defined in this case $\mathbf{r}_i = \{\text{In1}, \text{In2}, \text{Co}, \text{Co'}\}_i$.

Once features were extracted from optical and SAR images, we analysed their potential use for urban change detection. For that purpose, we represented scatter plots between features in the different dates (1995 and 1999), see Figure 6.3. As can be observed, the high

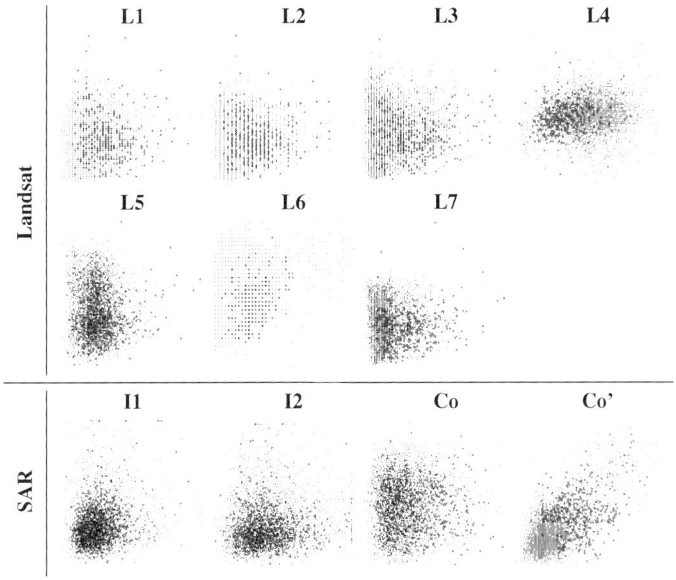

Figure 6.3 Scatter plots of the different considered features (both from Landsat and SAR data sources) for the Rome image; 1995 data is represented versus 1999 data. Big gray dots represent 'class change' and small black dots represent both 'no-change' or 'unknown class' in the image. These representations are very usual in the literature since one can promptly identify changes due to natural variability (ellipse-shaped distribution), atmospheric haze or sensor drift (skew of the ellipsoid), or the appearance of brighter/darker pixels (isolated areas in the plot).

degree of overlapping of change and no-change pixels, along with the wide spread out of the plot, indicate an extremely difficult change detection problem and suggests that nonlinear methods should be deployed. One can also observe that the extracted feature Co' can help in discriminating between changed and unchanged pixels. We also computed spatial and textural features from these optical and SAR features. Specifically, the spatial features for the optical images were the average of all pixels in the surrounding 7×7 window, and the textural features for the SAR data were Gabor-filtered (Clausi and Yue 2004) versions of $\mathbf{r}_i$ at different scales ($\theta = 1, \ldots, 4$) and orientations ($\{0^\circ, 45^\circ, 90^\circ, 135^\circ\}$) thus yielding the textural radar features.

In the following experiments, subset images from Rome and Naples scenes containing 200×200 pixels in areas with substantial urban changes were selected. For the Rome scene, 1392, 780 and 2978 pixels changed to 'non-urban', to 'urban', and to the 'unknown' status, respectively, in this 4-year period. For the case of the Naples scene, 1826, 215 and 1973 pixels changed to 'urban', to 'non-urban', and to 'unknown', respectively. Pixels in the unknown class were not considered, hence becoming a classical binary problem of change vs. no-change identification. In both cases, we randomly selected 25% of changed pixels for training, and five-fold cross-validation was used for free parameter tuning. The built classifier was finally tested on the whole image.

6.4.3 Multi-temporal image classification

Table 6.1 shows the results for supervised classifiers. In particular, we compared SVM and SVDD under multi-temporal classification (left) and change detection (right) using different temporal, spatio-spectral and multi-source composite kernels for the scenes of Rome and Naples. The best overall accuracy (OA)[%] and estimated κ values are provided. We also analysed class-by-class accuracies for selected cases, and statistical differences between classifiers are assessed with Wilcoxon's rank sum test at 95% confidence interval.

The following conclusions can be obtained from Table 6.1. In all cases and scenes, the use of the RBF kernel provides much better results than the linear kernel. In turn, the linear kernel is an upper bound of model's performance for (change detection) thresholding methods. For the Rome scene (Table 6.1[top]), and with unlabelled information of the prediction date image (1999), i.e. $t < t_T$, only labelled samples in 1995 can be used to train a classifier and predict in 1999. In this complex situation, a purely supervised approach like SVM yields poor solutions (OA[%]$<$ 70% and $\kappa < 0.6$), since there is no information on the change. However, SVDD still offers good results in this situation, since rather than building a separating hyperplane 'urban'/'non-urban', the method tries to model the 'urban' class . In all cases, the best composite kernel for integrating the spatial and the different data sources was the *summation kernel*, i.e. dedicating different kernels for the Landsat bands, SAR features, contextual Landsat features, and textural Gabor-filtered SAR features. This best method yielded a maximum accuracy of 84.2%, but with biased classifications, i.e., $\kappa = 0.51$, which was confirmed by looking at the individual class accuracies (90.3% for urban and 53.4% for non-urban, respectively). Note also that solutions are much sparser for the SVDD (average of 22% of SVs) than for the SVM (average of 59% of SVs). For the Naples scene, similar results are obtained (Table 6.1[bottom]). Again, when no information is available at time t_T, the SVDD is a better approximation, either with RBF or with linear kernel embedding. Differences were statistically significant (see star symbols in Table 6.1, $p < 0.05$).

In the case with available labelled information for t_T, several composite kernels dramatically improved the results in both scenes, because of the use of labelled samples from the t_T image. In these cases, SVM classifiers exhibit the best results, but SVDD classifiers also yield stable and robust outcomes, which confirms its suitability to application scenarios in which incomplete or partially complete information is available. Similar results have been observed in Muñoz Marí *et al.* (2007). The same behaviour can be also seen for the neural network, which provides lower accuracies than the nonlinear SVM and SVDD classifiers (both numerical and statistical). The best overall results were obtained by using simple summation kernels for integrating the spatio-spectral information and, in some cases, more complex cross-information kernels to process the temporal information. This classifier yielded a maximum OA = 94.3%, statistically compensated model ($\kappa = 0.78$), and good individual classification accuracies (97.1% for urban and 82.5% for non-urban, respectively) in the Rome image, and a maximum OA = 96.8% ($\kappa = 0.64$), and individual classification accuracies of 98.3% for urban and 83.3% for non-urban in the Naples image.

6.4.4 Change detection

Table 6.1 (right) shows the results for the *difference and ratio kernels* in the change detection problem. Labelled information for t_T is now provided in the form of 'change' vs. 'no-change' for the 1999 image and, hence, it can be seen as a supervised learning strategy. In general, a

Table 6.1 Results for the Rome (top) and Naples (bottom) scenes. Overall accuracy (OA[%]) and kappa statistic (κ) for the different scenarios ($t < t_T$ and $t \leq t_T$), time integration, multi-source fusion and classifier algorithms. Average results over ten realizations are shown for SVM and SVDD using both a linear and RBF kernel. Best results (bold) and second best (italics) are highlighted for each column and kernel type. Statistically different results, either in terms of OA[%] (tested through paired Wilcoxon's rank sum test at 95% confidence interval) or κ (considering it is normally distributed) from the best classifier are marked with a star '$\star$'

			Multi-temp. class. $t < t_T$	Multi-temp. class. $t \leq t_T$			Change detection $t \leq t_T$	
	Spatio-Spectral	Multi-source	Summation Equation (6.7)	Summation Equation (6.7)	Cross-terms Equation (6.8)	Weighted Equation (6.11)	Kernel diff. Equation (6.13)	Kernel ratio Equation (6.15)
				Rome scene				
SVM LIN	Sum	Stack	55.0 (0.20)	83.2 (0.45)	68.2 (0.61)	70.4 (0.64)	81.1 (0.70)	80.2 (0.70)
	Cross	Stack	54.1 (0.22)	81.4 (0.49)	69.2 (0.62)	71.4 (0.63)	**82.2 (0.68)**	*81.3 (0.71)*
	Sum	Sum	57.1 (0.31)	**84.1 (0.51)**	70.2 (0.63)	**73.4 (0.72)**	74.1 (0.72)*	79.4 (0.71)
SVDD LIN	Sum	Stack	*58.1 (0.33)*	71.5 (0.52)*	*73.4 (0.63)*	68.1 (0.62)*	82.1 (0.71)	80.2 (0.71)
	Cross	Stack	58.3 (0.34)	77.6 (0.55)	**74.1 (0.62)**	69.1 (0.54)*	81.3 (0.73)	78.3 (0.72)
	Sum	Sum	**66.1 (0.40)**	*78.2 (0.55)*	69.1 (0.62)	*78.3 (0.68)*	*82.2 (0.61)*	**83.1 (0.72)**
SVM RBF	Sum	Stack	61.1 (0.51)*	91.4 (0.67)	83.1 (0.70)	89.5 (0.78)	95.3 (0.81)	95.1 (0.77)
	Cross	Stack	66.5 (0.43)*	92.1 (0.69)	89.2 (0.71)	88.8 (0.77)	**96.1 (0.79)**	*95.8 (0.80)*
	Sum	Sum	68.3 (0.60)	**93.2 (0.77)**	**94.3 (0.78)**	**93.3 (0.81)**	94.1 (0.83)	93.3 (0.80)
SVDD RBF	Sum	Stack	*77.3 (0.63)*	83.3 (0.66)*	85.3 (0.68)	88.1 (0.71)	92.3 (0.80)	91.1 (0.79)
	Cross	Stack	75.1 (0.64)	88.7 (0.68)	84.5 (0.70)	79.0 (0.55)*	91.9 (0.82)	90.2 (0.79)
	Sum	Sum	**81.4 (0.70)**	*92.1 (0.75)*	*92.2 (0.75)*	*91.1 (0.79)*	*95.6 (0.80)*	**96.0 (0.81)**

(*continued*)

Table 6.1 (*Continued*)

			Multi-temp. class. $t < t_T$	Multi-temp. class. $t \leq t_T$			Change detection $t \leq t_T$	
	Spatio-Spectral	Multi-source	Summation Equation (6.7)	Summation Equation (6.7)	Cross-terms Equation (6.8)	Weighted Equation (6.11)	Kernel diff. Equation (6.13)	Kernel ratio Equation (6.15)
				Naples scene				
SVM LIN	Sum	Stack	58.3 (0.39)	**79.8 (0.48)**	*80.8 (0.51)*	81.0 (0.55)	85.1 (0.60)	81.3 (0.61)
	Cross	Stack	56.2 (0.41)	77.1 (0.45)	80.3 (0.50)	*82.9 (0.55)*	*87.4 (0.64)*	*86.3 (0.62)*
	Sum	Sum	59.1 (0.41)	*77.8 (0.45)*	**81.3 (0.52)**	**86.1 (0.55)**	87.4 (0.64)	86.1 (0.62)
SVDD LIN	Sum	Stack	60.1 (0.44)	77.8 (0.45)	71.8 (0.43)	71.1 (0.49)*	85.9 (0.60)	85.9 (0.61)
	Cross	Stack	*61.4 (0.52)*	72.6 (0.45)	70.6 (0.41)*	72.8 (0.44)*	**87.8 (0.64)**	85.1 (0.62)
	Sum	Sum	**64.3 (0.54)**	75.8 (0.55)	75.8 (0.42)	76.0 (0.51)	84.5 (0.62)	**87.3 (0.62)**
SVM RBF	Sum	Stack	70.1 (0.40)*	90.3 (0.71)	92.0 (0.59)	93.3 (0.60)	95.1 (0.71)	95.8 (0.75)
	Cross	Stack	68.9 (0.45)*	94.1 (0.60)	88.1 (0.59)	94.4 (0.60)	**98.0 (0.75)**	96.3 (0.76)
	Sum	Sum	66.4 (0.55)*	*95.0 (0.73)*	**96.8 (0.64)**	**97.5 (0.66)**	96.9 (0.73)	*96.9 (0.76)*
SVDD RBF	Sum	Stack	80.2 (0.50)	89.1 (0.57)	91.2 (0.56)	91.1 (0.59)	95.9 (0.70)	95.8 (0.71)
	Cross	Stack	*82.6 (0.50)*	92.7 (0.55)	90.4 (0.60)	92.2 (0.57)	97.2 (0.74)	95.3 (0.71)
	Sum	Sum	**84.2 (0.51)**	**95.3 (0.67)**	*95.8 (0.61)*	*96.0 (0.65)*	*97.6 (0.71)*	**97.6 (0.71)**

(both numerical and statistical) significant difference is given by RBF-based kernel classifiers (e.g. accuracy is about 12% higher). SVMs yield very good results in terms of accuracy (OA $> 90\%$, $\kappa > 0.7$), but SVDD provides better kappa values (but no significant statistical differences), which indicates well-balanced classifications with reduced false detections. For the best SVM (SVDD) classifier, individual accuracies were 97% (98%) for the change class and 69% (74%) for the unchange class in the Rome image dataset. For the Naples dataset, results between SVM and SVDD did not differ significantly (98% vs. 97% for the change class, and 75% vs. 74% for the unchange class), probably because this constitutes an easier problem and no particular guiding to learn a specific class is included in the methods. Note, however, that this is a different (and much easier) experimental setup than the multi-temporal approach, as the classifier only has to detect whether the pixel labels changed or not, rather than to estimate the class label.

6.4.5 Classification maps

Figure 6.4 represents the classification maps from the best methods (bold type in Table 6.1), for Rome (top) and Naples (bottom). The previously discussed numerical results are in general confirmed by visual inspection. For instance, results obtained by the SVDD method for $t < t_T$ are much better than the SVM (more homogenous areas and lower number of false detections), also observed in the case of using the difference or ratio kernels for change detection, even though (slightly) better accuracies are obtained using the SVM. In the case of $t \leq t_T$, SVDD does not outperform SVM, mainly because it poorly integrates the spatial/textural information, e.g. southern parts of the Rome scene or Naples middle east, where evident changes occur. Also, the neural network gives noisier classification maps, specially in change detection.

6.5 Conclusions

In this chapter we have introduced a general framework based on composite kernel methods for multi-temporal classification of remote sensing images that simultaneously take into account spectral, spatial and multi-sensor information. We also introduced composite kernel versions for the well-known difference and ratioing methods for change detection. All methods used, as core learners, classifiers based on statistical learning theory: the binary SVM and the one-class SVDD classifier. Experimental results show that binary SVM was more suitable than the SVDD when labelled data from the prediction instant was available. However, this is not the common situation, thus revealing the one-class SVDD as a particularly well-fitted tool for learning the change detection problem.

Results have also demonstrated that the use of the basic composite kernels approach yields good results in the particular application domain of urban monitoring, outperforming the traditional stacked vector approach in all cases. More sophisticated composite kernels that explicitly include cross-relations between different information sources show better performance at the expense of an increased computational burden. This might imply that more data need to be available for validation purposes, something to be confirmed in the future.

Further work will also consider the formulation of semi-supervised kernel-based techniques in the multi-temporal and multi-source change detection framework. At present, good results have been obtained using the graph-based method approach presented in

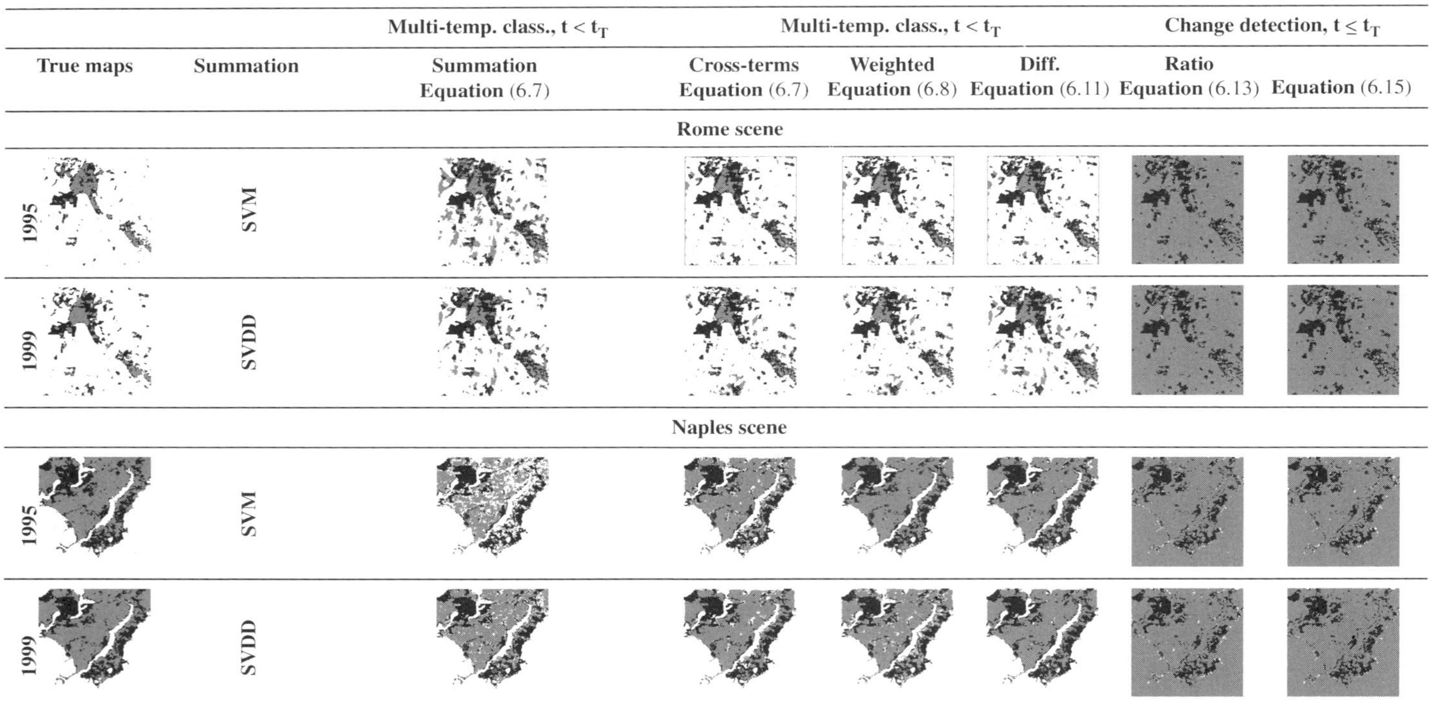

Figure 6.4 *Best classification maps for the Rome (top) and Naples (bottom) scenes. In the multi-temporal classification maps, white pixels represent the class 'non-urban', black pixels are 'unknown class', and grey pixels are 'urban'. For the change detection maps, we have plotted in grey the no-change classification, in white the change, and in black the unknown class.*

Bandos *et al.* (2006) and Camps-Valls *et al.* (2007), but other transductive and one-class approaches could be developed for this purpose. Chapters 9 and 10 are devoted to the analysis of semi-supervised methods.

Acknowledgments

This work was partly supported by the Spanish Ministry of Science projects CICYT-FEDER TEC2006-13845, AYA2008-05965-C04-03 and CSD2007-00018.

References

Bandos, T., Zhou, D. and Camps-Valls, G. (2006) Semi-supervised hyperspectral image classification with graphs. *IEEE International Geoscience and Remote Sensing Symposium, IGARSS'06*, pp. 3883–3886, Denver, CO.

Bosc, M., Heitz, F., Armspach, J.P., Namer, I., Gounot, D. and Rumbach, L. (2003) Automatic change detection in multimodal serial MRI: application to multiple sclerosis lesion evolution. *Neuroimage*, **20**, 643–656.

Boucher, A., Seto, K. and Journel, A. (2006) A novel method for mapping land cover changes: incorporating time and space with geostatistics. *IEEE Transactions on Geoscience and Remote Sensing*, **44**(11(2)), 3427–3435.

Bovolo, F. and Bruzzone, L. (2006) A theoretical framework for unsupervised change detection based on change vector analysis in the polar domain. *IEEE Transactions on Geoscience and Remote Sensing*, **45**(1), 218–236.

Bruzzone, L. and Cossu, R. (2002) A multiple-cascade-classifier system for a robust and partially unsupervised updating of land-cover maps. *IEEE Transactions on Geoscience and Remote Sensing*, **40**(9), 1984–1996.

Bruzzone, L. and Fernandez-Prieto, D. (2002) An adaptive semiparametric and context-based approach to unsupervised change detection in multi-temporal remote-sensing images. *IEEE Transactions on Image Processing*, **11**(4), 452–466.

Bruzzone, L. and Serpico, S. (1997) An iterative technique for the detection of land-cover transitions in multi-temporal remote-sensing images. *IEEE Transactions on Geoscience and Remote Sensing*, **35**(4), 858–867.

Camps-Valls, G. and Bruzzone, L. (2005) Kernel-based methods for hyperspectral image classification. *IEEE Transactions on Geoscience and Remote Sensing*, **43**(6), 1351–1362.

Camps-Valls, G., Bandos, T. and Zhou, D. (2007) Semi-supervised graph-based hyperspectral image classification. *IEEE Transactions on Geoscience and Remote Sensing*, **45**(10), 2044–3054.

Camps-Valls, G., Gómez-Chova, L., Calpe, J., Soria, E., Martín, J.D., Alonso, L. and Moreno, J. (2004) Robust support vector method for hyperspectral data classification and knowledge discovery. *IEEE Transactions on Geoscience and Remote Sensing*, **42**(7), 1530–1542.

Camps-Valls, G., Gomez-Chova, L., Muñoz Marí, J., Alonso, L., Calpe-Maravilla, J. and Moreno, J. (2006a) Multi-temporal image classification and change detection with kernels. *SPIE International Symposium Remote Sensing XII*, vol. 6365, pp. 63650H.1–63650H.12, Stockholm, Sweden.

Camps-Valls, G., Gómez-Chova, L., Muñoz Marí, J., Vila-Francés, J. and Calpe-Maravilla, J. (2006b) Composite kernels for hyperspectral image classification. *IEEE Geoscience and Remote Sensing Letters*, **3**(1), 93–97.

Camps-Valls, G., Gómez-Chova, L., Muñoz Marí, J., Martínez-Ramón, M., Rojo-Álvarez, J.L. (2008) Kernel-based framework for multi-temporal and multi-source remote sensing data classification and change detection. *IEEE Transactions on Geoscience and Remote Sensing*, **46**(6), 1822–1835.

Carincotte, C., Derrode, S. and Bourennane, S. (2006) Unsupervised change detection on SAR images using fuzzy hidden Markov chains. *IEEE Transactions on Geoscience and Remote Sensing*, **44**(2), 432–441.

Carlotto, M. (1997) Detection and analysis of change in remotely sensed imagery with application to wide area surveillance. *IEEE Transactions on Image Processing*, **6**(1), 189–202.

Castracane, P., Iavaronc, F., Mica, S., Sottile, E., Vignola, C., Arino, O., Cataldo, M., Fernandez-Prieto, D., Guidotti, G., Masullo, A. and Pratesi, I. (2003) Monitoring urban sprawl and its trends with EO data. UrbEx, a prototype national service from a WWF-ESA joint effort. *2nd GRSS/ISPRS Joint Workshop on Remote Sensing and Data Fusion over Urban Areas*, pp. 245–248.

Civco, D. (1993) Artificial neural networks for land cover classification and mapping. *International Journal of Geographic Information Systems*, **7**(2), 173–186.

Clausi, D. and Yue, B. (2004) Comparing co-occurrence probabilities and Markov random fields for texture analysis of SAR sea ice imagery. *IEEE Transactions on on Geoscience and Remote Sensing*, **42**(1), 215–228.

Collins, J.B. and Woodcock, C.E. (1996) An assessment of several linear change detection techniques for mapping forest mortality using multi-temporal Landsat TM data. *Remote Sensing of Environment*, **56**, 66–77.

Collins, R., Lipton, A. and T. Kanade, T. (2000) Introduction to the special section on video surveillance. *IEEE Transactions on Pattern Analysis and Machine Intelligence*, **22**(8), 745–746.

Coppin, P. and Bauer, M. (1996) Digital change detection in forest ecosystems with remote sensing imagery. *Remote Sensing Reviews*, **13**(6), 207–234.

G. Camps-Valls, J.L. Rojo-Álvarez and M. Martínez-Ramón (eds), (2006c) *Kernel Methods in Bioengineering, Signal and Image Processing*. Idea Group Inc., Harshey, PA (USA).

Fanelli, A., Santoro, M., Vitale, A., Murino, P. and Askne, J. (2000) Understanding ERS coherence over urban areas. In ESA-SP-461 (ed), *ERS-Envisat Symposium*.

Fang, C.Y., Chen, S.W. and Fuh, C.S. (2003) Automatic change detection of driving environments in a vision-based driver assistance system. *IEEE Transactions on Neural Networks*, **14**(3), 646–657.

Gamba, P., Dell'Acqua, F. and Lisini, G. (2006) Change detection of multi-temporal SAR data in urban areas combining feature-based and pixel-based techniques. *IEEE Transactions on Geoscience and Remote Sensing*, **44**(11(1)), 2820–2827.

Gómez-Chova, L., Fernández-Prieto, D., Calpe, J., Soria, E., Vila, J. and Camps-Valls, G. (2004) Partially supervised hierarchical clustering of SAR and multi-spectral imagery for urban areas. *SPIE Europe*, vol. 5573, pp. 138–149, Gran Canaria, Spain.

Gómez-Chova, L., Fernández-Prieto, D., Calpe, J., Soria, E., Vila, J. and Camps-Valls, G. (2006) Urban monitoring using multi-temporal SAR and multi-spectral data. *Pattern Recognition Letters, Special Issue on 'Pattern Recognition in Remote Sensing'*, **27**(4), 234–243.

Gopal, S. and Woodcock, C. (1996) Remote sensing of forest change using artifical neural networks. *IEEE Transactions on Geoscience and Remote Sensing*, **34**(2), 189–202.

Hagan, M. and Menhaj, M. (1994) Training feed-forward networks with the Marquardt algorithm. *IEEE Transactions on Neural Networks*, **5**(6), 989–993.

Hughes, G.F. (1968) On the mean accuracy of statistical pattern recognizers. *IEEE Trans. Information Theory*, **14**(1), 55–63.

Inglada, J. and Mercier, G. (2007) A new statistical similarity measure for change detection in multi-temporal SAR images and its extension to multiscale change analysis. *IEEE Transactions on Geoscience and Remote Sensing*, **45**(5(2)), 1432–1445.

Kushardono, D., Fukue, K., Shimoda, H. and Sakata, T. (1995) Comparison of multi-temporal image classification methods. *Proceedings of the International Conference on Geoscience and Remote Sensing*, pp. 1282–1284.

Lemieux, L., Wieshmann, U., Moran, N., Fish, D. and Shorvon, S. (1998) The detection and significance of subtle changes in mixed-signal brain lesions by serial MRI scan matching and spatial normalization. *Medical Image Analysis*, **2**(3), 227–242.

Li, J. and Narayanan, R.M. (2003) A shape-based approach to change detection of lakes using time series remote sensing images. *IEEE Transactions on Geoscience and Remote Sensing*, **41**(11), 2466–2477.

Lillesand, T.M., Kiefer, R.W. and Chipman, J.W. (2004) *Remote Sensing and Image Interpretation*, 5th edn John Wiley & Sons, New York.

Liu, D., Kelly, M. and Gong, P. (2005) Classifying multi-temporal Landsat TM imagery using Markov random fields and support vector machines. *3rd International Workshop on the Analysis of Multi-temporal Remote Sensing Images*.

Melgani, F. (2004) Classification of multi-temporal remote-sensing images by a fuzzy fusion of spectral and spatio-temporal contextual information. *International Journal of Pattern Recognition and Artificial Intelligence*, **18**(2), 143–156.

Melgani, F. and Serpico, S.B. (2003) A Markov random field approach to spatiotemporal contextual image classification. *IEEE Trans. Geosci. Remote Sens.*, **41**(11), 2478–2487.

Mercier, G. and Girard-Ardhuin, F. (2006) Partially supervised oil-slick detection by SAR imagery using kernel expansion. *IEEE Transactions on Geoscience and Remote Sensing*, **44**(10(1)), 2839–2846.

Morisette, J.T. and Khorram, S. (1997) An introduction to using generalized linear models to enhance satellite-based change detection. *Proceedings of the International Conference on Geoscience and Remote Sensing*, pp. 1282–1284.

Moser, G. and Serpico, S. (2006) Generalized minimum-error thresholding for unsupervised change detection from SAR amplitude imagery. *IEEE Transactions on Geoscience and Remote Sensing*, **44**(10(2)), 2972–2982.

Muñoz Marí, J., Bruzzone, L. and Camps-Valls, G. (2007) A support vector domain description approach to supervised classification of remote sensing images. *IEEE Transactions on Geoscience and Remote Sensing*, **45**(8), 2683–2692.

Pellizzeri, T., Gamba, P., Lombardo, P. and Dell'Acqua, F. (2003) Multi-temporal/Multiband SAR classification of urban areas using spatial analysis: statistical versus neural kernel-based approach. *IEEE Transactions on Geoscience and Remote Sensing*, **41**(10), 2338–2353.

Potin, D., Vanheeghe, P., Duflos, E. and Davy, M. (2006) An abrupt change detection algorithm for buried landmines localization. *IEEE Transactions on Geoscience and Remote Sensing*, **44**(2), 260–272.

Radke, R.J.S.A., Al-Kofahi, O. and Roysam, B. (2005) Image change detection algorithms: a systematic survey. *IEEE Transactions on Image Processing*, **14**(3), 294–307.

Schölkopf, B. and Smola, A.J. (2002) *Learning with Kernels*, MIT Press, Cambridge, MA.

Shawe-Taylor, J. and Cristianini, N. (2004) *Kernel Methods for Pattern Analysis*. Cambridge University Press.

Singh, A. (2003) Digital change detection techniques using remotely-sensed data. *International Journal of Remote Sensing*, **10**(6), 989–1003.

Tax, D. and Duin, R.P. (1999) Support vector domain description. *Pattern Recognition Letters*, **20**, 1191–1199.

Wang, F. (1993) A knowledge-based vision system for detecting land changes at urban fringes. *IEEE Transactions on Geoscience and Remote Sensing*, **31**(1), 136–145.

7

Target detection with kernels

Nasser M. Nasrabadi
US Army Research Laboratory

This book chapter provides a performance comparison of various linear and nonlinear anomaly detection techniques based on kernels. Three different subspace anomaly detectors based on Principal Component Analysis (PCA), Fisher Linear Discriminant (FLD) Analysis, and the Eigenspace Separation Transform (EST) are reviewed and their nonlinear versions (kernel extensions) are discussed. In addition to the subspace-based anomaly detectors, the well-known Reed–Xiaoli (RX) anomaly detector and its kernel version are also implemented. Comparisons between all linear and nonlinear anomaly detectors are made using receiver operating characteristics (ROC) curves for simulated toy data and real hyperspectral imagery.

7.1 Introduction

Hyperspectral imagery can be used in reconnaissance and surveillance applications where objects of interest are detected and identified. This chapter describes kernel-based techniques that are used to detect anomalies. These anomalies could be possible military targets or camouflaged objects. Hyperspectral imagery provides a significant amount of information about the spectral characteristics of the materials in the scene that can be used for anomaly and target detection (Chang *et al.* 1998; Harsanyi and Chang 1994; Healey and Slater 1999; Kwon *et al.* 2001; Manolakis and Shaw 2002; Reed and Yu 1990; Scharf and Friedlander 1994; Schweizer and Moura 2001; Stein *et al.* 2002).

The process of detecting and identifying a target in hyperspectral imagery can be considered as consisting of two stages. The first stage is an anomaly detector that identifies spectral anomalies or a localized spectral difference. The second stage is to identify whether or not the anomaly is a target or natural clutter. This stage can be achieved if the spectral signature of

Kernel Methods for Remote Sensing Data Analysis Edited by Gustavo Camps-Valls and Lorenzo Bruzzone

the target is known, which can be obtained from a spectral library or using a spectral subspace matched filter designed from a set of training data (Healey and Slater 1999; Manolakis and Shaw 2002).

Almost all the anomaly and target detectors are based on a linear process that exploits the first and second order moments of the data to identify anomalies or targets. For example, a well known spectral anomaly detection algorithm was developed in Reed and Yu (1990) called the RX algorithm, which is now considered as a benchmark anomaly detector. The RX algorithm is based on exploiting the difference between the spectral signature of an input pixel with its surrounding neighbours (Chang and Chiang 2002; Kwon *et al.* 2003; Stein 2001; Stein *et al.* 2002; Yu and Reed 1993). This distance comparison is very similar to the Mahalanobis distance measure, which is done by comparing the corresponding wavelengths (spectral bands) of two measurements normalized by the covariance matrix of background statistics. The conventional RX distance measure does not take into account the higher order relationships (higher order correlation) between the spectral bands at different wavelengths. The nonlinear relationships between different spectral bands within the target or clutter spectral signature need to be exploited in order to distinguish between target and background better. Similarly, most of the target detection algorithms are based on linear matched (subspace) filters where the spectral characteristics of a target or a target subspace representing target information is assumed to be known. In spectral matched filters (Manolakis *et al.* 2000; Robey *et al.* 1992) a linear mixing model is assumed where a known target spectral signature is used in conjunction with the covariance matrix of the background data to identify a specific target. In matched subspace detection (Scharf and Friedlander 1994; Thai and Healey 2002) a subspace linear mixture model is used for target detection where the target and background signatures are represented by their corresponding linear subspaces. Two other detectors, adaptive subspace detection (Kraut *et al.* 2001) and orthogonal subspace projection (Harsanyi 1993) have also been developed; they are based on using different linear models. All these target-based matched filter like detectors are based on linear mixture models that ignore the higher order correlations between the spectral bands.

In this chapter, we consider only the subspace-based anomaly detectors and their nonlinear versions using a kernel approach (Müller *et al.* 2001; Schölkopf and Smola 2002; Vapnik 1999). Kernel methods have emerged as new nonlinear-based learning techniques that implicitly exploit the dot product of feature vectors generated by the nonlinear mapping of the input vectors using kernel representations. The implicit exploitation of nonlinear features through kernels provides crucial information about given data that, in general, the learning methods based on linear models cannot achieve. In kernel methods the learning is performed in a high dimensional feature space where the complexity of the given data can be possibly reduced, subsequently generating simpler decision rules and improving generalization performance.

Many linear techniques have been extended to their corresponding kernel versions and promising results have been reported for many applications. For example, PCA, FLD and EST have been extended to their kernel versions corresponding to Kernel Principal Component Analysis (KPCA) (Schölkopf *et al.* 1999), Kernel Fisher Discriminant (KFD) analysis (Mika *et al.* 2003) and Kernel Eigenspace Separation Transform (KEST) (Goldberg *et al.* 2007), respectively. In Camps-Valls and Bruzzone (2005) several regularized kernel-based methods such as Support Vector Machine (SVM) (Vapnik 1998), KFD analysis, radial basis function neural network (Haykin 1999) and regularized AdaBoost (Rätsch *et al.* 2004) were implemented and their performance was compared for classification using hyperspectral imagery. In Melgani and Bruzzone (2004) the problem of multiclass discrimination was addressed by using binary SVMs for hyperspectral classification. The use of a single-class SVM anomaly detection in hyperspectral imagery and the use of KFD for outlier detection of faces and characters were reported in Banerjee *et al.* (2006) and Roth (2006), respectively. A kernel-based nonlinear subspace method using KPCA was implemented in Maeda and Murase (2002) for classification of Japanese characters. In Kwon and Nasrabadi (2005b) the well-known hyperspectral RX anomaly detector (Reed and Yu 1990) was extended to its nonlinear kernel version, Kernel RX (KRX). In Kwon and Nasrabadi (2007a) several target detection techniques, such as Kernel SMF (KSMF) (Kwon and Nasrabadi 2007b), Kernel MSD (KMSD) (Kwon and Nasrabadi 2006b), Kernel OSP (KOSP) (Kwon and Nasrabadi 2005a), and Kernel ASD (KASD) (Kwon and Nasrabadi 2006a) were reviewed and their performance was compared with their corresponding linear versions.

This chapter is organized in the following manner. Section 7.2 contains a brief overview of kernel-based learning techniques. In 7.3 the subspace anomaly detection methods based on PCA, FLD and EST are described. Their kernel extensions using their corresponding nonlinear subspaces generated by the KPCA, KFD and KEST, respectively are also reviewed. RX and KRX are also discussed for comparison with the subspace-based anomaly detectors. Results and analysis of the anomaly and target detectors that are reviewed in this chapter with their kernel versions can be found in Section 7.4 as applied to simulated data as well as multiple hyperspectral data sets. Finally, concluding remarks are made in Section 7.5.

7.2 Kernel learning theory

Suppose the input data set lies in the data space ($\mathcal{X} \subset \mathbb{R}^J$) and let $\mathcal{F}$ be a feature space (also known as a Hilbert space) associated with $\mathcal{X}$ by some nonlinear mapping function Φ. In particular,

$$\begin{aligned} \Phi &: \mathcal{X} \to \mathcal{F} \\ \mathbf{x} &\mapsto \Phi(\mathbf{x}), \end{aligned} \tag{7.1}$$

where $\mathbf{x}$ is an input vector ($\mathbf{x} \in \mathcal{X}$) that is mapped into a much higher dimensional feature space. Mapping the data using Φ into $\mathcal{F}$ is useful in many ways. The most significant benefit is that it is possible to define a similarity measure using the dot product in $\mathcal{F}$ in terms of a function of the corresponding data in the input space. Thus, it is possible to write

$$k(\mathbf{x}_i, \mathbf{x}_j) = \langle \Phi(\mathbf{x}_i), \Phi(\mathbf{x}_j) \rangle. \tag{7.2}$$

Equation (7.2), which is commonly referred to in the machine learning literature as the *kernel trick* (Schölkopf and Smola 2002) states that all dot products in $\mathcal{F}$ (a task that is otherwise computationally infeasible) can be implicitly computed by simply using a kernel function defined on the input data. Moreover, all of this can be accomplished without actually mapping the input vectors into $\mathcal{F}$. Hence, conveniently, the mapping Φ does not even need to be identified or defined. In other words, Equation (7.2) illustrates that all dot products in $\mathcal{F}$ can be replaced by an appropriately chosen *Mercer* kernel function k. For a more comprehensive discussion about the properties of various types of kernels and for more information on kernel-based learning in general, see one of the many references devoted to kernel methods (Schölkopf and Smola 2002; Shawe-Taylor and Cristianini 2004).

7.3 Linear subspace-based anomaly detectors and their kernel versions

Anomaly detectors (Ahlberg and Renhorn 2004; Banerjee *et al.* 2006; Kwon and Nasrabadi 2005b; Kwon *et al.* 2003; Manolakis and Shaw 2002; Ranney and Soumekh 2006; Reed and Yu 1990; Stein *et al.* 2002; Willis 2005) are pattern recognition schemes that are used to detect objects of interest. Almost all anomaly detectors attempt to locate anything that looks different either spatially or spectrally from its surroundings using a dual-window approach. Its use is predicated on the fact that it exploits both spatial variability in the image as well as spectral variability among different materials. In the dual-window approach, two concentric windows centred at each test pixel are opened creating two disjoint regions: an inner-window region (IWR) and an outer-window region (OWR). Hence, the local pixel neighbourhood is separated into two smaller regions. The size of the inner window is generally set so that the inner window can fully enclose a target. In most anomaly detectors another concentric window centred at the test pixel known as the 'guard band' is utilized as well. An example of a dual-window with guard band is seen in Figure 7.1. The guard band is slightly larger in size than the IWR, yet still smaller than the OWR and is used to reduce the probability that some target spectra will contaminate the OWR statistics and hence affect the background model (Ranney and Soumekh 2006).

One way of designing an anomaly detector is by projecting the input spectra on to a subspace whose bases are defined by some projection vectors. In Kwon *et al.* (2003) researchers compared subspace-based anomaly detection algorithms using projection vectors that were generated using three common pattern recognition techniques – PCA (Joliffe 1986), FLD Analysis (Duda *et al.* 2001) and EST (Torrieri 1999). In addition, they compared the performance of the RX algorithm with the performances of the aforementioned subspace-based anomaly detectors. In this chapter, we extend these linear subspace-based anomaly detectors proposed in Kwon *et al.* (2003) to their nonlinear kernel versions.

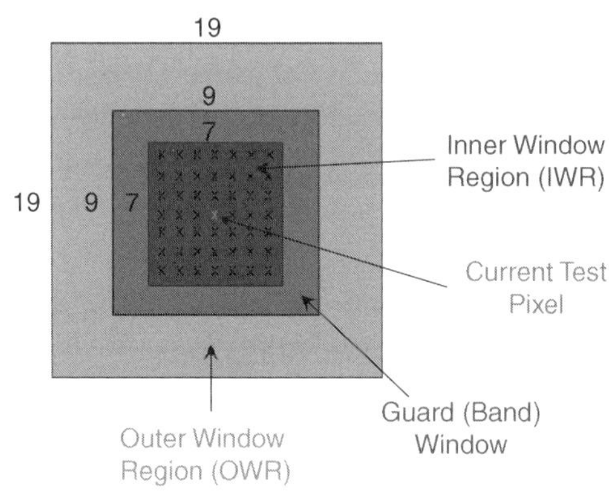

Figure 7.1 An example of a dual window with guard band. The numbers represent the length in pixels making up the side of each window. Each $\times$ in the IWR represents one pixel. The pixel in the middle of the figure represents the current test pixel. The figure is not necessarily drawn to scale.

In subspace-based anomaly detection techniques, projection (basis) vectors are generated using the statistical properties of the IWR and OWR covariance matrices. For example, using the eigenvalue decomposition of the covariance matrices of IWR or OWR spectra it is possible to generate basis vectors for a subspace onto which vectors from the IWR or OWR are projected for discrimination. Denote a spectral vector within the IWR of a dual window centred at a test pixel by $\mathbf{x}_k = (x_k(1), x_k(2), \ldots, x_k(J))^\top$ where J is the number of spectral bands and $k = 1, \ldots, N_{in}$. Assuming that there are a total of N_{in} pixels in the IWR, the matrix $\mathbf{X} = [\mathbf{x}_1, \mathbf{x}_2, \ldots, \mathbf{x}_{N_{in}}]$ is of size $J \times N_{in}$ and contains the spectra of each one of these samples as one of its N_{in} columns. Similarly, let a spectral vector within the OWR of the same dual window be denoted by $\mathbf{y}_l$ where $l = 1, \ldots, N_{out}$. Given that there are N_{out} pixels in the OWR, the $J \times N_{out}$ matrix $\mathbf{Y} = [\mathbf{y}_1, \mathbf{y}_2, \ldots, \mathbf{y}_{N_{out}}]$ is one whose columns are the spectral vectors of the pixels in the OWR representing the background clutter samples. The background clutter covariance statistics are estimated using the spectra of the pixels in the OWR. The covariance matrices of the IWR and OWR spectra are given by

$$\mathbf{C}_X = \frac{1}{N_{in}} \sum_{i=1}^{N_{in}} (\mathbf{x}_i - \hat{\boldsymbol{\mu}}_X)(\mathbf{x}_i - \hat{\boldsymbol{\mu}}_X)^\top, \tag{7.3}$$

$$\mathbf{C}_Y = \frac{1}{N_{out}} \sum_{i=1}^{N_{out}} (\mathbf{y}_i - \hat{\boldsymbol{\mu}}_Y)(\mathbf{y}_i - \hat{\boldsymbol{\mu}}_Y)^\top, \tag{7.4}$$

where $\hat{\boldsymbol{\mu}}_X$ and $\hat{\boldsymbol{\mu}}_Y$ represent the means of the IWR and OWR spectra, respectively.

The projection separation statistic for an input test pixel, $\mathbf{r}$, is calculated using

$$s' = (\mathbf{r} - \hat{\boldsymbol{\mu}}_Y)^\top \mathbf{W}\mathbf{W}^\top (\mathbf{r} - \hat{\boldsymbol{\mu}}_Y), \tag{7.5}$$

where $\mathbf{W} = [\mathbf{w}_1\ \mathbf{w}_2\ \ldots\ \mathbf{w}_m]$ is a matrix whose columns are m projection vectors. The product $\mathbf{W}\mathbf{W}^\top$ is known as a projection operator and represents a subspace characterizing the spectral region represented by the projection vectors $\mathbf{w}_i$. An anomaly is detected if the projection separation, s', is greater than some threshold, η. It is also possible to project the difference $(\mathbf{r} - \hat{\boldsymbol{\mu}}_Y)$ onto the complement subspace $(\mathbf{I} - \mathbf{W}\mathbf{W}^\top)$ given by

$$s'' = (\mathbf{r} - \hat{\boldsymbol{\mu}}_Y)^\top (\mathbf{I} - \mathbf{W}\mathbf{W}^\top)(\mathbf{r} - \hat{\boldsymbol{\mu}}_Y). \tag{7.6}$$

Several methods could be used to generate the projection vectors, $\mathbf{w}_i$, for Equations (7.5) or (7.6). In this chapter, three different methods PCA, FLD and EST as well as their kernel versions are implemented and their performance are compared.

7.3.1 Principal component analysis

PCA is one of the most commonly used methods for feature extraction and dimensionality reduction. The underlying goal is to find a projection that best represents some input data in the least-square sense (Joliffe 1986). In order to generate the PCA projection vectors, $\mathbf{W}_{PCA}$, the background clutter covariance matrix, $\mathbf{C}_Y$, is written in terms of its eigenvectors $\mathbf{V}$ and their corresponding eigenvalues $\mathbf{\Lambda}$ as

$$\mathbf{C}_Y = \mathbf{V}\mathbf{\Lambda}\mathbf{V}^\top, \tag{7.7}$$

the first m eigenvectors with the highest corresponding eigenvalues form the projection vectors. Thus,

$$\mathbf{W}_{PCA} = [\mathbf{v}_1, \mathbf{v}_2, \cdots, \mathbf{v}_m], \tag{7.8}$$

where m is a configurable constant. Altering the value of m (i.e. changing the number of eigenvectors used) will change the performance of the anomaly detector as shown by the experimental results. Using (7.8) as the projection vectors and substituting this result into (7.5) and (7.6) we obtain the corresponding projections of the input onto the PCA-based subspace anomaly detector, which is given by

$$\mathbf{PCA}(\mathbf{r}) = (\mathbf{r} - \hat{\boldsymbol{\mu}}_Y)^\top \left(\mathbf{W}_{PCA}\mathbf{W}_{PCA}^\top\right)(\mathbf{r} - \hat{\boldsymbol{\mu}}_Y), \tag{7.9}$$

and the projection onto its complement subspace is given by

$$\mathbf{PC\acute{A}}(\mathbf{r}) = (\mathbf{r} - \hat{\boldsymbol{\mu}}_Y)^\top \left(\mathbf{I} - \mathbf{W}_{PCA}\mathbf{W}_{PCA}^\top\right)(\mathbf{r} - \hat{\boldsymbol{\mu}}_Y). \tag{7.10}$$

The idea behind using the PCA eigenvectors lies in the fact that since these eigenvectors are optimal (i.e. they minimize the mean-square error) in terms of their representation of the spectral vectors of the OWR, the projection of the difference between the test pixel, $\mathbf{r}$, and the OWR mean, $\hat{\boldsymbol{\mu}}_Y$, should ideally be large if the dual window is centred on an anomalous target.

The algorithm outlined above can also be developed using samples collected from the IWR. In this case, Equations (7.9) and (7.10) remain the same with the exception that the projection vectors $\mathbf{W}_{PCA}$ are generated using the spectral information contained in the IWR rather than the OWR.

7.3.2 Kernel PCA subspace-based anomaly detection

In this section the linear PCA subspace-based method in Section 7.3.1 is extended into the kernel feature space $\mathcal{F}$ and then kernelized by replacing all dot products in the feature space by kernel functions using the kernel trick (7.2). Using a nonlinear mapping Φ, the original data, $\mathbf{X}$ and $\mathbf{Y}$, in the input space are mapped into the feature space $\mathcal{F}$ denoted by

$$\mathbf{X}_\Phi = \Phi(\mathbf{x}) = \left[\Phi(\mathbf{x}_1)\,\Phi(\mathbf{x}_2)\,\ldots\,\Phi(\mathbf{x}_{N_{in}})\right] \tag{7.11}$$

$$\mathbf{Y}_\Phi = \Phi(\mathbf{Y}) = \left[\Phi(\mathbf{y}_1)\,\Phi(\mathbf{y}_2)\,\ldots\,\Phi(\mathbf{y}_{N_{out}})\right]. \tag{7.12}$$

This means that $\mathbf{X}_\Phi$ represent the mapped IWR spectra and $\mathbf{Y}_\Phi$ represents the mapped OWR spectra. The statistical means of the mapped data in $\mathcal{F}$ are represented by $\hat{\boldsymbol{\mu}}_{X_\Phi}$ and $\hat{\boldsymbol{\mu}}_{Y_\Phi}$, respectively. For many of the following methods, it is assumed that the mapped data is mean-removed (centred) in $\mathcal{F}$. Thus, denote each centred vector for the IWR in $\mathcal{F}$ as $\Phi_c(\mathbf{x}_i) = \Phi(\mathbf{x}_i) - \hat{\boldsymbol{\mu}}_{X_\Phi}, i = 1, \ldots, N_{in}$ and similarly for the OWR spectra (i.e. $\Phi_c(\mathbf{y}_j) = \Phi(\mathbf{y}_j) - \hat{\boldsymbol{\mu}}_{Y_\Phi}$,

$j = 1, \ldots, N_{out}$). Then, let $\mathbf{x}_{c\Phi}$ and $\mathbf{Y}_{c\Phi}$ be matrices whose columns are the centred IWR and OWR in the feature space, respectively. Also, let $\mathbf{C}_{X_\Phi}$ and $\mathbf{C}_{Y_\Phi}$ be the covariance matrices of the centred spectra in the feature space. The projection of the mapped test pixel spectra $\Phi(\mathbf{r})$ onto a linear subspace in the feature space that is equivalent to a nonlinear subspace in the original input domain is given by

$$s' = (\Phi(\mathbf{r}) - \hat{\boldsymbol{\mu}}_{Y_\Phi})^\top \mathbf{W}_\Phi \mathbf{W}_\Phi^\top (\Phi(\mathbf{r}) - \hat{\boldsymbol{\mu}}_{Y_\Phi}), \tag{7.13}$$

where $\mathbf{W}_\Phi = \left[\mathbf{w}_\Phi^1 \, \mathbf{w}_\Phi^1 \, \ldots \, \mathbf{w}_\Phi^m\right]$ is a matrix whose columns are the set of m projection vectors in $\mathcal{F}$. Similarly, projection onto the complement subspace is given by

$$s' = (\Phi(\mathbf{r}) - \hat{\boldsymbol{\mu}}_{Y_\Phi})^\top \left(\mathbf{I}_\Phi - \mathbf{W}_\Phi \mathbf{W}_\Phi^\top\right) (\Phi(\mathbf{r}) - \hat{\boldsymbol{\mu}}_{Y_\Phi}). \tag{7.14}$$

In order to find the PCA eigenvectors in the feature space, simply solve the mapped version of Equation (7.7); specifically, the eigenvalues and eigenvectors of $\mathbf{C}_{Y_\Phi}$ in $\mathcal{F}$ can be found by solving

$$\mathbf{C}_{Y_\Phi} = \mathbf{V}_\Phi \boldsymbol{\Lambda}_\Phi \mathbf{V}_\Phi^\top, \tag{7.15}$$

where $\boldsymbol{\Lambda}_\Phi = diag\left(\lambda_\Phi^1 \, \lambda_\Phi^2 \, \ldots \, \lambda_\Phi^p\right)$ and $\mathbf{V}_\Phi = \left[\mathbf{v}_\Phi^1 \, \mathbf{v}_\Phi^2 \, \ldots \, \mathbf{v}_\Phi^p\right]$ contain only the p nonzero eigenvalues and corresponding eigenvectors of $\mathbf{C}_{Y_\Phi}$. All eigenvectors in the feature space lie in the span of the vectors in $\mathbf{Y}_{c\Phi}$. Therefore, they can be represented as

$$\mathbf{V}_\Phi = \boldsymbol{\Lambda}_\Phi{}^{-1/2} \mathbf{Y}_{c\Phi} \mathbf{A}, \tag{7.16}$$

where $\mathbf{A} = [\boldsymbol{\alpha}_1 \, \boldsymbol{\alpha}_2 \, \cdots \, \boldsymbol{\alpha}_{N_{out}}]$ as shown in Schölkopf *et al.* (1999) is a matrix whose columns are the nonzero eigenvectors of the centred Gram (kernel) matrix $\mathbf{K}_c$ and $\boldsymbol{\Lambda}_\Phi$ contains the associated nonzero eigenvalues. The centred kernel matrix can be calculated by $\mathbf{K}_c = (\mathbf{K} - \mathbf{1}_{N_{out}}\mathbf{K} - \mathbf{K}\mathbf{1}_{N_{out}} + \mathbf{1}_{N_{out}}\mathbf{K}\mathbf{1}_{N_{out}})$ where $\mathbf{K}$ is the kernel matrix whose elements are $(\mathbf{K})_{ij} = k(\mathbf{y}_i, \mathbf{y}_j)$ with $i, j = 1, \ldots, N_{out}$ and $(\mathbf{1}_{N_{out}})$ is an $N_{out} \times N_{out}$ with each element equal to $1/N_{out}$. It is shown in Schölkopf *et al.* (1999) that the eigenvalue decomposition of the centred kernel matrix is given by

$$\mathbf{K}_c = \mathbf{A} \boldsymbol{\Lambda}_\Phi \mathbf{A}^\top. \tag{7.17}$$

Utilizing only the m most significant eigenvectors in the features space they can be written as $\tilde{\mathbf{V}}_\Phi = \mathbf{Y}_{c\Phi} \tilde{\mathbf{A}}_{\text{KPCA}}$, where $\tilde{\mathbf{A}}_{\text{KPCA}} = [\tilde{\boldsymbol{\alpha}}_1 \, \tilde{\boldsymbol{\alpha}}_2 \, \cdots \, \tilde{\boldsymbol{\alpha}}_m]$ are the m most significant eigenvectors of $\mathbf{K}_c$ normalized by the square roots of their respective eigenvalues. Using vectors $\tilde{\mathbf{V}}_\Phi$ as the projection vectors in the feature space ($\mathbf{W}_\Phi$) in (7.13) we obtain KPCA-based anomaly detector

$$\begin{aligned}\mathbf{KPCA}(\mathbf{r}) &= (\Phi(\mathbf{r}) - \hat{\boldsymbol{\mu}}_{Y_\Phi}))^\top \left(\tilde{\mathbf{V}}_\Phi \tilde{\mathbf{V}}_\Phi^\top\right) (\Phi(\mathbf{r}) - \hat{\boldsymbol{\mu}}_{Y_\Phi}) \\ &= (\Phi(\mathbf{r}) - \hat{\boldsymbol{\mu}}_{Y_\Phi})^\top \mathbf{Y}_{c\Phi} \tilde{\mathbf{A}}_{\text{KPCA}} \tilde{\mathbf{A}}_{\text{KPCA}}^\top \mathbf{Y}_{c\Phi}^\top (\Phi(\mathbf{r}) - \hat{\boldsymbol{\mu}}_{Y_\Phi}).\end{aligned} \tag{7.18}$$

For notational simplicity, let $\mathbf{k}_{Yr}^{\top} = \Phi(\mathbf{r})^{\top}\mathbf{Y}_{c\Phi}$ and $\mathbf{k}_{Y\hat{\mu}}^{\top} = \hat{\boldsymbol{\mu}}_{Y_\Phi}^{\top}\mathbf{Y}_{c\Phi}$; these are commonly referred to as empirical kernel expansions (Schölkopf and Smola 2002). Substituting these expressions into (7.18) results in

$$\mathbf{KPCA(r)} = \left(\mathbf{k}_{Yr}^{\top} - \mathbf{k}_{Y\hat{\mu}}^{\top}\right)^{\top} \tilde{\mathbf{A}}_{\mathrm{KPCA}}\tilde{\mathbf{A}}_{\mathrm{KPCA}}^{\top} \left(\mathbf{k}_{Yr}^{\top} - \mathbf{k}_{Y\hat{\mu}}^{\top}\right). \tag{7.19}$$

As in the linear case, we can also project onto the complement subspace $(\mathbf{I} - \tilde{\mathbf{V}}_{\Phi}\tilde{\mathbf{V}}_{\Phi}^{\top})$ in the feature space. As mentioned above, the KPCA algorithm can also be formulated using the IWR spectra rather than the OWR spectra to generate the projection vectors $\mathbf{W}_{\Phi}$ in the feature space.

7.3.3 Fisher linear discriminant analysis

Although PCA has proven to be very useful for efficient representation of data, it does not exploit the information in the IWR and OWR at the same time in order to generate the target or background subspaces. However, the FLD analysis attempts to seek an optimal direction for discriminating between IWR and OWR data samples. First, the between-class scatter matrix is defined as $\mathbf{S}_B = (\hat{\boldsymbol{\mu}}_X - \hat{\boldsymbol{\mu}}_Y)(\hat{\boldsymbol{\mu}}_X - \hat{\boldsymbol{\mu}}_Y)^{\top}$ while the within-class scatter matrix can be written as $\mathbf{S}_W = \mathbf{C}_X + \mathbf{C}_Y$ where $\mathbf{C}_X$ and $\mathbf{C}_Y$ are the covariance matrices of the samples in the IWR and OWR defined by Equations (7.3) and (7.4), respectively. The matrix $\mathbf{S}_B$ is a measure of how well the means of the two classes are separated while the matrix $\mathbf{S}_W$ is a measure of the compactness of each class cluster.

In order to calculate the optimal discrimination direction, $\mathbf{w}^*$, the criterion function

$$\mathbf{w}^* = \max_{\mathbf{w}} \mathrm{J}(\mathbf{w}) = \frac{|\mathbf{w}^{\top}\mathbf{S}_B\mathbf{w}|}{|\mathbf{w}^{\top}\mathbf{S}_W\mathbf{w}|} \tag{7.20}$$

needs to be maximized over all possible $\mathbf{w}$ and has been shown (Duda *et al.* 2001) to be given by

$$\mathbf{w}_{\mathrm{FLD}} = \mathbf{w}^* = \mathbf{S}_W^{-1}(\hat{\boldsymbol{\mu}}_X - \hat{\boldsymbol{\mu}}_Y). \tag{7.21}$$

Using (7.21) as the projection vector and substituting this result into (7.5) gives

$$\mathbf{FLD(r)} = (\mathbf{r} - \hat{\boldsymbol{\mu}}_Y)^{\top}\left(\mathbf{w}_{\mathrm{FLD}}\mathbf{w}_{\mathrm{FLD}}^{\top}\right)(\mathbf{r} - \hat{\boldsymbol{\mu}}_Y). \tag{7.22}$$

The idea behind using FLD is that it will produce a large projection separation if the spectral means of the IWR and OWR are sufficiently dissimilar while the spectral vectors in each region are tightly clustered. In this chapter, Equation (7.22) is used as the FLD-based anomaly detector and referred to as the FLD detector.

7.3.4 Kernel fisher discriminant analysis

It has been shown in Baudat and Anouar (2000) and Mika *et al.* (2003) how to extend FLD analysis to its nonlinear version by using the kernel trick to compute the Fisher discriminant in the feature space (see Chapter 5). Defining FLD in the feature space is equivalent to maximizing the cost function given by

$$\mathrm{J}(\mathbf{w}_\Phi) = \frac{|\mathbf{w}_\Phi^\top \mathbf{S}_B^\Phi \mathbf{w}_\Phi|}{|\mathbf{w}_\Phi^\top \mathbf{S}_W^\Phi \mathbf{w}_\Phi|}, \tag{7.23}$$

where $\mathbf{w}_\Phi$ is the projection vector, $\mathbf{S}_W^\Phi = \mathbf{C}_{X_\Phi} + \mathbf{C}_{Y_\Phi}$ and $\mathbf{S}_B^\Phi = (\hat{\boldsymbol{\mu}}_{X_\Phi} - \hat{\boldsymbol{\mu}}_{Y_\Phi})(\hat{\boldsymbol{\mu}}_{X_\Phi} - \hat{\boldsymbol{\mu}}_{Y_\Phi})^\top$ are the within-class and between-class scatter matrices, respectively, in $\mathcal{F}$.

Finding an optimal $\mathbf{w}_\Phi$ by maximizing (7.23) is not mathematically tractable considering the simple fact that the feature space is of high (possibly infinite) dimensionality. Fortunately, we can reformulate this problem in terms of dot products in the feature space and then replace them with kernel functions. Based on reproducing kernel theory, any solution $\mathbf{w}_\Phi$ to Equation (7.23) can be expanded as

$$\mathbf{w}_\Phi = \sum_{i=1}^{N_t} \alpha_i \Phi(\mathbf{z}_i) = \mathbf{Z}_\Phi \boldsymbol{\alpha}, \tag{7.24}$$

where $\mathbf{Z}_\Phi = [\mathbf{z}_1, \mathbf{z}_2, \ldots, \mathbf{z}_{N_t}] = [\Phi_c(\mathbf{x}_1), \ldots, \Phi_c(\mathbf{x}_{N_{in}}), \Phi_c(\mathbf{y}_1), \ldots, \Phi_c(\mathbf{y}_{N_{out}})]$ is a matrix whose columns are the mapped vectors in $\mathcal{F}$ of the corresponding spectra in both the IWR and OWR concatenated together, $N_t = N_{in} + N_{out}$ and $\boldsymbol{\alpha} = (\alpha_1, \alpha_2, \ldots, \alpha_{N_t})$ is the KFD vector in the kernel space satisfying (7.25).

It has been shown in Mika *et al.* (2003) that maximizing the Fisher's discriminant (7.23) in $\mathcal{F}$ is equivalent to maximizing a dual representation given by

$$\mathrm{J}(\boldsymbol{\alpha}) = \frac{\boldsymbol{\alpha}^\top \mathbf{A} \boldsymbol{\alpha}}{\boldsymbol{\alpha}^\top \mathbf{B} \boldsymbol{\alpha}}, \tag{7.25}$$

where

$$\mathbf{A} = (\mathbf{M}_{in} - \mathbf{M}_{out})(\mathbf{M}_{in} - \mathbf{M}_{out})^\top, \; (\mathbf{M}_{in})_j = \frac{1}{N_{in}} \sum_{l=1}^{N_{in}} k(\mathbf{x}_j, \mathbf{x}_l),$$

$$(\mathbf{M}_{out})_j = \frac{1}{N_{out}} \sum_{l=1}^{N_{out}} k(\mathbf{y}_j, \mathbf{y}_l) \quad \text{and} \quad \mathbf{B} = \mathbf{K}_{in}(\mathbf{I} - \mathbf{1}_{N_{in}})\mathbf{K}_{in}^\top + \mathbf{K}_{out}(\mathbf{I} - \mathbf{1}_{N_{out}})\mathbf{K}_{out}^\top,$$

$\mathbf{I}$ is the identity matrix, $\mathbf{K}_{in}$ is an $N_t \times N_{in}$ Gram matrix, $\mathbf{K}_{out}$ is also an $N_t \times N_{out}$ Gram matrix, $\mathbf{1}_{N_{in}}$ and $\mathbf{1}_{N_{out}}$ are matrices with each entry equal to $1/N_{in}$ and $1/N_{out}$, respectively. Each element of $\mathbf{K}_{in}$ and $\mathbf{K}_{out}$ are defined to be $(\mathbf{K}_{in})_{mn} = k(\mathbf{x}_n, \mathbf{x}_m)$ and $(\mathbf{K}_{out})_{mn} = k(\mathbf{y}_n, \mathbf{y}_m)$, respectively. As in the solution to the analogous problem in the input space (Equation (7.20)), (7.25) can be solved simply by finding the leading eigenvector, $\boldsymbol{\alpha}_{\mathrm{KFD}}$, of $\mathbf{B}^{-1}\mathbf{A}$. Thus, the identity in Equation (7.24) becomes $\mathbf{w}_{\mathrm{FLD}_\Phi} = \mathbf{w}_\Phi = \mathbf{Z}_\Phi \boldsymbol{\alpha}_{\mathrm{KFD}}$. Substituting this result into (7.13), gives

$$\begin{aligned} \mathbf{KFD}(\mathbf{r}) &= (\Phi(\mathbf{r}) - \hat{\boldsymbol{\mu}}_{Y_\Phi})^\top \left(\mathbf{w}_{\mathrm{FLD}_\Phi} \mathbf{w}_{\mathrm{FLD}_\Phi}^\top\right)(\Phi(\mathbf{r}) - \hat{\boldsymbol{\mu}}_{Y_\Phi}) \\ &= (\Phi(\mathbf{r}) - \hat{\boldsymbol{\mu}}_{Y_\Phi})^\top \mathbf{Z}_\Phi \boldsymbol{\alpha}_{\mathrm{KFD}} \boldsymbol{\alpha}_{\mathrm{KFD}}^\top \mathbf{Z}_\Phi^\top (\Phi(\mathbf{r}) - \hat{\boldsymbol{\mu}}_{Y_\Phi}). \end{aligned} \tag{7.26}$$

To simplify (7.26), let

$$\Phi(\mathbf{r})^\top \mathbf{Z}_\Phi = \Phi(\mathbf{r})^\top \left[\Phi(\mathbf{z}_1)\ \Phi(\mathbf{z}_2)\ \ldots\ \Phi(\mathbf{z}_{N_t})\right] = \mathbf{k}\left(\mathbf{Z}, \mathbf{r}\right)^\top = \mathbf{K}_{Zr}, \tag{7.27}$$

and similarly

$$\hat{\boldsymbol{\mu}}_{Y_\Phi}^\top \mathbf{Z}_\Phi = \frac{1}{N_{out}} \sum_{\mathbf{y} \in OWR} \mathbf{k}\left(\mathbf{Z}, \mathbf{y}\right)^\top = \mathbf{K}_{Z\hat{\mu}}. \tag{7.28}$$

Using (7.27) and (7.28), we can rewrite (7.26) as

$$\mathbf{KFD}(\mathbf{r}) = \left(\mathbf{K}_{Zr}^\top - \mathbf{K}_{Z\hat{\mu}}^\top\right)^\top \boldsymbol{\alpha}_{\text{KFD}} \boldsymbol{\alpha}_{\text{KFD}}^\top \left(\mathbf{K}_{Zr}^\top - \mathbf{K}_{Z\hat{\mu}}^\top\right). \tag{7.29}$$

Equation (7.29) is the expression used for the Kernel Fisher Discriminant anomaly detector in this chapter, referred to as the KFD detector.

7.3.5 Eigenspace separation transform

The EST was developed by Torrieri (Torrieri 1999) as a preprocessing technique to extract features for neural network classifiers and has been successfully used by researchers for automatic clutter rejection (Chan *et al.* 2001). Like PCA, EST aims to extract features from a training set by projecting the input patterns onto a lower-dimensional orthogonal subspace. In this chapter, it is used to generate projection vectors in order to separate anomaly pixels from background clutter.

In the EST algorithm we first compute the $J \times J$ Difference Correlation (DCOR) matrix

$$\hat{\mathbf{R}} = \mathbf{R}_X - \mathbf{R}_Y = \frac{1}{N_{in}} \mathbf{X}\mathbf{X}^\top - \frac{1}{N_{out}} \mathbf{Y}\mathbf{Y}^\top, \tag{7.30}$$

where $\hat{\mathbf{R}}$ is simply the difference of the correlation matrices of the IWR ($\mathbf{R}_X$) and OWR ($\mathbf{R}_Y$), which represents the second order statistic differences between the two regions. The eigenvalue decomposition of DCOR can be rewritten in block-matrix form in terms of its positive and negative eigenvalues and eigenvectors as

$$\hat{\mathbf{R}} = \begin{bmatrix} \mathbf{V}_+ & \mathbf{V}_- \end{bmatrix} \begin{bmatrix} \boldsymbol{\Lambda}_+ & \mathbf{0} \\ \mathbf{0} & \boldsymbol{\Lambda}_- \end{bmatrix} \begin{bmatrix} \mathbf{V}_+^\top \\ \mathbf{V}_-^\top \end{bmatrix}, \tag{7.31}$$

where the columns of $\mathbf{V}_+$ and $\mathbf{V}_-$ are the eigenvectors with their corresponding nonzero positive ($\boldsymbol{\Lambda}_+$) and negative ($\boldsymbol{\Lambda}_-$) eigenvalues, respectively.

The matrix $\mathbf{W}_{EST}$ is then chosen to be the set of m positive or negative eigenvectors associated with $\hat{\mathbf{R}}$. The choice of which set to use hinges on which set of eigenvalues (positive or negative) has the largest absolute sum. Thus, the EST projection vectors are given as

$$\mathbf{W}_{EST} = [\mathbf{v}_1, \mathbf{v}_2, \ldots, \mathbf{v}_m], \tag{7.32}$$

where $\mathbf{v}_i$ $(i = 1, \ldots, m)$ are the m most significant (either positive or negative) eigenvectors for $\hat{\mathbf{R}}$. Using (7.32) as the projection vectors and substituting this result into (7.5) gives

$$\mathbf{EST}(\mathbf{r}) = (\mathbf{r} - \hat{\mu}_Y)^\top \left(\mathbf{W}_{EST}\mathbf{W}_{EST}^\top\right)(\mathbf{r} - \hat{\mu}_Y). \tag{7.33}$$

It is also possible to project onto the complement subspace, $\left(\mathbf{I} - \mathbf{W}_{EST}\mathbf{W}_{EST}^\top\right)$. Thus, substituting (7.32) into (7.6) yields

$$\mathbf{ES\acute{T}}(\mathbf{r}) = (\mathbf{r} - \hat{\mu}_Y)^\top \left(\mathbf{I} - \mathbf{W}_{EST}\mathbf{W}_{EST}^\top\right)(\mathbf{r} - \hat{\mu}_Y). \tag{7.34}$$

Since it is possible to use either the positive eigenvectors or the negative eigenvectors, there are four possible equations that can be used. This is outlined in the result section 7.4. In this chapter, Equations (7.33) and (7.34) are referred to as the EST-based anomaly detector or simply as the EST detector.

7.3.6 Kernel eigenspace separation transform

In Goldberg *et al.* (2007) EST is kernelized by first defining it in the feature space $\mathcal{F}$ and then reformulating it solely in terms of dot products where the kernel trick is utilized to convert it into its kernel version. The difference correlation matrix (DCOR) $\mathbf{R}_\Phi$ in the feature space can be written as

$$\begin{aligned}\mathbf{R}_\Phi = \mathbf{R}_{X_\Phi} - \mathbf{R}_{Y_\Phi} &= \frac{1}{N_{in}}\Phi(\mathbf{X})\Phi(\mathbf{X})^\top - \frac{1}{N_{out}}\Phi(\mathbf{Y})\Phi(\mathbf{Y})^\top \\ &= \begin{bmatrix}\Phi(\mathbf{x}) & -\Phi(\mathbf{Y})\end{bmatrix}\begin{bmatrix}\Phi(\mathbf{x})^\top/N_{in} \\ \Phi(\mathbf{Y})^\top/N_{out}\end{bmatrix},\end{aligned} \tag{7.35}$$

where $\mathbf{R}_{X_\Phi} = \Phi(\mathbf{X})\Phi(\mathbf{X})^\top/N_{in}$ and $\mathbf{R}_{Y_\Phi} = \Phi(\mathbf{Y})\Phi(\mathbf{Y})^\top/N_{out}$ are the correlation matrices in the feature space for IWR and OWR, respectively.

In order to diagonalize the DCOR matrix $\mathbf{R}_\Phi$ we must find all the eigenvectors $\mathbf{V}_\Phi = [\mathbf{V}_{+\Phi}\ \mathbf{V}_{-\Phi}]$ (both positive $\mathbf{V}_{+\Phi}$ and negative $\mathbf{V}_{-\Phi}$) and all the nonzero eigenvalues $\mathbf{\Lambda}_\Phi = [\mathbf{\Lambda}_{+\Phi}\ \mathbf{\Lambda}_{+\Phi}]^\top$ (both positive $\mathbf{\Lambda}_{+\Phi}$ and negative $\mathbf{\Lambda}_{-\Phi}$) that satisfy the equation

$$\mathbf{\Lambda}_\Phi \mathbf{V}_\Phi = \mathbf{R}_\Phi \mathbf{V}_\Phi. \tag{7.36}$$

Each eigenvector $\mathbf{v}_\Phi^k$ in the feature space can be written as a linear combination of the centred input data as

$$\begin{aligned}\mathbf{v}_\Phi^k &= \frac{1}{\sqrt{N_{in}}}\sum_{i=1}^{N_{in}} \alpha_i^k \Phi(\mathbf{x}_i)\lambda_i^{-\frac{1}{2}} - \frac{1}{\sqrt{N_{out}}}\sum_{j=1}^{N_{out}} \beta_j^k \Phi(\mathbf{y}_j)\lambda_j^{-\frac{1}{2}} \\ &= \frac{1}{\sqrt{N_{in}}}\Phi(\mathbf{x})\boldsymbol{\alpha}^k \mathbf{\Lambda}_{+\Phi}^{-\frac{1}{2}} - \frac{1}{\sqrt{N_{out}}}\Phi(\mathbf{Y})\boldsymbol{\beta}^k \mathbf{\Lambda}_{-\Phi}^{-\frac{1}{2}},\end{aligned} \tag{7.37}$$

where the expansion coefficients, $\boldsymbol{\alpha}^k$ and $\boldsymbol{\beta}^k$, are defined as $\boldsymbol{\alpha}^k = (\alpha_1^k, \alpha_2^k, \ldots, \alpha_{N_{in}}^k)^\top$ and $\boldsymbol{\beta}^k = (\beta_1^k, \beta_2^k, \ldots, \beta_{N_{out}}^k)^\top$ for $k = 1, \ldots, N_t$ where $N_t = N_{in} + N_{out}$. In Goldberg *et al.* (2007) it has been shown that each concatenated expansion coefficients $\mathbf{d}_k = [\boldsymbol{\alpha}^k \ \boldsymbol{\beta}^k]^\top$ for $k = 1, \ldots, N_t$ is an eigenvector of the KEST kernel matrix

$$\mathbf{K}_{KEST} = \begin{bmatrix} \frac{\mathbf{K}_{XX}}{N_{in}} & -\frac{\mathbf{K}_{XY}}{N_{in}} \\ \frac{\mathbf{K}_{YX}}{N_{out}} & -\frac{\mathbf{K}_{YY}}{N_{out}} \end{bmatrix}, \tag{7.38}$$

where $\mathbf{K}_{XX} = \Phi(\mathbf{X})^\top \Phi(\mathbf{X})$ is an $N_{in} \times N_{in}$ kernel matrix, $\mathbf{K}_{YY} = \Phi(\mathbf{Y})^\top \Phi(\mathbf{Y})$ is an $N_{out} \times N_{out}$ kernel matrix, $\mathbf{K}_{XY} = \Phi(\mathbf{X})^\top \Phi(\mathbf{Y})$ is an $N_{in} \times N_{out}$ kernel matrix, and $\mathbf{K}_{YX} = \Phi(\mathbf{Y})^\top \Phi(\mathbf{X})$ is an $N_{out} \times N_{in}$ kernel matrix. Each of the entries in all four matrices is obtained in terms of the kernel function k.

Let the KEST projection vectors, $\mathbf{W}_{KEST}$ be either the first m positive or negative eigenvectors of (7.38) normalized by the square roots of their corresponding eigenvalues. Thus, either

$$\begin{aligned} \mathbf{W}_{KEST} &= \mathbf{W}_{KEST}^{+} = \left[\tilde{\mathbf{d}}_1 \, \tilde{\mathbf{d}}_2 \, \cdots \, \tilde{\mathbf{d}}_m\right] \\ \mathbf{W}_{KEST} &= \mathbf{W}_{KEST}^{-} = \left[\tilde{\mathbf{d}}_{N_t} \, \tilde{\mathbf{d}}_{N_t-1} \, \cdots \, \tilde{\mathbf{d}}_{N_t-m+1}\right], \end{aligned} \tag{7.39}$$

where, as with KPCA, m is a configurable constant. The choice of using most positive significant or most negative significant is a data dependent choice and is determined using the same procedure outlined for the linear EST method in Section 7.3.5. Using (7.39) and (7.37) we can obtain the projection vectors, $\mathbf{V}_\Phi$, substituting this result in (7.13) yields

$$\begin{aligned} \mathbf{KEST}(\mathbf{r}) &= \left(\Phi(\mathbf{r}) - \hat{\boldsymbol{\mu}}_{Y_\Phi}\right)^\top \left(\mathbf{V}_\Phi \mathbf{V}_\Phi^\top\right) \left(\Phi(\mathbf{r}) - \hat{\boldsymbol{\mu}}_{Y_\Phi}\right) \\ &= \left(\Phi(\mathbf{r}) - \hat{\boldsymbol{\mu}}_{Y_\Phi}\right)^\top \Phi(\mathbf{Z}) \mathbf{W}_{KEST} \mathbf{W}_{KEST}^\top \Phi(\mathbf{Z})^\top \left(\Phi(\mathbf{r}) - \hat{\boldsymbol{\mu}}_{Y_\Phi}\right), \end{aligned} \tag{7.40}$$

where $\Phi(\mathbf{Z}) = [\Phi(\mathbf{X}) \ \ -\Phi(\mathbf{Y})]$. For notational convenience, let

$$\begin{aligned} \Phi(\mathbf{r})^\top \Phi(\mathbf{Z}) &= \Phi(\mathbf{r})^\top \left[\Phi(\mathbf{x}_1), \ldots, \Phi(\mathbf{x}_{N_{in}}), -\Phi(\mathbf{y}_1), \ldots, -\Phi(\mathbf{y}_{N_{out}})\right] \\ &= \left[k(\mathbf{x}_1, \mathbf{r}), \ldots, k(\mathbf{x}_{N_{in}}, \mathbf{r}), -k(\mathbf{y}_1, \mathbf{r}), \ldots, -k(\mathbf{y}_{N_{out}}, \mathbf{r})\right] \\ &= \mathbf{k}(\mathbf{Z}, \mathbf{r})^\top = \mathbf{K}_{Zr}. \end{aligned} \tag{7.41}$$

Similarly, define

$$\hat{\boldsymbol{\mu}}_{Y_\Phi}^\top \Phi(\mathbf{Z}) = \frac{1}{N_{out}} \sum_{\mathbf{y} \in OWR} \mathbf{k}(\mathbf{Z}, \mathbf{y})^\top = \mathbf{K}_{Z\hat{\mu}}. \tag{7.42}$$

Using (7.41) and (7.42), Equation (7.40) becomes

$$\mathbf{KEST}(\mathbf{r}) = \left(\mathbf{K}_{Zr}^\top - \mathbf{K}_{Z\hat{\mu}}{}^\top\right)^\top \mathbf{W}_{KEST} \mathbf{W}_{KEST}^\top \left(\mathbf{K}_{Zr}{}^\top - \mathbf{K}_{Z\hat{\mu}}{}^\top\right). \tag{7.43}$$

As in the case of linear EST in Section 7.3.5, it is also possible to obtain a kernel version projecting onto the complement subspace $(\mathbf{I} - \mathbf{W}_{KEST}\mathbf{W}_{KEST}^{\top})$ in the feature space.

7.3.7 RX algorithm

In Reed and Yu (1990) the RX algorithm was developed; this was based on a generalized likelihood ratio test (GLRT) for multi-dimensional image data assuming that the spectrum of the received signal (target) $\mathbf{s}$ and the covariance of the background clutter, $\mathbf{C}$, are unknown. In the conventional RX algorithm a non-stationary local mean is subtracted from each spectral pixel. The local mean $\hat{\mu}$ is estimated by sliding a dual window over every spectral pixel in the image and calculating the mean of the spectral pixels falling within the OWR. The size of the IWR is assumed to be the size of the typical target of interest in the image. The residual signal after mean subtraction is assumed to approximate a zero-mean pixel-to-pixel independent Gaussian random process.

Let each input spectral signal consisting of J spectral bands be denoted by $\mathbf{x}_n = (x_n(1), x_n(2), \ldots, x_n(J))^{\top}$. Define $\mathbf{X}$ to be a $J \times N$ matrix of the N reference background clutter pixels (or pixels in the OWR when the dual-window approach is used). Each observation spectral pixel is represented as a column in the sample matrix $\mathbf{X} = [\mathbf{x}_1\ \mathbf{x}_2\ \ldots\ \mathbf{x}_N]$. The model assumes that the data arises from two normal PDFs ($\mathbf{H}_0$ and $\mathbf{H}_1$) with the same covariance matrix but different means. Under $\mathbf{H}_0$ the data (background clutter) is modelled as $\mathcal{N}(0, \mathbf{C})$ and under $\mathbf{H}_1$ it is modelled as $\mathcal{N}(\mathbf{s}, \mathbf{C})$. The background covariance, $\mathbf{C}$, is estimated from the reference background clutter data given by

$$\hat{\mathbf{C}} = \frac{1}{N}\sum_{i=1}^{N}(\mathbf{x}_i - \hat{\boldsymbol{\mu}}_b)(\mathbf{x}_i - \hat{\boldsymbol{\mu}}_b)^{\top}, \tag{7.44}$$

where $\hat{\boldsymbol{\mu}}_b$ is the estimated background clutter sample mean

$$\hat{\boldsymbol{\mu}}_b = \frac{1}{N}\sum_{i=1}^{N}\mathbf{x}_i. \tag{7.45}$$

Assuming a single pixel target $\mathbf{r}$ as the observation test vector and using the GLRT the RX algorithm is formulated as

$$RX(\mathbf{r}) = (\mathbf{r} - \hat{\boldsymbol{\mu}}_b)^{\top}\mathbf{C}^{-1}(\mathbf{r} - \hat{\boldsymbol{\mu}}_b). \tag{7.46}$$

Using eigenvalue value decomposition of the background covariance matrix $\hat{\mathbf{C}} = \mathbf{V}\boldsymbol{\Lambda}\mathbf{V}^{\top}$, where the columns of $\mathbf{V}$ represent eigenvectors and the diagonal matrix Λ represents the corresponding non-zero eigenvlaues, we can rewrite the RX algorithm as

$$RX(\mathbf{r}) = (\mathbf{r} - \hat{\boldsymbol{\mu}}_b)^{\top}\mathbf{V}\boldsymbol{\Lambda}^{-1}\mathbf{V}^{\top}(\mathbf{r} - \hat{\boldsymbol{\mu}}_b) = (\mathbf{r} - \hat{\boldsymbol{\mu}}_b)^{\top}\tilde{\mathbf{V}}\tilde{\mathbf{V}}^{\top}(\mathbf{r} - \hat{\boldsymbol{\mu}}_b), \tag{7.47}$$

where $\tilde{\mathbf{V}} = \boldsymbol{\Lambda}^{-1/2}\mathbf{V}$ are the normalized eigenvectors with their corresponding eigenvalues. Therefore, the RX algorithm Equation (7.47) can be considered as a special case of the subspace-based anomaly detectors with subspace defined by the projection operator $\tilde{\mathbf{V}}\tilde{\mathbf{V}}^{\top}$. We can also use a regularized RX algorithm where only the m significant eigenvectors are used to define the subspace. Therefore, the RX algorithm is very similar to the PCA-based anomaly detector and the difference is that the basis in the regularized RX algorithm are weighted by their corresponding eigenvalues.

7.3.8 Kernel RX algorithm

Modelling the RX algorithm in the feature space by assuming that the input data has already been mapped into a high dimensional feature space via a nonlinear mapping Φ, the corresponding RX algorithm in the feature space is now represented as

$$RX(\Phi(\mathbf{r})) = (\Phi(\mathbf{r}) - \Phi(\hat{\boldsymbol{\mu}}_b))^\top \hat{\mathbf{C}}_\Phi^{-1}(\Phi(\mathbf{r}) - \Phi(\hat{\boldsymbol{\mu}}_b)), \tag{7.48}$$

where $\hat{\mathbf{C}}_\Phi$ and $\hat{\boldsymbol{\mu}}_b$ are the estimated covariance and mean of the background clutter pixels in the feature space, respectively. The estimated covariance matrix for the mapped data $\Phi(\mathbf{X}) = \mathbf{X}_\Phi := [\Phi(\mathbf{x}_1)\ \Phi(\mathbf{x}_2)\ \dots\ \Phi(\mathbf{x}_N)]$ is given by

$$\hat{\mathbf{C}}_\Phi = \frac{1}{N}\sum_{i=1}^{N}(\Phi(\mathbf{x}_i) - \Phi(\hat{\boldsymbol{\mu}}_b))(\Phi(\mathbf{x}_i) - \Phi(\hat{\boldsymbol{\mu}}_b))^\top, \tag{7.49}$$

where $\Phi(\hat{\boldsymbol{\mu}}_b)$ is the estimated background clutter sample mean given by

$$\Phi(\hat{\boldsymbol{\mu}}_b) = \frac{1}{N}\sum_{m=1}^{N}\Phi(\mathbf{x}_n). \tag{7.50}$$

The RX algorithm given by Equation (7.48) is now in the feature space that cannot be implemented explicitly due to the nonlinear mapping Φ that produces a data space of high dimensionality. In order to avoid implementing Equation (7.48) directly we need to kernelize it.

The estimated background covariance matrix can be represented by its eigenvector decomposition given by

$$\hat{\mathbf{C}}_\Phi = \mathbf{V}_\Phi \boldsymbol{\Lambda}_\Phi \mathbf{V}_\Phi{}^\top, \tag{7.51}$$

where $\boldsymbol{\Lambda}_\Phi$ is a diagonal matrix consisting of the eigenvalues and $\mathbf{V}_\Phi$ is a matrix whose columns are the eigenvectors of $\hat{\mathbf{C}}_\Phi$ in the feature space:

$$\mathbf{V}_\Phi = [\mathbf{v}_\Phi^1, \mathbf{v}_\Phi^2, \dots, \mathbf{v}_\Phi^N], \tag{7.52}$$

where N is the maximum number of the eigenvectors with nonzero eigenvalues.

The pseudoinvese of the estimated background covariance matrix can also be written in terms of its eigenvectors and eigenvalues as

$$\hat{\mathbf{C}}_\Phi^\# = \mathbf{V}_\Phi \boldsymbol{\Lambda}_\Phi^{-1} \mathbf{V}_\Phi{}^\top. \tag{7.53}$$

All the eigenvectors in the feature space span in the space of the input vectors $\mathbf{X}_\Phi$. Therefore, they can be represented as

$$\mathbf{V}_\Phi = \mathbf{\Lambda}_\Phi^{-1/2}\mathbf{X}_\Phi\mathcal{B}, \tag{7.54}$$

where $\mathcal{B} = (\beta^1, \beta^2, \ldots, \beta^N)^\top$ are the eigenvectors of the centred kernel matrix $\hat{\mathbf{K}}$ as shown in Schölkopf and Smola (2002). Substituting Equation (7.54) into (7.53) yields

$$\hat{\mathbf{C}}_\Phi^\# = \mathbf{X}_\Phi\mathcal{B}\mathbf{\Lambda}_\Phi{}^{-2}\mathcal{B}^\top\mathbf{X}_\Phi{}^\top. \tag{7.55}$$

Inserting Equation (7.55) into (7.48), it can be rewritten as

$$RX(\Phi(\mathbf{r})) = (\Phi(\mathbf{r}) - \Phi(\hat{\boldsymbol{\mu}}_b))^\top\mathbf{X}_\Phi\mathcal{B}\mathbf{\Lambda}_\Phi^{-2}\mathcal{B}^\top\mathbf{X}_\Phi{}^\top(\Phi(\mathbf{r}) - \Phi(\hat{\boldsymbol{\mu}}_b)). \tag{7.56}$$

The dot product term $\Phi(\mathbf{r})^\top\mathbf{X}_\Phi$ in the feature space can be represented in terms of the kernel function

$$\Phi(\mathbf{r})^\top\mathbf{X}_\Phi = (k(\mathbf{r}, \mathbf{x}_1)\ \ k(\mathbf{r}, \mathbf{x}_2)\ \ \ldots, k(\mathbf{r}, \mathbf{x}_N)) = \mathbf{k}(\mathbf{r}, \mathbf{X})^\top = \mathbf{k}_r^\top. \tag{7.57}$$

Similarly $\Phi(\hat{\boldsymbol{\mu}}_b)^\top\mathbf{X}_\Phi$,

$$\Phi(\hat{\boldsymbol{\mu}}_b)^\top\mathbf{X}_\Phi = (k(\hat{\boldsymbol{\mu}}_b, \mathbf{x}_1)\ \ k(\hat{\boldsymbol{\mu}}_b, \mathbf{x}_2)\ \ \ldots, k(\hat{\boldsymbol{\mu}}_b, \mathbf{x}_N)) = \mathbf{k}(\hat{\boldsymbol{\mu}}_b, \mathbf{X})^\top = \mathbf{k}_{\hat{\boldsymbol{\mu}}_b}^\top. \tag{7.58}$$

Also the inverse kernel matrix, as shown in Schölkopf and Smola (2002), can be written as

$$\hat{\mathbf{K}}^{-1} = \mathcal{B}\mathbf{\Lambda}_\Phi^{-1}\mathcal{B}^\top, \tag{7.59}$$

where we denote the centred background kernel matrix $\hat{\mathbf{K}} = (\mathbf{K} - \mathbf{1}_N\mathbf{K} - \mathbf{K}\mathbf{1}_N + \mathbf{1}_N\mathbf{K}\mathbf{1}_N)$ where $(\mathbf{1}_N)_{ij} = 1/N$ is an $N \times N$ matrix and $\mathbf{K} = \mathbf{K}(\mathbf{X}, \mathbf{X}) = (\mathbf{K})_{ij}$ a $N \times N$ kernel matrix whose entries $\mathbf{K}(\mathbf{x}_i, \mathbf{x}_j)$ are the dot products $< \Phi(\mathbf{x}_i), \Phi(\mathbf{x}_j) >$. Substituting (7.57), (7.58), and (7.59), into (7.56) and replacing $\mathbf{X}_\Phi$ with $\mathbf{X}_\Phi - \Phi(\hat{\boldsymbol{\mu}}_b)$ in (7.57), and (7.58) (due to centring) the kernelized version of the RX algorithm (Kwon and Nasrabadi 2005b) is given by

$$\begin{aligned} RX(\mathbf{k}(\mathbf{r})) &= (\hat{\mathbf{k}}(\mathbf{X}, \mathbf{r}) - \hat{\mathbf{k}}(\mathbf{X}, \hat{\mu}_b))^\top\hat{\mathbf{K}}_b^{-2}(\hat{\mathbf{k}}(\mathbf{X}, \mathbf{r}) - \hat{\mathbf{k}}(\mathbf{X}, \hat{\mu}_b)) \\ &= (\hat{\mathbf{k}}_r - \hat{\mathbf{k}}_{\hat{\boldsymbol{\mu}}_b})^\top\hat{\mathbf{K}}_b^{-2}(\hat{\mathbf{k}}_r - \hat{\mathbf{k}}_{\hat{\boldsymbol{\mu}}_b}), \end{aligned} \tag{7.60}$$

where $\hat{\mathbf{k}}_r^\top = \mathbf{k}_r^\top - \sum_{i=1}^N (\mathbf{x}_i, \mathbf{r})$ and $\hat{\mathbf{k}}_{\hat{\boldsymbol{\mu}}_b}^\top = \mathbf{k}_{\hat{\boldsymbol{\mu}}_b}^\top - \sum_{i=1}^N \mathbf{k}(\mathbf{x}_i, \hat{\mu}_b)$, respectively. The expression (7.60) can now be implemented with no knowledge of the mapping function Φ. The only requirement is a good choice for the kernel function k.

7.4 Results

In this section, the proposed subspace-based anomaly detectors are implemented using both simulated toy data as well as real hyperspectral imagery from the Hyperspectral Digital Imagery Collection Experiment (HYDICE) data set. In this chapter, we use a Gaussian Radial

Basis Function (RBF) kernel which takes the form $k(\mathbf{x}, \mathbf{y}) = \exp(-\|\mathbf{x} - \mathbf{y}\|^2/2\sigma^2)$ where $\sigma > 0$ is a critical kernel parameter representing the width of the Gaussian kernel. This parameter must be chosen so that the RBF function can fully exploit the data variations. One method of automatically selecting this parameter is to set it to a multiple of the variance of the data similarity difference (mean of the Euclidean distance between the data points). In this chapter, the value of σ was initially set to the variance of the data similarity difference for each image and the best value was determined experimentally for each algorithm using a cross-validation technique. Performance results using ROC curve analysis for each of the methods and their kernel versions are provided and compared.

7.4.1 Simulated toy data

The toy data set, shown in Figure 7.2(a), consists of two nonlinear Gaussian mixtures. Class 1 is represented by the red (*) points; Class 2 is represented by the blue (o) points. It is clear from this figure that no linear separating hyperplane can be placed that perfectly separates the two data classes. In order to implement the algorithms, Class 1 and Class 2 were defined as the two sets corresponding to the data in the IWR and OWR of a fictional dual window, respectively. Extending the problem to an anomaly or target detection setting, Class 1 represents the target data and Class 2 represents the background data. The results using each of the methods on the simulated data set are shown in Figures 7.2(b)–(i). To improve visual quality, the points in Class 2 are now yellow (o). The green lines in Figures 7.2(c)–(e) are the projection vectors

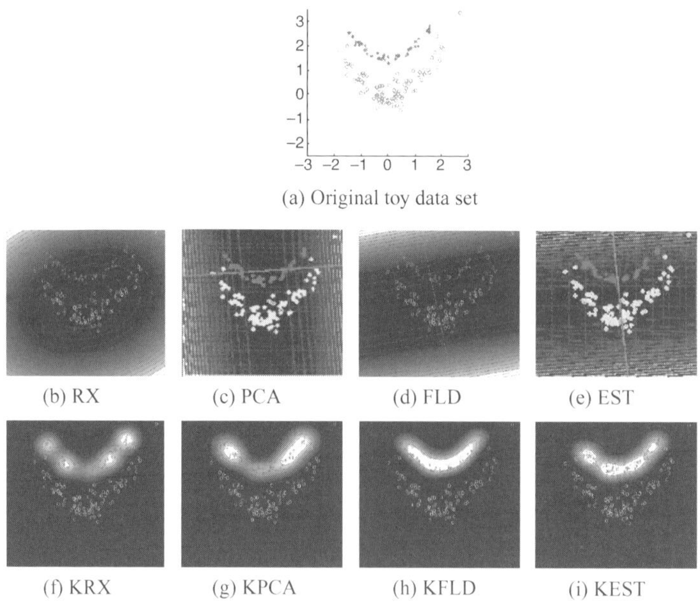

Figure 7.2 (a) Original simulated 2-D data set. A mixture of two nonlinear Gaussian distributions. The red points (*) represent the data in Class 1 and the blue (o) represent the data in Class 2. Contour and surface plots for the 2-D simulated data set using (b) RX, (c) PCA, (d) FLD, (e) EST, (f) KRX, (g) KPCA, (h) KFD and (i) KEST. (See plate 2)

used in each case. The blue contour lines are decision boundaries at different thresholds. The shading defines relative projection separation values; lighter shading means a larger projection separation value, which in turn implies a higher likelihood that points will be classified as anomalies. Similarly, darker areas correspond to points that are more likely to be classified as background clutter. It is clear that all four of the nonlinear methods have significantly better discrimination abilities than their linear counterparts. Each of the nonlinear methods generates a decision boundary which very nicely conform to the overall shape of the distribution. While it is difficult actually to compare the performances of the four nonlinear algorithms, it is nonetheless easy to see that the nonlinear methods perform better detection than do the linear methods.

7.4.2 Hyperspectral imagery

Two real hyperspectral images were used to compare the performance of all the algorithms in this chapter. The images are from the HDICE data set. The HYDICE sensor collects radiance information over a spectral range spanning the VNIR and SWIR frequency ranges (0.4–2.5μm). Each band is approximately 10 nm wide generating a spectral resolution consisting of 210 spectral bands. Owing to water absorption and low signal-to-noise ratio (SNR), only 150 of those bands are actually used here: bands 1–22, 102–108, 137–151 and 195–210 have been removed. The two HYDICE images used in this chapter are the Desert Radiance (DR-II) and Forest Radiance (FR-I) data sets. The DR-II image consists of six 'targets of interest' on a dirt road running through a dusty terrain with light vegetation. The FR-I image has fourteen 'targets of interest' in a grassy field situated near a densely wooded area. The DR-II and FR-I images are shown in Figures 7.3(a) and 7.4(a), respectively.

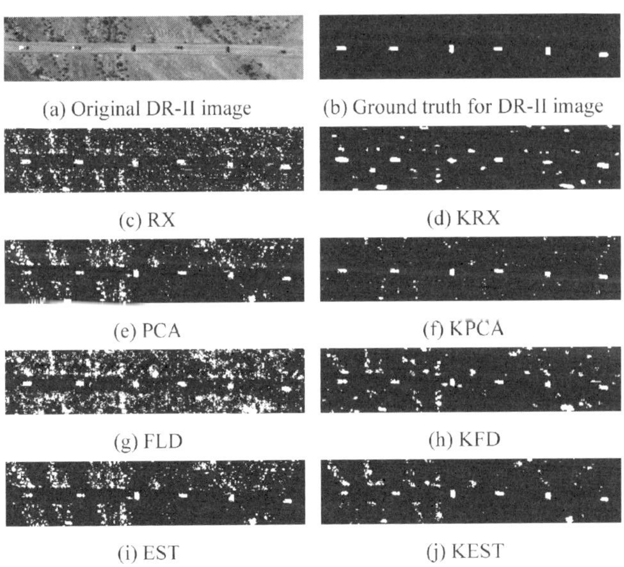

Figure 7.3 (a) Original DR-II HYDICE image. (b) Ground truth for the DR-II HYDICE image. Output results at 80% detection rate using (c) RX, (d) KRX, (e) PCA, (f) KPCA, (g) FLD, (h) KFLD (i) EST, (j) KEST.

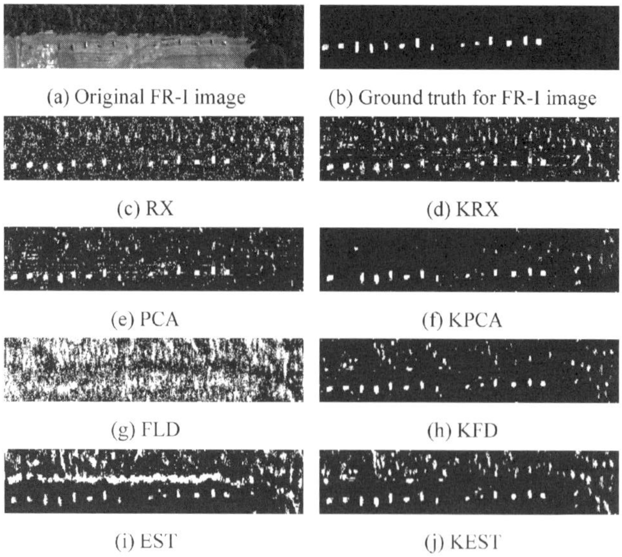

Figure 7.4 *(a) Original FR-I HYDICE image. (b) Ground truth for the FR-I HYDICE image. Output results at 80% detection rate using (c) RX, (d) KRX, (e) PCA, (f) KPCA, (g) FLD, (h) KFLD (i) EST, (j) KEST.*

Before any processing is done, all spectra in each image are normalized so that all values in the data cube lie between zero and one. The normalization factor is calculated as the largest value of all the spectral components in each hyperspectral image. This normalization helps to use the dynamic range of the RBF kernel effectively (Kwon and Nasrabadi 2005b). In all algorithms a dual-window approach was used to process input data. To provide consistency, an IWR of 7×7 pixels, a guard band of 9×9 pixels, and an OWR of 19×19 pixels were used with all algorithms and for all images. It was stated that the IWR size should be about as large as the biggest target in the image. This is more or less the case for all images. The size of the OWR was chosen such that there are a sufficient number of vectors available for calculating the background local statistics (background covariance matrix with a stable pseudo-inverse) and for obtaining meaningful empirical kernel expansions for the nonlinear techniques.

Whenever a small set of highly correlated background samples are used to represent the data statistics the kernel matrices need to be regularized (Nasrabadi 2007) in order to obtain a numerically stable pseudo-inverse. In this chapter, the regularized version of the kernel matrix is obtained by discarding all the eigenvectors with eigenvalues below a small threshold, 10^{-5}.

In order to compare the performance of each of the methods discussed in this chapter, receiver operating characteristic (ROC) curves were generated based on ground truth information obtained from each image. The ROC curves provide a visual quantitative comparison by plotting the probability of correct detection, P_D, versus the false alarm rate (FAR), R_{FA}. For each hyperspectral image, ground truth was obtained by determining the locations of all pixels in the image that correspond to a target to be detected. The probability of detection is defined as $P_D = N_{hit}/N_T$ and the false alarm rate is calculated by $R_{\mathrm{FA}} = N_{miss}/N_{TP}$ where, at each threshold T, N_{hit} is the number of pixels correctly identified as target, N_T is the total number

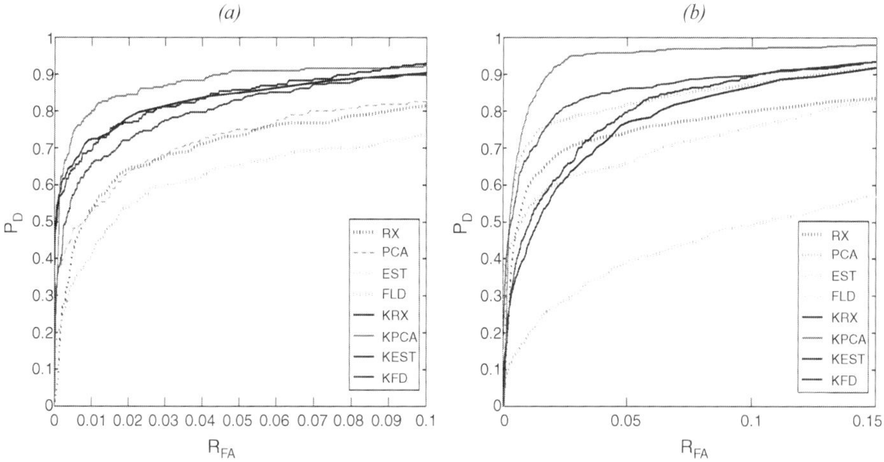

Figure 7.5 ROC curves obtained by conventional anomaly detectors and the corresponding kernel versions for (a) the DR-II image and (b) the FR-I image.

of target pixels in the ground truth for that image, N_{miss} is the number of pixels incorrectly labelled as targets, and N_{TP} is the total number of pixels in the image. For visual purposes, all outputs shown below have been binary thresholded at a value that corresponds to an 80% detection rate for that image.

The ground truth for the DR-II image is shown in Figure 7.3(b). It clearly shows the location of the six 'targets of interest'. All the anomaly detectors and their kernel versions were implemented for this image and the best outputs for each can be seen in Figures 7.3(c)–(j). For PCA-based anomaly detector, the first six eigenvectors using the OWR spectra were used in Equation (7.10). For KPCA, the first six eigenvectors using the OWR spectra were used in the complement subspace form of (7.19). For EST and KEST, the first three positive eigenvectors were used. The ROC curves for DR-II image at low FAR for each of the eight methods can be seen in Figure 7.5(a). From these results, it appears that each of the four nonlinear methods performs better than its respective linear counterpart. In addition, all four nonlinear anomaly detectors aggregately exhibit better results than all four linear detectors. At low FAR, KPCA performs the best of all methods followed by KRX, KEST and KFD. Of the linear methods, PCA, EST and RX all perform about the same, with FLD clearly performing the worst of all the anomaly detectors.

The ground truth for the FR-I HYDICE image is shown in Figure 7.4(b). It clearly shows the location of the fourteen 'targets of interest'. All eight algorithms were implemented for this image and the best outputs for each can be seen in Figures 7.4(c)–(j). The results shown are the best results obtained using the eight detectors outlined above. For PCA-based anomaly detector, the first six eigenvectors using the OWR spectra were used in (7.10) and for KPCA, the first six eigenvectors using the OWR spectra were used in the complement subspace form of (7.19). For EST and KEST, the first three negative eigenvectors were used. The ROC curves for each of the eight anomaly detectors at low FAR are shown in Figure 7.5(b). From these results, it is clear that KPCA performs the best of all the eight algorithms for this image since it detects very few background clutter regions. The results for KFD, KEST and PCA also appear to perform very well with slightly more false alarms appearing at this detection rate. At very low R_{FA}, PCA outperforms all algorithms except KPCA and KFD. KEST appears to perform much better than EST for this image as EST exhibits a large number of false alarms around the tree line region. A region similar to this one could prove to be problematic for anomaly detectors as there is an abrupt change from foliage material to a shadowed grassy region. Once again, the result using FLD is the poorest of all the detectors. The results for this image indicate that each of the nonlinear algorithms performs better than its respective linear version. However, since PCA performs well for this image, it cannot be said that all nonlinear versions as a whole perform this task better than the four linear methods.

7.5 Conclusion

This chapter provides a performance comparison of linear and nonlinear subspace-based anomaly methods for hyperspectral target detection. To generate the subspace for each of the anomaly detectors, four different algorithms (RX, PCA, FLD, EST) are proposed. In each algorithm the test pixel is projected onto its subspace by an appropriate projection operator. Each of these algorithms is then kernelized in an attempt to exploit the higher-order correlation between the spectral characteristics of the pixels. All eight anomaly detection algorithms are briefly explained and implemented using a toy example as well as two real hyperspectral data cubes containing a varying number of 'targets of interest'. It is shown that the nonlinear kernel-based anomaly detection techniques outperform their linear versions.

References

Ahlberg, J. and Renhorn, I. (2004) *Multi- and Hyperspectral Target and Anomaly Detection*. Technical report FOI-R-1526-SE, FOI -, Swedish Defence Research Agency.

Banerjee, A., Burlina, P. and Diehl, C. (2006) A support vector method for anomaly detectino in hyperspectral imagery. *IEEE Trans. Geosci. Remote Sensing*, **44**(8), 2282–2291.

Baudat, G. and Anouar, F. (2000) Generalized discriminant analysis using a kernel approach. *Neural Computation*, (12), 2385–2404.

Camps-Valls, G. and Bruzzone, L. (2005) Kernel-based methods for hyperspectral image classification. *IEEE Trans. Geosci. Remote Sensing*, **43**(6), 1351–1362.

Chan, L.A., Nasrabadi, N.M. and Torrieri, D. (2001) Eigenspace transform for automatic clutter rejection. *Optical Engineering*, **40**(4), 564–573.

Chang, C.I. and Chiang, S.S. (2002) Anomaly detection and classification for hyperspectral imagery. *IEEE Trans. Geosci. Remote Sensing*, **40**(6), 1314–1325.

Chang, C.I., Zhao, X.L., Althouse, M. and Pan, J.J. (1998) Least squares subspace projection approach to mixed pixel classification for hyperspectral images. *IEEE Trans. Geosci. Remote Sensing*, **36**(3), 898–912.

Duda, R.O., Hart, P.E. and Stork, D.G. (2001) *Pattern Classification*, 2 edn. John Wiley & Sons, Inc., New York.

Goldberg, H., Kwon, H. and Nasrabadi, N.M. (2007) Kernel eigenspace separation transform for subspace anomaly detection in hyperspectral imagery. *IEEE Trans. Geosci. Remote Sensing Letters*, **4**(4), 581–585.

Harsanyi, J.C. (1993) *Detection and Classification of Subpixel Spectral Signatures in Hyperspectral Image Sequences*. Ph.D. dissertation, Department & Electrical Engineering, University of Maryland, Baltimore County.

Harsanyi, J.C. and Chang, C.I. (1994) Hyperspectral image classification and dimensionality reduction: an orthogonal subspace projection approach. *IEEE Trans. Geosci. Remote Sensing*, **32**(4), 779–785.

Haykin, S. (1999) *Neural Networks: A Comprehensive Foundation*, Prentice-Hall, Englewood Cliffs, NJ.

Healey, G. and Slater, D. (1999) Models and methods for automated material identification in hyperspectral imagery acquired under unknown illumination and atmospheric conditions. *IEEE Trans. Geosci. Remote Sensing*, **37**(6), 2706–2717.

Joliffe, I.T. (1986) *Principal Component Analysis*. Springer-Verlag, Berlin, Germany.

Kraut, S., Scharf, L.L. and McWhorter, T. (2001) Adaptive subspace detectors. *IEEE Trans. Signal Process.* **49**(1), 208–216.

Kwon, H. and Nasrabadi, N.M. (2005a) Kernel orthogonal subspace projection for hyperspectral target classification. *IEEE Trans. Geosci. Remote Sensing*, **43**(12), 2952–2962.

Kwon, H. and Nasrabadi, N.M. (2005b) Kernel RX-algorithm: a nonlinear anomaly detector for hyperspectral imagery. *IEEE Trans. Geosci. Remote Sensing*, **43**(2), 388–397.

Kwon, H. and Nasrabadi, N.M. (2006a) Kernel adaptive subspace detector for hyperspectral imagery. *IEEE Trans. Geosci. Remote Sensing Letters*, **3**(2), 271–275.

Kwon, H. and Nasrabadi, N.M. (2006b) Kernel matched subspace detectors for hyperspectral target detection. *IEEE Trans. Pattern Anal. Machine Intell.*, **28**(2), 178–194.

Kwon, H. and Nasrabadi, N.M. (2007a) A comparative analysis of kernel subspace target detectors for hyperspectral imagery. *EURASIP Journal of Advances in Signal Processing*, **2007**, Article ID 29250.

Kwon, H. and Nasrabadi, N.M. (2007b) Kernel spectral matched filter for hyperspectral imagery. *Int. J. of Computer Vision*, **71**(2), 127–141.

Kwon, H., Der, S.Z. and Nasrabadi, N.M. (2001) An adaptive unsupervised segmentation algorithm based on iterative spectral dissimilarity measure for hyperspectral imagery. *Proc. SPIE*, vol. 4310, pp. 144–152.

Kwon, H., Der, S.Z. and Nasrabadi, N.M. (2003) Adaptive anomaly detection using subspace separation for hyperspectral images. *Optical Engineering*, **42**(11), 3342–3351.

Maeda, E. and Murase, H. (2002) Kernel-based nonlinear subspace method for pattern recognition. *Systems and Computers in Japan*, **33**(1), 38–52.

Manolakis, D. and Shaw, G. (2002) Detection algorithms for hyperspectral imaging applications. *IEEE Signal Processing Magazine*, **19**(1), 29–43.

Manolakis, D., Shaw, G. and Keshava, N. (2000) Comparative analysis of hyperspectral adaptive matched filter detector. *Proc. SPIE*, vol. 4049, pp. 2–17.

Melgani, F. and Bruzzone, L. (2004) Classification of hyperspectral remote sensing images with support vector machine. *IEEE Trans. Geosci. Remote Sensing Letters*, **42**(8), 1778–1790.

Mika, S., Rätsch, G., Weston, J., Schölkopf, B. and Müller, K.R. (2003) Constructing descriptive and discriminative nonlinear feature: Rayleigh coefficients in kernel feature space. *IEEE Trans. Pattern Anal. Machine Intell.*, **25**(5), 623–628.

Müller, K.R., Mika, S., Rätsch, G., Tsuda, K. and Schökopf, B. (2001) A introduction to kernel-based learning algorithms. *IEEE Trans. Neural Networks.*, **12**(2), 181–202.

Nasrabadi, N.M. (2007) Regularized spectral matched filter for target recognition in hyperspectral imagery. *IEEE Trans. Signal Process. Lett.*, **15**, 317–320.

Ranney, K.I. and Soumekh, M. (2006) Hyperspectral anomaly detection within the signal subspace. *IEEE Trans. Geosci. Remote Sensing Letters*, **3**(3), 312–316.

Rätsch, G., Onoda, T. and Müller, K.R. (2004) Soft margins for ADABOOST. *Machine Learning*, **42**(3), 287–320.

Reed, I.S. and Yu, X. (1990) Adaptive multiple-band CFAR detection of an optical pattern with unknown spectral distribution. *IEEE Trans. Acoustics, Speech and Signal Process.*, **38**(10), 1760–1770.

Robey, F.C., Fuhrmann, D.R. and Kelly, E.J. (1992) A CFAR adaptive matched filter detector. *IEEE Trans. on Aerospace and Elect. Syst.*, **28**(1), 208–216.

Roth, V. (2006) Kernel fisher discriminants for outlier detection. *Neural Computation*, **18**, 942–960.

Scharf, L.L. and Friedlander, B. (1994) Matched subspace detectors. *IEEE Trans. Signal Process.*, **42**(8), 2146–2157.

Schölkopf, B. and Smola, A.J. (2002) *Learning with Kernels*. The MIT Press.

Schölkopf, B., Smola, A.J. and Müller, K.R. (1999) Kernel principal component analysis. *Neural Computation*, (10), 1299–1319.

Schweizer, S.M. and Moura, J.M.F. (2001) Efficient detection in hyperspectral imagery. *IEEE Trans. Image Process.*, **10**(4), 584–597.

Shawe-Taylor, J. and Cristianini, N. (2004) *Kernel Methods for Pattern Analysis*. Cambridge University Press, Cambridge, UK.

Stein, D.W.J. (2001) Stochastic compositional models applied to subpixel analysis of hyperspectral imagery. *Proc. SPIE*, vol. 4480, pp. 49–56.

Stein, D.W.J., Beaven, S.G., Hoff, L.E., Winter, E. M , Schaum, A.P. and Stocker, A.D. (2002) Anomaly detection from hyperspectral imagery. *IEEE Signal Processing Mag.*, **19**(1), 58–69.

Thai, B. and Healey, G. (2002) Invariant subpixel material detection in hyperspectral imagery. *IEEE Trans. Geosci. Remote Sensing*, **40**(3), 599–608.

Torrieri, D. (1999) The eigenspace transform for neural network classifiers. *Neural Networks*, **12**(3), 419–427.

Vapnik, V.N. (1998) *Statistical Learning Theory*. John Wiley & Sons, New York.

Vapnik, V.N. (1999) *The Nature of Statistical Learning Theory*. Springer.

Willis, C.J. (2005) Comparison of anomaly detection methods for hyperspectral. In G.W. Kamernman and D.V. Willetts (eds), *Proc. SPIE, Electro-Optical Remote Sensing*, Vol. 5988, pp. B1-B12, July, pp. B1–B12. *Proc. SPIE*.

Yu, X. and Reed, I.S. (1993) Comparative performance analysis of adaptive multi-spectral detectors. *IEEE Trans. Signal Process.*, **41**(8), 2639–2656.

8

One-class SVMs for hyperspectral anomaly detection

Amit Banerjee, Philippe Burlina and Chris Diehl

Applied Physics Laboratory, The Johns Hopkins University, USA

A support vector framework for hyperspectral anomaly detection is developed in this chapter. Conventional methods for detecting anomalies in hyperspectral images are based on the popular Reed–Xiaoli (RX) detector. However, these algorithms typically suffer from a large numbers of false alarms, due to the assumptions that the background is Gaussian and homogeneous. In practice, these assumptions are often violated, especially when the neighbourhood of a pixel contains multiple types of terrain.

To remove these assumptions, a novel anomaly detector is proposed and derived that incorporates a nonparametric background model based on the Support Vector Data Description (SVDD). The SVDD is a one-class support vector classifier that can model the support of a distribution. Expanding on prior work, a geometric interpretation of the SVDD is developed to propose a decision rule that utilizes a new test statistic and shares some of the properties of constant false-alarm rate (CFAR) detectors. Two versions of the algorithm are presented to detect either local or global anomalies and an analysis of their computational requirements is provided. Using receiver operating characteristic (ROC) curves, the improved performance and reduction in the false alarm rate when using the SVDD-based detector are demonstrated on Wide Area Airborne Mine Detection (WAAMD) and Hyperspectral Digital Imagery Collection Experiment (HYDICE) hyperspectral imagery.

8.1 Introduction

Recent advances in hyperspectral sensors with high spectral and spatial resolution have led to increased interest in exploiting spectral imagery for target detection. Given the availability of

Kernel Methods for Remote Sensing Data Analysis Edited by Gustavo Camps-Valls and Lorenzo Bruzzone

spectral libraries for a wide range of materials, detection algorithms that exploit a known target signature have been widely investigated. It has been shown (Scharf and Friedlander 1994) that such algorithms are dependent on the degree of signal mismatch between the spectral libraries and the spectra observed in an image. Complications arising from (a) accurate spectral calibration, (b) compensating for atmospheric effects to convert radiance spectra to the reflectance domain, or (c) the availability of reliable atmospheric data to convert reflectance values to the radiance spectra can lead to errors that hinder the performance of known signature detectors.

Such complications can be avoided by using an anomaly detector. Detecting anomalies in hyperspectral imagery entails the task of locating pixels with spectral signatures that deviate significantly from the local background. Anomaly detectors have the advantage of not requiring *a priori* knowledge of the target's spectral signature, and therefore can process images completely in-scene. However, anomaly detectors usually suffer from a high false-alarm rate due to simplifying assumptions imposed on the background signature distribution. This chapter presents a new kernel-based approach that removes these assumptions in order to detect spectral anomalies while mitigating the false alarm rate.

Lacking prior target signatures, anomaly detection methods model the background and find pixels in the scene that are not well described by the background model. This is typically achieved by identifying the region of the given feature space that contains most of the background pixels. If the pixel under test falls in this region, it is labelled as part of the background; if it lies outside of the background's region of support, it is detected as an outlier and declared a target.

In general, there are two types of methods for computing the region of support for the background. One is to estimate the underlying probability density function (PDF) for the background signature and threshold the result. The form of the PDF resolves the shape and the threshold value determines the size of the support region. The most commonly used models for the density function $f(\mathbf{x})$ are the local Gaussian model, the global Gaussian mixture model, and the global linear mixture model. The second approach is to estimate the region of support directly without making assumptions about the distribution.

The RX algorithm, which is the benchmark anomaly detector for hyperspectral imagery, uses the local Gaussian model (Lu *et al.* 1997). With this approach, the background pixels in a local neighbourhood around the pixel under test are assumed to be independent, identically distributed Gaussian random variables. After estimating the background mean vector and covariance matrix, the Mahalanobis distance between the pixel under test and the background mean vector is compared with a threshold to detect an anomaly (Lu *et al.* 1997).

There are two issues with the RX algorithm that limit its performance. In many environments, it has been shown empirically that the local normal model provides an inadequate representation of the underlying distribution (Stein *et al.* 2002) leading to poor false alarm performance. This is especially true when the local background contains multiple classes of terrain. Using a goodness-of-fit test statistic for hyperspectral imagery based upon the Barringhaus, Henze, Epps and Pully (BHEP) test, it has been shown that the normal model is not valid in most situations (Henze and Wagner 1997). In addition, the RX algorithm is computationally intensive when operating on hyperspectral imagery. This is due to the need to estimate and invert large covariance matrices.

If the local background contains multiple types of terrain, the background cannot be properly modelled by a unimodal distribution. To more properly characterize non-homogeneous backgrounds, researchers have employed the mixture of Gaussians model (Beaven *et al.* 2000).

This approach models the background signature distribution as a linear combination of Gaussian distributions. The number of distributions, their weights, and the parameters of the normal PDFs are estimated using a stochastic Expectation Maximization (SEM) method (Stocker 1999). Given the parameters of the mixture of Gaussian model, a generalized likelihood ratio test (GLRT) is applied to detect outliers. While this Gaussian mixture model provides improved performance over the RX algorithm, it still requires the estimation and inversion of large covariance matrices and is further limited by the need to know or estimate *a priori* the number of classes of terrain in the image.

In Schweizer and Moura (2001), the use of Gauss–Markov random fields (GMRFs) is introduced for hyperspectral anomaly detection. The spectral and spatial correlations of background clutter in hyperspectral imagery are described by a GMRF. The estimated parameters of the GMRF are then used in a GLRT to detect outliers. The principal advantages of this method are that it is computationally efficient, even for high dimensional data, and uses spatial information that is usually ignored by most detectors. However, the model assumes that the background is locally homogeneous, which can lead to poor performance when targets are located along clutter boundaries.

To perform anomaly detection, the ultimate goal is to estimate the shape and size of the support region. Since the values of the PDF are ultimately not of interest, estimating the density function across the entire feature space is unnecessary. In keeping with Vapnik's (Vapnik 1998) principle of avoiding more difficult estimation problems along the path to the desired solution, one can estimate the region of support directly for the background and avoid the problem of estimating the underlying PDF.

Large margin techniques such as support vector machines (SVMs) have received considerable attention for classification tasks with high-dimensional, non-Gaussian data. SVMs yield good generalization performance on such problems by directly estimating a decision boundary with maximal separability. Motivated by the success of SVMs in classifying pixels in hyperspectral imagery (Gualtieri *et al.* 1999; Gualtieri and Cromp 1998) this work seeks to extend the SVM approach to detect spectral anomalies. For the particular problem of anomaly detection, several one-class classifiers have been derived in the literature, including the support vector data description (SVDD) (Tax *et al.* 1999; Tax and Duin 1999) and the ν-SVM (Scholkopf *et al.* 1999) methods. These classifiers are able to estimate directly the support region for a given dataset. The SVDD is a technique that has been used in several domains such as faulty machine part detection (Tax *et al.* 1999) and image retrieval (Lai *et al.* 2004). In the following, the SVDD is utilized to detect spectral anomalies that lie outside the region of support for the background pixels.

For anomaly detection, the SVDD approach has the following benefits: *Non-parametric*: it is a data-driven method that avoids prior assumptions about the distribution of the data; *Sparsity*: fewer training samples are needed, and thus fewer pixels are needed to accurately characterize the background; *Good generalization*: the method avoids overfitting and yields good generalization results when compared with other classical methods (Tax *et al.* 1999; Vapnik 1998); *Use of kernels*: by exploiting the 'kernel trick', the SVDD method is able to model accurately the support of non-trivial, multi-modal distributions. Therefore, the SVDD-based anomaly detector yields a more powerful detection performance for targets embedded in non-Gaussian and non-homogeneous backgrounds.

The recently introduced kernel RX algorithm (Kwon and Nasrabadi 2006) is a related technique for anomaly detection. It is a nonlinear version of the RX detector that uses kernels to model non-Gaussian distributions. The kernel RX detector has two key differences from the SVDD approach. While the SVDD avoids the problem of density estimation, the kernel RX assumes a Gaussian distribution in the kernel space. This assumption is the equivalent to a Parzen estimate for the distribution in the original input space (Kwon and Nasrabadi 2006). The second difference is that the kernel RX requires the estimation and inversion of a large covariance matrix, a computational burden that the SVDD avoids.

This chapter is organized as follows: Section 8.2 provides an overview of the derivation of the SVDD one-class classifier. A discussion on optimizing the SVDD function is given in Section 8.3. The algorithm to detect hyperspectral anomalies using the SVDD is presented in Section 8.4, along with a discussion of a new SVDD test statistic. Experimental results are provided in Section 8.5 to evaluate the RX and SVDD anomaly detectors. Finally, concluding remarks are given in Section 8.6.

8.2 Deriving the SVDD

Following Tax and Duin (1999), this section describes the derivation of the SVDD. The linear SVDD that models the support of the distribution as the minimal enclosing hypersphere containing the data in the original input space is introduced. Then it is generalized through the use of kernel functions to the nonlinear SVDD, which first maps the input space to a high-dimensional feature space and then estimates the minimum enclosing hypersphere in the feature space.

Before proceeding, it should be noted that the one-class classifiers mentioned above, the SVDD and the ν-SVM, are related. While the SVDD computes an optimal hypersphere that contains the data, the ν-SVM estimates a large margin hyperplane that separates the data and the origin of the space in which the data resides (Scholkopf *et al.* 2000). It has been shown (Scholkopf *et al.* 1999, 2000) that these methods are equivalent when the radial basis function (RBF) is chosen as the kernel function, as is common practice.

8.2.1 The linear SVDD

To determine the minimum enclosing hypersphere $S = \mathbf{x} : \| \mathbf{x} - \mathbf{a} \|^2 \leq R^2$ that contains the training set $T = \{\mathbf{x}_i | i = 1, \ldots, M\}$, the following constrained optimization problem must be solved:

$$\min(R) \text{ subject to } \mathbf{x}_i \in S, i = 1, \ldots, M. \tag{8.1}$$

The centre $\mathbf{a}$ and radius R of the minimum enclosing hypersphere can be found by optimizing the following Lagrangian:

$$L(R, \mathbf{a}, \alpha) = R^2 - \sum_i \alpha_i [R^2 - \langle \mathbf{x}_i, \mathbf{x}_i \rangle - 2\langle \mathbf{a}, \mathbf{x}_i \rangle + \langle \mathbf{a}, \mathbf{a} \rangle]. \tag{8.2}$$

The first term in (8.2) is the radius that is to be minimized. The second term constrains each training point $\mathbf{x}_i$ to lie within the sphere with centre $\mathbf{a}$ and radius R. The optimal solution must satisfy the Karush–Kuhn–Tucker (KKT) conditions. Taking the partial derivatives of L with respect to R and $\mathbf{a}$ and setting them to 0 yields:

$$\frac{\partial L}{\partial R} = 0 \Rightarrow \sum_i \alpha_i = 1 \tag{8.3}$$

$$\frac{\partial L}{\partial \mathbf{a}} = 0 \Rightarrow \mathbf{a} = \frac{\sum_i \alpha_i \mathbf{x}_i}{\sum_i \alpha_i}. \tag{8.4}$$

Combining Equations (8.3) and (8.4), a simple expression for the centre of the sphere is found:

$$\mathbf{a} = \sum_i \alpha_i \mathbf{x}_i. \tag{8.5}$$

The centre of the sphere is the centre of mass of all training points, with the weights equal to the Lagrange multipliers. Furthermore, the inequality $\| \mathbf{x}_i - \mathbf{a} \|^2 \leq R^2$ implies $\alpha_i \geq 0$, according to the non-negativity constraint of the KKT. Substituting Equations (8.3) and (8.5) into (8.2) yields another expression for the Lagrangian function to be maximized with respect to the α_i:

$$L = \sum_i \alpha_i \langle \mathbf{x}_i, \mathbf{x}_i \rangle - \sum_{i,j} \alpha_i \alpha_j \langle \mathbf{x}_i, \mathbf{x}_j \rangle. \tag{8.6}$$

After optimizing L with respect to α_i, it is typical to discover that a large fraction of the α_i is equal to zero. The training examples with nonzero α_i are called support vectors and lie on the boundary of the support region. This is a result of the complementary slackness condition of the KKT that states that $(\| \mathbf{x}_i - \mathbf{a} \|^2)\alpha_i = 0$ for optimality. Hence, the support vectors must lie on the boundary of the hypersphere. Therefore the SVDD yields a sparse representation of the support expressed entirely in terms of the support vectors.

Once the minimum enclosing hypersphere has been found, outliers are identified by testing whether or not the test examples lie within or outside the hypersphere. When the decision rule $\| \mathbf{y} - \mathbf{a} \|^2 > R^2$ holds true for a test example $\mathbf{y}$, the example is labelled as an outlier. Expanding

$$\| \mathbf{y} - \mathbf{a} \|^2 = (\mathbf{y} - \mathbf{a})^\top (\mathbf{y} - \mathbf{a}) \tag{8.7}$$

$$= \left(\mathbf{y} - \sum \alpha_i \mathbf{x}_i)^\top (\mathbf{y} - \sum \alpha_i \mathbf{x}_i\right) \tag{8.8}$$

$$= \langle \mathbf{y}, \mathbf{y} \rangle - 2 \sum_i \alpha_i \langle \mathbf{y}, \mathbf{x}_i \rangle + \sum_{i,j} \alpha_i \alpha_j \langle \mathbf{x}_i, \mathbf{x}_j \rangle, \tag{8.9}$$

the decision rule becomes:

$$\langle \mathbf{y}, \mathbf{y} \rangle - 2 \sum_i \alpha_i \langle \mathbf{y}, \mathbf{x}_i \rangle + \sum_{i,j} \alpha_i \alpha_j \langle \mathbf{x}_i, \mathbf{x}_j \rangle > R^2, \tag{8.10}$$

which is referred to as the linear SVDD decision rule.

8.2.2 The kernel-based SVDD

In most instances, a hypersphere does not provide a tight representation of the support of the data in the original input space. Additional flexibility is needed to model arbitrarily complex distributions. To address this, the nonlinear SVDD maps the data from the input space to a higher-dimensional feature space through the use of a mapping and models the support of the distribution as a minimum enclosing hypersphere in the feature space. This hypersphere corresponds to a tighter boundary for the support region in the original input space.

Proceeding as in the previous section, the algorithm seeks the smallest hypersphere in the induced feature space $S = \boldsymbol{\phi}(\mathbf{x}) : \| \boldsymbol{\phi}(\mathbf{x}) - \mathbf{c} \|^2 \leq R^2$ including the entire set of mapped training examples $T = \boldsymbol{\phi}(\mathbf{x}_i), i = 1, \ldots, M$. Therefore, the following constrained optimization problem needs to be solved:

$$\min(R) \text{ subject to } \boldsymbol{\phi}(\mathbf{x}_i) \in S, i = 1, \ldots, M. \tag{8.11}$$

The corresponding Lagrangian is expressed as:

$$L(R, \mathbf{a}, \alpha) = R^2 - \sum_i \alpha_i \left(R^2 - \langle \boldsymbol{\phi}(\mathbf{x}_i), \boldsymbol{\phi}(\mathbf{x}_i) \rangle - 2\langle \mathbf{c}, \boldsymbol{\phi}(\mathbf{x}_i) \rangle + \langle \mathbf{c}, \mathbf{c} \rangle \right) \tag{8.12}$$

with Lagrange multipliers α_i. Setting the partial derivatives of L with respect to R and $\mathbf{a}$ to zero and substituting the results into L yields:

$$L = \sum_i \alpha_i \langle \boldsymbol{\phi}(\mathbf{x}_i), \boldsymbol{\phi}(\mathbf{x}_i) \rangle - \sum_{i,j} \alpha_i \alpha_j \langle \boldsymbol{\phi}(\mathbf{x}_i), \boldsymbol{\phi}(\mathbf{x}_j) \rangle \tag{8.13}$$

with $\alpha_i \geq 0, \sum_i \alpha_i = 1$. This also gives an expression for the centre of the sphere:

$$\mathbf{c} = \sum_i \alpha_i \boldsymbol{\phi}(\mathbf{x}_i). \tag{8.14}$$

As above, the hypersphere centre is the centre of gravity of the support vectors given the optimal weights α_i. The corresponding decision rule to detect an anomaly for a test pixel $\mathbf{y}$ is $\text{SVDD}(\mathbf{y}) = \| \boldsymbol{\phi}(\mathbf{y}) - \mathbf{c} \|^2 > R^2$, which expands to

$$\begin{aligned}
\text{SVDD}(\mathbf{y}) &= \| \boldsymbol{\phi}(\mathbf{y}) - \mathbf{c} \|^2 \\
&= \| \boldsymbol{\phi}(\mathbf{y}) - \sum_{i=1}^{N} \alpha_i \boldsymbol{\phi}(\mathbf{x}_i) \|^2 \\
&= \langle \boldsymbol{\phi}(\mathbf{y}), \boldsymbol{\phi}(\mathbf{y}) \rangle - 2 \sum_i \alpha_i \langle \boldsymbol{\phi}(\mathbf{y}), \boldsymbol{\phi}(\mathbf{x}_i) \rangle + \sum_{i,j} \alpha_i \alpha_j \langle \boldsymbol{\phi}(\mathbf{x}_i), \boldsymbol{\phi}(\mathbf{x}_j) \rangle.
\end{aligned}$$

Note that the above expression consists of inner products of the mapping function $\boldsymbol{\phi}$. Using the well-known 'kernel trick', these inner products can be represented by a kernel function $K(\mathbf{x}, \mathbf{y})$. Kernel functions provide a computationally efficient technique to map implicitly the data into the induced feature space and compute the inner product. The only requirement imposed on the kernel function is that it satisfy Mercer's theorem (Vapnik 1998). Mercer's theorem states that in order for a kernel function $K(\mathbf{x}, \mathbf{y}) : \mathbb{R}^N \times \mathbb{R}^N \rightarrow \mathbb{R}, (\mathbf{x}, \mathbf{y}) \mapsto z = K(\mathbf{x}, \mathbf{y})$ to admit an eigenfunction expansion of the form

$$K(\mathbf{x}, \mathbf{y}) = \langle \boldsymbol{\phi}(\mathbf{x}_i), \boldsymbol{\phi}(\mathbf{y}_i) \rangle = \sum_{k=1}^{\infty} b_k \boldsymbol{\phi}(\mathbf{x}) \boldsymbol{\phi}(\mathbf{y}), \tag{8.15}$$

it must be positive and semi-definite:

$$\int K(x, y) g(x) g(y) dx dy \geq 0 \quad \forall g(\cdot) \in L_2. \tag{8.16}$$

Under these conditions, the SVDD statistic can be simply expressed as:

$$\mathrm{SVDD}(\mathbf{y}) = K(\mathbf{y}, \mathbf{y}) - 2 \sum_i \alpha_i K(\mathbf{y}, \mathbf{x}_i) + \sum_{i,j} \alpha_i \alpha_j K(\mathbf{x}_i, \mathbf{x}_j) > R^2. \tag{8.17}$$

Such continuous symmetric positive semi-definite kernel functions play the role of a dot product, thereby forming a new Hilbert space where the problem of determining the minimum enclosing hypersphere can be posed in a sensible fashion.

This work employs the popular Gaussian radial basis function (RBF) as the kernel function, defined as:

$$K(\mathbf{x}, \mathbf{y}) = \exp\left(\frac{- \parallel \mathbf{x} - \mathbf{y} \parallel^2}{\sigma^2} \right). \tag{8.18}$$

The choice of the RBF simplifies the SVDD(y) test statistic to:

$$\mathrm{SVDD}(\mathbf{y}) = 1 - 2 \sum_i \alpha_i K(\mathbf{y}, \mathbf{x}_i) + \sum_{i,j} \alpha_i \alpha_j K(\mathbf{x}_i, \mathbf{x}_j) > R^2, \tag{8.19}$$

since $K(\mathbf{y}, \mathbf{y}) = 1$. The RBF has one free parameter, which is the scale parameter σ. This parameter affects the tightness-of-fit for the training data (and therefore boundary smoothness), as shown in Figure 8.1 (Tax and Duin 1999). By varying the scale parameter of the RBF, the

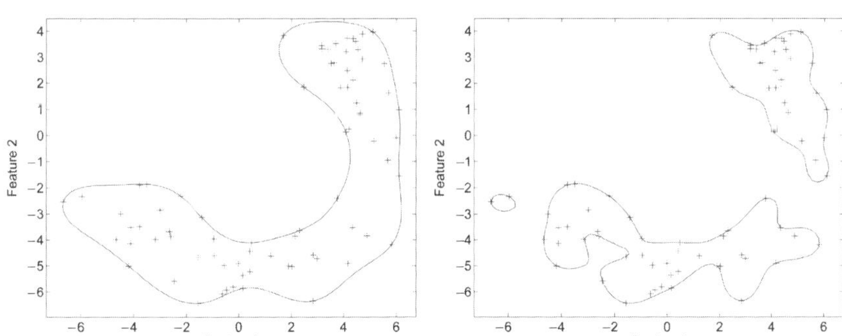

Figure 8.1 SVDD decision boundary using σ = 5 (left) and σ = 3 (right). Note how using a smaller value for sigma leads to a tighter decision boundary. (Modified from Banerjee, A., Burlina, P. and Diehl, C.P. (2006) A support vector method for anomaly detection in hyperspectral imagery, IEEE Trans. on Geoscience and Remote Sensing, 44(8), 2282–2291)

SVDD can determine multiple regions of support for a dataset. As can be seen Figure 8.1, this allows the SVDD to model multi-modal distributions. For anomaly detection, this implies that the SVDD can detect targets embedded in multiple types of clutter with fewer false alarms.

8.3 SVDD function optimization

Finding the SVDD decision boundary requires the following optimization:

$$\min_{\sigma} \frac{1}{2} \sum_{i,j} \alpha_i \alpha_j K(\mathbf{x}_i, \mathbf{x}_j) \text{ subject to } 0 \leq \alpha_i \leq 1, \sum \alpha_i = 1. \tag{8.20}$$

Traditional support vector algorithms use standard quadratic programming (QP) techniques such as active set and conjugate gradient methods to minimize the function. These methods have a complexity of $\mathcal{O}(N^3)$, where N is the number of training samples. However, given a large number of training samples, standard QP techniques cannot solve the problem efficiently. They involve a matrix whose size scales quadratically with the number of active constraints (non-zero α_is) and therefore require a lot of memory and a large number of iterations to arrive at a solution. For hyperspectral anomaly detection, this is a critical issue since many training pixels are needed to characterize the statistics of high-dimensional spectra.

To address this issue, a 'chunking' method was introduced in Vapnik (1982) to solve the dense SVM QP problem. The chunking algorithm uses the fact that the value of the quadratic function is the same if you remove the rows and columns of the matrix that correspond to zero Lagrange multipliers. Therefore, the large QP problem can be broken down into a series of smaller QP problems, whose ultimate goal is to identify and remove all of the nonzero Lagrange multipliers. Chunking reduces the size of the matrix from the number of training examples squared to approximately the number of nonzero Lagrange multipliers squared. However, chunking still cannot handle large-scale training problems, since even this reduced matrix cannot fit into memory.

In Osuna *et al.* (1997), a new class of QP algorithms was proposed by Osuna *et al*. for SVMs. They demonstrate that the large QP problem can be broken down into a series of smaller QP sub-problems. Sequential Minimal Optimization (SMO) is an example of such a fast method to find the solution for a very large QP optimization problem. SMO breaks this large QP problem into a series of QP problems involving only two examples. These small QP problems are solved analytically, which avoids using a time-consuming numerical QP optimization as an inner loop. The amount of memory required for SMO is linear in N, which allows SMO to handle very large training sets.

The results presented in Section 8.5 utilizes the SVDD implementation found in the NTU SVM software package (Chang and Lin 2001). This implementation uses the SMO minimization method to find the optimal hypersphere. To illustrate the increase in speed when using the SMO method, the SVDD algorithm is run using both the standard QP and SMO methods on the same data with the same value for sigma while varying the number of training samples. Plots of the run-times as a function of the number of samples are given for both methods in Figure 8.2. The plots show that the runtime for the standard QP increases exponentially with the number of training sample, while SMO increases linearly. Furthermore, SMO is an order-of-magnitude faster. This is a key benefit to using SMO for local hyperspectral anomaly detection, since the optimal SVDD decision boundary is computed for every pixel in the image.

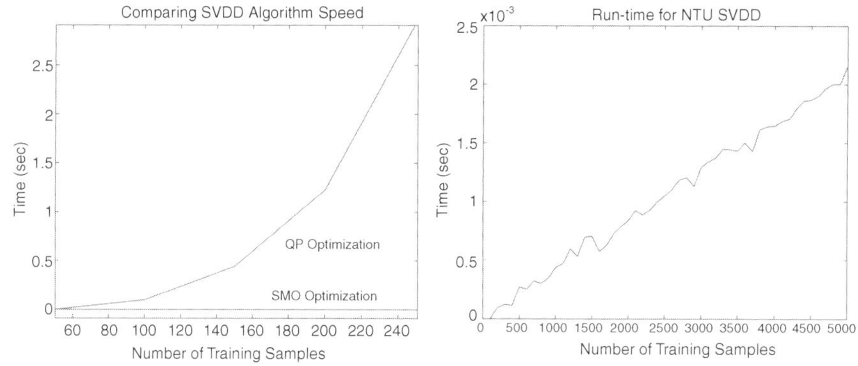

Figure 8.2 Comparing the speed of QP and SMO optimization methods. (Left) The runtime of the standard QP solver increases exponentially with the number of training samples, while SMO increases linearly. (Right) SMO runtime increases linearly with the number of samples.

Finally, it should be noted that incremental learning/update techniques found in the machine learning literature can be used to further improve the speed of local anomaly detectors (Syed *et al.* 1999). The local background of neighbouring pixels greatly overlap and share a large percentage of the same training data. Instead of retraining the SVDD at each pixel, incremental learning methods can efficiently update the SVDD model by considering only the new background spectra in the local neighbourhood.

8.4 SVDD algorithms for hyperspectral anomaly detection

This section presents local and global anomaly detection algorithms based upon the SVDD. A local anomaly detector finds pixels in the image that have different spectra from the clutter background in the local neighbourhood. A global anomaly detector finds pixels whose spectra are different from the other spectra in the entire image. The RX algorithm can also be used either as a local or global anomaly detector, depending on how the background samples are collected.

Both approaches have complementary advantages and disadvantages. Local anomaly detectors can use simpler models to describe the local background. However, they are computationally intensive and are susceptible to false alarms that are isolated spectral anomalies. For global background modelling, the background is estimated once for the entire hypercube. Since the global background contains many materials, the modelling required is much more complex than that required locally. They are often faster than the local anomaly detectors since the background clutter model is not computed for each pixel. Furthermore, global methods also provide the capability for enhanced ROC performance due to: (a) more accurate background characterization in the presence of closely spaced targets that is corrupted when a local model is used, and (b) the rejection of local anomalies that are not truly anomalous in the scene (e.g. isolated trees and shadows).

8.4.1 Outline of algorithms

The multi-modal background characterization of SVDD is virtually automatic and does not require any a priori knowledge of background characteristics (e.g. the number of components to be used in a mixture model). As a result, SVDD is well suited for global background characterization and is able to achieve high detection performance and low false alarm rates with extremely rapid computational processing.

The steps for the global SVDD anomaly detector are as follows.

1. Randomly select a set of N background pixels from the image as the training set.
2. Given the set of background spectra, estimate an optimal value for σ, the scale parameter of the RBF kernel.
3. Using the spectra from the training pixels, estimate the SVDD parameters $(\mathbf{a}, \alpha_i, R)$ to model the region of support for the background clutter.
4. For each pixel in the image perform the decision test:
    - If SVDD($\mathbf{y}$), the SVDD test statistic for pixel $\mathbf{y}$, is less than the detection threshold t, the pixel is part of the background.
    - Else, declare the pixel as an anomaly.

Note that since this is a global method, only the last step needs to be performed for each pixel, as the training is done only once for the entire image. To reduce the computational expense of the algorithm further, a closer look at the SVDD($\mathbf{y}$) (8.19) yields:

$$\mathrm{SVDD}(\mathbf{y}) = \kappa - c(\mathbf{y}) \sum_i \alpha_i C_i \exp\left(\frac{-2\mathbf{y}^\top \mathbf{x}}{\sigma^2}\right), \tag{8.21}$$

where $\kappa = 1 + \sum_{i,j} \alpha_i \alpha_j K(\mathbf{x}_i, \mathbf{x}_j)$, $c(\mathbf{y}) = \exp(-\mathbf{y}^\top \mathbf{y}/\sigma^2)$ and $C_i = \exp(-\mathbf{x}_i^\top \mathbf{x}_i/\sigma^2)$ may be pre-computed at the beginning after training when the support vectors $\mathbf{x}_i$ have been found. Therefore, only the inner-product in the exponential term on the right-hand side of the above equation needs to be computed at every pixel.

The local SVDD anomaly detector is similar to the RX algorithm. The SVDD statistic is computed at each pixel to determine if the pixel lies within the support region of the background pixels. This approach replaces the Mahalanobis distance used in the RX detector with the SVDD($\mathbf{y}$) measure. The steps for the local SVDD based anomaly detector are as follows.

1. Select the dimensions of the hollow background window.
2. Estimate σ, the scale parameter of the RBF kernel.
3. Given the window size and scale parameter, for each pixel:
    - Sample pixels from the local neighbourhood using the background window.
    - Using these pixels, compute the SVDD parameters $(\mathbf{a}, \alpha_i, R)$ to model the region of support for the background clutter.
    - Decision:
        - If SVDD($\mathbf{y}$), the SVDD test statistic for pixel $\mathbf{y}$ is less than the detection threshold T, the pixel is part of the background.
        - Else, declare the pixel as a target.

Details for steps 1 and 2 of the algorithm are given below.

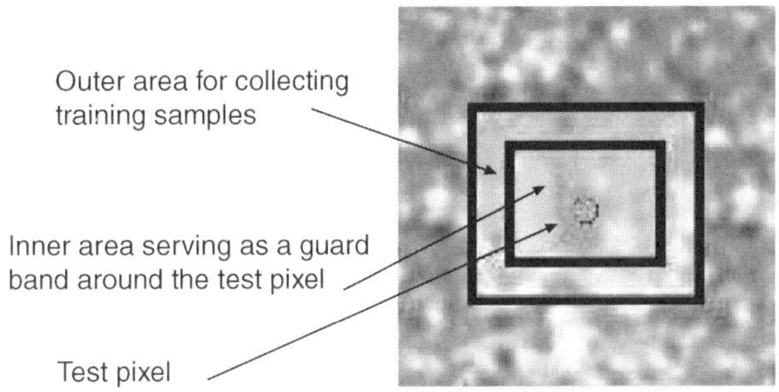

Figure 8.3 Example of a hollow window used to collect pixels for background estimation.

8.4.2 Dimensions for the background window

Local background samples around the pixel under test are used to estimate the background SVDD parameters. The samples are collected from a hollow rectangular window centred at the pixel. The window consists of two regions, the inner and outer window regions, as shown in Figure 8.3.

The dimensions of each region are based on the spatial resolution of the image and the expected size of the targets. The size of the inner area should accommodate the typical or largest expected target in the scene. The size of the outer window should be large enough to collect a sufficient number of training samples for background estimation.

8.4.3 Estimating sigma

Choosing an appropriate value for the RBF kernel scale parameter sigma is critical for acceptable detection performance. Typically, the performance of anomaly detectors is commonly characterized by two measures of error: the false alarm rate (classifying a background pixel as an outlier), and the false acceptance rate (accepting a target pixel as part of the background). Since the target spectra are unknown, it is not possible to estimate the false acceptance rate from the training samples directly.

To reliably estimate sigma, a Neyman–Pearson framework is developed based on two observations. First, it has been shown (Tax *et al.* 1999; Scholkopf *et al.* 2000; Tax and Duin 1999) that an upper bound on the false alarm rate for the SVDD is,

$$P_{FA} \leq \mathbb{E}\left[\frac{\#SV}{N}\right]. \tag{8.22}$$

In this expression, P_{fa} is the false alarm rate, $\#SV$ is the number of support vectors in the SVDD model and N is the total number of training samples.

The second observation is that anomaly detectors implicitly assume that the target space is simply the complement of the support region of the background training samples. Without knowledge of the target spectra and their distribution, anomaly detectors attempt to model the support region of the background spectra. They consider any pixels whose spectra lie outside the support region as targets.

Analogous to the Neyman–Pearson framework, an estimate for sigma that seeks to maximize the probability of detection while maintaining a desired false alarm rate can be defined as follows:

$$\hat{\sigma} = \max_{\sigma}(P_D) \text{ such that } P_{FA} \leq \tau, \tag{8.23}$$

where P_D is the probability of detection and τ is the desired false alarm rate. Utilizing the two observations above,

$$\hat{\sigma} = \max_{\sigma} \frac{1}{C} \int_{R^C} \mathbf{1}(\mathbf{x}) d\mathbf{x} \text{ such that } \mathbb{E}\left[\frac{\#SV}{N}\right] \leq \tau \tag{8.24}$$

$$= \max_{\sigma} volume(R^c) \text{ such that } \mathbb{E}\left[\frac{\#SV}{N}\right] \leq \tau \tag{8.25}$$

$$= \min_{\sigma} volume(R) \text{ such that } \mathbb{E}\left[\frac{\#SV}{N}\right] \leq \tau, \tag{8.26}$$

where $\mathbf{1}(\cdot)$ is the indicator function and R is the support region for the background pixels, and C is a normalizing constant. The probability of detection is approximated as the volume of the feature space that lies outside of the support region (R^c). Thus, the estimator seeks the value of sigma that (a) minimizes the size of the support region for the background spectra and (b) generates an average fraction of support vectors that is less than the desired false alarm rate.

As shown in Tax *et al.* (1999), Scholkopf *et al.* (2000) and Tax and Duin (1999), the choice of sigma affects the size of the support region and the number of support vectors. Figure 8.1 illustrates the SVDD support regions for a set of training data with two different values for sigma. It shows that if the value of sigma is small, the resulting support region is small and the decision boundary overfits the data. If sigma is big, the support region is large and the classifier underfits the data with a possibly trivial decision boundary. Since sigma is proportional to the volume of the support region, the estimate for sigma can be expressed as:

$$\hat{\sigma} = \min_{\sigma} \sigma \text{ such that } \mathbb{E}\left[\frac{\#SV}{N}\right] \leq \tau. \tag{8.27}$$

Furthermore, sigma is inversely proportional to the number of support vectors (Scholkopf *et al.* 2000; Tax and Duin 1999). Hence, the choice for sigma should be small enough to minimize the volume of the support region while large enough to maintain the desired false alarm rate.

Given the expectation in the above expression is over training sets of size N, the expectation from a single training set cannot be estimated. Therefore the resulting number of support vectors for the SVDD model is substituted in place of the expectation to obtain an approximate upper bound.

To avoid the computational expense of estimating sigma for every pixel, one can choose to estimate a value of sigma that minimizes the average false alarm rate across the entire image, i.e.

$$\frac{1}{M}\sum_{i=1}^{M} P_{FA_i} \approx \frac{1}{M}\sum_{i=1}^{M} \mathbb{E}\left[\frac{\#SV_i}{N}\right] \approx \frac{1}{M}\sum_{i=1}^{M} \frac{\#SV_i}{N} \leq \tau, \tag{8.28}$$

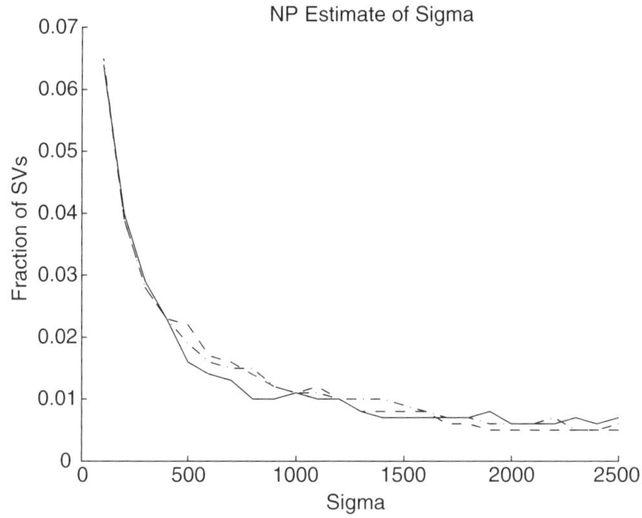

Figure 8.4 Fraction of support vectors as a function of sigma for three SVDD models. Sigma can be estimated using a Neyman–Pearson criterion.

where M is the number of training sets, N is the number of examples in each training set, P_{FA_i} is the ith training set, and $\#SV_i$ is the number of support vectors found for the ith training set. This approximation results in the following estimate for sigma:

$$\hat{\sigma} = \min_{\sigma} \sigma \text{ such that } \frac{1}{M}\sum_{i=1}^{M}\frac{\#SV_i}{N} \leq \tau. \tag{8.29}$$

The algorithm to obtain this global estimate for sigma is as follows. Generate multiple sets of training data by randomly selecting pixels from the image. For each set of training data, the SVDD decision boundary is determined using different values for sigma. For each value of sigma, the average fraction of support vectors is computed over all of the training sets. The smallest sigma such that the average fraction of support vectors is less than τ is the estimated value.

Figure 8.4 provides an example of the σ estimate as described above. In this example, three sets of 1000 training samples are randomly selected from a hyperspectral image. The SVDD classifier is trained on all three datasets with varying values for sigma. In this example, the smallest value for sigma that satisfies the desired false alarm rate of 0.01 is 1400, which is used as the estimate.

8.4.4 Normalized SVDD test statistic

Anomaly detectors compute a statistic that provides a measure of similarity to the background for each pixel in the image. The statistic can be used to compare the pixels directly and identify those that are more likely to be anomalies. This allows for the use of a single threshold for the entire image, which is a key consideration for local anomaly detectors. By varying this global threshold, an empirical receiver operating characteristic (ROC) curve can be generated to evaluate the detector's performance.

One class of such detectors is the constant false alarm rate (CFAR) detectors. The CFAR property is useful because it allows for the determination of a single detection threshold that (a) generates a desired number of false alarms, (b) is independent of the estimated parameters, and (c) can be used for every pixel in the image.

The RX algorithm is an example of a CFAR detector (Beaven *et al.* 2000). Using the Mahalanobis distance, each pixel is normalized to yield a zero-mean Gaussian random variable with an identity covariance matrix. This allows the algorithm to compute a detection threshold via

$$\int_{\mathbf{x}:f(\mathbf{x})<T} f(\mathbf{x})d\mathbf{x} = P_{FA}, \tag{8.30}$$

where $f(\mathbf{x}) \sim N(\mathbf{0}, I)$, and P_{FA} is the desired false alarm rate. Note that T is computed independently of the estimated background parameters. Furthermore, if the pixels are Gaussian distributed, the threshold will generate the desired number of false alarms.

For the SVDD, a normalized test statistic can be similarly derived. In contrast to the RX algorithm, the following normalization procedure is motivated geometrically and strictly avoids imposing assumptions about the underlying distribution. Recall from Section 8.2 that the nonlinear SVDD estimates the minimum enclosing hypersphere centred at $\mathbf{a}$ with radius R in the feature space as the support region for the background pixels. From Equation (8.11), the distance between the mapped example $\boldsymbol{\phi}(\mathbf{y})$ and the centroid of the hypersphere $\mathbf{c}$ is:

$$\| \boldsymbol{\phi}(\mathbf{y}) - \mathbf{c} \|^2 = 1 - 2\sum_i \alpha_i K(\mathbf{y}, \mathbf{x}_i) + \sum_{i,j} \alpha_i \alpha_j K(\mathbf{x}_i, \mathbf{x}_j), \tag{8.31}$$

which is the original SVDD statistic. Since the radius of the optimal hypersphere will vary from pixel to pixel, the distance to the centroid cannot be used to compare the similarities of multiple pixels to their respective backgrounds. Hence, a normalized version of the SVDD test statistic is given as:

$$\mathrm{SVDD}_N(\mathbf{y}) = \frac{\| \boldsymbol{\phi}(\mathbf{y}) - \mathbf{c} \|^2}{R^2} = \frac{1 - 2\sum_i \alpha_i K(\mathbf{y}, \mathbf{x}_i) + \sum_{i,j} \alpha_i \alpha_j K(\mathbf{x}_i, \mathbf{x}_j)}{R^2}. \tag{8.32}$$

By dividing the original statistic by the squared radius, the $\mathrm{SVDD}_N(\mathbf{y})$ function effectively transforms the feature space so that the minimum enclosing hypersphere encompassing the data has unit radius. This normalization allows for direct comparison of the SVDD function values for different pixels in the image. The detection threshold T can now be chosen independently of the estimated SVDD parameters.

Selecting the detection threshold analytically is an important consideration. CFAR detectors, such as the RX, allow for the estimation of a threshold based on a desired false alarm rate. However, if the data is non-Gaussian, the CFAR properties of the RX detector no longer hold. As the SVDD avoids estimating a PDF, it is difficult to compute the threshold analytically. Instead, its detection threshold can be determined empirically by using nonparametric approaches for data modelling. One example is to use extreme value theory (EVT) (Pickands 1975), a branch of statistics dealing with extreme deviations from the mean. EVT methods attempt to model the tails of the distribution empirically and determine a detection threshold based upon a desired false alarm rate. As this is an emerging area of study, its usefulness will be examined in future work.

8.5 Experimental results

In this section, the computational and detection performance of the SVDD anomaly detectors are evaluated. The first set of experiments compares the local anomaly detectors, in particular the RX, SVDD and normalized SVDD algorithms. Examples of how the false alarm rate is affected by the signal-to-noise ratio and the background clutter statistics are given. An analysis of how algorithm speed varies with the number of spectral bands and the number of training samples is provided. The analysis offers some insight into selecting the appropriate anomaly detector for different situations.

The second set of experiments compares these local anomaly detectors with the global SVDD algorithm. The computational complexity of the algorithms is discussed, and their detection capabilities are analysed using ROC curves.

Data overview

The detectors are compared using images taken from the Wide Area Airborne Mine Detection (WAAMD) and Hyperspectral Digital Imagery Collection Experiment (HYDICE) datasets. The WAAMD imagery is provided by the Night Vision Electronic Systems Directorate (NVESD). The COMPASS sensor, providing 256 bands in the VIS/NIR/SWIR part of the spectrum (400–2350 nm), was used to image the minefields. For the images used in this study, the sensor was flown at an altitude of 1000 and 2000 feet, with a GSD of approximately 4 and 8 inches, respectively. The images contain two types of mines whose sizes are approximately 2×2 pixels.

The HYDICE sensor was used to acquire the Forest Radiance II data collect. The sensor provides 210 spectral bands in the VIS/NIR/SWIR wavelengths with approximately 1 metre spatial resolution. Two images that contain the targets from detection experiments 2, 3 and 4 are considered. There are over 20 target vehicles in the scenes, including many that are partially hidden along tree lines. There are a handful of additional ‘transient’ non-military vehicles in the images that are not marked as ground truth, but used in the evaluation.

Pre-processing

The detection algorithms are evaluated on both multi-spectral and hyperspectral versions of the imagery. For hyperspectral data, the only pre-processing step involved manually removing the water and CO_2 absorption bands. The number of remaining bands is 145 for the WAAMD data and 133 for the HYDICE imagery. To generate multi-spectral data, an additional step was taken reduce the total number of bands to seven by averaging adjacent spectral bands. The spectrum was divided into seven regions where the averaging was performed; the regions correspond to red, green, blue, two near IR and two short-wave IR regions.

For all of the detectors, a hollow window is used to estimate the local background model. The window dimensions are determined by the largest expected target size in the scene, which can be determined by the pixel resolution of the data. A hollow window is used to avoid incorporating target pixels in the estimation of the background statistics. For the RX detector, a mean vector and covariance matrix is estimated for each pixel from the pixels in the hollow window. For the SVDD detectors, the samples from the background window are used to train the SVDD classifier at each pixel. The dimensions of the window are the same for all detectors. However, a larger window is used for hyperspectral data, since more training samples are required to estimate the covariance matrix for the RX algorithm. The actual window sizes used in the experiments are given below.

Performance evaluation of local anomaly detectors

The detectors are evaluated by comparing their ROC curves using groundtruthed imagery. The locations of the target pixels are represented by rectangular boxes. To generate the ROC curves, multiple thresholds are used for the RX, SVDD($\mathbf{y}$) and the $SVDD_N(\mathbf{y})$ test statistics. After applying a threshold, the detected pixels are clustered via a connected components algorithm; each cluster is counted as one detection or false alarm. If any part of the cluster falls within the target box, it is considered to be a detection, else it is a false alarm. In addition, detections arising from bad lines in the image are removed manually.

In Figures 8.5–8.8, ROC curves are provided to evaluate the ability of the algorithms to detect mines using the WAAMD data. A total of 44 mines were placed in a highly cluttered dirt field, resulting in a non-homogeneous background. The mine field was imaged with the COMPASS sensor at altitudes of 1000 and 2000 feet, with a GSD of 4 and 8 inches, respectively. The size of the 1000 and 2000 ft. altitude images are 1400 × 256 pixels and 1200 × 256 pixels, respectively. For multi-spectral processing, the dimensions of the inner and outer regions for the hollow background window are 7 × 7 and 13 × 13, respectively. For hyperspectral, the dimensions of the inner and outer regions for the hollow background window are 7 × 7 and 15 × 15, respectively.

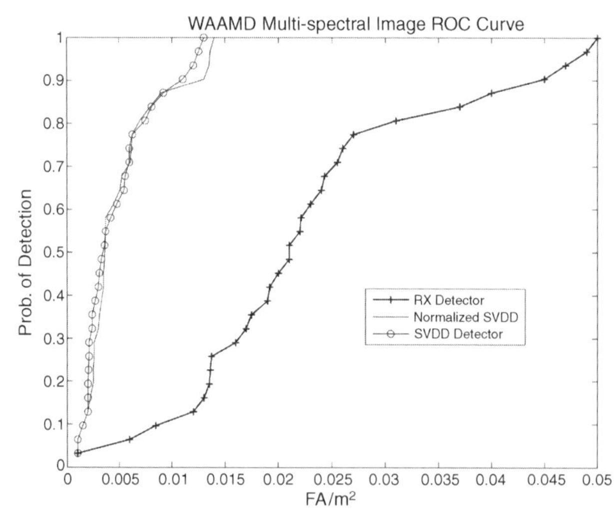

Figure 8.5 ROC curve comparing SVDD and RX detectors on WAAMD imagery using seven bands. The 1400 × 256 image contains 44 live mines in a dirt field and was taken at 1000 ft altitude.

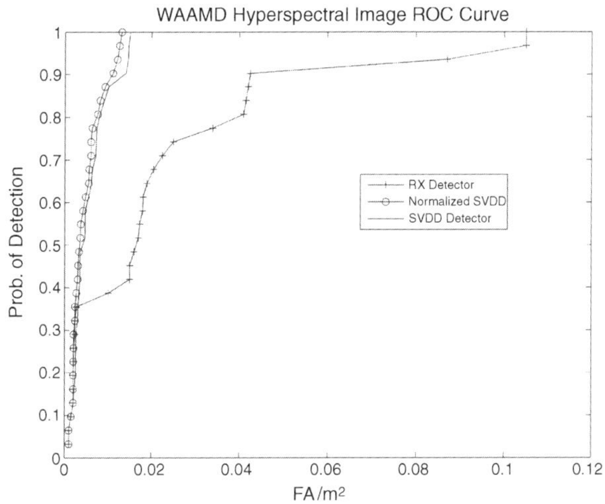

Figure 8.6 ROC curve comparing SVDD and RX detectors on WAAMD imagery using 145 bands. The 1400 × 256 image contains 44 live mines in a dirt field and was taken at 1000 ft altitude.

The ROC curves comparing the detectors on the 1000 ft. altitude image are shown in Figures 8.5 and 8.6. In the multi-spectral example of Figure 8.5, the SVDD detectors clearly outperform the RX algorithm throughout the curve. For 100% detection, the SVDD detectors reduce the false alarm rate by a factor of five. In the hyperspectral case (Figure 8.6), the SVDD detectors generate approximately the same false alarm rate as with the multi-spectral

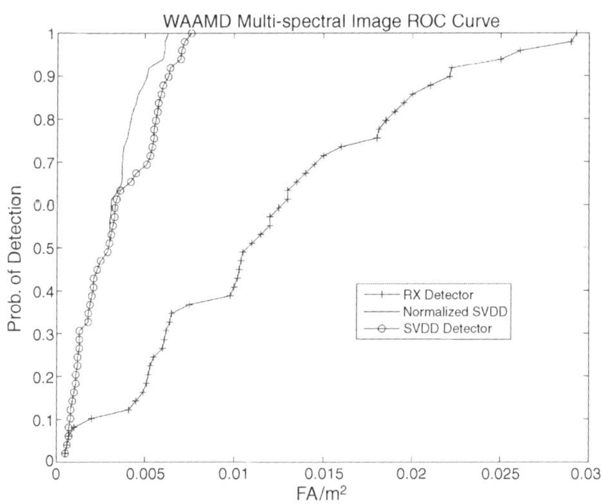

Figure 8.7 ROC curve comparing SVDD and RX detectors on WAAMD imagery using seven bands. The 1200 × 256 image contains 44 live mines in a dirt field and was taken at 2000 feet altitude.

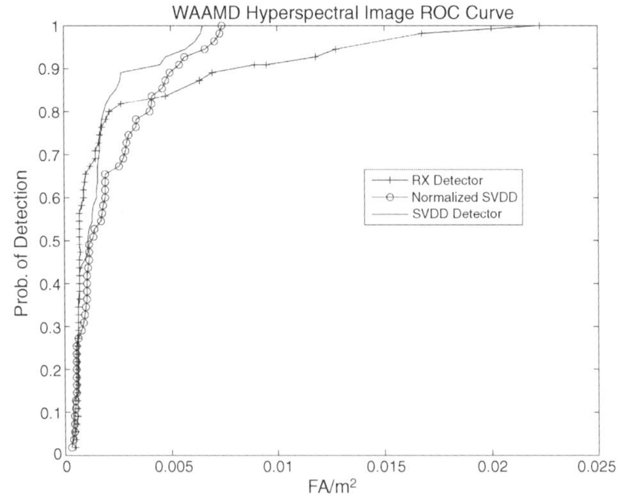

Figure 8.8 ROC curve comparing SVDD and RX detectors on WAAMD imagery using 145 bands. The 1200 × 256 image contains 44 live mines in a dirt field and was taken at 2000 feet altitude.

data (nearly 0.015 FA/m^2). However, the false alarm rate of the RX detector nearly doubles when processing hyperspectral data. In the hyperspectral representation of the image, the false alarm rate of the SVDD detectors is nearly an order of magnitude less than the RX algorithm. One explanation for the performance difference is the distribution of the background spectra. With a 4 inch GSD, the rocks and variations in the dirt field are more pronounced, making the background highly cluttered and non-homogeneous. Hence, the Gaussian assumption of the RX detector results in a high false alarm rate. The SVDD is better able to model the non-Gaussian and possibly multi-modal support of the background pixels and reduce the number false alarms.

From the curves in Figures 8.7 and 8.8, similar results are observed for the image at 2000 feet altitude. Processing the multi-spectral data, the SVDD detectors outperform the RX detector throughout the ROC curve. For 100% detection, they generate half of the false alarms of the RX. When analysing hyperspectral data, the ROC curves for the SVDD detectors remain consistent, yielding a false alarm rate of 0.01 FA/m^2 for 100% detection. For this image, the SVDD reduces the false alarm rate by a factor of nearly 2.5, instead of 10. One possible explanation for the difference is that the background for the higher altitude image is more homogeneous and therefore does not violate the unimodal assumption of the RX algorithm. It is also instructive to examine the Signal-to-Interference Noise Ratio (SINR) for the targets in the two images. The SINR for each target is computed by using the target spectrum and its covariance matrix estimated from a local neighbourhood of pixels. For the 1000 feet image, the range of SINR values is between 23 dB and 28 dB for multi-spectral data, and 20 to 26 dB for hyperspectral data. For the 2000 feet image, the SINR ranges between 26 dB and 29 dB for both multi- and hyperspectral versions of the image. These numbers suggest that for lower SINR levels, the SVDD detectors offer significant improvement over the RX algorithm. As the SINR increases, the performance gain between the algorithms appears to lessen.

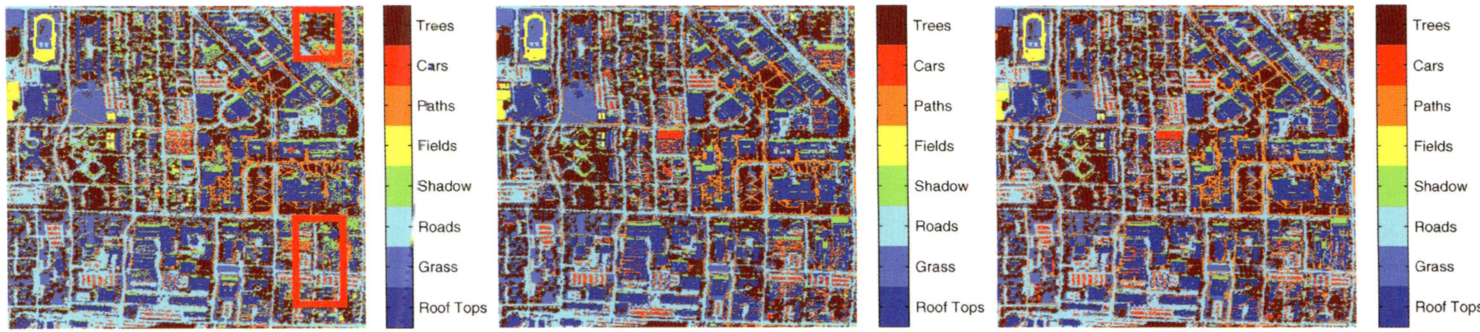

Plate 1 *Classification maps obtained for LFD, KFD and AKFDR for* $r = 0.04$, *with test accuracies of 86.5%, 90.5% and 90.6%, respectively.* *(See page 122.)*

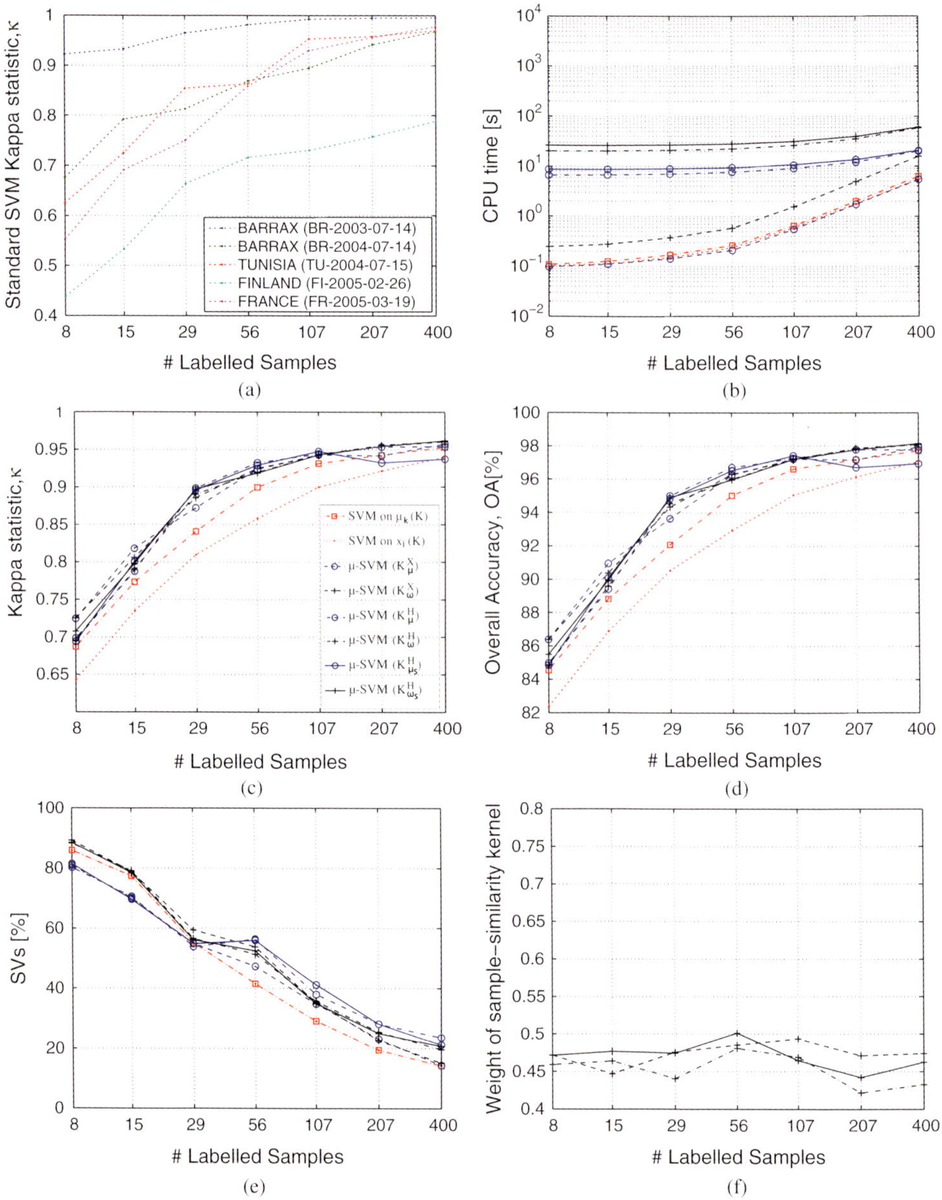

Plate 4 (a) Standard SVM classification results for the five MERIS images. (b)–(f) Average classification results over the five images training the models with labelled and unlabelled (800) samples from the image to be classified (single-image case)*: (b) CPU time [s], (c) κ, (d) OA, (e) SVs [%], and (f) weight ν of the sample-similarity kernel of labelled samples* K. *(See page 237.)*

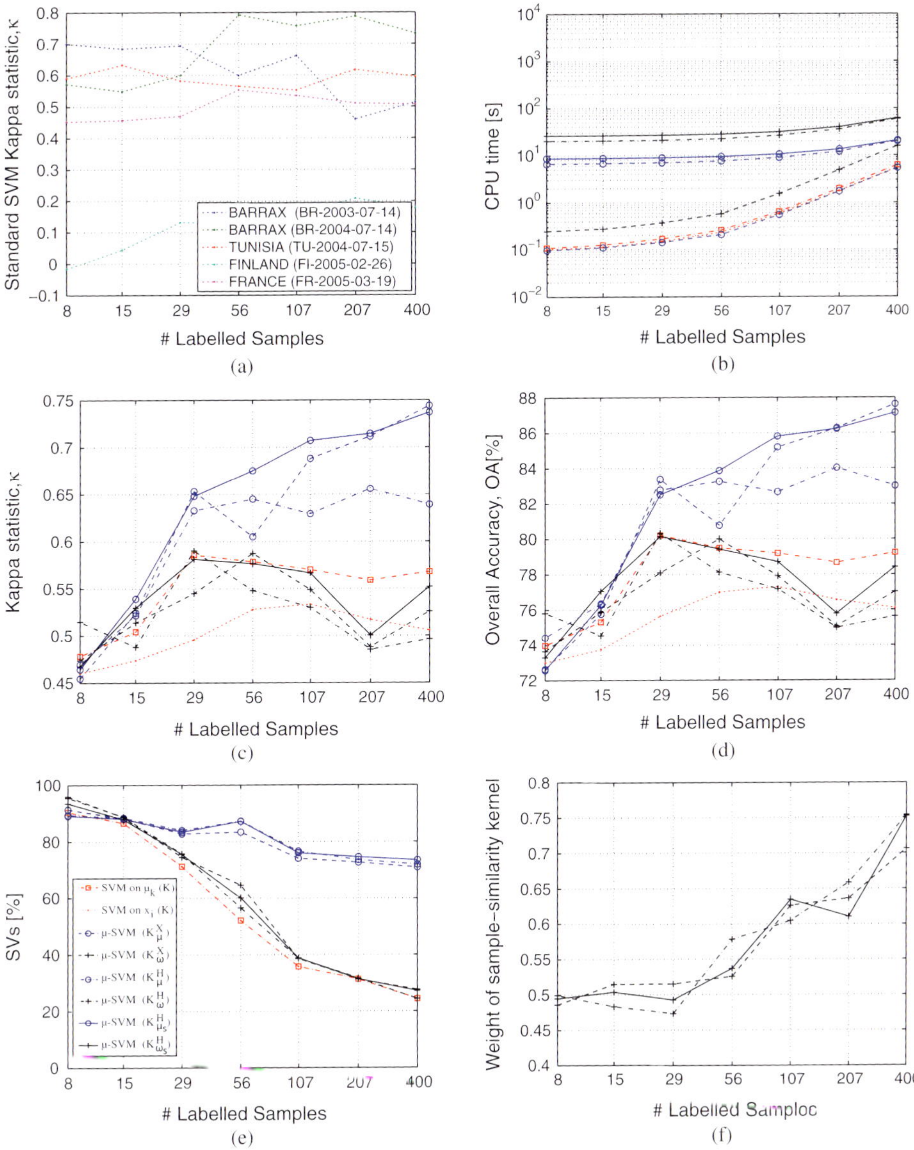

Plate 5 *(a) Standard SVM classification results for the five MERIS images. (b)–(f) Average classification results over the five images with labelled samples from the other four images and* 800 *unlabelled samples from the image to be classified* (image-fold case)*: (b) CPU time [s], (c) κ, (d) OA, (e) SVs[%], and (f) weight ν of the sample-similarity kernel (K). (See page 239.)*

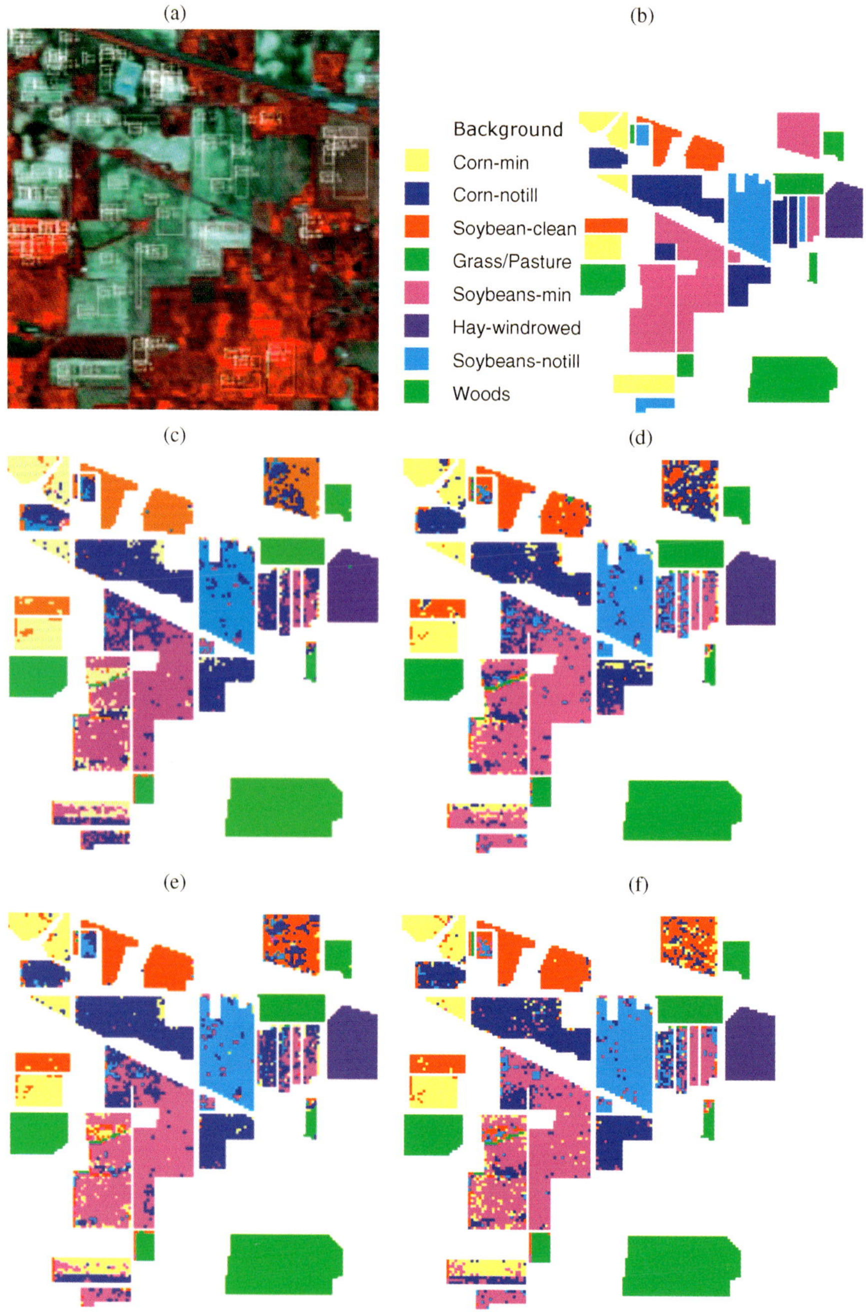

Plate 10 (a) *Simulated greyscale IR image of the Indian Pine Site dataset.* (b) *Ground truth of the area with eight classes.* (c) *Classification map by KPCA-poly1 with ML classifier ($n_i = 300$, $p = 13$),* (d) *GDA-poly2 with SVM-poly1 classifier ($n_i = 300$, $p = 7$),* (e) *KNWFE-poly1 with ML classifier ($n_i = 300$, $p = 11$), and* (f) *KFFE-poly1 with 1NN classifier ($n_i = 300$, $p = 13$).* *(See page 391.)*

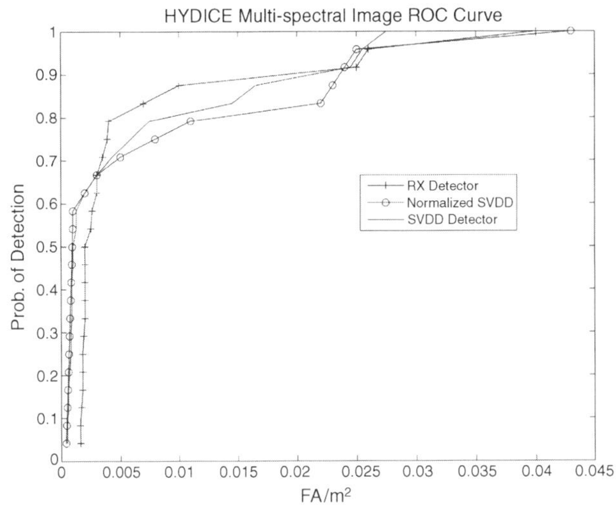

Figure 8.9 ROC curve comparing SVDD and RX detectors on HYDICE multi-spectral imagery. The 1280 × 320 image contains vehicles that are on roads and are partially camouflaged near tree lines.

The algorithms are also evaluated using the Forest Radiance II dataset to detect vehicles in various degrees of concealment. The HYDICE sensor was used to image vehicles on dirt roads in a heavily forested region at 1m GSD. The background consists of foliage, grass and roads, and is smoothly varying and homogeneous. The size of the image used in this study is 1280 × 320 pixels. For multi-spectral processing, the dimensions of the inner and outer regions for the hollow background window are 10 × 10 and 13 × 13, respectively. For hyperspectral, the dimensions of the inner and outer regions for the hollow background window are 10 × 10 and 17 × 17, respectively.

The ROC curves in Figures 8.9 and 8.10 show that all of the detectors work equally well in the low false-alarm region. In this area of the curve, the vehicles are predominantly in the open, so they are easier to detect. Using hyperspectral data (Figure 8.10), there is clear separation in the ability of the RX and SVDD algorithms to detect targets partially occluded along tree lines. Since the local background of these targets contains multiple types of terrain, a multi-modal representation for the background is required. In such cases, the SVDD is able to compute a more robust detection threshold than the RX detector. Furthermore, the SINR values for the targets in the multi-spectral image are above 30 dB. In the hyperspectral image, the SINR ranges from 27 dB to 33 dB. At such high signal-to-noise ratios, the targets are well-separated from the background. Therefore detectors, such as the RX, that use linear decision boundaries perform similarly to nonlinear detectors, such as the SVDD (as seen in Figure 8.9).

The results are summarized with two observations. First, the SVDD detectors are particularly useful for detecting low SINR targets. The spectra for these targets may not be linearly separable from the background, so a nonlinear decision boundary is required. Second, the SVDD detectors work well with both multi-spectral and hyperspectral data. They are able to incorporate the extra information provided by the addition spectral bands to reduce the number of false alarms for 100% detection. In contrast, the performance of the RX detector

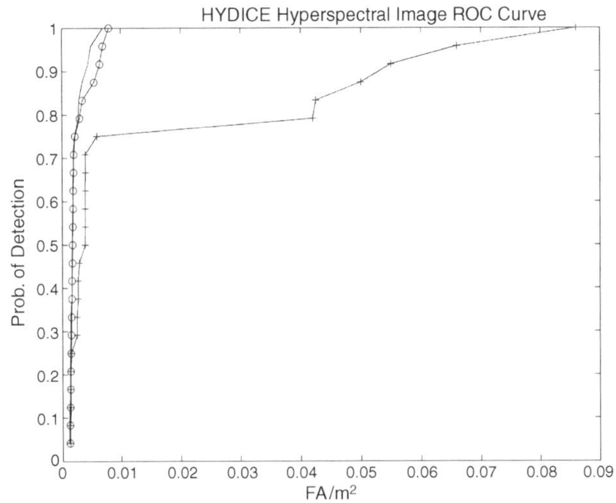

Figure 8.10 ROC curve comparing SVDD and RX detectors on HYDICE hyperspectral imagery. The 1280 × 320 image contains vehicles that are on roads and are partially camouflaged near tree lines.

appears to be sensitive to the number of bands. Among the reasons for the difference are (a) the Gaussian assumption of the RX may be less valid for higher dimensional data, as explained in Landgrebe (2002), and (b) robust estimation of the covariance matrix is more difficult as the number of bands increases.

Computational complexity

The ROC curves in Figures 8.5–8.10 illustrate the increase in detection performance offered by the SVDD approach. In addition to reducing the false alarm rate, anomaly detectors should also be computationally efficient in order to process a datacube in near real-time. Thus, the computational complexity of the RX and SVDD algorithms are considered.

The two critical factors that impact the speed of the detectors are the number of spectral bands and the number of pixels in the hollow window used to estimate the background model. For the RX detector, it has been shown (Schweizer 2001) that its computational complexity is linear with respect to the number of pixels used to estimate the background statistics. Since it needs to estimate and invert a $B \times B$ covariance matrix (where B is the number of spectral bands), the number of floating point operations per second (FLOPS) for the RX detector is approximately quadratic with respect to the number of bands (Schweizer and Moura 2001).

For the SVDD, the converse is true. By avoiding the need to invert large covariance matrices, the complexity of the SVDD increases linearly with the number of bands, as only dot products need to be computed. However, the speed of the SVDD detectors scales with the training set size (Lai *et al.* 2004). They require the inversion of an $N \times N$ kernel matrix, where N is the number of background pixels used to train the classifier at each pixel. Hence, their complexity is exponential with respect to the size of the background window.

In Figure 8.11, the processing times for both algorithms as function of the number of spectral bands and number of training samples are shown. The algorithms are implemented

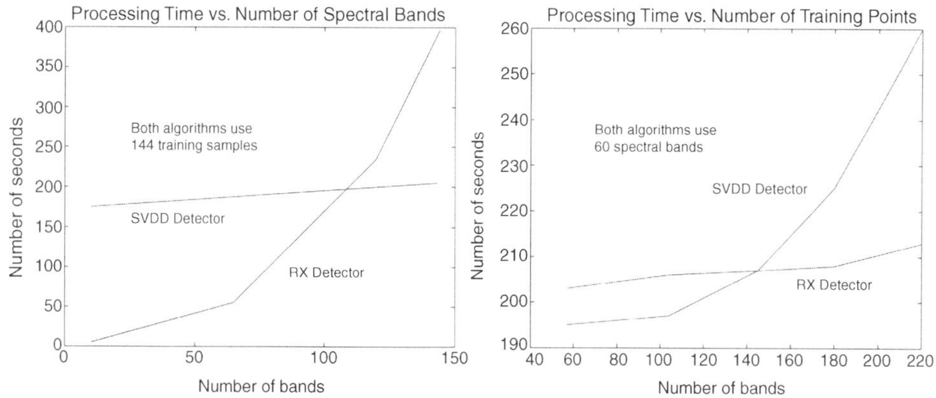

Figure 8.11 Illustration of how the computational complexity of the RX and SVDD anomaly detectors vary with the number of spectral bands and the number of training samples. Note that the speed of the SVDD in this example is based on the use of standard QP optimization.

in C++ on a 2.8 GHz Pentium machine, and the code has not been fully optimized. The CPU time required to process a 256×100 pixel image are measured. Figure 8.11(left) shows that for a fixed number of training samples (144 local background pixels), the processing time for the SVDD increases linearly with the number of spectral bands, while the RX increases quadratically. In Figure 8.11(right), it can be seen that the converse is true. For a fixed number of bands (60 in this example), the SVDD increases quadratically with the number of training samples, while the RX increases linearly.

Performance evaluation of global anomaly detectors

Finally, the performance of the global SVDD algorithm is compared with the local SVDD and RX anomaly detectors using the WAAMD imagery. Figure 8.12 shows their ROC

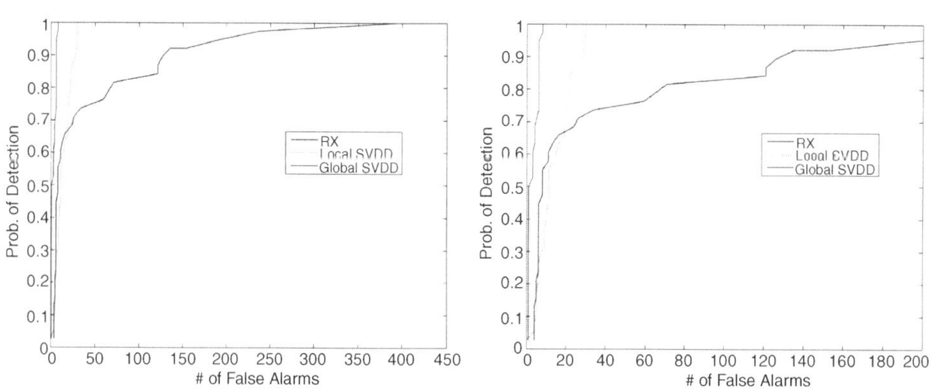

Figure 8.12 (Left) ROC performance for the RX, local SVDD and global SVDD detectors. (Right) Zoomed in view of upper left region of ROC plots providing a clearer view of global SVDD performance.

Table 8.1 Processing complexity and algorithm characteristics affecting detector performance

	Processing complexity	Clutter model/Notes
RX	$\mathcal{O}(N_p N_b^3)$	Local, Unimodal, Normal
SVDD (local)	$\mathcal{O}(N_p N_c^3)$	Local, Multimodal
SVDD (global)	$\mathcal{O}(N_c^3)$	Global, Multimodal

Key: N_p, N_b, N_c: Number of pixels, spectral bands, clutter training samples

performance on the WAAMD hyperspectral image captured at 2000 foot altitude containing 38 targets.

The global SVDD results in enhanced false alarm suppression due to the rejection of candidates that are locally anomalous but not globally anomalous (e.g. isolated trees). In addition, it provides enhanced detection due to enhanced background characterization that is not forced to include target spectra when targets are in close proximity to each other.

The computational complexity of the RX method stems from the inversion of the covariance matrix that is required for the background estimation, which in general is of order $\mathcal{O}(N_b^3)$ where N_b is the number of bands. Owing to the local nature of the estimation, this inversion is required at every pixel, so the actual processing time is of order $O(N_p N_b^3)$, where N_p is the number of spatial pixels in the hyperspectral image. By using a global background estimate for RX, N_p can be removed from the order of magnitude of the processing, and thereby achieve rapid processing of the image. However, the anomaly detector will not work well since the background is generally not globally unimodal.

The SVDD approach does not require a covariance matrix inverse computation and is also linear with respect to spectral dimensionality. Processing is commensurate with the inversion of the system matrix kernel, which is of order $\mathcal{O}(N_c^3)$. Since SVMs are effective with few exemplars, the background modelling is very robust to low training so N_c may be relatively low. In addition, since the background modelling is multi-modal, automatic and does not require any a priori knowledge of background characteristics (e.g. the number of components to be used in a mixture model), global background characterization may be reliably performed. As a result, SVDD is able to achieve high detection performance and a low false alarm rate with rapid computational processing. Computational complexity and associated characteristics affecting algorithm performance are summarized in Table 8.1.

8.6 Conclusions

A geometric interpretation of the SVDD is used to derive a normalized metric that is appropriate for anomaly detection in spectral imagery. ROC curves comparing the performance of the RX and the proposed detector show a marked reduction in the number of false alarms when using the SVDD. Further analysis of the background clutter and SINR for the targets to explain the differences in performance is also provided.

Other algorithms, such as the mixture of Gaussian anomaly detector and the sub-space RX (SSRX) produce similar reductions in the false alarm rate. However, the proposed SVDD approach for anomaly detection has several key differences.

- It is nonparametric and has the ability to model non-Gaussian background clutter.
- It can model nonhomogeneous backgrounds using multi-modal distributions. without making any prior assumptions (i.e. number of modes).
- It avoids the need to invert large covariance matrices when processing hyperspectral data.

References

Banerjee, A., Burlina, P. and Chellappa, R. (1999) Adaptive target detection in foliage-penetrating SAR images using alpha-stable models. *IEEE Trans. Image Processing*, **8**, 1823–1831.

Beaven, S.G., Stein, D. and Hoff, L.E. (2000) Comparison of Gaussian mixture and linear mixture models for classification of hyperspectral data. *IEEE Proc. International Geoscience and Remote Sensing Symposium*, 1597–1599.

Chang, C-C. and Lin, C-J. (2001) LIBSVM : a library for support vector machines. Software available at http://www.csie.ntu.edu.tw/ cjlin/libsvm.

Comaniciu D. and Meer, P. (1997) Robust analysis of feature spaces: colour image segmentation. *Proc. IEEE Conf. Computer Vision and Pattern Recognition*, 750–755.

Finn, H.M. (1986) A CFAR design for a window spanning two clutter fields. *IEEE Trans. on Aerospace and Electronic Systems*, **22**.

Fukunaga, K. and Hostetler, L.D. (1975) The estimation of the gradient of a density function. *IEEE Trans. Information Theory*, **21**, 32–40.

Gandhi, P. and Kassam, S. (1998) Analysis of CFAR processors in non-homogeneous backgrounds. *IEEE Trans. Aerospace and Electronic Systems*, **24**.

Gualtieri, J.A. and Cromp, R.F. (1998) Support vector machines for hyperspectral remote sensing classification. *27th AIPR Workshop, Proceedings of the SPIE*, **3584**, 221–232.

Gualtieri, J.A., Chettri, S.R., Cromp, R.F. and Johnson, L.F. (1999) *Support Vector Machines Classifiers as Applied to AVIRIS Data, Summaries of the Eighth JPL Airborne Earth Science Workshop*: **99-17**, 217–227.

Henze, N. and Wagner, T. (1997) A new approach to the BHEP tests for multivariate normality. *Journal Multivariate Analysis*, **62**, 1–23.

Kwon, H. and Nasrabadi, N. (2006) Kernel RX-algorithm: a nonlinear anomaly detector for hyperspectral imagery. *IEEE Trans. Geoscience Remote Sensing*, **43**.

Lai, C., Tax, D.M.J., Duin, R.P.W., Pekalska, E. and Paclik, P. (2004) A study on combining image representations for image classification and retrieval. *International Journal of Pattern Recognition and Artificial Intelligence*, **18**, 867–890.

Landgrebe, D. (2002) Hyperspectral image data analysis. *IEEE Signal Processing Magazine*, **19**, 17–28.

Lu, X., Hoff, L.E., Reed, I.S., Chen, M. and Stotts, L.B. (1997) Automatic target detection and recognition in multiband imagery: a unified ML detection and estimation approach. *IEEE Trans. Image Processing*, **6**, 143–156.

Osuna, E., Freund, R. and Girosi, F. (1997) Improved training algorithm for support vector machines. *Proc. IEEE Neural Networks in Signal Processing*, pp. 276–285.

Pickands, J. (1975) Statistical inference using extreme order statistics, *Annals of Statistics*, Vol. 3, pp. 119–131.

Scharf, L.L. and Friedlander, B. (1994) Matched subspace detectors. *IEEE Trans. Signal Proc.*, **42**, 2146–2157.

Scholkopf, B., Platt, J., Shawe-Taylor, J., Smola, A.J. and Williamson, R.C. (1999) Estimating the support of a high-dimensional distribution. *Tech Report Microsoft Research*, 99.

Scholkopf, B., Williamson, R., Smola, A.J., Shawe-Taylor, J. and Platt, J. (2000) Support vector method for novelty detection. *Neural Information Processing Systems*, 582–588.

Schweizer, S. and Moura, J.M. (2001) Efficient detection in hyperspectral imagery. *IEEE Trans. Image Processing*, **10**, 584–597.

Stein, D., Beaven, S., Hoff, L.E., Winter, E., Shaum, A. and Stocker, A.D. (2002) Anomaly detection from hyperspectral imagery. *IEEE Signal Processing Magazine*, **19**, 58–69.

Stocker, A.D. (1999) Stochastic expectation maximization (SEM) algorithm. *Proc. DARPA Adaptive Spectral Reconnaissance Algorithm Workshop.*

Syed, N., Liu, H. and Sung, K. (1999) Incremental learning with support vector machines. In *Proceedings of the Workshop on Support Vector Machines at the International Joint Conference on Artificial Intelligence (IJCAI-99),* Stockholm, Sweden 1999.

Tax, D.M.J. and Duin, R.P.W. (1999) Data domain description using support vectors. *Proc. European Symposium on Artificial Neural Networks*, 251–256.

Tax, D.M.J, Ypma, A. and Duin, R.P.W. (1999). Support vector data description applied to machine vibration analysis. *Proc. 5th Annual Conference of the Advanced School for Computing and Imaging*, 398–405.

Vapnik, V. *Estimation of Dependences Based on Empirical Data*. Springer-Verlag, 1982.

V. N. Vapnik, *Statistical Learning Theory*, John Wiley & Sons, New York, 1998.

III

Semi-supervised image classification

9

A domain adaptation SVM and a circular validation strategy for land-cover maps updating

Mattia Marconcini and Lorenzo Bruzzone

Dept. Information Engineering and Computer Science
University of Trento, Italy

In this paper, we present a novel domain adaptation classifier based on SVMs (DASVM) for land-cover maps updating, which can be employed in real operational situations when ground-truth labels are available only for a reference image acquired over the investigated geographical area before the one to classify. In addition, we also propose a circular validation strategy for the accuracy assessment of the classification results when no labelled samples are available for the image to be categorized.

9.1 Introduction

In the last few years, automatic image classification has become a challenging and important task in the framework of remote sensing thanks to the technological development of new sensors and the growing importance of related applications. In this context, two main families of learning methods are commonly employed, depending on whether true labels are available for the considered remote-sensing image or not: *supervised learning* methods and

Some parts of this chapter have been derived from papers currently under consideration in IEEE journals (Bruzzone and Marconcini 2008, 2009).

Kernel Methods for Remote Sensing Data Analysis Edited by Gustavo Camps-Valls and Lorenzo Bruzzone

unsupervised learning methods. The default assumption of supervised learning methods is that training and test data are drawn from the same domain (i.e., a domain D is defined as a distribution $p(\mathbf{x}, y) = p(\mathbf{x})p(y|\mathbf{x}), \mathbf{x} \in \tilde{\mathcal{X}}, y \in \Omega$, that governs the classification problem under investigation, where $\tilde{\mathcal{X}}$ and Ω represent all possible instances and all possible information classes for the considered problem, respectively). In this scenario, a *representative* set of labelled samples for the training of the classifier is assumed to be available and the learning of the algorithm is associated with the solution of a *well-posed* problem (i.e., the solution is stable with respect to small perturbations of the data). In unsupervised learning methods, instead, no training data are available and the problem is addressed according to the solution of *ill-posed* clustering problems (i.e., small changes in data can arbitrarily cause large unpredictable changes in results).

Nonetheless, in many applications, hybrid conditions can be encountered where, even if prior ground-truth information is available, it is not sufficient to define a training set representative of the distribution to which the trained model should be applied. These kinds of *ill-posed* problems can be addressed according to *transfer learning* methods. In particular, transfer learning refers to the problem of retaining and applying the knowledge available for one or more tasks, domains, or distributions to develop efficiently an effective hypothesis for a new task, domain or distribution. Depending on whether training and test data are drawn from the same remote-sensing image or not, two specific transfer learning sub-problems can be encountered: (i) learning under *sample selection bias*; and (ii) *domain adaptation*.

Learning under *sample selection bias* is associated with cases in which unlabelled test data are assumed to be drawn from the same remote-sensing image (i.e., the same domain) of training data. However, although training patterns are available (thus completely unsupervised methods are not optimal to solve the classification problem properly), they do not allow for reliable learning of supervised algorithms (*unrepresentative* training set). In particular, since the number and/or the quality of labelled samples is assumed to be inadequate for properly training the classifier, the estimated distribution $\hat{p}(\mathbf{x}, y) = \hat{p}(\mathbf{x})\hat{p}(y|\mathbf{x})$ does not correctly model the true underlying distribution $p(\mathbf{x}, y)$. On the one hand, a relatively small number of training samples (with respect to the number of free parameters of the classifier) may lead to a poor estimation $\hat{p}(\mathbf{x})$ and results, generally, in the well-known Hughes phenomenon (Hughes 1968). On the other hand, unrepresentative training patterns (e.g., correlated patterns violating the assumption of independence, samples that do not properly model all the realizations of the information classes, etc.) introduce a bias in the estimated class prior distribution (i.e., $\hat{p}(y) \neq p(y)$) and may cause a poor estimation of the conditional distribution (i.e., $\hat{p}(y|\mathbf{x}) \neq p(y|\mathbf{x})$).

Domain adaptation is related to more critical situations in which no training patterns are available for the considered dataset, but where labelled samples are known for a different dataset that refers to the same scenario under investigation. This is the case of land-cover maps updating when training samples are only provided for a reference image acquired at one date, but no true labels are available for a new image acquired over the same area at a different date (gathering reliable ground truth for each new acquisition is not realistic and is generally very expensive both in terms of time and economic cost). Although the considered images refer to the same area, it is reasonable to expect that relevant changes might occur between the information class distributions characterizing each specific acquisition date for several possible reasons (e.g., different acquisition system state, changes occurred on the ground, alterations in the phenologic state of vegetation, dissimilar illumination conditions, etc.). Therefore, unlabelled test patterns of the new image belong to a target domain D_t different from the source domain D_s of training samples of the reference image. In this context, the key idea is to infer a good approximation of $p^t(\mathbf{x}, y)$ by exploiting $p^s(\mathbf{x}, y)$, where $p^s(\mathbf{x}, y) = p^s(y|\mathbf{x}) \cdot p^s(\mathbf{x})$ and $p^t(\mathbf{x}, y) = p^t(y|\mathbf{x}) \cdot p^t(\mathbf{x})$ represent the true underlying distributions for D_s and D_t, respectively. In the framework of domain adaptation, most of the learning methods are inspired by the idea that, although different, the two considered domains are intimately related (Ben-David *et al.* 2007; Daumé III and Marcu 2006; Jiang and Zhai 2007; Satpal and Sarawagi 2007). In particular, it is intuitive to observe that considering the unlabelled data of the target domain in the training phase could improve the performances over the case when this information source is ignored.

It is worth noting that, in the context of sample selection bias the available training set, even if unrepresentative, provides a raw description of the analysed image, whereas in domain adaptation the available labelled samples are not related to the specific problem under investigation. This aspect involves an important difference in the learning strategies: in the context of sample selection bias, the unlabelled samples allow one to integrate and complete the information provided by training patterns; in domain adaptation, the available labelled samples can be used only for driving the learning problem, whose solution should reflect only the structure of the unlabelled patterns.

Recently, in the remote-sensing community great attention has been focused on solving problems with the sample selection bias hypothesis. In such situations, transductive and semi-supervised learning methods proved capable of recovering satisfactory discrimination performances by exploiting unlabelled data (taken from the image to classify) along with the unrepresentative training data. In particular, while *transductive* classifiers are specifically designed to handle only the unlabelled samples used in the learning process, *semi-supervised classifiers* allow one to handle any unlabelled data of the considered image. In several applications, these approaches proved to be particularly effective and resulted in a relevant increase of the classification accuracy with respect to supervised approaches (Zhu 2005).

On the other hand, due to its intrinsic higher complexity, less attention has been devoted to the development of domain adaptation classifiers for solving the challenging task of land-cover maps updating when ground truth is available only for a reference image. In this framework, the authors have defined and developed partially-unsupervised classification techniques in previous works (Bruzzone and Fernández Prieto 2001, 2002; Bruzzone and Cossu 2002; Bruzzone *et al.* 2002). However, even if such approaches proved to be effective, their parametric nature prevents them from being effectively employed in cases where it is not possible to explicitly model the type of distribution that governs the investigated classification problem. In order to overcome this drawback, in this chapter we present a novel domain adaptation support vector machine (DASVM) technique that extends support vector machines (SVMs) to the domain adaptation framework. From a general perspective, available labelled data from the reference image are used for determining an initial unreliable solution for the new image to be categorized. Then, the proposed technique iteratively selects and labels the unlabelled patterns of the new image that are most likely to be correctly classified. At the same time, original labelled samples are gradually erased, as they refer to an image that is different from the one to be classified.

Besides the intrinsic complexity of the considered problem, another important limitation in developing operational classification methods for addressing land-cover maps updating is related to the lack of validation strategies that allow assessing the effectiveness of the classification results. Under the assumption that no ground-truth labels are available for the image to classify, in fact, standard statistical validation methods cannot be employed. In this context, as a second contribution of this chapter we present a circular validation strategy for automatically estimating the correctness of a land-cover map obtained with the proposed DASVM technique. In particular, by assuming the existence of an intrinsic structure intimately relating the source and target domains, our strategy (which can be also used for validating the classification results of any domain adaptation method) indirectly analyses the robustness of the classification map obtained for the new image at the end of a circular (i.e., forward and backward) procedure by exploiting available labelled samples for the reference image.

Several experimental results obtained on a multi-temporal and multi-spectral dataset related to Lake Mulargia (Sardinia Island, Italy) confirmed the effectiveness and the reliability of the proposed system.

The chapter is organized into six sections. In Section 9.2 a survey on both transductive and semi-supervised methods is reported, as well as an overview of partially-unsupervised (domain adaptation) classifiers. Section 9.3 and Section 9.4 present the proposed DASVM technique and the circular validation strategy defined for addressing automatic land-cover maps updating, respectively. In Section 9.5 experimental results are reported. Finally, Section 9.6 draws the conclusions of the chapter.

9.2 Literature survey

In this section, we present a brief overview of the main contributions that have been proposed in the remote-sensing literature for addressing transfer learning classification problems. In particular, we first focus attention on both transductive and semi-supervised methods defined for solving problems under the hypothesis of sample selection bias. We then describe partially-unsupervised methods devised for addressing automatic updating of land-cover maps in the context of domain adaptation.

9.2.1 Learning under sample selection bias: transductive and semi-supervised methods

In many remote-sensing applications, several transductive and semi-supervised methods have been proposed so far for recovering satisfactory classification accuracies in the presence of unrepresentative training samples.

An effective family of approaches involves the use of generative models, which assume that data are drawn from an identifiable mixture distribution $p(\mathbf{x}|y)$. Under this hypothesis, it has been demonstrated that, in the presence of a large number of unlabelled patterns, it is possible to identify the mixture components; thus, only a few labelled samples are actually needed to estimate $\hat{p}(\mathbf{x}, y)$ correctly. In this context, several works presented in the literature exploit the Expectation-Maximization (EM) algorithm (Dempster *et al.* 1977) for a maximum likelihood estimation of the parameters that characterize the distributions of the considered information classes. Shahshahani and Landgrebe (1994) first introduced the EM algorithm to estimate model parameters of a Gaussian Maximum Likelihood classifier with both labelled and unlabelled samples for the classification of hyperspectral remote-sensing images. To limit the risk that semi-labelled samples (i.e., originally unlabelled samples associated with estimated labels during the learning process) could negatively affect the learning process, an effective weighting strategy was presented by Tadjudin and Landgrebe (2000). In particular, while full weights are associated to original labelled samples, reduced weights are assigned to semi-labelled samples. Since the covariance matrices are generally highly variable when only very few labelled patterns are available, Jackson and Landgrebe (2002) proposed an adaptive strategy that allows one to improve the performances by considering also semi-labelled patterns in the estimation phase.

In the last few years, in the machine learning community growing attention has been focused on the development of discriminative models, directly aimed at estimating the conditional distribution $\hat{p}(y|\mathbf{x})$. In this framework, Transductive SVMs (TSVMs) (Vapnik 1998, 1999) and Semi-supervised SVMs (S^3VMs) (Bennett and Demiriz 1998) proved particularly effective in many remote-sensing applications. These approaches are implemented under the assumption that each cluster of samples belongs to a specific information class; therefore, the decision boundary is defined between clusters in low density regions of the feature space. Based on the analysis of the properties of TSVMs, the authors first proposed a transductive classifier specifically designed for handling ill-posed problems in the context of remote-sensing image classification (Bruzzone *et al.* 2006). In particular, this technique: (i) exploits a transductive procedure that takes advantage of a strategy for weighting semi-labelled samples according to a time-dependent criterion and (ii) permits one to address multiclass problems. Also Dundar and Landgrebe (2004) presented a similar method that exploits a modified version of the Kernel Fisher's discriminant using both labelled and unlabelled data. In particular, the proposed classifier is obtained through an optimization of a quadratic programming problem that minimizes the total cost of misclassified training samples while maximizing the Rayleigh coefficient in the kernel space. Chi and Bruzzone (2007) introduced in remote sensing other two effective S^3VM algorithms for the classification of hyperspectral data implemented and optimized in the primal formulation. The constraints of the labelled and unlabelled samples are directly included in the cost function in order to obtain an unconstrained minimization problem as proposed by Chapelle (2007). The first presented algorithm optimizes the unconstrained objective function by the gradient descent technique, leading to the formulation of the ∇S^3VM. The second classifier combines a graph-based kernel matrix with the ∇S^3VM.

A third group of methods for addressing learning under sample selection bias includes graph-based semi-supervised algorithms, which define a graph where the nodes are both labelled and unlabelled patterns and edges reflect the similarity of the instances. Most of the graph-based classifiers proposed in the literature aim at estimating an objective function on the graph, which is expected to be smooth on the whole graph, as well as sufficiently close to the available labelled samples. In this context, an effective approach for hyperspectral image classification has been presented by Camps-Valls *et al.* (2007), who also integrated contextual information in the formulation of the graph-based method presented by Zhou *et al.* (2004). Another interesting technique has been recently presented by Gómez *et al.* (2008). In particular, the authors extended to the remote-sensing framework the Laplacian SVM technique proposed by Belkin *et al.* (2006), which introduces an additional regularization term on the geometry of both labelled and unlabelled samples by using the graph Laplacian (Jordan 1999). In contrast to most transductive learning methods, the proposed classifiers follows a non-iterative optimization procedure; moreover, with respect to other graph-based approaches, it also provides out-of-sample predictions.

Recently, systems based on ensemble methods have also been defined. As an example, Chi and Bruzzone (2005, 2006) proposed the employment of semi-labelled sample driven bagging techniques. Available labelled samples are used to train a base classifier; then a subset of representative semi-labelled samples is merged with the original training samples. This new hybrid set is used to train a number of semi-supervised classifiers whose cost functions account for the presence of semi-labelled samples. Finally, an ensemble architecture is used to combine the results provided by each classifier in order to increase the robustness of the classification results.

9.2.2 Domain adaptation: partially-unsupervised methods

Although several algorithms have been proposed for tackling remote-sensing problems under the hypothesis of sample selection bias, currently only few methods have been defined for handling the challenging problem of land-cover maps updating in the framework of domain adaptation. To this aim, the authors introduced in previous works partially-unsupervised techniques (Bruzzone and Fernández Prieto 2001, 2002; Bruzzone and Cossu 2002; Bruzzone *et al.* 2002).

Bruzzone and Fernández Prieto (2001) first presented an effective approach for updating the parameters of a maximum-likelihood (ML) classifier trained on a reference image on the basis of the distribution of a new image for which no ground-truth information is available. In particular, the key idea consists in modelling the observed spaces by a mixture of distributions, whose components are estimated through the use of a specific version of the EM algorithm that exploits both labelled samples of the reference image and unlabelled samples of the new image. To take into account the existence of a temporal correlation between two images acquired over the same area at different times, the partially-unsupervised ML approach has been then reformulated in the framework of the Bayesian rule for cascade classification, which jointly considers the information contained in all the items of a temporal series (Bruzzone and Fernández Prieto 2002).

Another interesting approach based on a multiple-classifier system has been proposed by Bruzzone *et al.* (2002). In particular, the aforementioned unsupervised retraining technique for the ML classifier was extended also to radial basis function neural networks (RBF-NNs) and different strategies for the combination of the resulting classifiers have been considered. Successively, this approach has been further extended for defining a multiple-cascade-classifier system (MCCS) architecture where each classifier of the ensemble is developed in the framework of cascade classification (Bruzzone and Cossu 2002). Besides partially-unsupervised ML and RBF-NNs, hybrid ML and RBF-NNs cascade classifiers have also been devised in order to increase both the effectiveness and the robustness of the ensemble. In this way, the resulting MCCS is characterized by a higher reliability than single algorithms composing the ensemble.

Despite the relevant impact it would have in several applications, nowadays the problem of automatic land-cover map updating is still scarcely investigated. In order to address this critical issue, we developed the domain adaptation SVM classifier described in the next section.

9.3 Proposed domain adaptation SVM

In this section, we describe the proposed domain adaptation SVM (DASVM) classifier. The considered assumptions are discussed in the following; the complete formulation of the DASVM algorithm is then presented in detail.

9.3.1 DASVM: problem definition and assumptions

Let $\mathcal{I}_1$ and $\mathcal{I}_2$ represent two remote-sensing images acquired over the same area at times t_1 and t_2, respectively. Let $\mathcal{X}_1 = \{\mathbf{x}_i^1 | \mathbf{x}_i^1 \in \mathcal{I}_1\}_{i=1}^N$ and $\mathcal{X}_2 = \{\mathbf{x}_i^2 | \mathbf{x}_i^2 \in \mathcal{I}_2\}_{i=1}^M$ denote two subsets of $\mathcal{I}_1$ and $\mathcal{I}_2$ composed of N and M instances, respectively, where $\mathbf{x}_i^1$ and $\mathbf{x}_i^2$ represent the d-dimensional feature vectors associated with the ith sample of $\mathcal{I}_1$ and $\mathcal{I}_2$. In the formulation of the proposed DASVM technique, let us assume that:

1. the same set of L information classes, $\Omega = \{\omega_i\}_{i=1}^L$, characterizes the two remote-sensing images $\mathcal{I}_1$ and $\mathcal{I}_2$;
2. the set of ground-truth labels $\mathcal{Y}_1 = \{y_i^1 | y_i^1 \in \Omega\}_{i=1}^N$ for $\mathcal{X}_1$ is available, therefore it is possible to define the training set $\mathcal{T}_1 = \{\mathcal{X}_1, \mathcal{Y}_1\} = \{(\mathbf{x}_\ell^1, y_\ell^1)\}_{\ell=1}^N$ for $\mathcal{I}_1$;
3. a set of ground-truth labels $\mathcal{Y}_2 = \{y_i^2 | y_i^2 \in \Omega\}_{i=1}^M$ for $\mathcal{X}_2$ is not available, thus it is not possible to define any training set for $\mathcal{I}_2$.

The objective of the DASVM technique is to obtain an accurate classification of the new image $\mathcal{I}_2$ by jointly exploiting labelled training samples $\mathcal{T}_1$ from $\mathcal{I}_1$ and unlabelled samples $\mathcal{X}_2$ from $\mathcal{I}_2$.

Note that it is reasonable to assume the existence of two extreme conditions for the distributions $p^1(\mathbf{x}, y)$ and $p^2(\mathbf{x}, y)$ that govern $\mathcal{I}_1$ and $\mathcal{I}_2$, respectively. If $p^1(\mathbf{x}, y) \equiv p^2(\mathbf{x}, y)$, then adaptation is not necessary and standard supervised learning algorithms can be effectively employed. Instead, if $p^1(\mathbf{x}, y)$ and $p^2(\mathbf{x}, y)$ are independent, then the available training set $\mathcal{T}_1$ for $\mathcal{I}_1$ is useless for building a model for $\mathcal{I}_2$, therefore adaptation cannot be performed. Nevertheless, in real applications $p^1(\mathbf{x}, y)$ and $p^2(\mathbf{x}, y)$ are generally neither identical nor independent; thus, it is reasonable to assume the existence of an intrinsic relationship between $\mathcal{I}_1$ and $\mathcal{I}_2$ that makes adaptation possible. It is worth noting that the probability of succeeding in the adaptation process is expected to depend on the intrinsic similarity between the two distributions (i.e., the larger the difference between $p^1(\mathbf{x}, y)$ and $p^2(\mathbf{x}, y)$, the harder it is to relate one domain to the other).

9.3.2 DASVM: formulation

For simplicity, in the following we describe the proposed DASVM technique in the case of a two-class problem. With respect to standard supervised SVMs, DASVMs are formulated directly considering that unlabelled samples $\mathcal{X}_2$ are drawn from a new image $\mathcal{I}_2$ that is different from the reference image $\mathcal{I}_1$ of the training samples. Accordingly, labelled patterns are only exploited for initializing the discriminant function for $\mathcal{I}_2$, while successively they are gradually erased in order to obtain a final separation hyperplane defined only on the basis of the samples of the new image (Figure 9.1). DASVMs, by iteratively deleting original training samples and gradually adapting the discriminant function to the unlabelled instances, can recover useful information and properly seize the classification problem at time t_2.

The DASVM algorithm is characterized by three different phases: (i) initialization (only original training samples $\mathcal{T}_1$ are used for initializing the discriminant function); (ii) iterative domain adaptation (both original training samples $\mathcal{T}_1$ and unlabelled samples $\mathcal{X}_2$, respectively, are jointly used for gradually adapting the discriminant function to $\mathcal{I}_2$); and (iii) convergence (only unlabelled samples $\mathcal{X}_2$ are used for defining the final discriminant function). In the following, $\mathcal{T}^{(i)}$ and $\mathcal{X}_2^{(i)}$ will denote the training set and the unlabelled set (i.e., the set containing

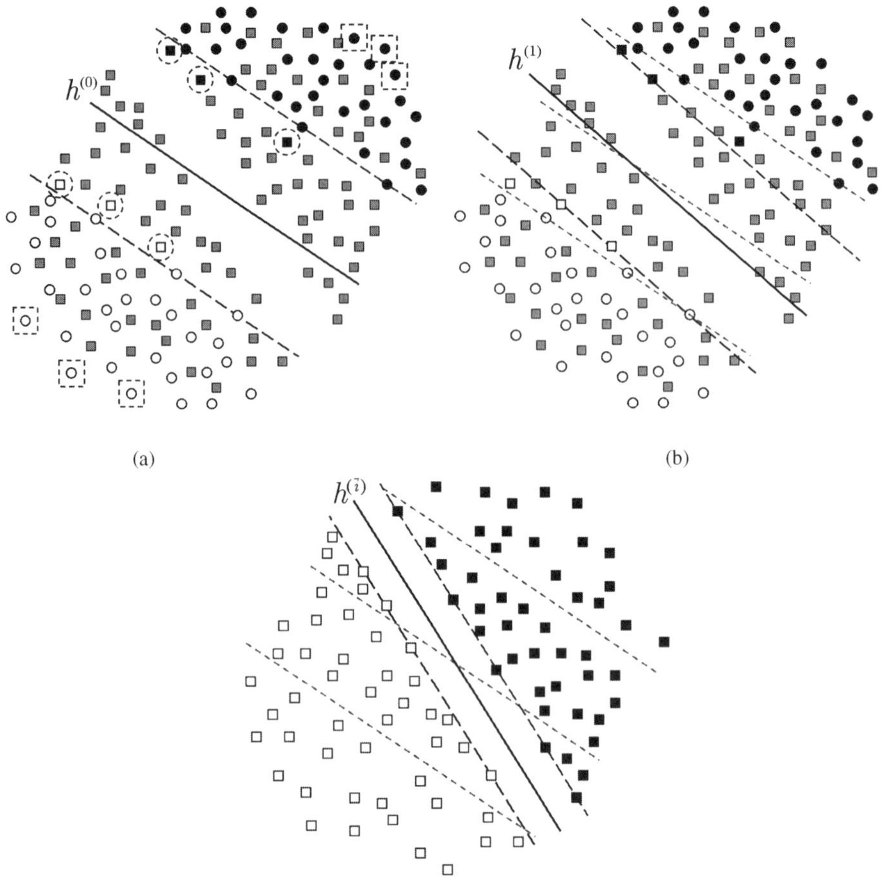

Figure 9.1 *Separation hyperplane (solid lines) and margin bounds (dashed lines) at different stages of the DASVM algorithm for a toy dataset. Original patterns at time t_1 are shown as white and black circles. Semi-labelled patterns are shown as white and black squares, respectively. Remaining unlabelled patterns at time t_2 are represented as grey squares. Feature space structure obtained: (a) at the first iteration (the dashed circles and squares highlight the current semi-labelled patterns and the current original training patterns to delete, respectively; in the example $\rho = 3$); (b) at the second iteration and (c) at the last iteration, respectively, in an ideal situation (the dashed grey lines represent both the separation hyperplane and the margin bounds at the beginning of the learning process).*

the unlabelled samples that have not been inserted into the training set $\mathcal{T}^{(i)}$) at the current iteration i, respectively.

Phase 1: Initialization In the first phase, as for supervised SVMs, an initial separation hyperplane $h^{(0)} : f^{(0)}(\mathbf{x}) = \langle \mathbf{w}^{(0)}, \mathbf{x} \rangle + b^{(0)} = 0$ is determined on the basis of the only original training patterns available for the reference image $\mathcal{I}_1$. We have that $\mathcal{T}^{(0)} \equiv \mathcal{T}_1 = \{(\mathbf{x}_\ell^1, y_\ell^1)\}_{\ell=1}^N$ and $\mathcal{X}_2^{(0)} = \{\mathbf{x}_u^2\}_{u=1}^M$. Accordingly, the functional to minimize is:

$$\begin{cases} \min_{\mathbf{w},b,\boldsymbol{\xi}} \left\{ \frac{1}{2} \left\| \mathbf{w}^{(0)} \right\|^2 + C \sum_{\ell=1}^{N} \xi_\ell^1 \right\} \\ y_\ell^1 \left(\left\langle \mathbf{w}^{(0)}, \mathbf{x}_\ell^1 \right\rangle + b^{(0)} \right) \geq 1 - \xi_\ell^1 \ \forall \ell = 1, \ldots, N \\ \xi_\ell^1 \geq 0, \end{cases} \tag{9.1}$$

where $\mathbf{w}$ is a vector normal to h, b is a constant such that $b/\|\mathbf{w}\|^2$ represents the distance of h from the origin, ξ_ℓ are slack variables allowing for (permitted) errors and C is the associated *regularization parameter*, which permits one to tune the generalization capability.

Phase 2: Iterative domain adaptation In the second phase, both labelled samples from $\mathcal{I}_1$ and unlabelled samples from $\mathcal{I}_2$ are jointly taken into consideration. At the generic iteration i, all the original unlabelled samples of the new image $\mathcal{I}_2$ are associated with an estimated label $\hat{y}_u^{2(i)} = sgn\left[f^{(i)}(\mathbf{x}_u^2)\right]$, where $f^{(i)}\left(\mathbf{x}_u^2\right) = \left\langle \mathbf{w}^{(i)}, \mathbf{x}_u^2 \right\rangle + b^{(i)}$ represents the current decision function. In order to seize the classification problem at time t_2, a subset of the remaining unlabelled samples $\mathcal{X}_2^{(i)}$ is iteratively selected and moved with the corresponding estimated labels into the training set at the following iteration $\mathcal{T}^{(i+1)}$. Let us consider that:

1. the current unlabelled patterns of $\mathcal{I}_2$ falling into the margin band $\mathcal{M}^{(i)} = \left\{\mathbf{x} \middle| -1 \leq f^{(i)}(\mathbf{x}) \leq 1\right\}$ have the highest probability of being associated with nonzero Lagrange multipliers once inserted into $\mathcal{T}^{(i+1)}$; accordingly, they have the highest chance of affecting the position of $h^{(i+1)}$ (samples lying outside the margin band, instead, are more likely to be associated with null multipliers);

2. the higher the distance of a given unlabelled sample from the separation hyperplane $h^{(i)} : \left\langle \mathbf{w}^{(i)}, \mathbf{x} \right\rangle + b^{(i)} = 0$, the higher is the chance that it is classified correctly.

In the light of these two observations, at each iteration of the learning process we progressively take into account the unlabelled samples into $\mathcal{M}^{(i)}$ that are closest to the margin bounds, as they are those further from the decision boundary $h^{(i)}$. Let us define the two following subsets:

$$\mathcal{K}_+^{(i)} = \{(\mathbf{x}_u^2, \hat{y}_u^{2(i)} = +1) | \mathbf{x}_u^2 \in \mathcal{X}_2^{(i)}, 1 \geq f^{(i)}(\mathbf{x}_u^2) \geq f^{(i)}(\mathbf{x}_{u+1}^2) \geq 0\}, \tag{9.2}$$

$$\mathcal{K}_-^{(i)} = \{(\mathbf{x}_u^2, \hat{y}_u^{2(i)} = -1) | \mathbf{x}_u^2 \in \mathcal{X}_2^{(i)}, -1 \leq f^{(i)}(\mathbf{x}_u^2) \leq f^{(i)}(\mathbf{x}_{u+1}^2) < 0\}. \tag{9.3}$$

The current unlabelled samples of $\mathcal{I}_2$ lying to the upper and lower side of $\mathcal{M}^{(i)}$ are inserted with their corresponding estimated labels into $\mathcal{K}_+^{(i)}$ and $\mathcal{K}_-^{(i)}$, respectively. In particular, samples of $\mathcal{K}_+^{(i)}$ and $\mathcal{K}_-^{(i)}$ are sorted in ascending order with respect to their distance from the upper (i.e., $h_+^{(i)} : \left\langle \mathbf{w}^{(i)}, \mathbf{x} \right\rangle + b^{(i)} = +1$) and lower bound (i.e., $h_-^{(i)} : \left\langle \mathbf{w}^{(i)}, \mathbf{x} \right\rangle + b^{(i)} = -1$) of the margin band, respectively. Then, at each iteration the first ρ unlabelled patterns of both $\mathcal{K}_+^{(i)}$ and $\mathcal{K}_-^{(i)}$ (where $\rho > 0$ is a free parameter defined *a priori* by the user) are selected (see Figure 10.1(a)).

Chen *et al.* (2003) proposed a similar strategy that iteratively considers the only two unlabelled patterns into $\mathcal{M}^{(i)}$ associated with the maximum and minimum value of the discriminant function. However, we assume that in real applications two patterns may not be representative enough for a proper tuning of the hyperplane. As the cardinality of $\mathcal{K}_+^{(i)}$ and $\mathcal{K}_-^{(i)}$ may be smaller than ρ, the set of semi-labelled patterns chosen at the generic ith iteration becomes:

$$\mathcal{K}^{(i)} = \{(\mathbf{x}_u^2, \hat{y}_u^{2(i)}) \in \mathcal{K}_+^{(i)} | 1 \leq u \leq \lambda^{(i)}\} \cup \{(\mathbf{x}_u^2, \hat{y}_u^{2(i)}) \in \mathcal{K}_-^{(i)} | 1 \leq u \leq \delta^{(i)}\}, \tag{9.4}$$

where $\lambda^{(i)} = \min(\rho, \#\mathcal{K}_+^{(i)})$ and $\delta^{(i)} = \min(\rho, \#\mathcal{K}_-^{(i)})$. Patterns of $\mathcal{K}^{(i)}$ are then merged into the current training set $\mathcal{T}^{(i)}$.

The main purpose of the proposed DASVM technique is to define and solve a bound minimization problem with respect only to the samples of the new image $\mathcal{I}_2$; hence, a strategy for iteratively erasing the original labelled patterns of the reference image $\mathcal{I}_1$ has been developed. In particular, we gradually delete the remaining original training samples that are furthest away from the current decision boundary. In fact, it is reasonable to expect that the higher the distance of a training sample from $h^{(i)}$, the lower is the chance that it can be misclassified after a new tuning of the hyperplane due to the insertion of semi-labelled samples into the training set at the following iteration $\mathcal{T}^{(i+1)}$. Let us define the two following subsets:

$$\mathcal{V}_+^{(i)} = \{(\mathbf{x}_\ell^1, y_\ell^1) \in \mathcal{T}^{(i)} | f^{(i)}(\mathbf{x}_\ell^1) \geq f^{(i)}(\mathbf{x}_{\ell+1}^1) \geq 0\}, \tag{9.5}$$

$$\mathcal{V}_-^{(i)} = \{(\mathbf{x}_\ell^1, y_\ell^1) \in \mathcal{T}^{(i)} | f^{(i)}(\mathbf{x}_\ell^1) \leq f^{(i)}(\mathbf{x}_{\ell+1}^1) < 0\}. \tag{9.6}$$

$\mathcal{V}_+^{(i)}$ and $\mathcal{V}_-^{(i)}$ include the remaining original labelled patterns at time t_1 that lie above and below $h^{(i)}$, respectively, sorted in descending order with respect to their distance from the separation hyperplane. For balancing the contribution of the new semi-labelled samples, at each iteration the number of patterns to delete from $\mathcal{V}_+^{(i)}$ and $\mathcal{V}_-^{(i)}$ is set equal to $\lambda^{(i)}$ and $\delta^{(i)}$ (i.e., the number of semi-labelled patterns selected from the upper and the lower side of the margin band), respectively. If none of the remaining unlabelled samples of $\mathcal{I}_2$ falls into the margin band (i.e., $\mathcal{K}^{(i)} = \varnothing$), the number of samples to delete is set equal to ρ. Accordingly, the final subset of patterns to delete is defined as:

$$\mathcal{V}^{(i)} = \{(\mathbf{x}_\ell^1, y_\ell^1) \in \mathcal{V}_+^{(i)} | 1 \leq \ell \leq \varsigma^{(i)}\} \cup \{(\mathbf{x}_\ell^1, y_u^1) \in \mathcal{V}_-^{(i)} | 1 \leq \ell \leq \kappa^{(i)}\}, \tag{9.7}$$

where

$$\varsigma^{(i)} = \begin{cases} \min(\lambda^{(i)}, \#\mathcal{V}_+^{(i)}) & \text{if } \mathcal{K}^{(i)} \neq \varnothing \\ \min(\rho, \#\mathcal{V}_+^{(i)}) & \text{if } \mathcal{K}^{(i)} = \varnothing \end{cases} \quad \text{and} \quad \kappa^{(i)} = \begin{cases} \min(\delta^{(i)}, \#\mathcal{V}_-^{(i)}) & \text{if } \mathcal{K}^{(i)} \neq \varnothing \\ \min(\rho, \#\mathcal{V}_-^{(i)}) & \text{if } \mathcal{K}^{(i)} = \varnothing \end{cases}.$$

Let $\upsilon^{(i)}$ and $\chi^{(i)}$ represent the number of remaining labelled samples of $\mathcal{I}_1$ and semi-labelled samples of $\mathcal{I}_2$ into the current training set $\mathcal{T}^{(i)}$, respectively. For $i \geq 1$, the optimization problem becomes:

$$\begin{cases} \min_{\mathbf{w},b,\boldsymbol{\xi}^1,\boldsymbol{\xi}^2} \left\{ \frac{1}{2} \left\| \mathbf{w}^{(i)} \right\|^2 + C^{(i)} \sum_{\ell=1}^{\upsilon^{(i)}} \xi_\ell^1 + \sum_{u=1}^{\chi^{(i)}} C_u^* \xi_u^2 \right\} \\ y_\ell^1 \left(\left\langle \mathbf{w}^{(i)}, \mathbf{x}_\ell^1 \right\rangle + b^{(i)} \right) \geq 1 - \xi_\ell^1 \ \forall \ell = 1, \ldots, \upsilon^{(i)} \\ \hat{y}_u^{2(i-1)} \left(\left\langle \mathbf{w}^{(i)}, \mathbf{x}_u^2 \right\rangle + b^{(i)} \right) \geq 1 - \xi_u^2 \ \forall u = 1, \ldots, \chi^{(i)} \\ \xi_\ell^1, \xi_u^2 \geq 0. \end{cases} \tag{9.8}$$

With respect to supervised SVMs, one can notice that:

1. semi-labelled samples $\left(\mathbf{x}_u^2, \hat{y}_u^{2(i-1)}\right) \in \mathcal{T}^{(i)}$ of the new image at time t_2 are associated with a specific regularization parameter $C_u^* \in \mathbb{R}^+$;
2. training patterns $\left(\mathbf{x}_\ell^1, y_\ell^1\right) \in \mathcal{T}^{(i)}$ of the reference image at time t_1 are associated with a regularization parameter $C^{(i)}$ that varies at each iteration.

The purpose of C_u^* and $C^{(i)}$ is to control the number of misclassified samples of the current training set $\mathcal{T}^{(i)}$ associated with $\mathcal{I}_2$ and $\mathcal{I}_1$, respectively. The larger their values, the higher is the influence of the associated samples in determining the position of the separation hyperplane. Accordingly, since it is reasonable to expect that the two distributions $p^1(\mathbf{x}, y)$ and $p^2(\mathbf{x}, y)$ could be rather different, in order to avoid instabilities a weighting strategy based on a temporal criterion is adopted.

On the one hand, we associate semi-labelled samples with a regularization parameter that grows depending on the number of iterations for which they have been assigned the same estimated label until iteration $i - 1$. The confidence of the system on a semi-labelled pattern is expected to increase depending on the number of iterations for which that pattern is associated with the same information class. Hence, by increasing the value of C_u^*, the influence of the associated sample in the definition of the decision boundary increases. In particular, we make C_u^* rise iteratively in a quadratic way (see Figure 9.2(a)); thus, for each semi-labelled sample

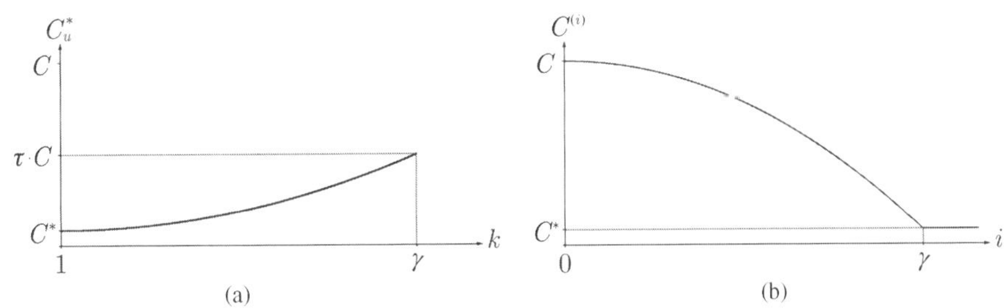

Figure 9.2 (a) Behaviour of C_u^* (regularization parameter for the semi-labelled patterns belonging to the current training set $\mathcal{T}^{(i)}$) versus k (index corresponding to the number of iterations for which they have been associated with the same label until iteration $i - 1$). (b) Behaviour of the regularization parameter $C^{(i)}$ for the original training patterns at time t_1 versus the number of iterations i.

$\left(\mathbf{x}_u^2, \hat{y}_u^{2(i-1)}\right)$ the corresponding regularization parameter is defined as:

$$C_u^* = \frac{\tau \cdot C - C^*}{(\gamma - 1)^2}(k-1)^2 + C^* \Longleftrightarrow \left(\mathbf{x}_u^2, \hat{y}_u^{2(i-1)}\right) \in \mathcal{J}_k^{(i)} \quad k = 1, \ldots, \gamma, \tag{9.9}$$

where $\mathcal{J}_k^{(i-1)}$ includes all the current semi-labelled patterns in $\mathcal{T}^{(i)}$ that have been assigned the same estimated label for k successive iterations, C^* is the initial regularization value for semi-labelled samples (this is a user defined parameter), and τ, $0 < \tau \leq 1$, tunes the maximum cost value of semi-labelled samples (i.e., a reasonable choice has generally proved to be $\tau = 0.5$). In (9.9), γ is defined as the maximum number of iterations for which the regularization parameter for semi-labelled samples is allowed to increase by the user.

On the other hand, we make the cost factor $C^{(i)}$ for the original labelled samples of $\mathcal{I}_1$ decrease in a quadratic way too depending on i (see Figure 10.2(b)):

$$C^{(i)} = \max\left(\frac{C^* - C}{\gamma^2} i^2 + C, C^*\right). \tag{9.10}$$

Such a strategy lets us balance the growing importance of semi-labelled samples in tuning the position of the separation hyperplane. In particular, while at the beginning the position of h mainly depends on labelled patterns available at time t_1, as the number of iteration increases (i.e., until $i = \gamma$) their influence always becomes less relevant.

The strategy adopted for deleting patterns of $\mathcal{I}_1$ might also result in the removal of mislabeled original training samples, which affect the position of the separation hyperplane as they are associated with nonzero Lagrange multipliers. Nonetheless, this aspect does not seem to be particularly critical for two main reasons.

1. If any mislabeled training samples of the reference image fall on the wrong side of the margin band or close to the wrong margin bound, they are expected to be erased when the learning process is close to convergence or, more generally, when the number of current semi-labelled samples of $\mathcal{I}_2$ is high. It is reasonable to assume that in such a scenario the decision boundary strongly depends on semi-labelled samples. Accordingly, the deletion of the aforementioned original training samples is not critical. It is also possible to notice that, since this is expected to happen after several iterations, training patterns would be associated with a small value of the regularization parameter.

2. If mislabeled original training samples fall far away from the separation hyperplane (i.e., $\xi_\ell \gg 1$), they can be reasonably assumed to be outliers (e.g., they could have been associated with a wrong label while defining the training set at time t_1). In general, at most only a few original training samples are expected to exhibit such behaviour. Therefore, even if they are still associated with a non-negligible value of the regularization parameter, their deletion would not significantly affect the position of the decision boundary.

Since the training set progressively changes, the position of h is also expected to change at each iteration (see Figure 9.1(b)). Hence, a dynamical adjustment has been designed to take into account that a given semi-labelled sample could be associated with different estimated labels between two successive iterations. Let

$$\mathcal{S}^{(i)} = \left\{ \left(\mathbf{x}_u^2, \hat{y}_u^{2(i-1)} \right) \in \mathcal{T}^{(i)} | \hat{y}_u^{2(i)} \neq \hat{y}_u^{2(i-1)} \right\} \tag{9.11}$$

represent the set of semi-labelled samples belonging to $\mathcal{T}^{(i)}$ whose labels at iteration i (i.e., $\hat{y}_u^{2(i)}$) are different from those at iteration $i-1$ (i.e., $\hat{y}_u^{2(i-1)}$). If $\hat{y}_u^{2(i)}$ differs from $\hat{y}_u^{2(i-1)}$, we assume that the system is no more confident on that sample. Accordingly, such a label is erased and the corresponding 'inconsistent' pattern is reset to the unlabelled state and moved back to $\mathcal{X}_2^{(i+1)}$. This makes it possible to reconsider the same sample in the following iterations of the learning process.

The second phase ends when the convergence criteria described in the next sub-section are satisfied.

Phase 3: Convergence As stated above, the goal of the proposed DASVM is to obtain a final discriminant function that depends only on samples of $\mathcal{I}_2$. Accordingly, a mandatory condition for the system to reach convergence is that all the original labelled samples of the reference image $\mathcal{I}_1$ must be completely erased (i.e., $\mathcal{V}^{(i)} = \varnothing$). Under this assumption, it can be assumed that convergence occurs if none of the remaining unlabelled samples at time t_2 fall into $\mathcal{M}^{(i)}$, i.e. $\mathcal{K}^{(i)} = \varnothing$ (see Figure 9.1(c)). However, such an approach might result in a high computational load, especially when the cardinality of $\mathcal{X}_2$ is high. Furthermore, even when the margin band is empty, it should be taken into consideration that the number of inconsistent semi-labelled patterns may not be negligible. In the light of these considerations, the following empirical stopping criteria have been defined:

$$\begin{cases} \mathcal{V}^{(i)} = \varnothing \\ \#\mathcal{K}^{(i)} \leq \lceil \vartheta \cdot M \rceil \\ \#\mathcal{S}^{(i)} \leq \lceil \vartheta \cdot M \rceil, \end{cases} \tag{9.12}$$

where ϑ is a user-defined parameter that allows one to tune the sensitivity of the learning process, and M is the number of original unlabelled samples drawn from $\mathcal{I}_2$. This means that convergence is reached if both the number of inconsistent semi-labelled patterns and remaining unlabelled patterns falling into the current margin band $\mathcal{M}^{(i)}$ is lower than or equal to $\lceil \vartheta \cdot M \rceil$. The final optimization problem at the last iteration $\bar{i}$ is defined as:

$$\begin{cases} \min\limits_{\mathbf{w},b,\boldsymbol{\xi}} \left\{ \frac{1}{2} \|\mathbf{w}\|^2 + \sum\limits_{u=1}^{\chi^{(\bar{i})}} C_u^* \xi_u \right\} \\ \hat{y}_u^{2(\bar{i}-1)} \left(\langle \mathbf{w}, \mathbf{x}_u^2 \rangle + b \right) \geq 1 - \xi_u \ \forall u = 1, \ldots, \chi^{(\bar{i})} \\ \xi_u \geq 0. \end{cases} \tag{9.13}$$

At the end, all the patterns of the new image $\mathbf{x}_u^2 \in \mathcal{I}_2$ are labelled according to the resulting separation hyperplane, i.e. $\hat{y}_u^2 = \text{sgn}[\langle \mathbf{w}, \mathbf{x}_u^2 \rangle + b]$.

At each iteration of the learning process the objective function to minimize is convex. In this context, since a unique solution exists (independently of the number of considered semi-labelled patterns), convergence always occurs. Nevertheless, note that it is not possible to guarantee that the system converges towards a final solution that is acceptable for the investigated task. Indeed, this strictly depends on the unlabelled samples iteratively considered in the training phase and, implicitly, on the similarity between source and target domains.

The DASVM algorithm described in this section is defined for two-class problems. For addressing multiclass problems, a one-against-all (OAA) strategy can be employed (Cristianini and Taylor 2000; Bruzzone *et al.* 2007).

9.4 Proposed circular validation strategy

In the framework of land-cover maps updating, when no labelled samples are available for the new image to classify, standard model selection and statistical validation approaches (e.g., holdout, cross-validation, leave-one-out, etc.) cannot be employed. In order to overcome this limitation, in the following we introduce a novel strategy that can be employed for assessing the validity of the results obtained with any domain adaptation classifier. The joint use of the proposed circular validation strategy and the DASVM classifier described in the previous section let us define an effective system for addressing real land-cover maps updating problems.

9.4.1 Circular validation strategy: rationale

The rationale of the proposed strategy is based on the two following observations.

1. Since the two images $\mathcal{I}_1$ and $\mathcal{I}_2$ are assumed to be acquired over the same geographical area at different times, the corresponding distributions $p^1(\mathbf{x}, y)$ and $p^2(\mathbf{x}, y)$ are reasonably neither identical or independent. Under this hypothesis, there exists a direct relationship between $\mathcal{I}_1$ and $\mathcal{I}_2$ that depends on the similarity between the two distributions.

2. Training samples are available only for the reference image $\mathcal{I}_1$. Accordingly, we are forced to define an indirect strategy that exploits such labelled samples at time t_1 for validating the accuracy of the classification results obtained for the new image $\mathcal{I}_2$ at time t_2.

Let us assume that the considered domain adaptation algorithm, starting from an acceptable accuracy for the reference image $\mathcal{I}_1$, is able to shape the structure of the problem at time t_2 and, thus, to obtain a reliable classification map for $\mathcal{I}_2$. In this hypothesis, it is reasonable to assume that the classifier seized the intrinsic relationship between $p^1(\mathbf{x}, y)$ and $p^2(\mathbf{x}, y)$. Therefore, we expect that it is still possible to obtain an acceptable classification map for $\mathcal{I}_1$ by again applying the same domain adaptation algorithm in the reverse sense (i.e., using the estimated classification labels in place of the missing true labels at time t_2, keeping the same learning parameters, and considering $\mathcal{I}_1$ as the new image to be categorized).

In contrast, if the classifier does not provide a classification that is satisfactory for the investigated task at time t_2, it means that it was unable to recover a reliable estimate $\hat{p}^2(\mathbf{x}, y)$, thus converging to a wrong solution for $\mathcal{I}_2$. In this situation, we assume that it is never possible to obtain a reliable classification map for $\mathcal{I}_1$ again at the end of the backward process from time t_2 to time t_1, due to the unreliability of the estimated training set for $\mathcal{I}_2$.

According to these expected properties, it is possible to exploit the classification accuracy evaluated on the original labelled samples $\mathcal{T}_1$ to make inferences about the correctness of the classification maps obtained for the new image after a circular (forward and backward) application of the considered domain adaptation algorithm.

In the following, we describe the formulation of the proposed validation strategy in the specific case of DASVM classifiers.

9.4.2 Circular validation strategy: formulation

Let $\Lambda(\mathcal{Y}_k, \hat{\mathcal{Y}}_k)$ represent a classification accuracy measure (e.g., the overall accuracy, the kappa coefficient of accuracy, etc.) that evaluates the similarity between a set of estimated labels $\hat{\mathcal{Y}}_k$ (i.e., a solution) predicted by a generic classifier and the corresponding set of true labels $\mathcal{Y}_k$. Let Λ_{th} represents a threshold for Λ. If $\Lambda(\mathcal{Y}_k, \hat{\mathcal{Y}}_k) \geq \Lambda_{\text{th}}$, we assume that the solution $\hat{\mathcal{Y}}_k$ is *consistent* with $\mathcal{Y}_k$ (i.e., it is acceptable for the investigated problem). Accordingly, let us define the following four sets.

1. $\mathcal{A}$ contains all the solutions consistent with the reference image $\mathcal{I}_1$, i.e. $\Lambda(\mathcal{Y}_1, \hat{\mathcal{Y}}_1) \geq \Lambda_{\text{th}}$.
2. $\mathcal{B}$ contains all the solutions consistent with the new image $\mathcal{I}_2$, i.e. $\Lambda(\mathcal{Y}_2, \hat{\mathcal{Y}}_2) \geq \Lambda_{\text{th}}$.
3. $\mathcal{C}$ contains all the solutions not consistent with the reference image $\mathcal{I}_1$, i.e. $\Lambda(\mathcal{Y}_1, \hat{\mathcal{Y}}_1) < \Lambda_{\text{th}}$.
4. $\mathcal{D}$ contains all the solutions not consistent with the new image $\mathcal{I}_2$, i.e. $\Lambda(\mathcal{Y}_2, \hat{\mathcal{Y}}_2) < \Lambda_{\text{th}}$.

Let us train a supervised SVM using the labelled samples available at time t_1, and select a solution for $\mathcal{I}_1$ that belongs to $\mathcal{A}$ (i.e., the system is in the state $\bar{A}$). Then, we train a DASVM using $\mathcal{T}_1$ and $\mathcal{X}_2$ as training and unlabelled sets, respectively, and keep the same values adopted in the aforementioned supervised learning for both the regularization parameter C and the kernel function free parameters. We assume that, if the choice of the values for the domain adaptation parameters (i.e., C^*, ρ, and γ) is adequate, then the accuracy of the classification map at time t_2 is acceptable and the solution belongs to $\mathcal{B}$ (i.e., the system moves to the state $\bar{B}$). On the other hand, with an improper choice for the parameters, we expect to obtain an unreliable estimate $\hat{p}^2(\mathbf{x}, y)$, thus the resulting solution belongs to $\mathcal{D}$ (i.e., the system moves to the state $\bar{D}$).

Let us now consider the backward domain adaptation problem from time t_2 to time t_1. By exploiting the set $\hat{\mathcal{Y}}_2$ of estimated labels for $\mathcal{X}_2$, we define the estimated training set $\hat{\mathcal{T}}_2 = \{\mathcal{X}_2, \hat{\mathcal{Y}}_2\}$ for $\mathcal{I}_2$. Then, we again train the DASVM classifier keeping the same learning parameters and jointly exploiting $\hat{\mathcal{T}}_2$ and the subset of instances $\mathcal{X}_1$ at time t_1 (considered without the corresponding ground-truth labels $\mathcal{Y}_1$) in the training phase.

Note that, since true labels $\mathcal{Y}_1$ are available, it is possible to evaluate the considered classification accuracy measure $\Lambda(\mathcal{Y}_1, \hat{\mathcal{Y}}_1)$ associated with the results obtained after the circular learning process. If $\Lambda(\mathcal{Y}_1, \hat{\mathcal{Y}}_1) \geq \Lambda_{\text{th}}$ the resulting solution is acceptable for $\mathcal{I}_1$ (i.e., it belongs to $\mathcal{A}$) and the system moves back to state $\bar{A}$. In contrast, if $\Lambda(\mathcal{Y}_1, \hat{\mathcal{Y}}_1) < \Lambda_{\text{th}}$, the classification accuracy at time t_1 cannot be considered acceptable and the solution belongs to $\mathcal{C}$ (i.e., the system moves to state $\bar{C}$). All the possible transitions of the system are shown in Figure 9.3.

It is worth noting that, when the system starting from state $\bar{A}$ is able to return again to state $\bar{A}$ at the end of the circular procedure, the solution obtained at time t_2 is always consistent. Thus, the classification map for $\mathcal{I}_2$ represents a reliable updating of the land-cover map obtained for the reference image $\mathcal{I}_1$. This issue is crucial, since it means that it is possible to validate the correctness of the updated classification map even in the absence of prior ground-truth labels at time t_2.

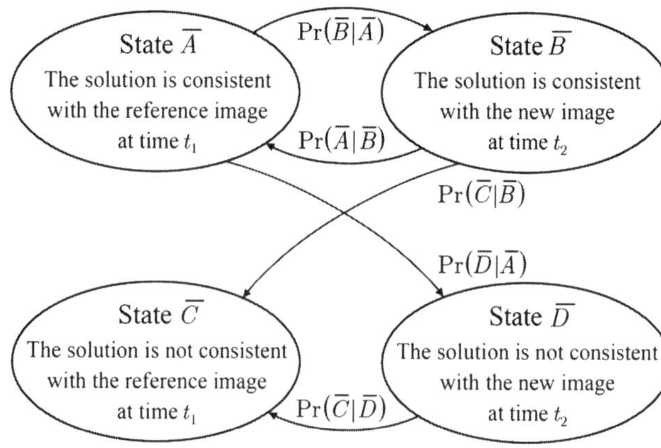

Figure 9.3 Diagram of all the possible state transitions of solutions exploited from the proposed circular validation strategy.

Let Pr $(\bar{Y}|\bar{X})$ represent the probability that, by applying the proposed DASVM algorithm, the system moves to the generic state $\bar{Y}$ starting from the generic state $\bar{X}$. Accordingly, the proposed validation technique is effective under the two following assumptions.

1. *Starting from the state $\bar{B}$ the system can return to the state $\bar{A}$.* If there exists a subset of solutions obtained for the unlabelled instances $\mathcal{X}_2$ in the forward sense (from time t_1 to time t_2) that are acceptable for $\mathcal{I}_2$ (i.e., $\mathcal{B} \neq \varnothing$), then by applying the DASVM in the backward sense (from time t_2 to time t_1) for at least one of them it must be possible to obtain a consistent solution for $\mathcal{I}_1$ (i.e., Pr $(\bar{A}|\bar{B}) > 0$).

2. *Starting from state $\bar{D}$ the system must never move back to state $\bar{A}$.* If the solution obtained in the forward sense (from time t_1 to time t_2) for unlabelled instances $\mathcal{X}_2$ is not consistent, then by applying the DASVM classifier algorithm in the backward sense (from time t_2 to time t_1) it must never be possible to obtain an acceptable solution at time t_1 (i.e., Pr $(\bar{A}|\bar{D}) = 0$). Instead, the final solution must always be inconsistent with the true labels $\mathcal{Y}_1$ available for the reference image $\mathcal{I}_1$ (i.e., Pr $(\bar{C}|\bar{D}) = 1$). Similarly to the behaviour of the trajectories that model transitions between different states in chaotic systems, it seems impossible to recover an acceptable solution for $\mathcal{I}_1$ starting from a wrong solution for $\mathcal{I}_2$.

Note that, according to the aforementioned assumptions, it may happen that in some cases solutions acceptable for $\mathcal{I}_2$ are actually rejected. However, this does not represent a critical issue, since the learning parameters are not optimized for the backward process. Instead, the relevant aspects are that: (i) a set of consistent solutions can be correctly validated; and (ii) the system never accepts solutions that are non-consistent at time t_2 (a situation that is potentially very critical in the operational application of DASVM).

9.5 Experimental results

The dataset used in our experiments refers to the Lake Mulargia area, Sardinia Island, Italy (see Figure 9.4). In particular, we considered two co-registered multi-spectral images composed

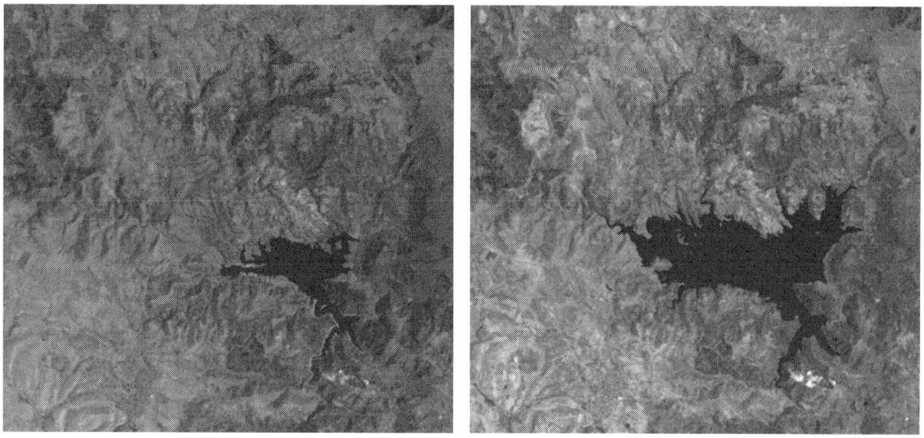

(a) (b)

Figure 9.4 Band 5 of the multi-spectral Landsat-5 Thematic Mapper images used in the experiments. (a) Image acquired in September 1995. (b) Image acquired in July 1996.

of 412 × 382 pixels (i.e., about 11.7 km ×10.8 km) acquired by the Thematic Mapper (TM) sensor of the Landsat-5 Satellite in September 1995 ($\mathcal{I}_1$) and in July 1996 ($\mathcal{I}_2$). In our analysis, the five land-cover classes that mainly characterized the area of interest at both the acquisition dates were taken into account, i.e. forest, pasture, urban area, water and vineyard. Note that, as commonly done in the literature, of the seven available TM spectral bands we did not consider band 6 (i.e., the low-resolution band associated with the thermal infrared channel). Moreover, in order to characterize the textural properties of the investigated information classes and to exploit the distribution-free nature of SVMs, in addition to the six remaining TM bands we also computed five texture features based on the grey-level co-occurrence matrix (GLCM), i.e. correlation, sum average, sum variance, difference variance and entropy. In particular, the GLCM was obtained by compressing the original dynamic of 256 levels to 32 levels with an equal-probability quantizing algorithm (Haralick *et al.* 1973) and using a 7 × 7 window with the inter-pixel distance equal to 1.

Table 9.1 Number of labelled patterns available for the September 1995 ($\mathcal{I}_1$) and the July 1996 ($\mathcal{I}_2$) images and Jensen–Shannon Divergence (D_{JS}) values

Classes	Available labelled patterns		D_{JS}
	$\mathcal{I}_1$	$\mathcal{I}_2$	
Pasture	554 (24.63%)	589 (30.22%)	0.517
Forest	304 (13.52%)	274 (14.06%)	0.278
Urban area	408 (18.14%)	418 (21.45%)	0.391
Water	804 (35.75%)	551 (28.27%)	0.423
Vineyard	179 (7.96%)	117 (6.00%)	0.567
Overall	2249	1949	0.391

By exploiting ground-truth information available for the investigated area, we defined two spatially uncorrelated sets of labelled samples at time t_1 (i.e., September 1995) and time t_2 (i.e., July 1996), respectively (see Table 9.1). However, note that labelled samples related to $\mathcal{I}_2$ were used only for a quantitative assessment of the performances of the proposed system.

According to the notation adopted in the previous sections of this chapter, in the following we will denote as $\mathcal{T}_1 = \{\mathcal{X}_1, \mathcal{Y}_1\}$ the training set available for $\mathcal{I}_1$ (i.e., $\mathcal{X}_1$ and $\mathcal{Y}_1$ represent the instances and the corresponding true labels considered at time t_1, respectively); whereas, we will refer to $\mathcal{X}_2$ as the subset of instances at time t_2 for which ground-truth labels were actually available but were not considered in the learning phase.

In order to assess the complexity of the investigated problem, we estimated the similarity between the pattern distributions $p^1(\mathbf{x})$ and $p^2(\mathbf{x})$ related to $\mathcal{I}_1$ and $\mathcal{I}_2$, respectively. To this aim, we computed the Jensen–Shannon divergence D_{JS} (Lin 1991), which is a symmetrized and smoothed version of the well-known Kullback–Leibler divergence (Kullback and Leibler 1951). D_{JS} is defined as:

$$D_{JS}\left[p^1(\mathbf{x}), p^2(\mathbf{x})\right] = \alpha \cdot \sum_n p_n^1 \log \frac{p_n^1}{p_n^3} + \beta \cdot \sum_n p_n^2 \log \frac{p_n^2}{p_n^3}, \tag{9.14}$$

where p_n^1 and p_n^2 are point probabilities of $p^1(\mathbf{x})$ and $p^2(\mathbf{x})$, respectively, whereas p_n^3 are point probabilities of $p^3(\mathbf{x}) = \alpha \cdot p^1(\mathbf{x}) + \beta \cdot p^2(\mathbf{x})$. In our analysis, we set $\alpha = \beta = 0.5$ (in the literature this particular case is referred to as *specific* D_{JS}), therefore it holds that $D_{JS}[p^1(\mathbf{x}), p^2(\mathbf{x})] \in [0; \log 2]$. This aspect is particularly important, since the existence of upper and lower bounds for D_{JS} let us guess to what extent the two distributions differ. In particular, if $D_{JS}[p^1(\mathbf{x}), p^2(\mathbf{x})] = 0$, then $p^1(\mathbf{x})$ and $p^2(\mathbf{x})$ can be considered identical, whereas if $D_{JS}[p^1(\mathbf{x}), p^2(\mathbf{x})] = \log 2 \simeq 0.693$, $p^1(\mathbf{x})$ and $p^2(\mathbf{x})$ can be considered independent.

Since $\mathcal{I}_1$ and $\mathcal{I}_2$ were acquired in different periods of the year, the resulting overall D_{JS} was significant (i.e., 0.391), as reported in Table 9.1. However, in order to stress further the complexity of the investigated domain adaptation problem, we also evaluated the divergence between conditional class distributions $D_{JS}[p^1(\mathbf{x}|\omega_i), p^2(\mathbf{x}|\omega_i)]$. This let us quantitatively estimate the distance between the pattern distributions related to corresponding land-cover classes in the two images. In particular, from the table it is possible to notice that both the distributions of the class vineyard (i.e., $D_{JS} = 0.567$) and the class pasture (i.e., $D_{JS} = 0.517$) experienced relevant changes between the two acquisition dates.

In all the experiments, we chose as reference classification accuracy measure Λ the percentage overall accuracy $OA\%$ (i.e., the percentage of correctly labelled patterns over the total number of considered samples). Moreover, we set $\Lambda_{\text{th}} = OA\%_{th} = 85$ (this value represents a reasonable lower bound for a solution to be acceptable in the investigated problem); therefore, solutions were assumed to be acceptable only if $OA\% \geq 85$.

In the learning phase of both DASVMs and supervised SVMs (used for comparison purposes) we employed Gaussian kernel functions (ruled by the free parameter σ), as they generally proved effective in addressing the classification of multi-spectral remote-sensing images. In both cases, for handling the multiclass problem, we employed the OAA strategy. The Sequential Minimal Optimization algorithm developed by Platt (1998) was used for training both supervised SVMs and, with proper modifications, also the proposed DASVMs.

At the beginning of the experimental analysis, we trained several supervised SVMs by exploiting the training patterns $\mathcal{T}_1$ available at time t_1. In order to identify models with good generalization capabilities, we adopted a ten-fold cross validation (CV) strategy (Cristianini and Shawe-Taylor 2000), which permits one to effectively estimate the quality of the predicted models using the same data exploited for building the models themselves. Accordingly, we were able to identify the values of the supervised learning parameters (i.e., σ and C) that resulted in solutions acceptable for the reference image $\mathcal{I}_1$.

Then we trained several DASVMs using $\mathcal{T}_1$ and $\mathcal{X}_2$ as labelled and unlabelled sets, respectively. Both σ and C were assigned the same values that resulted in solutions consistent at time t_1 with the supervised approach, whereas a grid search strategy was applied to the remaining domain adaptation parameters (i.e., C^*, ρ and γ). Concerning the convergence criterion, on the basis of the performances exhibited by the system on a variety of preliminary experimental trials on toy datasets, we set $\vartheta = 3 \times 10^{-2}$. Moreover, as regards the maximum weight for semi-labelled samples, we fixed in all the trials $\tau = 0.5$ (it is reasonable that semi-labelled samples are associated at most with a regularization parameter equal to one half of the initial regularization parameter for original training samples). On the one hand, γ must be lower than the number of total iterations to let the final regularization parameter for the most of semi-labelled samples achieve sensible values. On the other hand, the higher is ρ, the lower is expected to be the number of total iterations. Accordingly, there exists an intrinsic inverse proportion between these two parameters, as is confirmed by the behaviour exhibited in the experiments. A reasonable range for ρ was $[5 - 15]$. When $\rho = 5$, we obtained consistent solutions for $\gamma \in [45 - 55]$; whereas, when $\rho = 15$ we obtained consistent solutions for $\gamma \in [30 - 40]$. The model selection for C^* never proved to be critical. In fact, for very small values of C^* (i.e., in the range $[0.5 - 1]$) it was always possible to obtain good classification accuracies.

For assessing the effectiveness of the circular validation strategy described in Section 9.4, each of the aforementioned DASVMs was re-trained for addressing the backward domain adaptation problem from $\mathcal{I}_2$ to $\mathcal{I}_1$. In particular, we used $\hat{\mathcal{T}}_2 = \{\mathcal{X}_2, \hat{\mathcal{Y}}_2\}$ (where $\hat{\mathcal{Y}}_2$ is the set of estimated labels predicted for $\mathcal{X}_2$) as the labelled set and the set of instances $\mathcal{X}_1$ at time t_1 (considered without the associated true labels) as the unlabelled set. The free parameters were assigned the same values adopted in the forward learning process. Given the availability of true labels $\mathcal{Y}_1$ for $\mathcal{X}_1$, it was possible to evaluate the $OA\%$ of the solution $\hat{\mathcal{Y}}_1$ obtained at time t_1 at the end of the circular procedure and, thus, to indirectly evaluate the consistency of the solution $\hat{\mathcal{Y}}_2$ related to the new image $\mathcal{I}_2$.

Table 9.2 shows the results obtained in terms of $OA\%$ and both percentage producers' and users' accuracies (i.e., $PA\%$ and $UA\%$, respectively) for each of the considered land-cover classes at time t_2 by: (i) the supervised SVM trained according to the ten-fold CV strategy with the labelled samples available for $\mathcal{I}_1$ that provided the highest $OA\%$ at time t_1 ($\text{SVM}_{t_1}^{CV}$); and (ii) the proposed DASVM technique with supervised selection of domain adaptation parameters (DASVM^{best}). Furthermore, also the average accuracies associated with the consistent solutions obtained by the presented domain adaptation algorithm and correctly identified by the circular validation strategy (DASVM^{ave}) are reported (in order to obtain significant estimations, 400 backward DASVMs have been trained both starting from consistent and non-consistent solutions at time t_2). Note that these values are particularly important, since they represent an average measure for the quality of the solutions correctly identified as acceptable for $\mathcal{I}_2$ without using any prior ground-truth information at time t_2.

At the end of our analysis, the proposed system proved to be particularly effective, resulting in very good classification performances. The increase in the accuracies exhibited by the DASVM technique is noteworthy with respect to the supervised approach. In particular,

Table 9.2 Percentage Overall accuracy ($OA\%$), Producer's Accuracies ($PA\%$) and User's Accuracies ($UA\%$) obtained for the new image $\mathcal{I}_2$ by: (i) the supervised SVM trained according to the ten-fold CV strategy with the labelled patterns available for $\mathcal{I}_1$ that provided the highest $OA\%$ at time t_1 ($\text{SVM}_{t_1}^{CV}$); (ii) the proposed DASVM technique with supervised selection of domain adaptation parameters (DASVM^{best}). The average accuracies associated with the consistent solutions obtained by the proposed DASVM technique and correctly identified by the circular validation strategy (DASVM^{ave}) are also given

		$\text{SVM}_{t_1}^{CV}$	DASVM^{best}	DASVM^{ave}
$OA\%$		78.76	96.67 (+17.91)	93.12 (+14.36)
Pasture	$PA\%$	53.48	96.60 (+43.12)	89.81 (+36.33)
	$UA\%$	94.03	96.61 (+2.58)	96.36 (+2.33)
Forest	$PA\%$	99.27	99.63 (+0.36)	99.27 (+0.00)
	$UA\%$	61.82	94.46 (+32.64)	81.68 (+19.86)
Urban area	$PA\%$	80.34	97.85 (+17.51)	93.54 (+13.20)
	$UA\%$	98.25	99.51 (+1.26)	99.24 (+0.99)
Water	$PA\%$	100	100 (+0.00)	100 (+0.00)
	$UA\%$	99.28	100 (+0.72)	100 (+0.72)
Vineyard	$PA\%$	52.14	70.01 (+17.87)	61.54 (+9.40)
	$UA\%$	22.02	75.23 (+53.21)	59.02 (+37.00)

without exploiting ground-truth labels at time t_2, the average $OA\%$ exhibited by the solutions automatically identified as acceptable for $\mathcal{I}_2$ was equal to 93.12, which corresponds to a relevant increase with respect to $\text{SVM}_{t_1}^{CV}$ (i.e., $+14.36$). Moreover, with a supervised selection of the domain adaptation parameters it was possible to obtain a further increase of the $OA\%$ up to 96.67 (i.e., $+17.91$ with respect to $\text{SVM}_{t_1}^{CV}$), which represents a very good result in the light of the complexity of the considered problem.

The behaviour of DASVMs was significantly different with respect to the standard SVM approach, as confirmed by the confusion matrices reported in Table 9.3(a) and 9.3(b). Note that it seems reasonable that $\text{SVM}_{t_1}^{CV}$ provided low accuracies for both pasture and vineyard, since the corresponding distributions exhibited relevant changes at the two acquisition dates (see the values for the D_{JS} reported in Table 9.1). In particular, $\text{SVM}_{t_1}^{CV}$ often misclassified pasture patterns as forest and vineyard areas. The proposed DASVM technique, instead, rarely incurred these errors. As concerns pasture, the gain in the $PA\%$ is noteworthy (i.e., $+36.33$

Table 9.3 Confusion matrices obtained at time t_2 by: (a) the supervised SVM trained according to the ten-fold CV strategy with the labelled patterns available for $\mathcal{I}_1$ that provided the highest $OA\%$ at time t_1 ($\text{SVM}_{t_1}^{CV}$); and (b) the proposed DASVM technique with optimal selection of domain adaptation parameters (DASVM^{best})

(a)	Pasture	Forest	Urban area	Water	Vineyard
Pasture	315	0	2	0	18
Forest	59	272	80	0	29
Urban area	0	1	336	0	5
Water	0	0	0	551	4
Vineyard	215	1	0	0	61
(b)	**Pasture**	**Forest**	**Urban area**	**Water**	**Vineyard**
Pasture	569	0	2	0	18
Forest	1	273	0	0	15
Urban area	0	0	409	0	2
Water	0	0	0	551	0
Vineyard	19	1	7	0	82

on average, +43.12 in the best case). Concerning the vineyard class, the increase of DASVMs with respect to the supervised approach is significant both in terms of $PA\%$ (i.e., +9.40 on average, +17.87 in the best case) and $UA\%$ (i.e., +37.00 on average, +53.21 in the best case). It is worth noting that in this case the accuracies are smaller than those obtained for the other information classes; however, this is mainly because the vineyard proved to be the most critical class since it had the lowest prior probability.

$\text{SVM}_{t_1}^{CV}$ frequently also confused urban areas with forest. On the other hand, DASVMs never incurred this type of error, thus providing a significant improvement both in the $UA\%$ of forest (i.e., +19.86 on average, +32.64 in the best case) and $PA\%$ of urban area (i.e., +13.20 on average, +17.51 in the best case). As regards water, although the related D_{JS} was not negligible, the accuracies provided by $\text{SVM}_{t_1}^{CV}$ were also particularly good at time t_2, since the spectral signature of this class is rather different with respect to those of the other information classes.

For the sake of comparison, in Table 9.4 we report the results obtained for $\mathcal{I}_2$ by: (i) a supervised SVM trained according to a ten-fold CV strategy at time t_2 by exploiting the true labels available for $\mathcal{X}_2$ ($\text{SVM}_{t_2}^{CV}$); (ii) the partially-unsupervised retraining technique for maximum likelihood classifiers (denoted as $\text{ML}_{retrain}$) presented by Bruzzone and Fernández Prieto in 2001; and (iii) the partially-unsupervised maximum likelihood cascade classifier (denoted as $\text{ML}_{cascade}$) proposed by Bruzzone and Fernández Prieto in 2002. As expected, $\text{SVM}_{t_2}^{CV}$ resulted in very good classification accuracy at time t_2, due to the supervised selection of the values for the learning parameters performed according to the CV strategy. However, it is worth mentioning that for a sub-optimal model selection, the supervised approach provided results comparable to those exhibited by DASVMs. Moreover, the proposed DASVM algorithm once again confirmed its effectiveness, exhibiting accuracies higher than those provided by both the $\text{ML}_{retrain}$ and the $\text{ML}_{cascade}$ techniques.

In order to characterize the behaviour of the proposed circular validation strategy, in Figure 9.5 we reported the empirical cumulative distribution function (CDF) of the $OA\%$

Table 9.4 Percentage Overall accuracy ($OA\%$), Producer's Accuracies ($PA\%$) and User's Accuracies ($UA\%$) obtained at time t_2 by: (i) the supervised SVM trained according to the ten-fold CV strategy at time t_2 exploiting true labels for the patterns of $\mathcal{X}_2$ ($\mathrm{SVM}_{t_2}^{CV}$); (ii) the partially-supervised retraining technique for maximum likelihood classifiers ($\mathrm{ML}_{retrain}$) presented by Bruzzone and Fernández Prieto in 2001; and (iii) the partially-unsupervised maximum likelihood cascaded classifier ($\mathrm{ML}_{cascade}$) proposed by Bruzzone and Fernández Prieto in 2002

		$\mathrm{SVM}_{t_2}^{CV}$	$\mathrm{ML}_{retrain}$	$\mathrm{ML}_{cascade}$
$OA\%$		99.69	92.76	91.48
Pasture	$PA\%$	100	94.06	94.25
	$UA\%$	99.49	88.50	83.53
Forest	$PA\%$	100	87.22	90.51
	$UA\%$	100	88.19	97.45
Urban area	$PA\%$	99.28	93.06	81.48
	$UA\%$	100	97.49	95.69
Water	$PA\%$	100	100	100
	$UA\%$	100	100	100
Vineyard	$PA\%$	97.44	64.10	86.90
	$UA\%$	97.44	73.53	62.39

(denoted as $\hat{p}(OA\%)$) estimated from the solutions obtained for $\mathcal{I}_1$ at the end of the circular learning procedure. In particular, the grey line refers to the solutions obtained at time t_1 when the system started from the state $\bar{D}$ (i.e., non-acceptable classification accuracy for $\mathcal{I}_2$); whereas, the black line refers to the solutions obtained at time t_1 when the system started from the state $\bar{B}$ (i.e., acceptable classification accuracy for $\mathcal{I}_2$). The ordinate of each point of the considered CDFs (referred to as the *quantile*) represents the probability that the final $OA\%$ obtained at time t_1 at the end of the backward learning process is lower or equal to the $OA\%$ value of the corresponding abscissa.

The two most relevant considerations that emerge from the analysis of the aforementioned curves, are as follows.

1. The proposed validation strategy was always able to correctly reject solutions that were not acceptable for $\mathcal{I}_2$. Indeed, the distribution corresponding to solutions not consistent at time t_2 (grey line) saturates to 1 for $OA\% \approx 78.5$, which is lower than $OA\%_{th} = 85$. This means that, when DASVMs were not able to seize the structure of the classification problem for the new image, the system could not recover a correct solution for the reference image $\mathcal{I}_1$ at the end of the circular procedure, thus satisfying the most critical requirement for the operational employment of the proposed strategy (i.e., $\Pr(\bar{A}|\bar{D}) = 0$).

2. The quantile corresponding to $OA\%_{th} = 85$ [i.e., $q_{0.85}(OA\%)$] of the CDF related to the solutions acceptable for $\mathcal{I}_2$ (black line) is equal to 0.45. Accordingly, since the system moves to the state $\bar{C}$ if $OA\% < OA\%_{th}$, we have that $\Pr\left(\bar{C}|\bar{B}\right) = q_{0.85}(OA\%) = 0.45$, therefore it holds that $\Pr\left(\bar{A}|\bar{B}\right) = 1 - \Pr\left(\bar{C}|\bar{B}\right) = 0.55$. This aspect is particularly

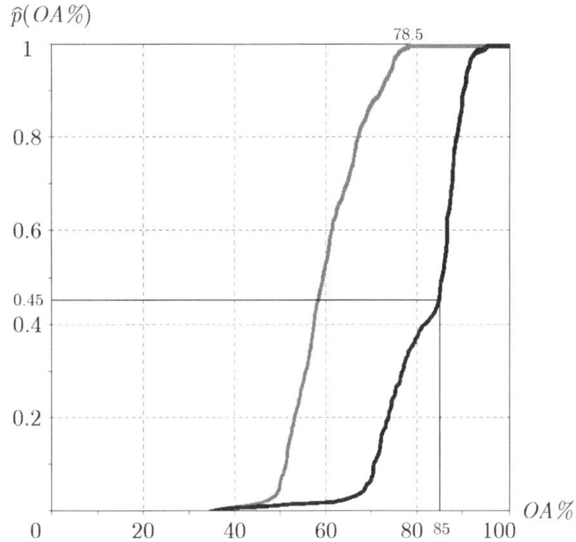

Figure 9.5 Empirical cumulative distribution function of the percentage overall accuracy ($OA\%$) estimated from the solutions obtained for the reference image $\mathcal{I}_1$ at the end of the backward learning procedure when the system started from both state $\bar{B}$ (i.e., solutions acceptable for the new image $\mathcal{I}_2$, black line) and state $\bar{D}$ (i.e., solutions not acceptable for the new image $\mathcal{I}_2$, grey line). If $OA\% < 85$ the system moved to the state $\bar{C}$ (i.e., solutions not acceptable for $\mathcal{I}_1$); if $OA\% \geq 85$ the system moved to the state $\bar{A}$ (i.e., solutions acceptable for $\mathcal{I}_1$).

important, as it means that it was possible to correctly identify 55% of the correct solutions obtained in our experiments without exploiting any prior ground-truth label at time t_2. Moreover, as discussed above, the average quality of these solutions is comparable to that obtained with supervised selection of domain adaptation parameters, thus confirming the effectiveness and the robustness of the proposed strategy.

Let us now focus attention on the computational load required by the proposed DASVM technique. It is possible to notice that, while the total number of semi-labelled patterns drawn from $\mathcal{I}_2$ increases, at the same time the number of remaining labelled patterns available for $\mathcal{I}_1$ decreases. As a consequence, since the cardinality of the training set does not vary significantly during the iterative domain adaptation process, the computational load grows almost linearly with the number of iterations. In our experiments, carried out on a PC mounting an Intel Core2 Duo processor at 2.6 GHz and a 4 Gb DDR2 RAM, training a supervised SVM took about 10 seconds. Some tens of iterations were necessary, instead, for the proposed DASVM to reach convergence. Therefore, since each iteration of the DASVM algorithm requires a time equivalent to that necessary for training a supervised SVM, the average learning time resulted in about 9 minutes. In the light of the complexity of the investigated problem and the huge increase of the classification accuracy with respect to the standard supervised approach, it is reasonable to consider this computational cost acceptable.

9.6 Discussions and conclusion

In this chapter we introduced a novel domain adaptation SVM (DASVM) technique for automatic updating land-cover maps when prior ground information is available only for a reference image acquired over the same geographical area before the one to classify. Furthermore, for assessing the correctness of the classification maps in the presence of no labelled patterns for the remote-sensing image to be categorized, we also presented a novel circular validation strategy.

The proposed DASVM algorithm directly takes into account that the distribution $p^2(\mathbf{x}, y)$ that governs the new image $\mathcal{I}_2$ at time t_2 is different from the distribution $p^1(\mathbf{x}, y)$ that characterizes the reference image $\mathcal{I}_1$ at time t_1. However, since the two images are acquired over the same geographical area, it is reasonable to assume the existence of an intrinsic relationship between $p^1(\mathbf{x}, y)$ and $p^2(\mathbf{x}, y)$. In this hypothesis, labelled samples available for $\mathcal{I}_1$ are used for obtaining an initial raw decision boundary for the classification problem at time t_2. Then, unlabelled samples of $\mathcal{I}_2$ are iteratively considered in the learning process for adapting the discriminant function to the new image. In order to improve the robustness and to better control the behaviour of the classifier, we defined an adaptive weighting strategy based on a temporal criterion for the regularization parameters of both original training samples available at time t_1 and semi-labelled samples (i.e., originally unlabelled samples associated with estimated labels during the learning process) drawn from $\mathcal{I}_2$. This allows one to obtain very good classification performances and to reduce the risk of converging towards an improper solution at time t_2.

The presented circular validation strategy let us bridge the lack of procedures for assessing the reliability of the classification results when labelled samples for the new image to classify are not available. Given the relationship existing between $p^1(\mathbf{x}, y)$ and $p^2(\mathbf{x}, y)$, we assume that it is anyhow possible to infer about the correctness of the updated land-cover map at time t_2 by exploiting the training samples for the reference image at time t_1. In particular, a solution for $\mathcal{I}_2$ obtained with a domain adaptation learning algorithm is identified as consistent if the solution obtained by applying the same algorithm in the reverse sense (i.e., using the estimated classification labels in place of missing true labels for the new image and considering the training patterns of the reference image as unlabelled) is acceptable for $\mathcal{I}_1$ (this can be evaluated thanks to the availability of true labels at time t_1).

Several experiments carried out on a dataset made up of two multi-spectral images acquired by the TM sensor of the Landsat-5 satellite confirmed the robustness and the effectiveness of both the proposed methods. The presented DASVM technique exhibited very good classification performances and proved capable of outperforming standard supervised SVMs resulting in high and satisfactory classification accuracies. Furthermore, the circular validation strategy was able to reject all the solutions that were not acceptable for the new image $\mathcal{I}_2$ (thus satisfying the most critical requirement for an operational employment) and correctly identified a significant set of solutions consistent at time t_2.

Note that if the initial decision boundary results in a poor discrimination capability for $\mathcal{I}_2$, most of the semi-labelled samples selected by the DASVM algorithm can be misclassified. In such cases, it becomes difficult for the proposed system to recover a reliable land-cover map. Thus, the convergence towards a consistent solution depends on the similarity between $p^1(\mathbf{x}, y)$ and $p^2(\mathbf{x}, y)$ (i.e., it can be effectively estimated by proper statistical distance measures). Therefore, when the two distributions are considerably different, it is not possible to guarantee that a solution acceptable for the investigated classification problem can be obtained.

Nevertheless, note that the proposed circular validation strategy is able to identify these critical situations and to reject all the non-consistent solutions automatically.

As a future development of this work, in order to reduce the computational load and to further speed up the learning process, we are studying an adaptive version of the proposed DASVM technique. In particular, similarly to what it is commonly done in the active learning framework, we are considering the possibility of iteratively taking into account only the semi-labelled samples selected at the current iteration and the patterns of the current training set identified as support vectors.

References

Belkin, M., Niyogi, P. and Sindhwani, V. (2006) Manifold regularization: A geometric framework for learning from labelled and unlabelled samples. *Journal of Machine Learning Research*, **7**, pp. 2399–434.

Ben-David, S., Blitzer, J., Crammer, K. and Pereira, F. (2007) Analysis of representation for domain adaptation. In *Advances in Neural Information Processing Systems* 19, *NIPS Conference*. Vancouver, British Columbia, Canada 4–9 December 2006. The MIT Press: Cambridge, Massachusetts.

Bennett, K. P. and Demiriz, A. (1998) Semi-supervised support vector machines. In *Advances in Neural Information Processing Systems* 10, *NIPS Conference*. Denver, Colorado, USA, 5–10 December 1997. The MIT Press: Cambridge, Massachusetts.

Bruzzone, L., Chi, M. and Marconcini, M. (2006) A novel transductive SVM for the semi-supervised classification of remote-sensing images. *IEEE Transactions on Geoscience and Remote Sensing*, **44** (11), pp. 3363–73.

Bruzzone, L., Chi, M. and Marconcini, M. (2007) Semi-supervised support vector machines for classification of hyperspectral remote sensing images. In C. Chang (ed), *Hyperspectral Data Exploitation Theory and Applications*. New York: John Wiley & Sons, Inc. Ch 11.

Bruzzone, L. and Cossu, R. (2002) A multiple-cascade-classifier system for a robust and partially unsupervised updating of land-cover maps. *IEEE Transactions on Geoscience and Remote Sensing*, **40** (9), pp. 1984–96.

Bruzzone, L., Cossu, R. and Vernazza, G. (2002) Combining parametric and non-parametric algorithms for a partially unsupervised classification of multi-temporal remote-sensing images. *Information Fusion*, **3** (4), pp. 289–97.

Bruzzone, L. and Fernández Prieto, D. (2001) Unsupervised retraining of a maximum-likelihood classifier for the analysis of multi-temporal remote-sensing images. *IEEE Transactions on Geoscience and Remote Sensing*, **39**, pp. 456–60.

Bruzzone, L. and Fernández Prieto, D. (2002) A partially unsupervised approach to the automatic classification of multi-temporal remote-sensing images. *Pattern Recognition Letters*, **23** (9), pp. 1063–71.

Bruzzone, L. and Marconcini, M. (2008) Domain adaptation problems: a DASVM classification technique and a circular validation strategy. *IEEE Transaction on Pattern Analysis and Machine Intelligence*, submitted.

Bruzzone, L. and Marconcini, M. (2009) Toward the automatic updating of land-cover maps by a DASVM classifier and a circular validation strategy. *IEEE Transactions on Geoscience and Remote Sensing*, in press.

Camps-Valls, G., Bandos T. and Zhou, D. (2007) Semi-supervised graph-based hyperspectral image classification. *IEEE Transactions on Geoscience and Remote Sensing*, **45** (10), pp. 3044–54.

Chapelle, O. (2007) Training a support vector machine in the primal. *Neural Computation*, **19** (5), pp. 1155–78.

Chen, Y., Wang, G. and Dong, S. (2003) Learning with progressive transductive support vector machine. *Pattern Recognition Letters*, **24** (12), pp. 1845–55.

Chi, M. and Bruzzone, L. (2005) A semi-labelled-sample-driven bagging technique for ill-posed classification problems. *IEEE Geoscience and Remote Sensing Letters*, **2** (1), pp. 69–73.

Chi, M. and Bruzzone, L. (2006) An ensemble-driven *k*-NN approach to ill-posed classification problems. *Pattern Recognition Letters*, **27** (4), pp. 301–7.

Chi, M. and Bruzzone, L. (2007) Semi-supervised classification of hyperspectral images by SVMs optimized in the primal. *IEEE Transactions on Geoscience and Remote Sensing*, **45** (6), pp. 1870–80.

Cover, T. M. (1965) Geometrical and statistical properties of systems of linear inequalities with applications in pattern recognition. *IEEE Transactions on Electronic Computers*, **14**, pp. 326–34.

Cristianini, N. and Shawe-Taylor, J. (2000) *An Introduction to Support Vector Machines*. Cambridge University Press.

Daumé III, H. and Marcu, D. (2006) Domain adaptation for statistical classifiers. *Journal of Artificial Intelligence Research*, **26**, pp. 101–26.

Dempster, A., Laird, N. and Rubin, D. (1977) Maximum likelihood from incomplete data via the EM algorithm. *Journal of the Royal Statistical Society. Series B (Methodological)*, **39** (1), pp. 1–38.

Dundar, M. M. and Landgrebe, D. A. (2004) A cost-effective semi-supervised classifier approach with kernels. *IEEE Transactions on Geoscience and Remote Sensing*, **42** (1), pp. 264–70.

Gómez, L., Camps-Valls, G., Muñoz-Marí, J. and Calpe-Maravilla, J. (2008) Semi-supervised image classification with laplacian support vector machines. *IEEE Geoscience and Remote Sensing Letters*, **5** (3), pp. 336–40.

Haralick, R. M., Shanmugan, K. and Dinstein, I. (1973) Textural features for image classification. *IEEE Transactions on Systems, Man, and Cybernetics*, **3**, pp. 610–21.

Hughes, G. F. (1968) On the mean accuracy of statistical pattern recognizers. *IEEE Transaction on Information Theory*, **14** (1), pp. 55–63.

Jackson, Q. and Landgrebe, D. A. (2002) An adaptive method for combined covariance estimation and classification. *IEEE Transactions on Geoscience and Remote Sensing*, **40** (5), pp. 1082–7.

Jiang, J. and Zhai, C. (2007) Instance weighting for domain adaptation in NLP. In *Proceedings of the 45th Annual Meeting of the Association for Computational Linguistics, ACL2007 Conference*. Prague, Czech Republic 23–30 June 2007.

Jordan, M. I. (1999) *Learning in Graphical Models*. 1st edn Cambridge, Massachusetts and London, England: MIT Press.

Kullback, S. and Leibler, R. (1951) On information and sufficiency. *Annals of Mathematical Statistics*, **22**, pp. 79–86.

Lin, J. (1991) Divergence measures based on the Shannon entropy. *IEEE Transactions on Information Theory*, **37**, pp.145–51.

Platt, J. (1998) Fast training of support vector machines using sequential minimal optimization. In B. Schölkopf, C. Burges, and A. Smola (ed), *Advances in Kernel Methods: Support Vector Learning*, Cambridge, Massachusetts and London, England: MIT Press. Ch. 12.

Satpal, S. and Sarawagi, S. (2007) Domain adaptation of conditional probability models via feature subsetting. In *Proceedings of the 11th European Conference on Principles and Practice of Knowledge Discovery in Databases, PKDD Conference*. Warsaw, Poland 17–21 September 2007.

Shahshahani, B. M. and Landgrebe, D. A. (1994) The effect of unlabelled samples in reducing the small sample size problem and mitigating the Hughes phenomenon. *IEEE Transactions on Geoscience and Remote Sensing*, **32** (5), pp. 1087–95.

Tadjudin, S. and Landgrebe, D. A. (2000) Robust parameter estimation for mixture model. *IEEE Transactions on Geoscience and Remote Sensing*, **38** (1), pp 439–45.

Vapnik, V. N. (1998) *Statistical Learning Theory*. New York: John Wiley & Sons, Inc.

Vapnik, V. N. (1999) The Nature of *Statistical Learning Theory*. 2nd edn Berlin: Springer-Verlag.

Zhou, D., Huang, J. and Schölkopf, B. (2004) Learning with local and global consistency. In *Advances in Neural Information Processing System 16, NIPS Conference*. Vancouver and Whistler, British Columbia, Canada 8–13 December 2003. The MIT Press: Cambridge, Massachusetts.

Zhu, X. (2005) *Semi-Supervised Learning Literature Survey*. Computer Sciences, University of Wisconsin-Madison. TR-1530.

10

Mean kernels for semi-supervised remote sensing image classification

Luis Gómez-Chova[1], Javier Calpe-Maravilla[1], Lorenzo Bruzzone[2] and Gustavo Camps-Valls[1]

[1] *Image Processing Laboratory (IPL) & Dept. Enginyeria Electrònica, Universitat de València, Spain*

[2] *Dept. Information Engineering and Computer Science, University of Trento, Italy*

This chapter presents a semi-supervised support vector machine (SVM) classifier based on the combination of the expectation-maximization (EM) algorithm for Gaussian mixture models (GMM) and the *mean map* kernel. The proposed method improves classification accuracy by reinforcing samples in the same cluster belonging to the same class through the use of composite kernels learned from the testing image. These sample and cluster similarities are included in the standard SVM by means of a linear combination of kernels. The cluster similarity is directly computed in the kernel space with a dedicated kernel that is based on the means of the feature vectors in this space. Moreover, only the most reliable samples in terms of likelihood values are used to compute a kernel function that accurately reflects the similarity between clusters in the feature space. Results show that the proposed approach is especially suited for situations where the available labelled information does not properly describe the classes in the test image.

Kernel Methods for Remote Sensing Data Analysis Edited by Gustavo Camps-Valls and Lorenzo Bruzzone

10.1 Introduction

In many remote sensing image classification problems, it is difficult to collect a sufficient number of statistically significant ground-truth samples to define a complete training set for developing robust supervised classifiers. In this setting, extrapolating a trained classifier to unseen scenes is a challenging problem when no training samples in the test image are available. In such situations, labelled data extracted from other images modelling similar problems can be used. However, the resulting problem is not well posed since training samples might not adequately describe the test data. This is known as the *sample selection bias* problem. In this kind of problem, unlabelled samples from the test image can be jointly used with the available training samples for increasing the reliability and accuracy of the learning machine.

On the one hand, applying unsupervised clustering methods to the whole image allows us to take advantage of the wealth of information and the high amount of spatial and spectral correlation of the image pixels. On the other hand, supervised kernel-based classification methods, such as supervised support vector machines (SVMs) excel in using the labelled information (Camps-Valls and Bruzzone 2005; Shawe-Taylor and Cristianini 2004) for image classification. These methods, nevertheless, obviate the potentially useful wealth of unlabelled data, and need to be reformulated to exploit it properly. Semi-supervised learning (SSL) is concerned with such situations (Chapelle *et al.* 2006; Zhu 2005), and several approaches have been carried out in the context of remotely sensed image classification (Bruzzone *et al.* 2006; Camps-Valls *et al.* 2007; Chi and Bruzzone 2007; Gómez-Chova *et al.* 2008b). However, when selecting or developing an SSL method, one has to verify that the imposed model assumptions fit the problem data structure. Before presenting the methods, some key issues must be considered:

Data structure

The main assumptions made about data structure in SSL are that data are organized into a number of groups (or clusters) according to a given distance measure in some representation space (*cluster assumption*) or that data hold a manifold forming a nonlinear complex global structure (*manifold assumption*) (Bachmann *et al.* 2005, 2006). In the case of most natural land covers, distribution of multi-spectral remote sensing data is smooth and the spectra of pixels of the same land cover are similar, making the cluster assumption suitable to model each class. For this reason, the proposed method assumes that data are organized into a number of groups (or clusters) according to a given distance measure in some representation space. If the correspondence between sample and cluster is known for some training samples, one can reinforce samples in the same cluster to belong to the same class by taking into account the similarity between clusters.

SSL model

In general, SSL methods aim to retrieve information from unlabelled data by estimating their conditional distribution, and one can distinguish between *generative models* and *discriminative models*. In multi-spectral image processing, the assumption that the data distribution can be approximated as a mixture of normally distributed samples is commonly accepted (Jackson and Landgrebe 2001; Shahshahani and Landgrebe 1994). Hence, generative models, which estimate the conditional distribution by modelling the class-conditional distributions explicitly, may provide a good performance and approximately meet the cluster assumption.

Training set representativeness

As mentioned above, in some applications, few labelled samples (with respect to the size of the input feature space) are available to train classifiers. In this setting, two main different conditions are usually considered: (1) few training samples are available for characterizing the image to be classified; and (2) no training samples are available for the test image to be classified. In the latter case, one-class classifiers can be used (Muñoz-Marí *et al.* 2007; Muñoz-Marí *et al.* 2008). We adopt a similar strategy by which training data extracted from other images modelling similar problems can be exploited. In both situations, unlabelled samples of test image can be jointly used with the available training samples for increasing the reliability and accuracy of the classifier.

All these issues are addressed in this chapter. Our aim is to take advantage of the benefits shown by both supervised and unsupervised methods. This chapter proposes a semi-supervised kernel-based classification method that takes into account the particularities and problems found in practical remote sensing applications. First, some recommendations about how to use clustering methods to extract information from unlabelled samples of the test image are given. The concept of mean kernel is then introduced to compute cluster similarities and it is then extended under the framework of composite kernels. Finally, the proposed solution is presented. For illustration purposes, we show results in the complex problem of detecting clouds from multi-spectral satellite images.

The chapter is organized as follows. Section 10.2 presents the proposed kernel method that combines both clustering and the *mean map kernel*. Section 10.3.1 describes the experimental setup and model development. Sections 10.3.2 and 10.3.3 are devoted to analysing the results of the proposed method in synthetic and real situations where the model assumptions are verified or broken. Section 10.4 concludes the chapter.

10.2 Semi-supervised classification with mean kernels

The proposed *semi-supervised* kernel method is based on three assumptions: (1) data is organized into a number of groups (or clusters) according to a distance measure in some representation space; (2) classes in the test image(s) present features inducing compact clusters; and (3) unlabelled data can help model these clusters. The key point is the definition of a kernel function that accurately reflects the similarity between samples and similarity between clusters in the kernel space (Gómez-Chova *et al.* 2008a), while performing the classification at pixel level (Gómez-Chova 2008). The method combines the expectation-maximization (EM) algorithm for fitting Gaussian mixture models (GMM) (Dempster *et al.* 1977), which has been extensively used in remote sensing image classification (Bruzzone and Fernández Prieto 2001; Gómez-Chova *et al.* 2007; Shahshahani and Landgrebe 1994), and the *mean map kernel*, which computes the similarity between clusters in the kernel space. This alleviates the *sample selection bias* problem and produces improved classification accuracy by reinforcing samples in the same cluster belonging to the same class.

10.2.1 Learning from labelled samples

In supervised classification problems, we are given a set of ℓ labelled (training) samples $\{\mathbf{x}_i, y_i\}_{i=1}^{\ell}$, where $\mathbf{x}_i \in \mathbb{R}^d$ represents the d-bands pixel defined in an input space $\mathcal{X}$, and $y_i \in \mathbb{N}$ belongs to the observation (output) space of land cover classes. SVMs attempt to separate samples belonging to different classes by tracing a maximum margin hyperplane in a high dimension (possibly infinite) space $\mathcal{H}$ where samples are mapped to through a nonlinear mapping function $\boldsymbol{\phi}$ (Schölkopf and Smola 2001; Shawe-Taylor and Cristianini 2004) (see also Chapters 1 and 3 of the present book). One can work in this *feature space* without even knowing the explicit form of the mapping $\boldsymbol{\phi}$, but only the kernel function formed by the dot product among them:

$$K(\mathbf{x}_i, \mathbf{x}_j) = \left\langle \boldsymbol{\phi}(\mathbf{x}_i), \boldsymbol{\phi}(\mathbf{x}_j) \right\rangle . \tag{10.1}$$

In any kernel method, a proper definition of the structural form of K reflecting signal relations is the crucial step to attaining good results.

10.2.2 Image clustering

Instead of working with individual samples in a supervised manner, we also include the data structure (characterized by the clusters of the analysed image) into the training process. In order to get cluster information, we first apply a clustering algorithm, which provides for each sample $\mathbf{x}_i$ a *crisp* or *soft* association, h_{ik}, to each cluster ω_k, $k = 1, \ldots, c$. In particular, the input image is considered as a mixture of normal distributions so the EM algorithm can be used to obtain the maximum likelihood estimation of the probability density function (pdf) of the Gaussian mixture. The EM algorithm estimates the mixture coefficient π_k, the mean $\boldsymbol{\mu}_k$, and the covariance matrix $\boldsymbol{\Sigma}_k$ for each component of the mixture. Then, the algorithm assigns each sample to the cluster with the maximum *a posteriori* probability (MAP); and the cluster membership h_{ik} represents the estimates of the posterior probabilities; that is, membership or probability value between [0, 1], and sum-to-one cluster memberships, $\sum_k h_{ik} = 1$. Hence, the optimal cluster label for each sample is found as $h_i = \text{argmax}_k \{h_{ik}\}$, i.e. $h_i = k$ if the sample $\mathbf{x}_i$ is assigned to the cluster ω_k. The ease of use and fast classification performance justifies the selection of this algorithm, even though other clustering algorithms and additional input data (e.g. spatial information) might be equally used in our method. The analysis of the available clustering algorithms, the clustering initialization, and the selection of the number of clusters is beyond the scope of this chapter (see Chapter 1 of the present book for further information) but, in order to make suitable the cluster assumption, the number of clusters should be high enough and the clusters should fairly approximate the class distribution. Once a clustering is done, one should compute cluster similarity.

10.2.3 Cluster similarity and the mean map

One of the basic ideas when working with kernel methods is that the mapping $\boldsymbol{\phi}(\mathbf{x})$ into the kernel-defined feature space $\mathcal{H}$ does not need to be explicitly known. Although we do not have access to the representation of samples in this space, it is possible to perform some elementary calculations in the kernel space implicitly, such as computing sums, products, means and distances (see Chapter 2 of the present book for some examples).

Given a finite subset of training samples $S = \{\mathbf{x}_1, \ldots, \mathbf{x}_n\}$ laying in an input space $\mathcal{X}$ and a kernel $K(\mathbf{x}_i, \mathbf{x}_j) = \left\langle \boldsymbol{\phi}(\mathbf{x}_i), \boldsymbol{\phi}(\mathbf{x}_j) \right\rangle$, let $\boldsymbol{\Phi}(S) = \{\boldsymbol{\phi}(\mathbf{x}_1), \ldots, \boldsymbol{\phi}(\mathbf{x}_n)\}$ be the representation of S under the map $\boldsymbol{\phi}$. Hence $\boldsymbol{\Phi}(S)$ is a subset of the inner product space $\mathcal{H}$. In particular, the centre of mass of the set S in the kernel space is the vector:

$$\boldsymbol{\phi}_\mu(S) = \frac{1}{n}\sum_{i=1}^{n}\boldsymbol{\phi}(\mathbf{x}_i), \tag{10.2}$$

where $\boldsymbol{\phi}_\mu(\cdot)$ denotes the *mean map*. We should stress that, in principle, there is not an explicit vector representation of the centre of mass, since, in this case, there may also not exist a point in the input space $\mathcal{X}$ whose image under $\boldsymbol{\phi}$ is $\boldsymbol{\phi}_\mu(S)$. In other words, we are now considering points that potentially lie outside $\boldsymbol{\phi}(\mathcal{X})$, that is, the image of the input space $\mathcal{X}$ under the mapping $\boldsymbol{\phi}$. However, computing the mean in a high dimensional feature space can report additional advantages.

Let us now consider two subsets of samples $S_1 = \{\mathbf{a}_1, \ldots, \mathbf{a}_m\}$ and $S_2 = \{\mathbf{b}_1, \ldots, \mathbf{b}_n\}$ belonging to two different clusters ω_1 and ω_2, respectively. We are interested in defining a *cluster similarity* function that estimates the proximity between them in a sufficiently rich feature space. A straightforward kernel function reflecting the similarity between clusters is obtained by evaluating the kernel function between the means of the clusters in the input space $\mathcal{X}$:

$$K_\mu^{\mathcal{X}}(S_1, S_2) \equiv \langle \boldsymbol{\phi}(\boldsymbol{\mu}_1), \boldsymbol{\phi}(\boldsymbol{\mu}_2)\rangle = K(\boldsymbol{\mu}_1, \boldsymbol{\mu}_2), \tag{10.3}$$

but then we loose the advantage of directly working in the feature space $\mathcal{H}$.

The centre of mass of the sets S_1 and S_2 in the kernel space are the vectors $\boldsymbol{\phi}_\mu(S_1) = \frac{1}{m}\sum_{i=1}^{m}\boldsymbol{\phi}(\mathbf{a}_i)$ and $\boldsymbol{\phi}_\mu(S_2) = \frac{1}{n}\sum_{i=1}^{n}\boldsymbol{\phi}(\mathbf{b}_i)$. Despite the apparent inaccessibility of the points $\boldsymbol{\phi}_\mu(S_1)$ and $\boldsymbol{\phi}_\mu(S_2)$ in the kernel space $\mathcal{H}$, we can compute the *cluster similarity* in $\mathcal{H}$ using only evaluations of the *sample similarity* contained in the kernel matrix:

$$\begin{aligned} K_\mu^{\mathcal{H}}(S_1, S_2) &\equiv \left\langle \boldsymbol{\phi}_\mu(S_1), \boldsymbol{\phi}_\mu(S_2)\right\rangle \\ &= \left\langle \frac{1}{m}\sum_{i=1}^{m}\boldsymbol{\phi}(\mathbf{a}_i), \frac{1}{n}\sum_{j=1}^{n}\boldsymbol{\phi}(\mathbf{b}_j)\right\rangle \\ &= \frac{1}{mn}\sum_{i=1}^{m}\sum_{j=1}^{n}K(\mathbf{a}_i, \mathbf{b}_j). \end{aligned} \tag{10.4}$$

Note how significant information about the cluster similarities can be obtained by using only the inner product information contained in the kernel matrix, $\mathbf{K}_{ij} = K(\mathbf{x}_i, \mathbf{x}_j)$, of kernel evaluations between all pairs of elements in S_1 and S_2:

$$\mathbf{K} = \left[\begin{array}{ccc|ccc} K(\mathbf{a}_1, \mathbf{a}_1) & \cdots & K(\mathbf{a}_1, \mathbf{a}_m) & K(\mathbf{a}_1, \mathbf{b}_1) & \cdots & K(\mathbf{a}_1, \mathbf{b}_n) \\ \vdots & \ddots & \vdots & \vdots & \ddots & \vdots \\ K(\mathbf{a}_m, \mathbf{a}_1) & \cdots & K(\mathbf{a}_m, \mathbf{a}_m) & K(\mathbf{a}_m, \mathbf{b}_1) & \cdots & K(\mathbf{a}_m, \mathbf{b}_n) \\ \hline K(\mathbf{b}_1, \mathbf{a}_1) & \cdots & K(\mathbf{b}_1, \mathbf{a}_m) & K(\mathbf{b}_1, \mathbf{b}_1) & \cdots & K(\mathbf{b}_1, \mathbf{b}_n) \\ \vdots & \ddots & \vdots & \vdots & \ddots & \vdots \\ K(\mathbf{b}_n, \mathbf{a}_1) & \cdots & K(\mathbf{b}_n, \mathbf{a}_m) & K(\mathbf{b}_n, \mathbf{b}_1) & \cdots & K(\mathbf{b}_n, \mathbf{b}_n) \end{array}\right], \tag{10.5}$$

which is reduced to $\mathbf{K}_\mu$ by applying (10.4):

$$\mathbf{K}_\mu^{\mathcal{H}} = \left[\begin{array}{c|c} K_\mu^{\mathcal{H}}(S_1, S_1) & K_\mu^{\mathcal{H}}(S_1, S_2) \\ \hline K_\mu^{\mathcal{H}}(S_2, S_1) & K_\mu^{\mathcal{H}}(S_2, S_2) \end{array}\right] \tag{10.6}$$

The concept of the *mean map kernel* has recently been extended to compare data distributions in the kernel space (Gretton *et al.* 2007).

10.2.4 Composite sample-cluster kernels

The concept of computing similarities between sets of vectors in the kernel space has been previously explored in several application fields. For example, in Gärtner *et al.* (2002), a kernel on sets is proposed to solve multi-instance problems, where individuals are represented by structured sets. In Kondor and Jebara (2003), the Bhattacharya's measure is computed in the Hilbert space between the Gaussians obtained after mapping the set of vectors into $\mathcal{H}$. In Jebara *et al.* (2004), kernel machines are combined with generative modelling using a kernel between distributions called Probability Product Kernel (PPK). In Zhou and Chellappa (2006), expressions for the most common probabilistic distance measures in the reproducing kernel Hilbert space are presented. Finally, in Li *et al.* (2007), the PPK is used to develop a Support Cluster Machine for large-scale classification problems, where the labelled samples are clustered and the obtained labelled clusters are used to train the model instead of using the training samples. However, all these works consider the sets of samples or distributions as a single entity with a given label and no information is provided for each individual sample. In our approach, classifying clusters is not the goal since we seek a detailed classification at a pixel level. This approach thus falls in the field of *cluster kernels*, which are focused on changing the representation given to a classifier by taking into account the structure described by the unlabelled data (Chapelle *et al.* 2003; Szummer and Jaakkola 2002; Weston *et al.* 2005; Zhu and Ghahramani 2002). Let us remember that the bottleneck for any kernel method is the definition of a suitable kernel function that accurately reflects the similarity between samples. Hence, the proposed algorithm should compute and combine both similarity between samples and similarity between clusters in the kernel space.

SSL methods assume that one has access to a set of unlabelled (test) samples and learn from both labelled and unlabelled samples. To fix notation, we are given a set of ℓ labelled samples, $\{\mathbf{x}_i, y_i\}_{i=1}^{\ell}$, where $\mathbf{x}_i \in \mathbb{R}^d$ and $y_i \in \{-1, +1\}$, and a set of u unlabelled samples $\{\mathbf{x}_i\}_{i=\ell+1}^{\ell+u}$. In the proposed semi-supervised method, the u unlabelled training samples coming from the test image are used to describe the clusters and to compute the similarity between clusters, which is used to weight the similarity between the ℓ labelled training samples that define the classes. The similarity between clusters is included through the use of a *composite kernel* (Camps-Valls *et al.* 2006) that balances both similarity distances

$$K_\omega(\mathbf{x}_i, \mathbf{x}_j) = \nu\, K(\mathbf{x}_i, \mathbf{x}_j) + (1 - \nu)\, K_\mu(S_{h_i}, S_{h_j}) \quad \forall i, j = 1, \ldots, \ell, \tag{10.7}$$

where ν is a positive real-valued free parameter ($0 \leqslant \nu \leqslant 1$), which is tuned in the training process and constitutes a trade-off between the sample and corresponding cluster information. It is worth noting that: (1) the number of training samples is $(\ell + u)$, because unlabelled samples are used to compute the cluster similarities by summing elements of the kernel matrix;

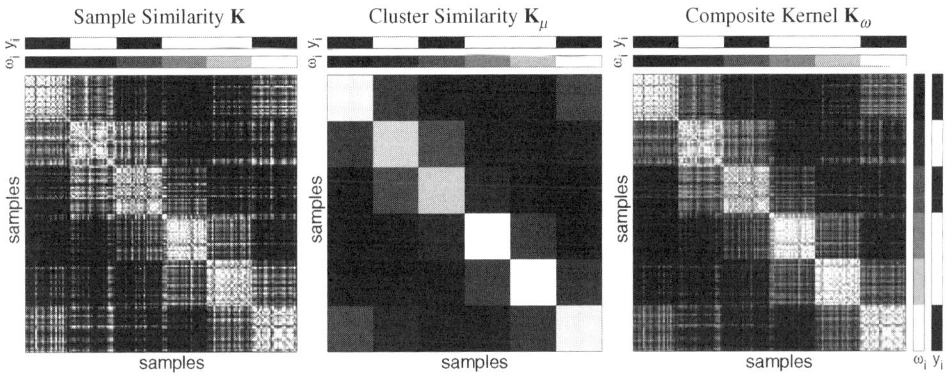

Figure 10.1 Illustrative example of the three involved kernel matrices ($\ell \times \ell$): sample similarity *accounted by the kernel of the training samples* $\mathbf{K}$ *($\nu = 1$);* cluster similarity *accounted by the mean map kernel of the clusters* $\mathbf{K}_\mu$ *($\nu = 0$); and the* composite kernel $\mathbf{K}_\omega$ *obtained by combining the sample and the cluster similarities for each sample ($0 < \nu < 1$). Note that samples are sorted by class y_i and by cluster ω_i for a proper interpretation.*

and (2) the number of clusters is c, thus one will obtain only $c \times c$ cluster similarities using K_μ. However, the size of the final kernel matrix $\mathbf{K}_\omega$ used to train the standard SVM is $\ell \times \ell$ (the first ℓ samples are labelled). Summarizing, each position (i, j) of matrix $\mathbf{K}_\omega$ contains the similarity between all possible pairs of the ℓ labelled training samples ($\mathbf{x}_i$ and $\mathbf{x}_j$) and their corresponding clusters (defined by h_i and h_j), which are measured with suitable kernel functions K and K_μ fulfilling Mercer's conditions (see Chapters 2 and 6 of the present book to find Mercer's kernels properties).

Figure 10.1 shows an illustrative example of the three kernel matrices ($\ell \times \ell$) involved in the proposed method: *sample similarity* accounted by the kernel of the training samples $\mathbf{K}$; *cluster similarity* accounted by the mean map kernel of the clusters $\mathbf{K}_\mu$; and the *composite kernel* $\mathbf{K}_\omega$ obtained by combining the sample and the cluster similarities for each sample. It is worth noting that the proposed composite kernel K_ω maintains the sample similarity at pixel level while making pixels in the same cluster more similar, thus reinforcing them to belong to the same class. It may be interpreted as a smoothing of K attending to the cluster structure in K_μ.

10.2.5 Sample selection bias and the soft mean map

So far we have assumed that training and test data are independently and identically distributed (i.i.d.) drawn from the same pdf, but actually training and test set distributions could not match, which is known in the literature as *sample selection bias* (Bickel *et al.* 2007; Heckman 1979; Huang *et al.* 2007) or *covariate shift* (Shimodaira 2000; Sugiyama and Müller 2005; Sugiyama *et al.* 2007, 2008). Obviously, if the training and the test data have nothing in common there is no chance of learning anything. Thus, we assume that both follow a similar conditional distribution $p(y|\mathbf{x})$ and the input distributions $p(\mathbf{x})$ differ, yet not completely (see Figure 10.2). In remote sensing, this is a likely situation since usually no training samples are available for the image to be classified and labelled data is extracted from other images

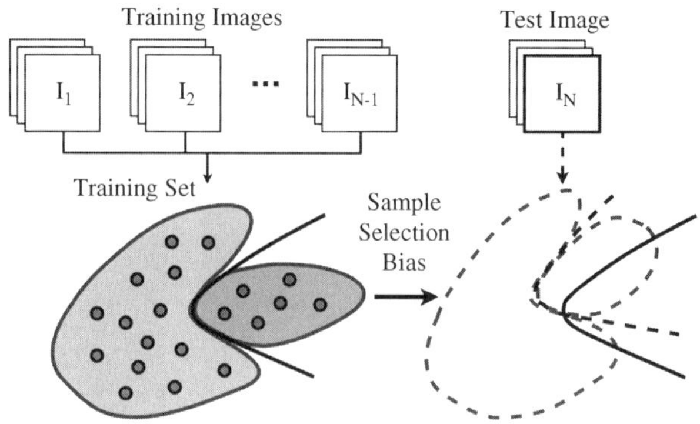

Figure 10.2 Illustrative example of the sample selection bias *produced when the distributions on training and test sets do not match.*

modelling similar problems. In these situations, not all training samples are equally reliable. In the literature, the training samples are weighted in different ways in order to avoid the sample selection problem: by deriving the conditional density to maximize the log-likelihood function (Shimodaira 2000); by changing the criterion to be maximized for learning, such as the nonparametric kernel mean matching method presented in Huang *et al.* (2007) that tries to match the first momentum of training and test sets in the kernel space; or by modifying the model selection criteria to obtain unbiased results (Sugiyama *et al.* 2007).

In the proposed method, the most reliable samples in terms of maximum likelihood in the input space are used to compute a kernel function that accurately reflects the similarity between clusters in the kernel space. Therefore, the relative reliability of training samples is trimmed by weighting the contribution of each sample $\mathbf{x}_i$ to the definition of the centre of mass of each cluster in the kernel space $\mathcal{H}$ with the EM estimated posterior probabilities h_{ik}, that is:

$$\boldsymbol{\phi}_{\mu_s}(S_k) = \frac{\sum_i h_{ik}\boldsymbol{\phi}(\mathbf{x}_i)}{\sum_i h_{ik}}, \tag{10.8}$$

which we call the *soft mean map*. The corresponding soft mean map kernel can be easily computed as:

$$\begin{aligned} K_{\mu_s}^{\mathcal{H}}(S_k, S_l) &= \left\langle \boldsymbol{\phi}_{\mu_s}(S_k), \boldsymbol{\phi}_{\mu_s}(S_l) \right\rangle \\ &= \left\langle \frac{\sum_i h_{ik}\boldsymbol{\phi}(\mathbf{x}_i)}{\sum_i h_{ik}}, \frac{\sum_j h_{jl}\boldsymbol{\phi}(\mathbf{x}_j)}{\sum_j h_{jl}} \right\rangle \\ &= \frac{\sum_i \sum_j h_{ik} h_{jl} K(\mathbf{x}_i, \mathbf{x}_j)}{\sum_i h_{ik} \sum_j h_{jl}}, \end{aligned} \tag{10.9}$$

and now, when computing cluster similarities, all samples contribute to all clusters but with different relative weights according to their a posteriori probability. The main advantage of the proposed method is that weights for the training samples are directly computed by taking advantage of the full statistical information of the test data distribution without modifying the classification problem to be solved, i.e. the same quadratic programming (QP) optimization problem of the SVM is solved with the available standard optimization tools. We should stress that, with this approach, the EM algorithm is applied to the entire image in order to properly characterize the test data distribution and compute the sample weights, but the number of unlabelled samples used to describe the clusters in the soft mean map can be selected by the user to reduce the size of the kernel matrix and thus control the computational effort.

Note that the mean map kernel in (10.4) is a particular case of the proposed soft mean map kernel in (10.9) when the training samples are associated only with one cluster (*crisp* association), i.e. when $h_{ik} = 1$ if $\mathbf{x}_i$ belongs to cluster ω_k and $h_{ik} = 0$ otherwise. The expression of the soft mean map kernel in (10.9) can be rewritten in matrix notation as follows:

$$\mathbf{K}_{\mu}^{\mathcal{H}} = \mathbf{D}\mathbf{H}^{\top}\mathbf{K}\mathbf{H}\mathbf{D}, \tag{10.10}$$

where $\mathbf{K}$ is the $(\ell + u) \times (\ell + u)$ kernel matrix of both labelled and unlabelled training samples; $\mathbf{H}$ is a $(\ell + u) \times c$ matrix containing the membership value h_{ik} of each training sample to each cluster of the analysed image; and $\mathbf{D}$ is a $c \times c$ diagonal matrix with normalization factors for each cluster $D_{kk} = 1/\sum_i h_{ik}$. Note that the computational effort is directly controlled by the number of unlabelled samples u included in $\mathbf{K}$ to describe the clusters.

The size of the matrix containing the similarity between clusters $\mathbf{K}_{\mu}$ is $c \times c$. Thus, it has to be expanded to match the number of labelled samples, in order to obtain the final $\ell \times \ell$ kernel matrix $\mathbf{K}_{\omega}$ in (10.7) used to train the classifier:

$$\mathbf{K}_{\omega} = \nu\,\mathbf{J}\mathbf{K}\mathbf{J}^{\top} + (1 - \nu)\,\mathbf{W}\mathbf{K}_{\mu}\mathbf{W}^{\top}, \tag{10.11}$$

where $\mathbf{J} = [\mathbf{I}\ \mathbf{0}]$ is an $\ell \times (\ell + u)$ matrix with $\mathbf{I}$ as the $\ell \times \ell$ identity matrix (the first ℓ samples are labelled); and $\mathbf{W}$ is a $\ell \times c$ sparse matrix that stores the cluster of each labelled sample h_i, i.e. $W_{ik} = 1$ if sample $\mathbf{x}_i$ belongs to cluster ω_k and $W_{ik} = 0$ otherwise.

10.2.6 Summary of composite mean kernel methods

The proposed method brings together the ideas of unsupervised clustering, mean map kernel, composite kernel, and SVM in a simple and natural way. Essentially, the method tries: (1) to reinforce both the local and global consistencies, and (2) to mitigate the sample selection bias problem. The method combines the expectation-maximization (EM) algorithm for fitting Gaussian mixture models (GMM) and the *mean map kernel*. The final classification model is obtained by solving a standard SVM (a convex optimization problem) but the kernel of the labelled training samples (local consistency) is previously deformed to take into account the similarities between clusters (global consistency), which are easily computed from the unlabelled samples of the analysed image.

Table 10.1 shows several particular cases of the proposed method (denoted by μ-SVM) depending on: (1) the balance between the sample similarity and the cluster similarity (free parameter ν), (2) in which space the cluster similarities are computed (input or kernel space), and (3) how the unlabelled training samples contribute to each cluster (crisp or soft association). In this table we indicate the kernel function used in the SVM, the mapping function whose dot product generates the corresponding composite kernel, and the value of ν that constitutes a trade-off between the sample ($\nu = 1$) and the cluster information ($\nu = 0$).

Table 10.1 Particular cases of the proposed method depending on: (1) the sample-cluster similarity balance (free parameter ν), (2) in which space the cluster similarities are computed (input or kernel space), and (3) how the unlabelled training samples contribute to each cluster (crisp or soft association)

Method	Kernel	Mapping	Similarity	Equation	Legend
SVM	K	$\phi(\mathbf{x})$	$\nu = 1$	(1.1)	···•···
μ-SVM in $\mathcal{X}$	$K_\omega^\mathcal{X} = \nu K + (1-\nu)K_\mu^\mathcal{X}$	$\{\sqrt{\nu}\phi^\top(\mathbf{x}), \sqrt{1-\nu}\phi^\top(\boldsymbol{\mu})\}^\top$	$0 < \nu < 1$	(1.7)	- + -
	$K_\mu^\mathcal{X}$	$\phi(\boldsymbol{\mu})$	$\nu = 0$	(1.3)	- ⊖ -
μ-SVM in $\mathcal{H}$	$K_\omega^\mathcal{H} = \nu K + (1-\nu)K_\mu^\mathcal{H}$	$\{\sqrt{\nu}\phi^\top(\mathbf{x}), \sqrt{1-\nu}\phi_\mu^\top(S)\}^\top$	$0 < \nu < 1$	(1.7)	·-+-
	$K_\mu^\mathcal{H}$	$\phi_\mu(S)$	$\nu = 0$	(1.4)	·-⊙-
μ_s-SVM in $\mathcal{H}$	$K_{\omega_s}^\mathcal{H} = \nu K + (1-\nu)K_{\mu_s}^\mathcal{H}$	$\{\sqrt{\nu}\phi^\top(\mathbf{x}), \sqrt{1-\nu}\phi_{\mu_s}^\top(S)\}^\top$	$0 < \nu < 1$	(1.7)	—+—
	$K_{\mu_s}^\mathcal{H}$	$\phi_{\mu_s}(S)$	$\nu = 0$	(1.9)	—⊖—

10.3 Experimental results

10.3.1 Model development

The proposed μ-SVM classifiers are benchmarked against: (1) the standard SVM applied to the image pixels (denoted by 'SVM on $\mathbf{x}_i$'), which is used as a reference for supervised methods; and (2) the standard SVM applied to each cluster centre in order to obtain its class label that is propagated to all the samples belonging to this cluster (denoted by 'SVM on $\boldsymbol{\mu}_k$'), which is the standard approach in unsupervised classification problems.

We used for all the experiments the radial basis function (RBF) kernel, $K(\mathbf{x}_i, \mathbf{x}_j) = \exp\left(-\|\mathbf{x}_i - \mathbf{x}_j\|^2/2\sigma^2\right)$, where $\sigma \in \mathbb{R}^+$ is the kernel width, which can be different for K and K_μ. Kernel width σ was tuned in the range $\{10^{-3}, \ldots, 10\}$. SVM regularization parameter C was varied in the range $\{10^{-1}, \ldots, 10^2\}$. Weight ν was tuned in the range $\{0.01, \ldots, 0.99\}$ for composite kernels. The selection of the best subset of free parameters was done through ten-fold cross-validation in the training set.

For both synthetic and real experiments, we generated training sets consisting of ℓ labelled samples (equal number of samples *per* class), and added u unlabelled (randomly selected) samples from the analysed test data for the training of the SSL methods. We varied the rate of labelled samples, i.e. $\{2, 4, 7, 14, 27, 52, 100\}\%$ of the labelled samples of the training set were used to train the models in each experiment. By decreasing the number of labelled training samples ℓ, one can analyse how the SSL methods efficiently exploit the information contained in the available unlabelled samples compared with the labelled samples. In order to avoid skewed conclusions, for each value of ℓ, the experiments were run for ten realizations using randomly selected training samples. Once classifiers are trained and adjusted, they are compared using the overall accuracy OA[%] and the kappa statistic κ (estimated in the classification of 5000 independent samples) as a measure of robustness in the classification over the validation set and the test image.

10.3.2 Results on synthetic data

This section illustrates with a synthetic example how the proposed method efficiently exploits the information contained in the available unlabelled samples to characterize the *marginal distribution* of data by means of the cluster structure. In order to demonstrate the robustness of the method in the sample selection bias problem, we used a two-dimenssional nonlinear classification problem where training and test data are generated by two slightly different mixture models composed of six Gaussian components (Figures 10.3(a)–(c)). Three different models are obtained in order to illustrate the trade-off between the sample ($\nu = 1$) and the cluster information ($\nu = 0$) provided by the proposed method (Figures 10.3(d)–(f)). In particular, we have depicted results of the pixel-based approach ($\nu = 1$) equivalent to a standard SVM trained with K; the cluster-based approach ($\nu = 0$) of the μ_s-SVM trained with $K^{\mathcal{H}}_{\mu_s}$; and the composite pixel-cluster approach ($0 < \nu < 1$) of the μ_s-SVM trained with $K^{\mathcal{H}}_{\omega_s}$.

Selected models are compared with the classification boundary of the maximum likelihood classifier (MLC) built with the true mixture models (upper-bound of the classification performance) of both the training distribution and the test distribution, respectively. In the three plots at the bottom part of the figure, we can observe from the sample-similarity based approach to the cluster-similarity based approach, how the classification boundary changes. On the one hand, Figure 10.3(d) corresponds to the standard SVM looking for a maximum margin classifier of the training samples. In this case, the classification boundary matches the MLC boundary of the training set (*gold line*). On the other hand, Figure 10.3(e) corresponds to the cluster-based approach that trains a SVM using as kernel matrix the cluster similarity $K^{\mathcal{H}}_{\mu_s}$ of the associated clusters. In this case, classification boundary is dominated by the cluster distribution of the unlabelled samples from the test set, and thus it matches the MLC boundary of the test set (*blue line*). It is worth noting that in the case where both sample and cluster similarities are combined (Figure 10.3(f)) we obtain an intermediate solution, that is: (1) in regions densely populated by labelled training samples, the μ_s-SVM follows the optimal MLC boundary of the training data (middle-left part of the plot); and (2) in regions without labelled training samples forcing the output class, μ_s-SVM follows the optimal MLC boundary of the test data (bottom-right corner of the plot). The value of ν can be tuned by the user in the training process or selected through cross-validation in the training set, as is commonly done with the other free parameters of SVMs, such as C and σ. However, we should note that the model selection in semi-supervised methods applied to problems affected by sample selection bias is still unsolved since, if the training and test distributions are significantly different, the selected model will be biased towards the training samples when performing cross-validation in the training set. Recently, some efforts have been made to alleviate this problem (Marconcini and Bruzzone 2007; Sugiyama *et al.* 2007).

10.3.3 Results on real data

This section presents the experimental results on real remote sensing images. For illustration purposes, we focus on the challenging *sample selection biased* scenario of cloud identification from multi-spectral images.

Cloud screening constitutes a clear candidate for the proposed SSL method, since very few labelled cloud pixels are typically available, and cloud features change to a great extent depending on the cloud type, thickness, transparency, height or background. We used data from the MEdium Resolution Imaging Spectrometer (MERIS) instrument on board the ESA ENVIronmental SATellite (ENVISAT) (Rast *et al.* 1999). We used as inputs six physically-inspired features extracted from the 15 MERIS multi-spectral bands (Gómez-Chova *et al.* 2007).

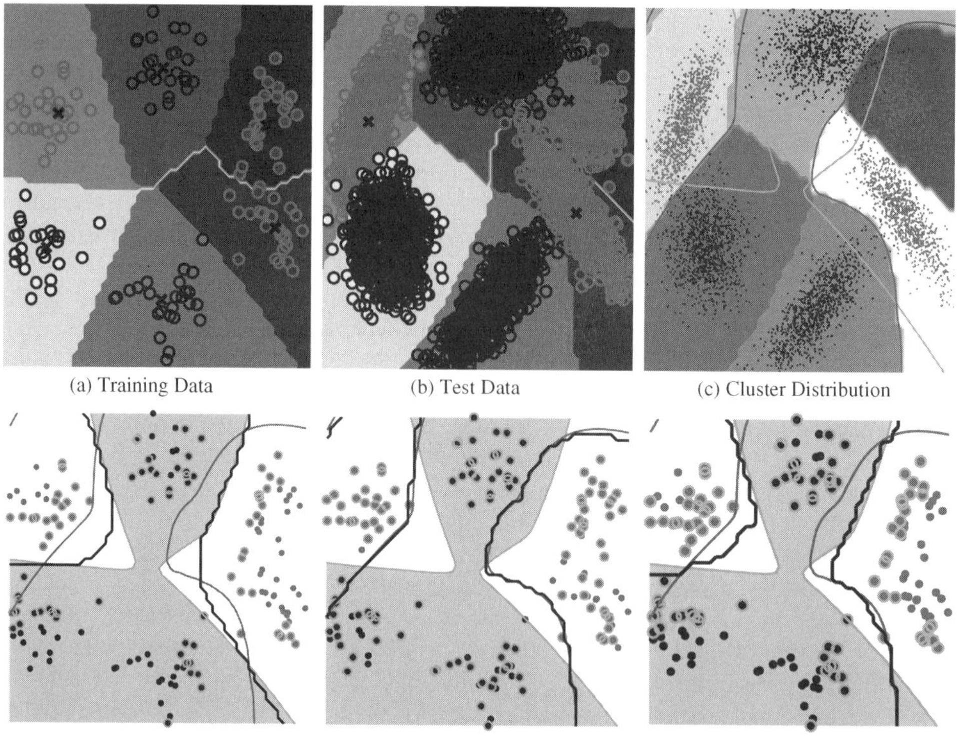
(a) Training Data (b) Test Data (c) Cluster Distribution

(d) SVM (K) (e) μ_s-SVM($K^{\mathcal{H}}_{\mu_s}$) (f) μ_s-SVM($K^{\mathcal{H}}_{\omega_s}$)

*Figure 10.3 Illustrative example of the use of composite mean kernel in a two-dimensional two-classes problem where training and test data are generated by two slightly different mixture models composed of six Gaussian components. Plots (a) and (b) show the samples (circles) and the distribution (coloured areas) of each Gaussian component of the the training (6 × 25 samples) and test (6 × 1000 samples) sets, respectively. Plot (c) shows test samples with the true class labels (*red *and* black points*) and the distribution (*shaded areas*) of the found clusters by the EM algorithm. Plots (d)–(f) show the classification results of the proposed composite kernel using the soft mean map for three different situations: (d) pixel-based approach ($\nu = 1$) equivalent to a standard SVM trained with K; (e) cluster-based approach ($\nu = 0$) of the μ_s-SVM trained with $K^{\mathcal{H}}_{\mu_s}$; and (f) composite pixel-cluster approach ($0 < \nu < 1$) of the μ_s-SVM trained with $K^{\mathcal{H}}_{\omega_s}$. In these plots,* red *and* black points *represent the training samples;* blue circles *indicate the samples used as support vectors in each model;* golden line *and* shaded areas *indicate the classification boundary of the maximum likelihood classifier (MLC) built with the true mixture model of the training data (upper-bound of the classification performance);* blue line *indicates the classification boundary of the MLC built with the true mixture model of the test data; and* black line *shows the classification boundary of the trained models. (See plate 3)*

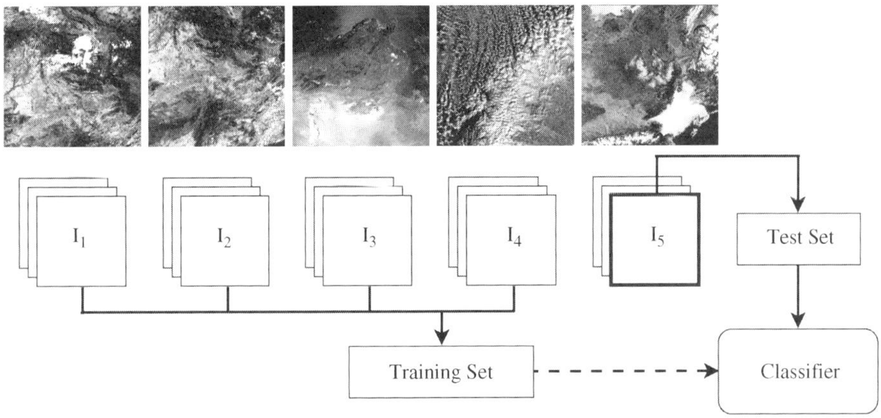

Figure 10.4 MERIS images acquired over Spain (BR-2003-07-14 and BR-2004-07-14), Tunisia (TU-2004-07-15), Finland (FI-2005-02-26), and France (FR-2005-03-19). The bottom part of the figure shows the image-fold case *followed in the experiments that induce the* sample selection bias *in the training process.*

Experiments were carried out using five MERIS images acquired over Spain (two images), Tunisia, Finland and France (Figure 10.4). In order to test the robustness of the algorithm to differences in the training and test distributions (*sample selection bias*), the proposed kernel methods (Table 10.1) are benchmarked in two different conditions: (1) few labelled training samples are available for characterizing the image to be classified; and (2) no labelled training samples are available for the test image to be classified. The first case is called the *single-image case*, since labelled and unlabelled training samples belong to the image to be classified. Note that, in this case, both the training and test data come from the same marginal distribution, and thus the classifiers are not affected by the sample selection bias problem. In the second case, results are obtained following an *image-fold cross-validation* strategy[1], that is, each test image was classified with a model built with labelled samples from the other images and unlabelled samples coming from the same test image (Figure 10.4). In this case, the available training set might not be representative enough for the test (sample selection bias problem) and the unlabelled samples from the image to be classified should help to improve the classification accuracy.

Single-image case results

First, we are going to analyse the classification results in the *single-image case*, i.e. each test image is classified with a model built with labelled and unlabelled samples coming from the same test image. This procedure is aimed at comparing the different algorithms in an ideal situation where both training and test data come from the same distribution. Hence, one can

[1] Classifiers are trained using both labelled samples from the $N-1$ out of N available images and unlabelled samples from the image to be classified. A different classifier is trained for each image and, finally, classification results on test are averaged for the N images to evaluate the robustness of the classifiers to the sample selection bias.

assess which method learns best from the labelled samples, and how the semi-supervised methods take advantage of the available unlabelled samples. In the following paragraphs, we will discuss results obtained for the single-image case. Average results over ten realizations are shown in Figure 10.5. Several conclusions are obtained from these plots.

Figure 10.5(a) shows the κ statistic versus the number of labelled samples obtained with the standard SVM for the five images. This plot provides us with a reference on how difficult is the classification problem or the cloud screening problem in each MERIS image (Figure 10.4). From the results one can conclude that the classification complexity of the images increases in the following order: Barrax (BR-2003-07-14) that presents a bright and thick cloud in the centre of the image; Barrax (BR-2004-07-14) that presents small clouds over land and sea in the right part of the image; Tunisia (TU-2004-07-15) that presents clouds and bright dessert areas; France (FR-2005-03-19) that presents opaque clouds at south and north France, but also snowy mountains at various altitudes; and, finally, Finland (FI-2005-02-26), which presents cirrus clouds over the sea and the icy coast of Finland. Therefore, we are including in the experiments both easy cloud screening problems, where few labelled samples are enough to obtain accurate classifications, and extremely complex cloud screening scenarios, where a relatively high number of labelled samples is required to correctly detect clouds when using a standard supervised classifier. In the remaining plots of Figure 10.5, the results obtained for these five scenes are averaged in order to discover which method performs best in most of the scenarios.

Figure 10.5(b) shows the average CPU time consumed by each method during the training phase. All experiments were carried out in a 64-bit dual-core Intel® Xeon™ CPU 2.80 GHz processor under Linux, and all methods are based on MATLAB implementations with a QP/SMO algorithm programmed in C++ (Chang and Lin 2001). In this plot, two groups of methods are easily distinguished. First, the best performance in terms of CPU time is obtained by the standard SVM and the μ-SVM in $\mathcal{X}$, which only require one to compute the kernel matrix for the labelled samples $K_{\ell\times\ell}$. In fact, the $K_\mu^{\mathcal{X}}$ method is slightly faster than the SVM since it only computes the kernel matrix over the cluster centres $\boldsymbol{\mu}_k$ in the input space ($K_\mu^{\mathcal{X}} = \langle\boldsymbol{\phi}(\boldsymbol{\mu}_1), \boldsymbol{\phi}(\boldsymbol{\mu}_2)\rangle$) and the number of clusters c in the image is usually much lower than the number of labelled samples ℓ. On the other hand, $K_\omega^{\mathcal{X}} = \nu K + (1-\nu)K_\mu^{\mathcal{X}}$ is slightly slower than these methods since the weighting parameter ν is also tuned during the training phase. Secondly, the proposed μ_s-SVM classifiers in $\mathcal{H}$ provide an acceptable performance but are slower than previous methods since, in order to compute the similarity between clusters in the kernel space $K_\mu^{\mathcal{X}}$, they have to compute the kernel matrix for the labelled and unlabelled samples $K_{(\ell+u)\times(\ell+u)}$. However, this difference is reduced when the number of labelled samples ℓ approaches the number of unlabelled samples $u = 5000$. Again, the weighted versions of the μ-SVM ('+' markers) are slower than the versions based on clusters exclusively ('o' markers) because of the tuning of ν.

Figures 10.5(c) and 10.5(d) show the average κ and OA for all the methods. The first conclusion extracted from the curves is that the proposed μ-SVM method clearly improves the results of the other methods. The mean kernel classifiers produce better classification results than the reference provided by the supervised SVM in all cases (note that SVM is a particular case of the μ-SVM for $\nu = 1$). These results are a consequence of taking into account the distribution of image data to define the clusters. In consequence, μ-SVM classifiers can be considered as a good trade-off between computational cost and classification accuracy. In addition, μ-SVM classifiers working in the kernel space provide slightly better results, supporting the idea that we can find a richer space $\mathcal{H}$ for separating classes. In ill-posed

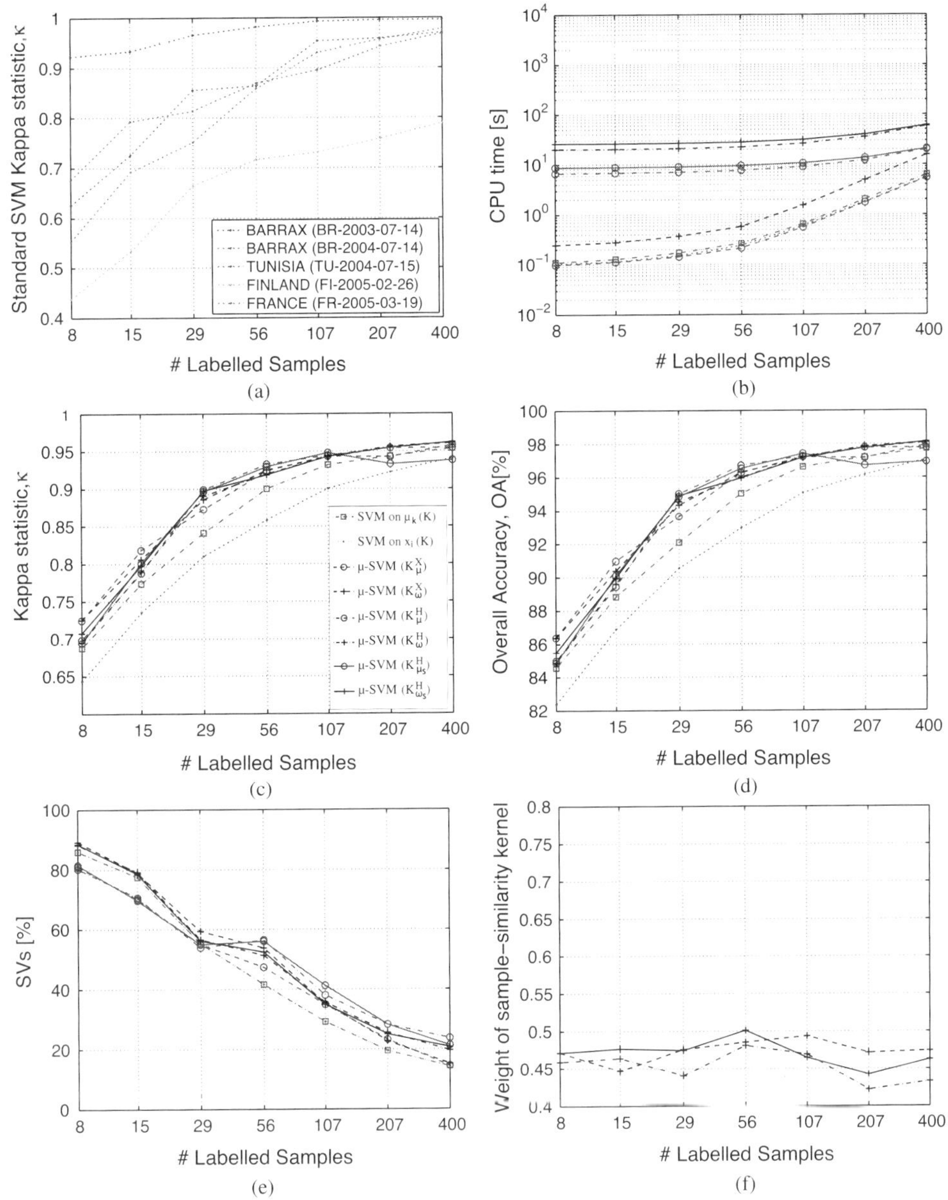

Figure 10.5 (a) Standard SVM classification results for the five MERIS images. (b)–(f) Average classification results over the five images training the models with labelled and unlabelled (800) samples from the image to be classified (single-image case)*: (b) CPU time [s], (c) κ, (d) OA, (e) SVs [%], and (f) weight ν of the sample-similarity kernel of labelled samples* K. *(See plate 4)*

situations, with a low number of labelled samples, the performance of μ-SVM in $\mathcal{H}$ is reversed and μ-SVM in $\mathcal{X}$ provide better results. This fact can be explained since, when working with a low number of labelled samples, v-fold cross-correlation techniques are less efficient at tuning the kernel width σ, which actually defines the mapping to $\mathcal{H}$. Therefore, the cluster similarity $K_{\mu}^{\mathcal{H}}$, computed only from the unlabelled samples in $\mathcal{H}$, is less meaningful than $K_{\mu}^{\mathcal{X}}$, computed by using the cluster centres $\boldsymbol{\mu}_k$ obtained when applying the EM algorithm to the whole image. We can also observe that the proposed method is not equivalent to a simple segmentation of the image by classifying the centres of the clusters (*red dash-doted line*), that is, classifying $\boldsymbol{\mu}_k$ is not a good option but it is still better than purely supervised SVM (*red dotted line*). This indicates that the EM clustering of the image provides a good image segmentation, which is mainly due to the physically-inspired extracted features.

Figure 10.5(e) shows the average percentage of support vectors (SVs) for each method, i.e. the number of labelled training samples used as SVs in the selected models. In these experiments, all SVM methods produce sparse models with a low number of support vectors. The trend of all methods is coherent, since as the number of labelled samples in the training set increases, the rate of samples (SVs) required to classify correctly decreases. The only significant difference between methods is that, in ill-posed situations with a low number of labelled samples, the classifiers based on cluster similarity require fewer SVs since the class distribution is approximated by the cluster distribution. However, when increasing the number of labelled samples, simple spaces (such as that of SVM) increase sparsity, but also worsen models in terms of kappa.

Finally, Figure 10.5(f) shows the relative weight ν of the sample-similarity kernel of labelled samples K with respect to the cluster-similarity kernel of the unlabelled samples in the selected $K_{\omega}^{\mathcal{X}}$, $K_{\omega}^{\mathcal{H}}$ and $K_{\omega_s}^{\mathcal{H}}$ models. The value of ν can be tuned by the user in the training process, but we have selected it through cross-validation in the training set. In these experiments, the sum of Hilbert spaces (i.e. different mapping and kernel functions) leads approximately to an average weighting as the optimal solution ($\nu \sim 0.5$). Intuitively, this means that both the labelled information and the cluster information show similar importance in the classification. Hence, taking into account the excellent classification accuracy obtained in the experiments, we can conclude that both the labelled samples included in the training set and the clusters from unlabelled samples properly describe the class distribution in the test image. This situation is coherent in the context of the *single-image case* followed to obtain the results presented in Figure 10.5.

Image-fold case results

Now, we analyse the classification results in the *image-fold case*, i.e. each test image is classified according to a model built with labelled samples from the other four images and unlabelled samples coming from the same test image. This procedure is aimed at testing the robustness of the algorithm to differences between the training and test distributions.

Figure 10.6(a) shows the κ statistic for the five images obtained with the standard SVM. This plot provides us with a reference of how difficult the classification problem is in each MERIS image, but now it also depends on how well the labelled samples from the other images in the training set represent the classes in the test image. Now, the order of the images depending on their classification complexity is very similar: Barrax (BR-2003-07-14), Barrax (BR-2004-07-14), Tunisia (TU-2004-07-15), France (FR-2005-03-19) and Finland (FI-2005-02-26). Also, as expected, poorer classification accuracy is obtained in all images. However,

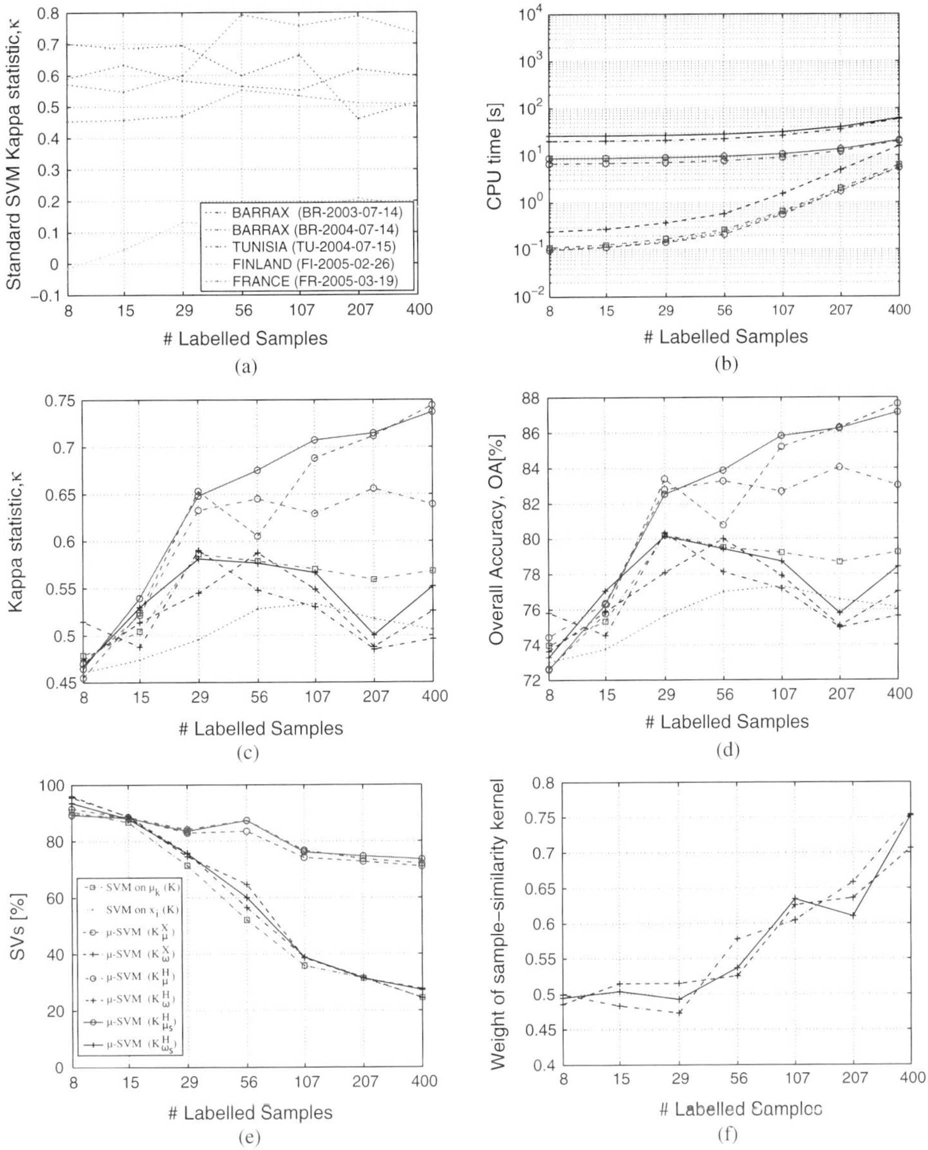

Figure 10.6 (a) Standard SVM classification results for the five MERIS images. (b)–(f) Average classification results over the five images with labelled samples from the other four images and 800 unlabelled samples from the image to be classified (image-fold case)*: (b) CPU time [s], (c) κ, (d) OA, (e) SVs[%], and (f) weight ν of the sample-similarity kernel (K). (See plate 5)*

it remains almost independent of the number of labelled samples, which indicates that, from the very beginning, labelled samples from other images roughly describe the type of clouds in the test image, and these situations are not improved by adding more labelled samples with the same information. Therefore, we can conclude that, following the proposed image-fold case, the experiments for the five images are affected by the sample selection bias problem. In the remaining plots of Figure 10.6, the results obtained for these five scenes are averaged in order to discover which method performs better in most of the scenarios.

Figure 10.6(b) shows the average CPU time consumed by each method during the training phase. The results in this plot are almost identical to the results in Figure 10.5, since the computational burden mainly depends on the amount and type of data, and the experiments were carried out on the same computer.

Figures 10.6(c) and 10.6(d) show the average κ and OA for all the methods respectively. The situation now is completely different from the previous experiments at the single-image case. Almost all the methods provide moderate classification accuracy, and all of them provide poor results in ill-posed situations. However, a large difference can be observed between the μ-SVM classifiers based on clusters exclusively ('o' markers) and the rest. Several conclusions can be obtained from these curves. The standard SVM is affected by the sample selection bias and this cannot be solved since it relies on the training labelled samples exclusively. When using the standard SVM to directly classify the centres of the clusters from the EM algorithm, classification results improve since cluster prototypes have a higher probability of being correctly classified by SVM than test samples. Again, the main impediment is the SVM classifier used to label the cluster centres. The μ-SVM classifiers based exclusively on cluster-based approaches $K_\mu^{\mathcal{X}}$, $K_\mu^{\mathcal{H}}$ and $K_{\mu_s}^{\mathcal{H}}$ provide excellent results when there are enough labelled samples to describe the class conditional distribution of the clusters (with few labelled samples a whole cluster can be misclassified). Of these three classifiers, $K_\mu^{\mathcal{H}}$ produce worse results. This can be explained since the selection of the mapping function to $\mathcal{H}$, i.e. the selection of kernel free parameters, depends on the classification accuracy in the training set. Therefore, an inappropriate training set produces an inappropriate mapping (in terms of class separability in the test). As a consequence, $K_\mu^{\mathcal{H}}$ is more affected by the sample selection bias since all the unlabelled samples in the training set are used to compute the cluster similarity in an inappropriate kernel space. On the other hand, $K_\mu^{\mathcal{X}}$ is more robust to the sample selection bias because it approximates the cluster similarity to the similarities of the cluster centres $\boldsymbol{\mu}_k$ already defined in the input space, and thus it is less dependent on how the unlabelled samples that represent the clusters are mapped into $\mathcal{H}$. In this context, $K_{\mu_s}^{\mathcal{H}}$ provides the best results and is also more robust to the sample selection bias because it uses the soft mean map to compute the cluster similarity in the kernel space. The soft mean map weights the contribution of each sample to the definition of the centre of mass of each cluster in the kernel space $\mathcal{H}$ with the EM estimated posterior probabilities. This is equivalent to eliminating the samples that do not properly represent the cluster in the input space, and thus the estimation of the cluster centre in $\mathcal{H}$ is less affected by the selection of an inappropriate mapping. Finally, the μ-SVM classifiers based on composite mean kernels $K_\omega^{\mathcal{X}}$, $K_\omega^{\mathcal{H}}$ and $K_{\omega_s}^{\mathcal{H}}$ (*black '+' lines*) produce significantly worse results than the cluster-based approaches K_μ. These approaches combine both sample and cluster similarities and, intuitively, should produce better classification results than standard SVM (K) and K_μ, which are a particular case of K_ω for $\nu = 1$ and $\nu = 0$, respectively. Therefore, the divergence in the results should be explained by the tuning of ν, and we have also to analyse the selected ν value for the different experiments.

Figure 10.6(f) shows the relative weight ν of the sample-similarity kernel of labelled samples K with respect to the cluster-similarity kernel K_{μ}. In this plot, we can observe that for a low number of labelled samples ($\ell \leqslant 30$) the sample-similarity and cluster-similarity have the same weight ($\nu = 0.5$), and then the weight of the sample-similarity kernel of labelled samples K increases exponentially with the number of labelled samples. As the number of labelled samples increases, K becomes more important than the cluster information (the exponential value of labelled samples in supervised classifiers is analysed in detail by Castelli and Cover (1995)). However, the opposite behaviour is observed in the classification accuracy (Figures 10.6(c) and 10.6(d)) that starts to decrease when ν increases. As mentioned above, the labelled samples in the training set used to compute K come from different images and are affected by the sample selection bias. For this reason, high values of ν (greater weight of K) produce worse values. The key point is to understand the reason for which high values of ν are selected during the training. The reason is that ν has been selected through ν-fold cross-validation on the labelled training samples, thus the selected model (ν, σ and C) will be biased towards the training samples, which produces worse results on the test set if the training and test distributions are significantly different. It should be noted that the model selection in semi-supervised methods applied to problems affected by sample selection bias is not well-solved yet. One solution would be to select the best model by cross-validation in a labelled validation set coming from the image to be classified, but this is an unrealistic situation since, in this case, these data can be used in the training. The results suggest that a simple and reasonable trade-off is to equally weight the sample and cluster similarity ($\nu = 0.5$), which provides good results in standard semi-supervised scenarios (Figure 10.5) and it is not biased towards the training set when dealing with sample selection biased problems (Figure 10.6).

Finally, Figure 10.6(e) shows the average percentage of support vectors (SVs) for each method. Again, most of the methods produce sparse models with a low number of support vectors. The only exception are the three cluster-based methods that require more SVs to correctly weight the cluster similarities. Here we can clearly observe the trade-off between sparsity and accuracy, with the over-sparse solutions that provide low classification accuracy and the moderately sparse (80% over 400 samples) models that provide better classification accuracy. The high number of SVs in cluster-based methods can be explained since in this image-fold experiment the information (similarities) in K and K_{μ} are contradictory because the class distribution in training and the cluster distribution in test do not match. Hence, a higher number of samples is needed to find a maximum-margin hyperplane in the kernel space.

Classification maps

Finally, the developed models are applied to all the pixels of the test images in order to obtain the classification maps. Results shown in the previous section have been obtained by using the validation set (5000 labelled samples from the test image used to compare the performance of the classifiers). In this section, a quantitative and a visual analysis of the corresponding classification maps of the test images are carried out.

Figure 10.7 shows the comparison of the μ_s-SVM methods (both composite $K_{\omega_s}^{\mathcal{H}}$ and cluster-based $K_{\mu_s}^{\mathcal{H}}$ classifiers) against the cloud screening reference. The images selected to illustrate the results are one image over Barrax (BR-2003-07-14) and the France image (FR-2005-03-19), which present different cloud screening problems and are differently affected by the sample selection bias problem. In the case of the unprojected MERIS FR images,

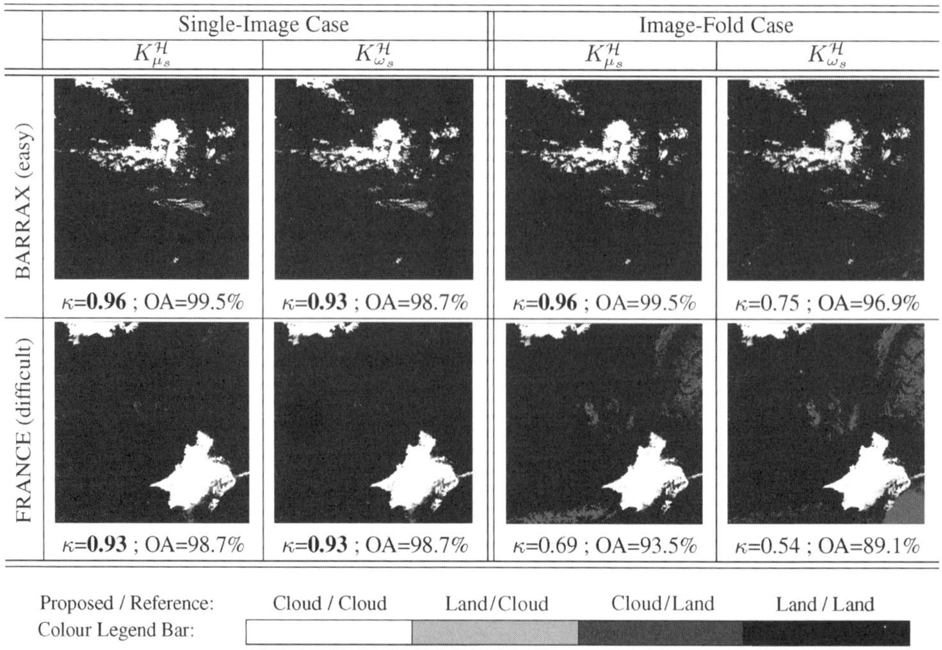

Figure 10.7 Comparison of the cloud mask of the proposed kernel methods against the ground truth for the MERIS images over Barrax (BR-2003-07-14) and France (FR-2005-03-19). Discrepancies are shown in red when proposed kernel methods detect cloud and in yellow when pixels are classified as cloud-free. (See plate 6)

the quarter scenes have 1153×1153 pixels (e.g. BR-2003-07-14) and the full scenes have 2241×2241 pixels (e.g. FR-2005-03-19). The selected images are classified using the best models (realization with best validation results) trained with 400 labelled samples for both the single-image case and the image-fold case (last point in Figures 10.5 and 10.6, respectively). Pixels where the algorithms and the ground truth agree are depicted in *white* for the cloudy pixels and in *blue* for the cloud free pixels; discrepancies are shown in *yellow* and *red*. The agreement between both maps is expressed in terms of the overall accuracy (OA) and kappa statistic (κ) for the whole image.

Classification accuracies above 90% are obtained for most cases, but the lower values of κ for some cases reflect that classification results are unbalanced due to the misclassification of a significant number of pixels of one class. Note that the overall accuracy is directly interpretable as the ratio of pixels being classified correctly, while the kappa coefficient allows for a statistical test of the significance of the divergence between two algorithms (Congalton and Green 1999). The best kappa result (>0.9) for each experiment is bold faced. Let us analyse the Barrax image (BR-2003-07-14), which represents an easy cloud screening problem. When comparing the two kernel methods with the reference, the cluster-based classifier $K^{\mathcal{H}}_{\mu_s}$ provides good results even in the image-fold case, which means that training samples from the other images are useful to classify correctly the clusters found in the test image. However, the composite kernel classifier $K_{\omega_s}^{\mathcal{H}}$ works properly in the single-image case, while results are worse for the image-fold case, which means that the model is biased towards the labelled samples of the other images instead of to the cluster structure (see explanation of Figure 10.6(e)).

Let us now consider the second image, which is the 'France image', that presents opaque clouds at the south and north and also snow in the Alps, the Pyrenees and the Massif Central. Attending to the image-fold experiments, μ_s-SVM methods agree with the cloud mask. In the image-fold case, neither the cluster-based classifier $K_{\mu_s}^{\mathcal{H}}$ nor the composite kernel classifier $K_{\omega_s}^{\mathcal{H}}$ produce an accurate cloud screening. This can be explained since, in this case, no training samples from the other images model the difference between clouds and snowy mountains, thus the classifier cannot learn this difference.

10.4 Conclusions

A semi-supervised SVM classification method based on both sample and cluster similarity was presented in this chapter. It assumes that some supervised information is available and it is used together with the unlabelled samples of the analysed image to develop a classifier that provides the class of each pixel taking also into account the image data distribution. The information from unlabelled samples of the test dataset is included in the standard SVM by means of a linear combination of kernels, and the cluster similarity is directly computed in the kernel space with a dedicated kernel, which is based on the means of the feature vectors in this space. From a methodological point of view, it has two main advantages: (1) the complexity of the problem is not increased (similar computational complexity) and the objective function is still convex; and (2) the mixture of kernels is much more flexible than an objective function and parameters are easier to tune.

In some real applications training samples are available, while in others they are not available for the image to be classified, and thus, labelled data extracted from other images modelling similar problems can be useful in improving the classification accuracy. In this scenario, not all the training samples are equally reliable and hence the presented soft mean map constitutes a suitable approach to facing the sample selection bias.

Good results have been obtained in different synthetic and real situations. The good results obtained for the cloud screening application suggest that, when a proper data assumption is made, the proposed semi-supervised method outperforms the standard supervised algorithm. Since the distribution of remote sensing data over natural land covers is usually smooth (spectra of pixels from the same land cover are similar), the cluster assumption imposed in the model is reasonable, and the EM algorithm with GMM efficiently implements it. However, although the proposed semi-supervised methods benefit from the inclusion of unlabelled samples by estimating the marginal data distribution, results have shown that these methods are limited by the quality of the available labelled information and cannot alleviate situations with a dramatic sample selection bias problem. This suggests that further developments might be focused on new validation methods for these situations, and on performing extensive experiments in other applications.

Acknowledgments

This work was partly supported by the Spanish Ministry of Science projects CICYT-FEDER TEC2006-13845, AYA2008-05965-C04-03 and CSD2007-00018.

References

Bachmann, C., Ainsworth, T. and Fusina, R. (2005) Exploiting manifold geometry in hyperspectral imagery. *IEEE Transactions on Geoscience and Remote Sensing*, **43**(3), 441–454.

Bachmann, C., Ainsworth, T. and Fusina, R. (2006) Improved manifold coordinate representations of large-scale hyperspectral scenes. *IEEE Transactions on Geoscience and Remote Sensing*, **44**(10), 2786–2803.

Bickel, S., Brückner, M. and Scheffer, T. (2007) Discriminative learning for differing training and test distributions *ICML '07: Proceedings of the 24th International Conference on Machine Learning*, pp. 81–88. ACM, New York, NY, USA.

Bruzzone, L. and Fernández Prieto, D. (2001) Unsupervised retraining of a maximum likelihood classifier for the analysis of multi-temporal remote sensing images. *IEEE Transactions on Geoscience and Remote Sensing*, **39**(2), 456–460.

Bruzzone, L., Chi, M. and Marconcini, M. (2006) A novel transductive SVM for the semi-supervised classification of remote-sensing images. *IEEE Transactions on Geoscience and Remote Sensing*, **44**(11), 3363–3373.

Camps-Valls, G. and Bruzzone, L. (2005) Kernel-based methods for hyperspectral image classification. *IEEE Transactions on Geoscience and Remote Sensing*, **43**(6), 1351–1362.

Camps-Valls, G., Bandos Marsheva, T.V. and Zhou, D. (2007) Semi-supervised graph-based hyperspectral image classification. *IEEE Transactions on Geoscience and Remote Sensing*, **45**(10), 3044–3054.

Camps-Valls, G., Gómez-Chova, L., Muñoz-Marí, J., Vila-Francés, J. and Calpe-Maravilla, J. (2006) Composite kernels for hyperspectral image classification. *IEEE Geoscience and Remote Sensing Letters*, **3**(1), 93–97.

Castelli, V. and Cover, T.M. (1995) On the exponential value of labelled samples. *Pattern Recogn. Lett.* **16**(1), 105–111.

Chang, C.C. and Lin, C.J. (2001) *LIBSVM: a library for support vector machines*. Software available at http://www.csie.ntu.edu.tw/ cjlin/libsvm.

Chapelle, O., Schölkopf, B. and Zien, A. (2006) *Semi-supervised Learning* 1st edn. MIT Press, Cambridge, MA, USA.

Chapelle, O., Weston, J. and Schölkopf, B. (2003) Cluster kernels for semi-supervised learning In *NIPS 2002* (eds. Becker, S., Thrun, S. and Obermayer, K.), vol. 15, pp. 585–592. MIT Press, Cambridge, MA, USA.

Chi, M. and Bruzzone, L. (2007) Semi-supervised classification of hyperspectral images by SVMS optimized in the primal. *IEEE Transactions on Geoscience and Remote Sensing*, **45**(6), 1870–1880.

Congalton, R. and Green, K. (1999) *Assessing the Accuracy of Remotely Sensed Data: Principles and Practices*, 1st edn. CRC Press, Boca Raton, FL, USA.

Dempster, A.P., Laird, N.M. and Rubin, D.B. (1977) Maximum likelihood from incomplete data via the EM algorithm. *Journal of the Royal Statistical Society, Series B*, **39**, 1–38.

Gärtner, T., Flach, P.A., Kowalczyk, A. and Smola, A.J. (2002) Multi-instance kernels *ICML'02: Proceedings of the Nineteenth International Conference on Machine Learning*, pp. 179–186. Morgan Kaufmann Publishers Inc., San Francisco, CA, USA.

Gómez-Chova, L. (2008) *Cloud screening algorithm for MERIS and CHRIS multi-spectral sensors*. PhD thesis Universitat de València, Spain.

Gómez-Chova, L., Bruzzone, L., Camps-Valls, G. and Calpe-Maravilla, J. (2008a) Semi-supervised remote sensing image classification based on clustering and the mean map kernel. *IEEE International Geoscience and Remote Sensing Symposium, IGARSS'2008*, vol. IV, pp. 391–394, Boston, USA.

Gómez-Chova, L., Camps-Valls, G., Calpe, J., Guanter, L. and Moreno, J. (2007) Cloud-screening algorithm for ENVISAT/MERIS multi-spectral images. *IEEE Transactions on Geoscience and Remote Sensing*, **45**(12), 4105–4118.

Gómez-Chova, L., Camps-Valls, G., Muñoz-Marí, J. and Calpe-Maravilla, J. (2008b) Semi-supervised image classification with laplacian support vector machines. *IEEE Geoscience and Remote Sensing Letters*, **5**(3), 336–340.

Gretton, A., Borgwardt, K.M., Rasch, M., Schölkopf, B. and Smola, A. (2007) A kernel method for the two-sample-problem *NIPS 2006*, vol. 19, pp. 1–8. MIT Press, Cambridge, MA, USA.

Heckman, J.J. (1979) Sample selection bias as a specification error. *Econometrica* **47**(1), 153–161.

Huang, J., Smola, A., Gretton, A., Borgwardt, K.M. and Schölkopf, B. (2007) Correcting sample selection bias by unlabelled data *NIPS 2006*, vol. 19, pp. 1–8. MIT Press, Cambridge, MA, USA.

Jackson, Q. and Landgrebe, D. (2001) An adaptive classifier design for high-dimensional data analysis with a limited training data set. *IEEE Transactions on Geoscience and Remote Sensing*, **39**(12), 2664–2679.

Jebara, T., Kondor, R. and Howard, A. (2004) Probability product kernels. *Journal of Machine Learning Research, JMLR, Special Topic on Learning Theory* **5**, 819–844.

Kondor, R. and Jebara, T. (2003) A kernel between sets of vectors *International Conference on Machine Learning, ICML03*.

Li, B., Chi, M., Fan, J. and Xue, X. (2007) Support Cluster Machine. In *ICML'07: Proceedings of the 24th International Conference on Machine Learning* (ed. Ghahramani Z), vol. 227 of *ACM International Conference Proceeding Series*, pp. 505–512. ACM, New York, USA.

Marconcini, M. and Bruzzone, L. (2007) Partially-supervised updating of land-cover maps: A P^2S^2VM technique and a circular validation strategy *IEEE International Geoscience and Remote Sensing Symposium, IGARSS'2007*, pp. 1079–1082.

Muñoz-Marí, J., Camps-Valls, G., Gómez-Chova, L. and Calpe-Maravilla, J. (2007) Combination of one-class remote sensing image classifiers *IEEE International Geoscience and Remote Sensing Symposium, IGARSS'2007*, pp. 1509–1512, Barcelona, Spain.

Muñoz-Marí, J., Gómez-Chova, L., Camps-Valls, G. and Calpe-Maravilla, J. (2008) Image classification with semi-supervised support vector domain description In *SPIE International Remote Sensing Symposium 2008* (ed. Bruzzone L), vol. 7109A, pp. 7109A–11. SPIE.

Rast, M., Bézy, J. and Bruzzi, S. (1999) The ESA Medium Resolution Imaging Spectrometer MERIS: a review of the instrument and its mission. *International Journal of Remote Sensing*, **20**(9), 1681–1702.

Schölkopf, B. and Smola, A. (2001) *Learning with Kernels – Support Vector Machines, Regularization, Optimization and Beyond*. MIT Press Series, Cambridge, MA, USA.

Shahshahani, B. and Landgrebe, D. (1994) The effect of unlabelled samples in reducing the small sample size problem and mitigating the Hughes phenomenon. *IEEE Transactions on Geoscience and Remote Sensing*, **32**(5), 1087–1095.

Shawe-Taylor, J. and Cristianini, N. (2004) *Kernel Methods for Pattern Analysis*. Cambridge University Press, Cambridge, MA, USA.

Shimodaira, H. (2000) Improving predictive inference under covariate shift by weighting the log-likelihood function. *Journal of Statistical Planning and Inference*, **90**(2), 227–244.

Sugiyama, M. and Müller, K.R. (2005) Input-dependent estimation of generalization error under covariate shift. *Statistics & Decisions*, **23**(4), 249–279.

Sugiyama, M., Krauledat, M. and Müller, K.R. (2007) Covariate shift adaptation by importance weighted cross validation. *Journal of Machine Learning Research*, **8**, 985–1005.

Sugiyama, M., Suzuki, T., Nakajima, S., Kashima, H., von, Bünau, P. and Kawanabe, M. (2008) Direct importance estimation for covariate shift adaptation. *Annals of the Institute of Statistical Mathematics*, **60**(4), 699–746.

Szummer, M. and Jaakkola, T. (2002) Partially labeled classification with Markov random walks In *NIPS 2001* (ed. et al. TD), vol. 14, pp. 945–952. MIT Press, Cambridge, MA, USA.

Weston, J., Leslie, C., Ie, E., Zhou, D., Elisseeff, A. and Noble, W.S. (2005) Semi-supervised protein classification using cluster kernels. *Bioinformatics*, **21**(15), 3241–3247.

Zhou, S. and Chellappa, R. (2006) From sample similarity to ensemble similarity: probabilistic distance measures in reproducing kernel Hilbert space. *IEEE, Transactions on Pattern Analysis and Machine Intelligence* **28**(6), 917–929.

Zhu, X. (2005) Semi-supervised learning literature survey. Technical Report 1530, Computer Sciences, University of Wisconsin-Madison, USA. Online document: http://www.cs.wisc.edu/jerryzhu/pub/ssl_survey.pdf. Last modified on July 19, 2008.

Zhu, X. and Ghahramani, Z. (2002) Learning from labelled and unlabelled data with label propagation. Technical Report CMU-CALD-02-107, Carnegie Mellon University.

IV

Function approximation and regression

11

Kernel methods for unmixing hyperspectral imagery

Joshua Broadwater, Amit Banerjee and Philippe Burlina
Applied Physics Laboratory, The Johns Hopkins University, USA

This chapter introduces the concept of kernel unmixing algorithms for hyperspectral imagery. The physics of linear and nonlinear mixtures are explored and shown to be suited for kernel approximations. Based on this evidence, a fully automatic kernel unmixing algorithm is developed for endmember extraction and abundance estimation. This kernel unmixing framework has the flexibility to account for both linear and nonlinear behavior by using different kernel functions. Additionally to support nonlinear mixtures, a new physics-based kernel is proposed based on the physics of intimate mixtures. Experimental results on both in-lab and real-world hyperspectral data show both qualitative and quantitative improvements when using the proposed kernel unmixing approach.

11.1 Introduction

Hyperspectral sensors provide unique data for remote sensing applications. These imaging spectrometers collect not only spatial information (pixels), but also a spectral response at each pixel (spectra). Imaging spectroscopy is a direct descendent of reflectance spectroscopy, which began in the 1920s (Wendtland and Hecht 1966) and radiative transfer theory which began in the 1940s (Chandrasekhar 1960). From these sciences, researchers learned how to measure a material's chemical property from the way light interacted with it. Perhaps more importantly, researchers learned how to identify different chemicals within the same sample by 'unmixing' the sample back into its constituent parts. These developments led to multi-spectral imagers

Kernel Methods for Remote Sensing Data Analysis Edited by Gustavo Camps-Valls and Lorenzo Bruzzone

such as Landsat (Landgrebe 1997) that ultimately led to the hyperspectral sensors in use today (Keshava and Mustard 2002).

Hyperspectral sensors provide data that allow researchers to spectrally unmix an entire scene into its different materials based on imaging spectroscopy. Besides obtaining estimates of the materials in a scene, spectroscopy also allows the measurement of the amount of each material both spatially across the entire image and also spectrally within each individual pixel. This unmixing of spectral signatures allows hyperspectral imagery to be used for applications varying from subpixel detection (Manolakis and Shaw 2002) to mineral mapping (Mustard and Pieters 1989).

Two types of models exist that describe how materials mix within a pixel. The first model is called the areal mixing model. It describes the cases where the materials are non-overlapping (Keshava and Mustard 2002) and can be mathematically expressed as a linear combination where the weights are the amounts of each material (Hapke 1993). The other model is the intimate mixture model and it is described by nonlinear functions. This nonlinear mixing model has been difficult to model, but can show up to 30% improvement in measurements over the linear mixing model when intimate mixtures are present (Herzog and Mustard 1996). The difficulty is how to model these nonlinearities when much of the necessary information is not available.

Kernel functions were introduced as a way to generalize linear algorithms to nonlinear data. In the case of detection and classification applications, kernel functions can induce high dimensional feature spaces. In these spaces, previously non-separable classes can be made linearly separable. Thus, linear methods can be applied in this new feature space that provides nonlinear boundaries back in the original data space. Another example is the kernel PCA method (Scholkopf *et al.* 1998), which can be found in Chapters 2, 7, 14 and 15. In this case, the kernel is not used to induce a high dimensional space, but is used to better match the data structure through nonlinear mappings. It is in this mode that kernels can be used to produce nonlinear mixing results while essentially using a linear mixture model. What is more appealing is that the physics suggest that such a method is ideal if one can model the kernel correctly.

Therefore, this chapter introduces the concept of a kernel unmixing algorithm for hyperspectral imagery. The chapter describes the proposed method by first describing the two mixing models for linear and nonlinear mixtures in Section 11.2. Section 11.3 describes the proposed kernel unmixing algorithm based on the SVDD method presented in Chapter 8 and a kernel fully constrained least squares abundance estimate. Section 11.4 presents results of the proposed unmixing algorithms on both in-lab and real-world hyperspectral data. Section 11.5 describes our initial work on a physics-based kernel that provides improved performance over a standard RBF kernel as shown in Section 11.6. Section 11.7 concludes with a summary of the results.

11.2 Mixing models

To develop algorithms for unmixing hyperspectral imagery, the physics of spectroscopy has to be understood first. Based on the work of early researchers in spectroscopy, Clark and Roush (1984) and Hapke (1993) have studied how spectral mixtures affect imaging spectrometers. The result is two well studied mixing models: areal mixtures and intimate mixtures. The next two sections describe the physics behind each mixing type.

11.2.1 Areal mixtures

Areal mixtures are the simplest of the two models and hence the most used in practice. The model is a simple linear combination of endmembers and abundances. Endmembers are the spectra representing the unique materials in an image such as water, soil or vegetation. Abundances are the percentage of each endmember within a pixel. At the physical level, the model assumes that the materials are arranged within a pixel such that they are non-overlapping (Hapke 1993). Assuming that multiple scattering effects between materials are negligible (Keshava and Mustard 2002), the photons interact with each material individually. Therefore, a linear model can be used to identify the materials and their amounts in this scenario.

This linear mixing model can be expressed mathematically as

$$\mathbf{x} = \sum_{q=1}^{Q} a_q \mathbf{m}_q, \quad a_q \geq 0, \quad \sum_{q=1}^{Q} a_q = 1, \tag{11.1}$$

where $\mathbf{x}$ is an $L \times 1$ vector that represents the spectral signature of the current pixel, Q is the number of endmembers within the image, $\mathbf{m}_q$ is an $L \times 1$ vector that represents the qth endmember, and a_q is the qth abundance value. Note that the linear mixing model includes two constraints on the abundance values: non-negativity and sum-to-one. These constraints are necessary to match the physics of the problem that define the abundances as the percentage of the pixel occupied by the different materials. Since they are percentages, the two constraints must hold.

11.2.2 Intimate mixtures

Intimate mixtures are much more difficult to model and require significant knowledge of the material properties and Sun–ground–sensor geometry. In this model, the materials are interspersed with one another at a particulate level. An example of such an intimate mixture is the mapping of materials on the lunar surface where impacts from meteors have pulverized the materials into particulate matter and mixed them (Mustard *et al.* 1998). Another example is the interspersion of different types of vegetation in tree canopies that can lead to nonlinear mixtures (Roberts *et al.* 1993).

In intimate mixtures, the photons are interacting with all the materials simultaneously. In areal mixtures, the assumption was that the photons scattered off one material at a time. Since intimate mixtures have multiple different materials in close relation to one another, the photons can bounce from one particle to another causing different absorption and reflection effects. The result is mixing that cannot be well captured by simple linear models. In Herzog and Mustard (1996), the researchers found that by not accounting for nonlinear mixing, the linear estimates could be substantially different from truth by as much as 30%.

It is well-known that reflectance is 'the ratio of reflected radiance to incident irradiance' (Schott 1996); however, how this ratio is measured leads to different types of reflectance measurements (Hapke 1993). For example, reflectance can be measured as bidirectional reflectance, hemispherical reflectance, hemispherical directional reflectance, spherical reflectance, or diffusive reflectance. For the remainder of this work, the hemispherical directional reflectance will be used since we are assuming Lambertian scattering.

This Lambertian reflectance (or simply reflectance as we will refer to it from now on) is a function of the material's ability to scatter and absorb light such that

$$\rho(\lambda) = \frac{w(\lambda)}{4(\mu + \mu_0)}\left((1 + B(g))P(g) + H(\lambda, \mu)H(\lambda, \mu_0) - 1\right), \tag{11.2}$$

where ρ is the reflectance in terms of wavelength λ, w is the average Single-Scattering Albedo (SSA), μ is the cosine of the angle of emergence, μ_0 is the cosine of the angle of incidence, g is the phase angle, $B(g)$ is the backscatter function $P(g)$ is the average single particle phase function, and H is Chandrasekhar's function for isotropic scattering (Mustard and Pieters 1989). The average SSA is mathematically modelled as

$$w(\lambda) = \frac{\sum_j (M_j w_j(\lambda))/(\rho_j d_j)}{\sum_j (M_j)/(\rho_j d_j)}, \tag{11.3}$$

where M_j is the mass fraction for material j, d_j is the diameter of the particles of material j, ρ_j is the solid density of the particles of material j, and w_j is the SSA of material j. The SSA is 'the ratio of the total amount of power scattered to the total power removed from the wave' (Hapke 1993).

Therefore, intimate mixtures are actually a linear mixture of single-scattering albedos in (11.3) nonlinearly mapped to reflectance using (11.2). Their abundances or mass fractions can be estimated from the relative geometric cross-section F_j where

$$F_j = \frac{(M_j)/(\rho_j d_j)}{\sum_j (M_j)/(\rho_j d_j)}. \tag{11.4}$$

Thus, in cases where we do not know the particle densities or sizes, the best estimate that can be obtained is the relative geometric cross-sections of the materials. If the particle density and sizes are known or can be inferred, then a direct measure of the abundance or mass fraction can be found. Note that, in either case, the measures must be non-negative and sum to one similar to the abundances in the linear mixing model.

11.3 Proposed kernel unmixing algorithm

A number of algorithms have been developed over the years for unmixing hyperspectral images. Unmixing algorithms generally require two steps: endmember extraction and abundance estimation. Previous algorithms for endmember extraction can suffer from several issues. One, they impose assumptions on the data that are often faulty and can lead to suboptimal performance. Many algorithms assume that the distribution of the spectra is Gaussian or can be modelled by a simplex. There is little theoretical or empirical physical evidence to support these assumptions. Second, they need to know the number of endmembers *a priori*. Finally, many algorithms provide non-physical endmembers, which can cause their corresponding abundances to be either negative or greater than one.

One class of such algorithms is based upon the statistical properties of the image cube. Examples include the Principal Components Analysis (PCA) and Minimum Noise Fraction (MNF) algorithms (Green *et al.* 1988). While useful for dimensionality reduction, they do not provide physical endmembers, since they consider only the statistical properties of the data. Therefore, they often generate endmembers with abundances that violate the non-negativity and sum-to-one constraints.

To determine physical endmembers, several algorithms have been developed, such as N-FINDR (Winter 1999), the pixel purity index (PPI) (Boardman *et al.* 1995), and the convex cone analysis (CCA) (Ifarraguerri and Chang 1999). For an image cube with L spectral bands, these algorithms attempt to find the endmembers as spectral convexities in an L-dimensional space. However, these methods: (1) require prior knowledge of the number of endmembers; (2) often make limiting assumptions about the spectral distribution; (3) may require user supervision, and (4) may be computationally complex.

The linear mixing model states that the spectra lie in a simplex. However, since hyperspectral imagery often contains nonlinear mixtures, the spectra may not lie in a simplex. Therefore, another method is needed to describe the boundary containing the data. The method should have the flexibility to generalize the simplex and fit the spectral distribution with greater accuracy.

Thus, a support vector approach for endmember selection is proposed. Specifically, we use the Support Vector Data Description (SVDD), developed by Tax and Duin (1999) for one-class classification. It is a novel approach that is motivated by the success of support vector algorithms for pixel classification (Gualtieri and Cromp 1998) and anomaly detection (Banerjee *et al.* 2006). Given a set of spectra from an image, the SVDD is able to determine the boundary of its region of support without making prior assumptions about its shape. The benefits of using the SVDD include:

- non-parametric modelling of the distribution of the spectra. The SVDD finds the convex boundary of the pixels using kernels without assuming the size or shape of the convex boundaries;
- readily selecting the spectra that lie on the convex boundary as the endmembers. The SVDD directly provides a set of support vectors that by definition are located along the edge of the distribution;
- automatically and quickly selecting the endmembers without any dimensionality reduction.

The problem of estimating the number of endmembers is addressed by using ideas from the field of data compression and rate-distortion. Since the spectra in the image are linear combinations of the endmembers in either the original input space or the kernel space, an automatic method to determine the number of endmembers using rate-distortion theory is proposed.

The SVDD algorithm generates endmembers that do not necessarily form a convex boundary in the original input space. This is a desirable property because the endmembers are allowed to model the data without imposing any assumptions about its underlying distribution. Current algorithms from researchers such as Boardman (1990), Settle and Drake (1993), Ashton and Schaum (1998), and Heinz and Chang (2001) cannot be used in this case as they will ignore those endmembers not on the convex boundary.

To overcome this problem and estimate abundance estimates in the correct feature space, we propose a kernel fully constrained least squares (KFCLS) algorithm. This algorithm generalizes previous abundance algorithms by using a kernel function. If we choose a linear kernel function (i.e., an inner product), the KFCLS algorithm is simply the FCLS algorithm (Heinz and Chang 2001). However, if we choose a more complex kernel like a radial basis function, the KFCLS algorithm can calculate abundances for those endmembers that form a non-convex boundary.

This section is organized as follows. After introducing the SVDD endmember extraction method in Section 11.3.1, a rate-distortion approach for estimating the number of endmembers is described in Section 11.3.2. Section 11.3.3 introduces the KFCLS algorithm. Finally, Section 11.3.4 provides an outline of the fully automatic kernel unmixing algorithm.

11.3.1 Support vector data description for endmember extraction

This section provides a brief derivation of the SVDD algorithm. Given a set of L-dimensional spectra, the linear version of the SVDD seeks the smallest hypersphere to model the region of support in the L-dimensional space. The nonlinear version of the SVDD uses kernels to construct non-spherical support regions (Tax and Duin 1999).

The SVDD is derived by considering the following geometric problem: find the smallest L-dimensional sphere with radius R centred at $\mathbf{a}$, $S = \left\{\mathbf{x} : ||\mathbf{x} - \mathbf{a}||^2 < \mathbb{R}^2\right\}$ enclosing the entire set of training exemplars $T = \left\{\mathbf{x}_i \in \mathbb{R}^L, i = 1, .., M\right\}$. This constrained optimization problem is solved by defining and minimizing a Lagrangian. It can be shown that, for the optimal solution, the following statements hold (Tax and Duin 1999).

- Only a small fraction of the Lagrangian multipliers, α_i, are non-zero.
- The sample vectors corresponding to non-zero α_i called support vectors, lie at the sphere's boundary and form a sparse representation of the training data.
- The Lagrange multipliers sum to one: $\sum_i \alpha_i = 1$.
- The centre of the sphere is the centre of mass of the support vectors: $\mathbf{a} = \sum_i \alpha_i \mathbf{x}_i$.

The functional form of the optimal sphere is given by the SVDD function:

$$\text{SVDD}(\mathbf{x}) = (\mathbf{x}, \mathbf{x}) - 2\sum_i \alpha_i (\mathbf{x}, \mathbf{x}_i) + \sum_{i,j} \alpha_i\alpha_j \left(\mathbf{x}_j, \mathbf{x}_i\right). \tag{11.5}$$

The SVDD method is generalized to allow for non-spherical support regions by using the well-known 'kernel-trick' (Tax and Duin 1999), where the SVDD function is now expressed as

$$\text{SVDD}(\mathbf{x}) = K(\mathbf{x}, \mathbf{x}) - 2\sum_i \alpha_i K(\mathbf{x}, \mathbf{x}_i) + \sum_{i,j} \alpha_i\alpha_j K\left(\mathbf{x}_j, \mathbf{x}_i\right). \tag{11.6}$$

In this work, we use the popular Gaussian radial basis function (RBF) as the kernel function, as is common practice. Since $K(\mathbf{y}, \mathbf{y}) = 1$, the SVDD function simplifies to

$$\text{SVDD}(\mathbf{x}) = 1 - 2\sum_i \alpha_i K(\mathbf{x}, \mathbf{x}_i) + \sum_{i,j} \alpha_i\alpha_j K\left(\mathbf{x}_j, \mathbf{x}_i\right),$$

$$\text{SVDD}(\mathbf{x}) = C - 2\sum_i \alpha_i K(\mathbf{x}, \mathbf{x}_i), \tag{11.7}$$

where C collects all of the constant terms.

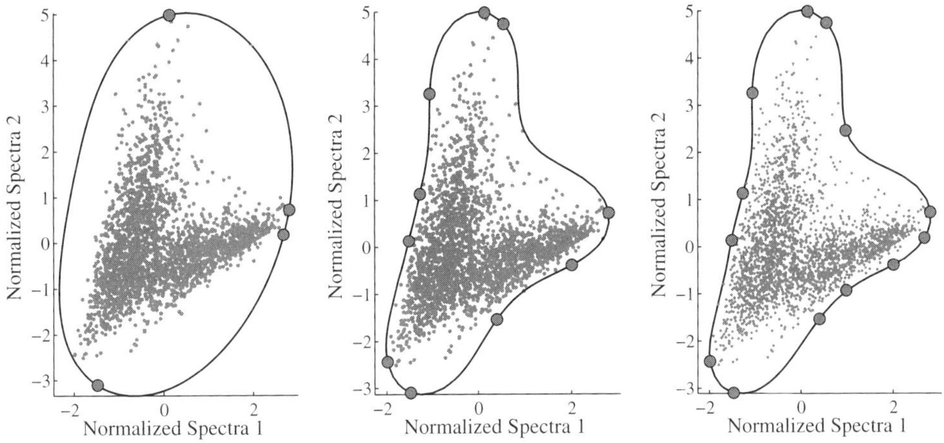

Figure 11.1 *Detecting the endmembers via SVDD. The SVDD boundaries for a 2 band scatter plot using decreasing sigma to achieve 4, 10 and 13 endmembers respectively. The black line represents the SVDD boundary and the large grey circles represent the endmembers.*

Finally, the RBF scale parameter σ^2 controls how tightly the SVDD models the support boundaries (Tax and Duin 1999). This allows the SVDD to model non-Gaussian, multi-modal data (Vapnik 1998). For the problem of endmember extraction, it determines the number of detected endmembers (support vectors). This is exemplified in Figure 11.1. Given a set of 5000 spectra, two bands are arbitrarily selected and their scatter plots are shown. The SVDD function, denoted by the solid line, defines the support region for the spectra for different values of sigma. As sigma is decreased, the SVDD determines a tighter fit to the data and finds additional support vectors (endmembers). Note that the SVDD defines a convex hull in the feature space (the optimal hypersphere), it can correspond to a non-convex hull in the input space.

11.3.2 Rate-distortion theory

Next, we consider the problem of automatically computing the number of endmembers in the scene. From simple observations of the linear mixture model, an algorithm to estimate the number of endmembers based on rate-distortion theory is developed (Cover and Thomas 2006). It is fully automatic and borrows from basic ideas in the field of data compression.

The linear mixture model states that the spectrum of every pixel is a linear combination of the endmembers. These endmembers form a set of basis functions that span the space of the spectra in the datacube. Since the spectra generally lie in a small subspace of the L-dimensional space, a relatively small number of endmembers is needed to span the subspace, and they can be used to approximate all of the spectra in the datacube according to the mixing model.

In general, it can also be shown that increasing the number of endmembers provides a better approximation to the spectra. In fact, if all of the pixels are considered as endmembers, the approximation is perfect. Of course, it is extremely unlikely that each of the pixels is a unique and pure material. Therefore, the goal is to find a compact set of endmembers that provides the desired approximation for the hyperspectral datacube. This problem is analogous to the rate-distortion optimization problem often encountered in data compression. For compression algorithms, the Rate-Distortion (R-D) function describes how the distortion between the original and reconstructed data varies as a function of rate (i.e bits/second, bits/pixel, number of vector quantizers, etc.). The function is often used to find an optimal quantizer, allocate the number of bits, or compare different compression algorithms.

The rate-distortion function can similarly be used to find the number of endmembers in a hyperspectral image. Since the datacube is approximated by a linear combination of the endmembers, the rate is defined as the number of endmembers, and the distortion is the mean-square error between the original datacube and its estimate:

$$D(Q) = \sum_{i=1}^{N} \|\mathbf{x}_i - \mathbf{E}\mathbf{a}_i\|^2 , \tag{11.8}$$

where Q is the number of endmembers (no. of columns of $\mathbf{E}$), D is the distortion, and N is the number of pixels. Note that the distortion is a function of Q since changing the number of endmembers changes the difference between the original spectra of the ith pixel $\mathbf{x}_i$ and its estimate $\mathbf{E}\mathbf{a}_i$. In general, the distortion decreases with increasing number of endmembers.

An optimal choice for the rate Q can be found via the following Lagrangian formulation:

$$\hat{Q} = \min_{Q} J(Q) = D + \lambda Q. \tag{11.9}$$

The cost function seeks to minimize the distortion subject to a constraint on the rate. The optimal Q is found as

$$\frac{\partial J}{\partial Q} = \frac{\partial D}{\partial Q} + \lambda = 0 \Rightarrow \lambda = -\frac{\partial D}{\partial Q},$$

$$J^*(Q) = D(Q) - \frac{\partial D}{\partial Q} Q. \tag{11.10}$$

Note that $\partial D/\partial Q$ is simply the slope of the R-D curve and can be computed using finite differences for any R-D function. Since D is in general a decreasing function of Q, the optimal λ is positive and will find the minimum of the cost function J^*. Thus, using the rate-distortion criterion, the optimal estimate for the number of endmembers is the Q that minimizes J^*.

11.3.3 Kernel fully constrained least squares abundance estimates

To estimate the abundances, a standard quadratic programming package can be used and easily kernelized in most cases. We use the FCLS algorithm as the base quadratic programming solution (Heinz and Chang 2001) because it is an efficient algorithm that reduces computational complexity and thus provides quick processing times.

The KFCLS derivation begins with the non-negativity constraints. Similar to other algorithms, we begin by minimizing the Least Squares Error (LSE) to estimate the non-negative abundance values. Mathematically this is expressed as

$$\min_{\mathbf{a}}\{(\mathbf{x}-\mathbf{Ea})^\top(\mathbf{x}-\mathbf{Ea})\} \text{ s.t. } a_i \geq 0 \ \forall i, \tag{11.11}$$

where $\mathbf{E}$ is a matrix of Q background endmembers. Knowing that the abundances have to be non-negative, we write a Lagrangian function L such that

$$L = \frac{1}{2}\left(\mathbf{x}^\top\mathbf{x} - 2\sum_i a_i\mathbf{x}^\top\mathbf{e}_i + \sum_{i,j} a_i a_j \mathbf{e}_i{}^\top\mathbf{e}_j\right) + \sum_i \lambda_i(a_i - c_i), \tag{11.12}$$

where $\mathbf{e}_i$ is the ith column from $\mathbf{E}$, $a_i = c_i$, and each c_i is non-negative to enforce the non-negativity constraint. By replacing the inner products with their kernel counterparts, the new Lagrangian is

$$L = \frac{1}{2}\left(K(\mathbf{x},\mathbf{x}) - 2\sum_i a_i K(\mathbf{x},\mathbf{e}_i) + \sum_{i,j} a_i a_j K(\mathbf{e}_i,\mathbf{e}_j)\right) + \sum_i \lambda_i(a_i - c_i). \tag{11.13}$$

To calculate the estimate of a, we take the partial derivative of L with respect to each a_i to obtain

$$\frac{\partial L}{\partial a_i} = -K(\mathbf{x},\mathbf{e}_i) + \sum_j a_j K(\mathbf{e}_i,\mathbf{e}_j) + \lambda_i. \tag{11.14}$$

The equation above contains a number of unknowns for each a_i and each λ_i. We can solve for each of these unknowns to produce the following equations:

$$\lambda_i = -\sum_j a_j K(\mathbf{e}_i,\mathbf{e}_j) + K(\mathbf{x},\mathbf{e}_j), \tag{11.15}$$

$$a_i = K(\mathbf{x}_i,\mathbf{x}_i)^{-1}\left(\sum_{i\neq j} -a_j K(\mathbf{e}_i,\mathbf{e}_j) + K(\mathbf{x},\mathbf{e}_j) - \lambda_i\right). \tag{11.16}$$

To solve for the non-negative abundances, we set all the Lagrange multipliers to zero and calculate the abundance using (11.16). From this solution, we identify those abundance values that are greater than zero and place them in the passive set P. The remaining non-positive abundance values are placed in the active set R. Equations (11.15) and (11.16) are iterated until all Lagrange multipliers in the passive set are zero and all Lagrange multipliers in the active set are either zero or negative. At this point, the Kuhn–Tucker conditions have been met and an optimal solution for the non-negative abundance values has been found.

To handle the sum-to-one constraints, an easy modification of the aforementioned algorithm was developed to retain the optimality guaranteed under the Kuhn–Tucker conditions for numerical optimization on a finite computing machine (Haskell and Hansen 1981). After performing some simple algebra, it can be shown that each kernel function should be replaced with

$$K_{\delta}(\mathbf{x},\mathbf{y}) = \delta^2 K(\mathbf{x},\mathbf{y}) + 1, \tag{11.17}$$

where δ is a small number (typically 10^{-5}). The δ variable controls how tightly the solution will sum to one so that smaller values provide a better solution, but may need longer convergence time. The new kernel matrix in (11.17) is then used in (11.15) and (11.16) to obtain an abundance solution that meets both the non-negativity and sum-to-one constraints.

The original purpose for creating the KFCLS algorithm was to guarantee that the abundance estimates are being calculated in the same feature space as the endmembers. This is a necessary condition because different kernel functions and even different kernel parameters generate different feature spaces. Therefore, the endmembers and abundances should be calculated in the same feature space to minimize errors in computation. Also, the SVDD creates a convex hull in the feature space where the abundance constraints hold. Because of these reasons, the KFCLS algorithm uses the same kernel and kernel parameters as the SVDD algorithm.

11.3.4 Outline of full algorithm

The SVDD, KFCLS and the rate-distortion estimate for the number of endmembers can be combined for a fully-automatic hyperspectral kernel unmixing algorithm. The steps are:

1. Choose an initial value for the RBF parameter sigma.
2. Given the set of spectra from the image, compute the SVDD function and select the support vectors.
3. Using the support vectors as the endmembers, compute their abundances using KFCLS at every pixel and reconstruct the datacube.
4. Measure the distortion for the current set of endmembers as the mean-square error between the original and reconstructed datacubes.
5. Using the values for the distortion and rate, compute the value for cost function $J^*(Q)$.
6. If sigma is less than threshold, then stop. Else, decrease sigma and go to step 2.

The value of Q that minimizes $J^*(Q)$ is the optimal estimate for the number of endmembers. Furthermore, the value for sigma corresponding to this R is also used as the estimate for the RBF kernel function.

11.4 Experimental results of the kernel unmixing algorithm

This section provides a quantitative and qualitative analysis of kernel unmixing algorithm performance. The first set of data was measured in a controlled laboratory environment so the number of endmembers, type of endmembers, and abundance of each endmember in a data sample could be precisely known. For this data, quantitative assessments could be made to compare algorithm performance.

The second set of data was taken in a real-world scenario using AVIRIS. For this data, the basic type and number of endmembers is well known, but there does not exist any absolute ground truth for abundance estimates. Therefore, the AVIRIS imagery is used to qualitatively show that kernel methods can identify physically meaningful endmembers and produce reasonable abundance planes in real-world hyperspectral data.

11.4.1 RELAB data results

The first set of data was acquired by Professor Carle Pieters and Professor John Mustard at the NASA Reflectance Experiment Laboratory (RELAB) at Brown University. The data includes binary and ternary nonlinearly mixed mineral samples with known endmembers and abundances. The data ware originally collected to measure the effectiveness of the physics models such as those presented in Section 11.2 (Mustard and Pieters 1989). Other researchers have used it to measure the effectiveness of their nonlinear abundance estimation algorithms against known samples, such as Guilfoyle *et al.* (2001) and Plaza *et al.* (2004). Thus, the RELAB data provides a good way to measure the effectiveness of the kernel unmixing algorithm for both endmember extraction and abundance estimation.

The sensor used to collect this data is the RELAB spectrometer. This is a bidirectional reflectance spectrometer that allows for different incident and emergence angles to be measured. The data was collected with at a resolution of 10 nm from 400 nm to 2500 nm (VNIR/SWIR). To calculate the reflectance of the material, the measurements were compared to pressed halon which is uniformly bright ($>96\%$) throughout the wavelengths of interest. More information can be found in (Mustard and Pieters 1989) on the sensor and its calibration as well as the creation of the various samples. The final measurements thus contain 211 wavelength measurements per sample.

From the RELAB database, we chose three sets of intimate mixtures. Each set of mixtures was formed into a small image. Each image contained the pure endmembers and their intimate mixtures. Thus, these images could be used to test not only the ability of an algorithm to estimate mass fractions, but also whether the unmixing algorithm can identify the true endmembers in the scene.

Before describing the three mixture sets, one last note needs to be mentioned. For all of these experiments, the particle densities and particle sizes of the various materials are approximately the same. In this case, the relative geometric cross-section and mass fraction estimates are roughly equivalent. Therefore, algorithms like FCLS and KFCLS that measure the relative geometric cross-section are also directly measuring the mass fraction estimates for these experiments; hence, we will refer to these estimates from now on as mass fractions for this chapter.

The first mixture set comprises the minerals enstatite and olivine. The corresponding test image was given to the kernel unmixing algorithm. The algorithm successfully found the two true endmembers of enstatite and olivine. For example in Figure 11.2, the distortion curve monotonically decreased as expected but levels when $Q = 2$. Therefore, when calculating the Lagrange cost function from the distortion, the Lagrange cost function forms a minimum at $Q = 2$, which correctly identifies that two endmembers are in the scene. In fact, the kernel unmixing algorithm found the correct type and number of endmembers in all three mixture sets.

As for the mass fraction estimates for the first set of mixtures, Table 11.1 provides the true mass fraction estimates, mass fraction estimates from using the standard FCLS algorithm,

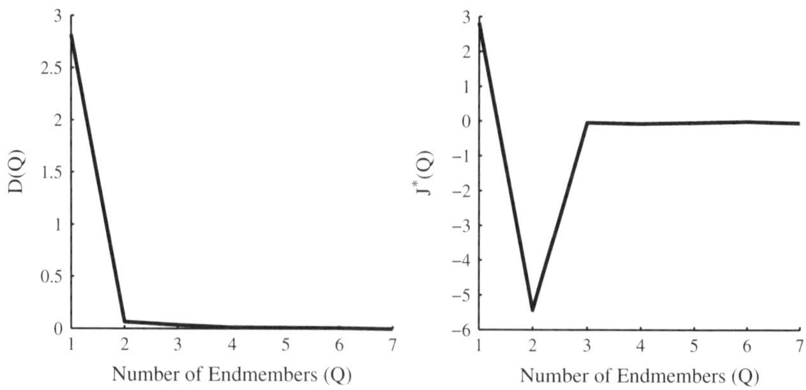

Figure 11.2 Distortion and Lagrangian curves for enstatite/olivine mixture.

and mass fraction estimates using the KFCLS algorithm. As can be seen in the table, this mixture is slightly nonlinear and thus the FCLS algorithm provides estimates that are near the true abundances with an RMSE of 0.0348. The KFCLS results show improvement over FCLS with an RMSE of 0.0280. While this is not a large improvement over FCLS, the KFCLS was able to model some nonlinear mixing effects to improve the overall estimate of the mass fractions.

The second set of mixtures comprises the endmembers magnetite and olivine. While the kernel unmixing algorithm was able to identify the two endmembers in the data correctly, the abundances were more difficult to model with either the FCLS or KFCLS algorithms. This set of mixtures has a high degree of nonlinearity, which results in an RMSE of 0.2100 for FCLS and 0.1770 for KFCLS. Again, KFCLS was able to outperform its linear counterpart because it could model some of the nonlinear behaviour; however, it still was on average 17.7 percentage points from the true abundance. Table 11.2 provides the abundance estimates for each algorithm.

The final set of mixtures is the most difficult and comprises the endmembers anorthite, enstatite and olivine. Again, the proposed unmixing algorithm identified the correct type and number of endmembers in the scene. Again, the abundance estimates for both FCLS and KFCLS were far from the truth. FCLS has a RMSE of 0.1266 and KFCLS has a RMSE of 0.1383. This is the only case where the FCLS estimates were slightly better than KFCLS. As

Table 11.1 Comparison of estimated mass fractions of enstatite/olivine mixtures

Truth	Enstatite	0.1000	0.2500	0.5000	0.7500	0.9000
	Olivine	0.9000	0.7500	0.5000	0.2500	0.1000
FCLS	Enstatite	0.1174	0.2249	0.4446	0.7061	0.8894
	Olivine	0.8826	0.7751	0.5554	0.2939	0.1106
RBF	Enstatite	0.1167	0.2076	0.4695	0.7617	0.9281
	Olivine	0.8833	0.7924	0.5305	0.2383	0.0719

Table 11.2 Comparison of estimated mass fractions of magnetite/olivine mixtures

Truth	Magnetite	0.0500	0.1000	0.2500	0.5000	0.7500
	Olivine	0.9500	0.9000	0.7500	0.5000	0.2500
FCLS	Magnetite	0.1570	0.3026	0.5437	0.7445	0.8982
	Olivine	0.8430	0.6974	0.4563	0.2555	0.1018
RBF	Magnetite	0.0916	0.2516	0.4818	0.7454	0.8838
	Olivine	0.9084	0.7484	0.5182	0.2546	0.1162

shown in Table 11.3, both algorithms tended to underestimate the true amount of enstatite present in the samples.

From these quantitative results, the kernel unmixing algorithm can successfully identify the true endmembers in a scene, but as the mixtures become more nonlinear, the mass fraction estimates tend to stray away from truth. The reason for this mismatch when it comes to abundance estimates is due to the choice of kernel. The RBF kernel is a general purpose kernel that can provide some degree of nonlinearity; however, it does not guarantee that the nonlinearity it introduces is going to match the nonlinear behaviour of an intimate mixture as detailed in Section 11.2. To match that type of nonlinear behaviour, a new kernel function is derived from the physics of intimate mixtures in Section 11.5.

11.4.2 AVIRIS data results

The second set of data was collected using the AVIRIS sensor developed at the California Institute of Technology Jet Propulsion Laboratory (JPL) (Vane *et al.* 1993). The AVIRIS sensor is a hyperspectral sensor that collects 224 contiguous spectral bands from 400 to 2500 nm. The ground sampling distance is typically 17 m per side. Imagery collected by AVIRIS can be received in the original raw radiance units or corrected reflectance units. In this analysis, we chose the reflectance units so we could compare the endmembers and abundances calculated with known laboratory spectra.

The AVIRIS image we used for our analyses is the well documented Cuprite, Nevada image set. This imagery has been studied in previous papers and has documented ground

Table 11.3 Comparison of estimated mass fractions of ternary mixtures

Truth	Anorthite	0.1616	0.1605	0.4192	0.3361	0.1613	0.4183	0.6781
	Enstatite	0.1624	0.4183	0.1611	0.3303	0.6785	0.4177	0.1599
	Olivine	0.6760	0.4212	0.4197	0.3336	0.1602	0.1640	0.1620
FCLS	Anorthite	0.2318	0.2474	0.5182	0.4536	0.2890	0.5761	0.7670
	Enstatite	0.0476	0.2214	0.0190	0.1415	0.4528	0.2074	0.0403
	Olivine	0.7206	0.5312	0.4628	0.4049	0.2582	0.2164	0.1927
RBF	Anorthite	0.1974	0.1810	0.5018	0.4007	0.1805	0.5090	0.7640
	Enstatite	0.0332	0.2200	0.0000	0.1178	0.4784	0.1771	0.0149
	Olivine	0.7694	0.5990	0.4982	0.4814	0.3411	0.3139	0.2211

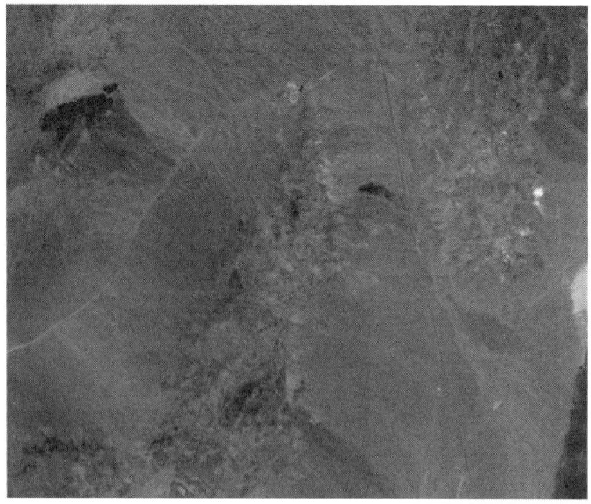

Figure 11.3 Broadband SWIR image of Cuprite, Nevada.

truth information. The Cuprite image is broken into five parts for easier processing. We chose the fourth image of the set which is used by most other researchers. As was done in Swayze *et al.* (1992) and Winter (1999), we also limited the spectrum to the short-wave infrared bands from 2.0 to 2.5 μm. This gave us a final image size of 512 × 614 × 57 covering approximately a 10 km^2 area. While this imagery is not ideal for abundance estimation, it does provide a real-world hyperspectral image for which the type and number of endmembers are well documented. Therefore, this experiment is designed to show that kernel methods can produce physically meaningful endmembers with reasonable abundance planes on real-world hyperspectral imagery. Figure 11.3 shows the broadband SWIR Cuprite AVIRIS image.

For this experiment, the SVDD algorithm found 24 endmembers using the RBF kernel. Figure 11.4 shows the Lagrange cost function curve showing a clear minimum at 24. This

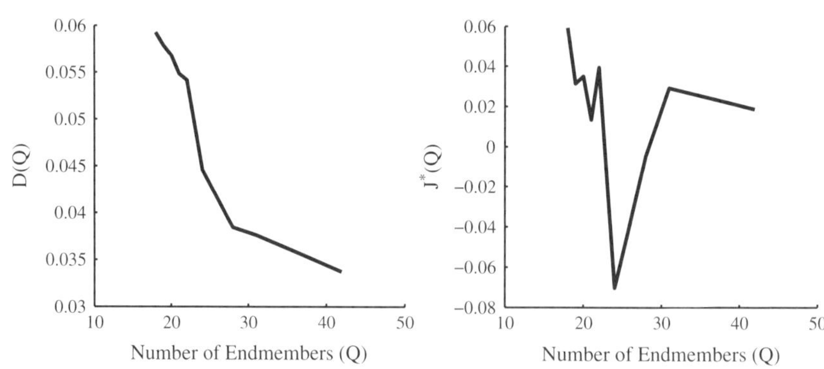

Figure 11.4 Rate distortion function and Lagrange function to estimate the number of endmembers for AVIRIS imagery.

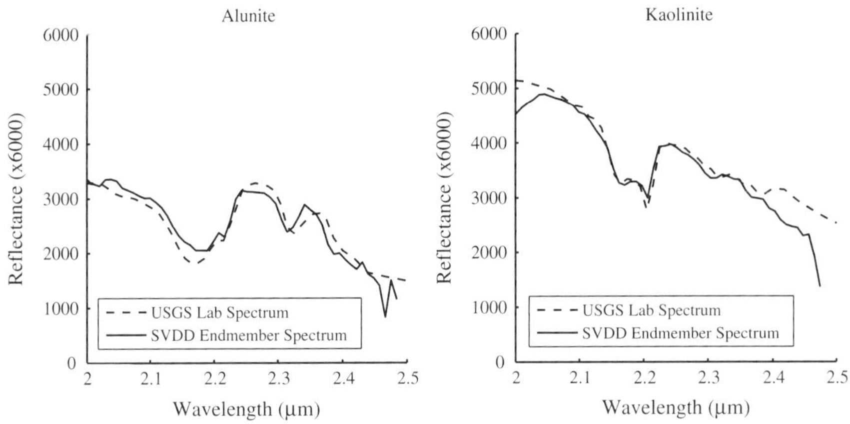

Figure 11.5 Comparison of estimated endmember with lab measured spectrum of alunite and kaolinite.

result agrees well with results reported in Ifarraguerri and Chang (1999) and Winter (1999), where PPI and N-FINDR found 22 endmembers for the Cuprite imagery.

In Figure 11.5, we compare several of the detected endmembers with the spectra of minerals known to be in the image. The laboratory reflectance spectra of the known minerals are publicly available at the USGS Spectroscopy Website (http://speclab.cr.usgs.gov/). We chose the two cases in Figure 11.5 because they are well studied and have been mapped in previous research by Resmini *et al.* (1997), Swayze *et al.* (1992) and Winter (1999). As hypothesized, the estimated endmembers show a high correlation with the known materials in the imagery. The differences are most likely due to artefacts from inverting the image from radiance to reflectance and from slight material variations.

From these endmembers, we calculated 24 abundance maps using the KFCLS algorithm. Abundance maps are images of the abundance values shown across all pixels in the image. The images are mapped such that dark pixels represent abundances of zero and white pixels represent abundances of one. Note that in this case we are using the words abundance and mass fraction interchangeably.

Four abundance maps are shown in Figure 11.6. The top row provides the abundance maps for alunite and kaolinite using the FCLS algorithm. The bottom row provides the abundance maps for alunite and kaolinite using the KFCLS algorithm. For alunite, the maps are similar and correspond well to the abundance maps of Swayze *et al.* (1992). The only difference is that the KFCLS abundance map does not suffer from the granularity seen in the FCLS map. For kaolinite, the FCLS and KFCLS maps are completely different, but the KFCLS map is the one that is most similar to the abundance maps in Swayze *et al.* (1992). This phenomenon is likely due to the ability of SVDD/KFCLS to model the nonlinear mixtures in the imagery that cannot be captured by NFINDR/FCLS. Also interesting is the fact that the endmembers found with SVDD and KFCLS produce more meaningful abundance maps than NFINDR with FCLS. In a number of cases, NFINDR finds 'noise' endmembers that produce speckle abundance images. No such abundance images can be found when using the proposed algorithm. In all cases, the

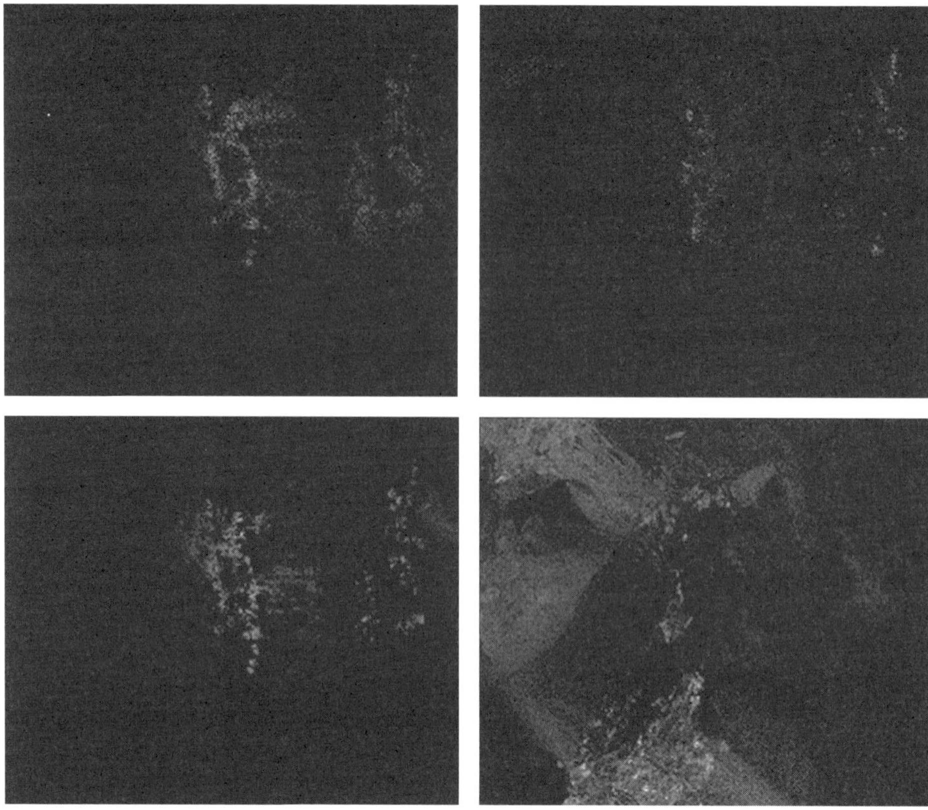

Figure 11.6 Abundance images for alunite (left) and kaolinite (right). Top row is using FCLS and bottom row is using KFCLS.

kernel unmixing algorithm produces abundance maps that follow the natural variation in the image.

11.4.3 Processing times

Despite the complexity of these methods, the processing times are short. An implementation of the kernel unmixing algorithm was created in C++ and integrated into ENVI. The C++ code was partially optimized to reduce computational complexity and checked against slower Matlab development code to validate the results. The result is that the C++ code can process the entire $512 \times 614 \times 57$ Cuprite image in less than two minutes. This estimate is for the fully automated kernel unmixing algorithm outlined in Section 11.3.4 that unmixes the image for many values of sigma. Therefore, while this method may seem time-consuming due to the quadratic optimization in both the SVDD and KFCLS algorithms, both of these algorithms were chosen because they are computationally efficient and thus have short processing times even when dealing with large hyperspectral images. For a computational analysis of SVDD, the interested reader is referred to Chapter 8.

11.5 Development of physics-based kernels for unmixing

Section 11.3 focused on developing a set of kernel algorithms to unmix hyperspectral imagery; however, no focus was placed on the kernel itself. While RBF kernels typically provide good performance in general, they may not be the best fit for unmixing intimate mixtures as shown in 11.5.1. Instead, we can use the physics shown in Section 11.2 to design a kernel more suited for unmixing purposes. The rest of this section is as follows. Section 11.5.1 simplifies the albedo to reflectance transform for ease of calculation. Section 11.5.2 develops a kernel from the simplified transform that can be used with the algorithm developed in Section 11.3.

11.5.1 Simplification of the albedo to reflectance transform

Mustard and Pieters recommend that a simplified model of albedo to reflectance would be good for remote sensing applications such as AVIRIS (Mustard and Pieters 1989). Following their work, (11.2) can be simplified by making some reasonable assumptions. First, they assume the average single particle phase function $P(g)$ is constant for all materials. While this may not be true in general, they shown in their paper that it provides a good fit nevertheless. Second, they assume that $g = 30^\circ$ which makes $P(30) = 1$. Third, they assume that the backscatter function is negligible and set $B(g) = 0$. These simplifications result in a new nonlinear mapping between reflectance and single-scattering albedo such that

$$\rho(\lambda) = \frac{w(\lambda)}{4(\mu + \mu_0)} \left(H(\lambda, \mu) H(\lambda, \mu_0) \right). \tag{11.18}$$

Chandrasekhar's function can be approximated by Hapke (1981) such that

$$H(\lambda, \mu) = \frac{1 + 2\mu}{1 + 2\mu\sqrt{1 - w(\lambda)}}. \tag{11.19}$$

Substituting (11.19) into (11.18) provides

$$\rho(\lambda) = \frac{w(\lambda)}{4(\mu + \mu_0)} \left(\frac{1 + 2\mu}{1 + 2\mu\sqrt{1 - w(\lambda)}} \right) \left(\frac{1 + 2\mu_0}{1 + 2\mu_0\sqrt{1 - w(\lambda)}} \right), \tag{11.20}$$

which is the function used in Guilfoyle *et al.* (2001) for their nonlinear estimation algorithms. Mustard and Pieters show this method can calculate reasonable abundances (mass fractions) for nonlinear mixtures within 7% of truth, but a method that better models the phase function and Chandrasekhar's function obviously provides improved results (Mustard and Pieters 1989). Despite the loss of accuracy, (11.20) is far more computationally efficient than calculating the full integral of Chandrasekhar's function.

11.5.2 Kernel approximation of intimate mixtures

In Section 11.2.2, nonlinear mixing was shown to be a nonlinear mapping of a linear mixture of single-scattering albedos. This is a perfect match for kernel functions. As noted in Chapter 2, kernel functions can apply a nonlinear transform to a linear method. For the physics of nonlinear mixtures, the nonlinear transform is (11.2) applied to the linear combination in (11.1).

To see this relation clearly, (11.1) is rewritten such that

$$\mathbf{w} = \sum_{q=1}^{Q} F_q \mathbf{w}_q, \quad F_q \geq 0, \quad \sum_{q=1}^{Q} F_q = 1. \tag{11.21}$$

Equation (11.21) is equivalent to (11.1) except that the combination is now based on single-scattering albedos and not pure endmember reflectances. Therefore, the reflectance of a nonlinear mixture is nothing more than a nonlinear function of (11.21) such that

$$\rho(\lambda) = \boldsymbol{\phi}\left(\sum_{q=1}^{Q} F_q w_q(\lambda)\right), \tag{11.22}$$

which shows that nonlinear mixing can be approximated by a nonlinear function $\boldsymbol{\phi}$ operating on a linear function. This is an excellent match for kernel functions: to take previously linear functions and make them nonlinear through a kernel function such that our new physics-based kernel is

$$K(\mathbf{x}, \mathbf{y}) = \left\langle \boldsymbol{\phi}\left(\sum_{q=1}^{Q} F_q \mathbf{x}_q\right), \boldsymbol{\phi}\left(\sum_{q=1}^{Q} F_q \mathbf{y}_q\right)\right\rangle. \tag{11.23}$$

The interesting twist to nonlinear mixing using kernels is that the function $\boldsymbol{\phi}$ is known via the full model in (11.2) or its simplified counterpart in (11.20). Kernels can be useful though for two reasons. First, as demonstrated in Section 11.4.3, algorithms based on kernel functions can be computationally efficient leading to small processing times. Second, kernels provide a framework to calculate either linear or nonlinear mixtures. By choosing a simple inner product, the kernel method will calculate a linear mixing model. By choosing (11.23), the kernel can be used to approximate nonlinear mixing effects. In this way, kernels generalize the mathematical model for both linear and nonlinear mixtures. For example, the core KFCLS algorithm can calculate linear mixtures if the kernel is an inner product and nonlinear mixtures if the kernel is (11.23) without changing the fundamental KFCLS algorithm. Currently, which kernels provide the best approximation to the nonlinear mixing model remains an open research question; however, the work presented in this section shows that kernels can be an efficient way both computationally and algorithmically to introduce physical models into hyperspectral unmixing.

11.6 Physics-based kernel results

For this set of experiments, we use the RELAB data described in Section 11.4.1. Instead of using an RBF kernel, we use the kernel in (11.23). Since the RELAB data were taken under conditions where the $P(g) = 1$ and $B(g) = 0$, the proposed kernel in (11.23) should match well with the ground truth. Additionally, we again mention that the RELAB data were created such that the particle densities and particle sizes were roughly equivalent; therefore, the KFCLS algorithm with physics-based kernel is providing an estimate of the mass fraction in this case.

Table 11.4 Augmented mass fractions estimates of enstatite/olivine mixtures

Truth	Enstatite	0.1000	0.2500	0.5000	0.7500	0.9000
	Olivine	0.9000	0.7500	0.5000	0.2500	0.1000
FCLS	Enstatite	0.1174	0.2249	0.4446	0.7061	0.8894
	Olivine	0.8826	0.7751	0.5554	0.2939	0.1106
RBF	Enstatite	0.1167	0.2076	0.4695	0.7617	0.9281
	Olivine	0.8833	0.7924	0.5305	0.2383	0.0719
Physics	Enstatite	0.1358	0.2569	0.5187	0.7712	0.9207
	Olivine	0.8642	0.7404	0.4813	0.2288	0.0793

Table 11.5 Augmented mass fractions estimates of magnetite/olivine mixtures

Truth	Magnetite	0.0500	0.1000	0.2500	0.5000	0.7500
	Olivine	0.9500	0.9000	0.7500	0.5000	0.2500
FCLS	Magnetite	0.1570	0.3026	0.5437	0.7445	0.8982
	Olivine	0.8430	0.6974	0.4563	0.2555	0.1018
RBF	Magnetite	0.0916	0.2516	0.4818	0.7454	0.8838
	Olivine	0.9084	0.7484	0.5182	0.2546	0.1162
Physics	Magnetite	0.0285	0.0733	0.2001	0.4070	0.6915
	Olivine	0.9715	0.9267	0.7999	0.5930	0.3085

Table 11.6 Augmented mass fractions estimates of ternary mixtures

	Anorthite	0.1616	0.1605	0.4192	0.3361	0.1613	0.4183	0.6781
Truth	Enstatite	0.1624	0.4183	0.1611	0.3303	0.6785	0.4177	0.1599
	Olivine	0.6760	0.4212	0.4197	0.3336	0.1602	0.1640	0.1620
	Anorthite	0.2318	0.2474	0.5182	0.4536	0.2890	0.5761	0.7670
FCLS	Enstatite	0.0476	0.2214	0.0190	0.1415	0.4528	0.2074	0.0403
	Olivine	0.7206	0.5312	0.4628	0.4049	0.2582	0.2164	0.1927
	Anorthite	0.1974	0.1810	0.5018	0.4007	0.1805	0.5090	0.7640
RBF	Enstatite	0.0332	0.2200	0.0000	0.1178	0.4784	0.1771	0.0149
	Olivine	0.7694	0.5990	0.4982	0.4814	0.3411	0.3139	0.2211
	Anorthite	0.1649	0.1574	0.4193	0.3238	0.1490	0.4009	0.6710
Physics	Enstatite	0.1838	0.4534	0.1802	0.3679	0.7078	0.4498	0.1788
	Olivine	0.6513	0.3892	0.4005	0.3093	0.1432	0.1493	0.1502

Tables 11.4–11.6 provide all of the different mass fraction estimates for the RELAB data. The last two rows show the mass fraction estimates when using the physics-based kernel. In all cases, the physics-based kernel outperforms either of the other methods as expected. What was not expected was the significant decrease in RMSE by up to an order of magnitude. The respective RMSE values for the three mixtures are 0.0228, 0.0561 and 0.0214. This experiment shows the ability of the physics-based kernel in (11.23) to model intimate mixtures accurately and estimate their mass fractions.

11.7 Summary

This chapter introduced the concept of kernel unmixing algorithms for hyperspectral imagery. The physics of linear and nonlinear mixtures are explored and shown to be suited for kernel approximations. Based on this evidence, a fully automatic kernel unmixing algorithm is developed for endmember extraction (SVDD) and abundance estimation (KFCLS). This kernel framework has the flexibility to account for both linear and nonlinear behaviour by using different kernel functions. Specifically, we have derived a kernel based on the physics of intimate mixtures.

The experimental results show both qualitative and quantitative improvements when using the kernel approach. On the AVIRIS data, physically meaningful endmembers were identified against known ground truth without any noise endmembers. Using the RELAB data, we provided quantitative comparisons of several algorithms. They demonstrate an order of magnitude improvement when using the proposed method with the physics-based kernel.

References

Ashton, E.A. and Schaum, A. (1998) Algorithms for the detection of sub-pixel targets in multispectral imagery. *Photogrammetric Engineering and Remote Sensing*, **64**, 723–31.

Banerjee. A., Burlina, P. and Diehl, C. (2006) A support vector method for anomaly detection in hyperspectral imagery. *IEEE Transactions on Geoscience and Remote Sensing*, **37**, 756–70.

Boardman, J. (1990) Inversion of high spectral resolution data. *Proceedings of the SPIE*, **1298**, 222–33.

Boardman, J., Kruse, F.A. and Green, R.O. (1995) Mapping target signatures via partial unmixing of AVIRIS data, *Summaries of the VI JPL Airborne Earth Science Workshop*, Pasadena, CA.

Chandrasekhar, S. (1960) *Radiative Transfer*. Dover, New York.

Clark, R. and Roush, T. (1984) Reflectance spectroscopy: quantitative analysis techniques for remote sensing applications. *Journal of Geophysical Research*, **89**, 6329–40.

Cover, T.M. and Thomas, J.A. (2006) *Elements of Information Theory*. 2nd edn. John Wiley & Sons, New York, NY.

Green, A.A., Berman, M., Switzer, P. and Craig, M.D. (1988) A transformation for ordering multispectral data in terms of image quality with implications for noise removal. *IEEE Transactions on Geoscience and Remote Sensing*, **26**, 65–74.

Gualtieri, J.A. and Cromp, R.F. (1998) Support vector machines for hyperspectral remote sensing classification, *Proceedings of the SPIE 27th AIPR Workshop*, **3584**, 221–32.

Guilfoyle, K.J., Althouse, M.L. and Chang, C.-I. (2001) A quantitative and comparative analysis of linear and nonlinear spectral mixture models using radial basis function neural networks. *IEEE Transactions on Geoscience and Remote Sensing*, **39**, 2314–8.

Hapke, B. (1981) Bidirectional reflectance spectroscopy, 1, Theory. *Journal of Geophysical Research*, **86**, 3039–54.

Hapke, B. (1993) *Theory of Reflectance and Emittance Spectroscopy*. Cambridge University Press, New York.

Haskell, K.H. and Hansen, R.J. (1981) An algorithm for linear least squares problems with equality and non-negativity constraints generalized. *Math Prog.*, **21**, 98–118.

Heinz, D.C. and Chang, C.-I. (2001) Fully constrained least squares linear spectral mixture analysis method for material quantification in hyperspectral imagery. *IEEE Transactions on Geoscience and Remote Sensing*, **39**, 529–45.

Herzog, S.G. and Mustard, J.F. (1996) Reflectance spectra of five-component mineral mixtures: Implications for mixture modelling. *Lunar and Planetary Science*, **27**, 535–536.

Ifarraguerri, A. and Chang, C.-I. (1999) Multis-pectral and hyperspectral image analysis with convex cones. *IEEE Transactions on Geoscience and Remote Sensing*, **37**, 756–70.

Keshava, N. and Mustard, J.F. (2002) Spectral unmixing. *IEEE Signal Processing Magazine*, **19**, 44–57.

Landgrebe, D. (1997) The evolution of landsat data analysis. *Photogrammetric Engineering and Remote Sensing*, **LXIII**, 859–67.

Manolakis, D. and Shaw, G. (2002) Detection algorithms for hyperspectral imaging applications. *IEEE Signal Processing Magazine*, **19**, 29–43.

Mustard, J.F., Li, *L.* and He, G. (1998) Compositional gradients across mare-highland contacts: The importance and geological implications of lateral mixing. *Journal of Geophysical Research–Planets*, **103**, 419–25.

Mustard, J.F. and Pieters, C. (1987) Quantitative abundance estimates from bidirectional reflectance measurements, *Proc. of 17th Lunar and Planetary Science Conf.*, (eds Ryder, G. and Schubert, G.), E617–26.

Mustard, J.F. and Pieters, C. (1989) Photometric phase functions of common geologic minerals and applications to quantitative analysis of mineral mixture reflectance spectra. *Journal of Geophysical Research*, **94**, 13619–34.

Plaza, J., Martinez, P., Perez, R. and Plaza, A. (2004) Nonlinear neural network mixture models for fractional abundance estimation in AVIRIS hyperspectral images, *Proc. XIII JPL Airborne Earth Science Workshop*, Pasadena, CA.

Resmini, R.G., Karpus, M.E., Aldrich, W.S., Harsanyi, J.C. and Anderson, M. (1997) Mineral mapping with hyperspectral digital imagery collection experiment (HYDICE) sensor data at Cuprite, Nevada. *International Journal of Remote Sensing*, **18**, 1553–70.

Roberts, D.A., Smith, M.O. and Adams, J.B. (1993) Green vegetation, nonphotosynthetic vegetation, and soils in AVIRIS data. *Remote Sensing of Environment*, **44**, 255–69.

Schölkopf, B. and Smola, A.J. (2002) *Learning with Kernels: Support Vector Machines, Regularization, Optimization, and Beyond*. MIT Press, Cambridge, MA.

Schölkopf, B., Smola, A.J. and Muller, K.-R. (1998) Kernel principal component analysis. *Neural Computation*, **10**, 1299–1319.

Schott, J.R. (1996) *Remote Sensing: The Image Chain Approach*. Oxford University Press, Oxford, UK.

Settle, J.J. and Drake, N.A. (1993) Linear mixing and estimation of ground cover proportions. *International Journal of Remote Sensing*, **14**, 1159–77.

Swayze, G., Clark, R.N., Kruse, F., Sutley, S. and Gallagher, A. (1992) Ground-truthing AVIRIS mineral mapping at Cuprite, Nevada. *Summaries of the Third Annual JPL Airborne Geoscience Workshop*, **1**, 47–9.

Tax, D.M.J. and Duin, R.P.W. (1999) Data domain description using support vectors, *Proc. European Symposium on Artificial Neural Networks*, Brussels, Belgium, 251–6.

Vane, G., Green, R.O., Chrien, T.G., Enmark, H.T., Hansen. E.G. and Porter, W.M. (1993) The airborne visible/infrared imaging spectrometer (AVIRIS). *Remote Sensing of Environment*, **44**, 127–43.

Vapnik, V.N. (1998) *Statistical Learning Theory*. John Wiley & Sons, New York.

Wendtland, W. and Hecht, H. (1966) *Reflectance Spectroscopy*. Wiley-Interscience, New York.

Winter, E.M. (1999) N-FINDR An algorithm for fast autonomous spectral end-member determination in hyperspectral data. *Imaging Spectrometry V*, pp. 266–277.

12

Kernel-based quantitative remote sensing inversion

Yanfei Wang[1], Changchun Yang[1] and Xiaowen Li[2]

[1] *Key Laboratory of Petroleum Geophysics, Institute of Geology and Geophysics, Chinese Academy of Sciences, Beijing, P.R. China*
[2] *Research Center for Remote Sensing and GIS, Beijing Normal University, Beijing, P.R. China*

To estimate structural parameters and spectral component signatures of Earth surface cover type, quantitative remote sensing seems to be an appropriate way of dealing with these problems. Since the real physical system that couples the atmosphere and the land surface is very complicated and should be continuous, sometimes it requires a comprehensive set of parameters to describe the system, so any practical physical model can only be approximated by a model that includes only a limited number of the most important parameters that capture the major variation of the real system.

The pivotal problem for quantitative remote sensing is the inversion. Inverse problems are typically ill-posed. The ill-posed nature is characterized by: (C1) the solution may not exist; (C2) the dimension of the solution space may be infinity; (C3) the solution is not continuous with the variation of the observed signals. These issues exist for all quantitative remote sensing inverse problems. For example, when sampling is poor, i.e., there are too few observations, or directions are poorly located, the inversion process would be underdetermined, which leads to the large condition number of the normalized system and significant noise propagation. Hence (C2) and (C3) would be the highlighted difficulties for quantitative remote sensing inversion.

This chapter will address the theory and methods from the viewpoint that the quantitative remote sensing inverse problems can be represented by kernel-based operator equations.

Kernel Methods for Remote Sensing Data Analysis Edited by Gustavo Camps-Valls and Lorenzo Bruzzone

12.1 Introduction

Both modelling and model-based inversion are important for quantitative remote sensing. Here, modelling mainly refers to data modelling, which is a method used to define and analyse data requirements; model-based inversion mainly refers to using physical or empirically physical models to infer unknown but interesting parameters. Hundreds of models related to vegetation and radiation have been established during past decades. The model-based inversion in atmospheric science has been well understood. However, the model-based inverse problems for land surface has only recently received much attention from scientists. Compared with modelling, model-based inversion is still at an exploration stage. This is because intrinsic difficulties exist in the application of *a priori* information, inverse strategy and inverse algorithms. The appearance of hyperspectral and multiangular remote sensors enhanced the means of exploration and have provided us with more spectral and spatial dimensional information than before. However, how to utilize this information to solve the problems faced in quantitative remote sensing to make remote sensing really enter the time of quantification is still an arduous and urgent task for remote sensing scientists. Remote sensing inversion for different scientific problems in different branches has attracted increasing attention in recent years. In a series of international study projections, such as the International Geosphere-Biosphere Programme (IGBP), the World Climate Research Programme (WCRP) and NASA's Earth Observing System (EOS), remote sensing inversion has become a focal point of study.

Remote sensing inversions are usually optimization problems with different constraints. Therefore, a way of incorporating the method developed in the operational research field into the remote sensing inversion field is very much needed. In quantitative remote sensing, since the real physical system that couples the atmosphere and the land surface is very complicated and should be continuous, sometimes it requires a comprehensive set of parameters to describe the system, so any practical physical model can only be approximated by a model that includes only a limited number of the most important parameters that capture the major variations of the real system. Generally speaking, a discrete forward model to describe such a system is in the form

$$y = h(\mathbf{x}, \mathbf{S}), \tag{12.1}$$

where y is single measurement, $\mathbf{x}$ is a vector of controllable measurement conditions such as the wave band, viewing direction, time, Sun position, polarization, etc., $\mathbf{S}$ is a vector of the state parameters of the system approximation, and h is a function that relates $\mathbf{x}$ to $\mathbf{S}$, which is generally nonlinear and continuous.

With the ability of satellite sensors to acquire multiple bands, multiple viewing directions, etc., while keeping $\mathbf{S}$ essentially the same, we obtain the following nonhomogeneous equations

$$\mathbf{y} = h(\mathbf{x}, \mathbf{S}) + \mathbf{n}, \tag{12.2}$$

where $\mathbf{y}$ is a vector in $\mathbb{R}^M$, which is an M dimensional measurement space with M values corresponding to M different measurement conditions and $\mathbf{n} \in \mathbb{R}^M$ is the vector of random noise with same vector length M. Assume that there are m undetermined parameters needing to be recovered. Clearly, if $M = m$, (12.2) is a determined system, so it is not difficult to develop some suitable algorithms to solve it. If more observations can be collected than the existing parameters in the model (Verstraete *et al.* 1996), i.e., $M > m$, the system (12.2) is over-determined. In this situation, the traditional solution does not exist. We must define its solution by some other meaning, for example, the least squares error (LSE) solution. However as Li in Li *et al.* (1998) pointed out, 'for physical models with about ten parameters (single band), it is questionable whether remote sensing inversion can be an over-determined one in the foreseeable future.' Therefore, the inversion problems in geosciences seem to be always underdetermined in some sense. Nevertheless the underdetermined system, in some cases, can always be converted to an overdetermined one by utilizing multiangular remote sensing data or by accumulating some *a priori* knowledge (Li *et al.* 2001).

Developed methods in literature for quantitative remote sensing inversion are mainly statistical methods with different variations from Bayesian inference. In this chapter, using kernel expression, we analyse from the algebraic point of view, the solution theory and methods for quantitative remote sensing inverse problems. The kernels mentioned in this chapter mainly refer to integral kernel operators (characterized by integral kernel functions) or discrete linear operators (characterized by finite rank matrices). It is closely related with the kernels of linear functional analysis, Hilbert space theory and spectral theory. The kernels are different from those in the support vector machine learning for data mining, where kernels must satisfy the Mercer theorem (see Chapter 2, Vapnik (1995) and Wang (2007) for details). But in essence, for support vector machine learning problems, they also require are to solve ill-posed problems. In particular, we present regularizing retrieval of parameters with *a posteriori* choice of regularization parameters, several cases of choosing scale/weighting matrices to the unknowns, numerically truncated singular value decomposition, and sparse inversion in l_1 space. These methods, as far as we know, are novel to literature.

The outline of this chapter is as follows: in Section 12.2, we list two typical kernel-based remote sensing inverse problems. One is aerosol size distributions from optical transmission or scattering measurements, which is a long existing problem and still an important topic today; the other is the linear kernel-based bidirectional reflectance distribution function (BRDF) model inverse problem, which is of great importance for land surface parameter retrieval. In Section 12.3 the conception of well-posed problems and ill-posed problems are introduced. In Section 12.4 the regularization theory and solution techniques for ill-posed quantitative remote sensing inverse problems are described. In Section 12.4.4 the regularization scheme formulated in the Bayesian statistical inference is introduced. In Section 12.5 optimization theory and solution methods are discussed for finding an optimized solution of a minimization model. In Section 12.6 the detailed regularizing solution methods for retrieval of ill-posed land surface parameters are discussed. In Section 12.7 the regularization and optimization methods for recovering aerosol particle size distribution functions are presented. Finally, in Section 12.8, some concluding remarks are given.

Throughout this chapter, we use the following notations: ':=' denotes 'defined as'; 'arg min' denotes 'minimization for an argument'; 'max' and 'min' denote 'maximizing' and 'minimizing' some functional respectively; $\mathbf{x}$ and x denote the vector and the continuous function respectively, as do other functions and their corresponding vectors; 'A^*' denotes the adjoint of an operator A, '$A^\top$' denotes the transpose of a matrix A, 's.t.' denotes 'subject to', and 'diag(·)' denotes a diagonal matrix.

12.2 Typical kernel-based remote sensing inverse problems

The kernel methods can increase the accuracy of remote sensing data processing, including specific land-cover identification, biophysical parameter estimation and feature extraction

(Camps-Valls 2008). Here we include two typical kernel-based remote sensing inverse problems. One is the problem of indirect detection of the distribution of the number of size distribution functions of atmospheric aerosols, which has long existed; the other is the kernel-based land surface parameter retrieval problem.

12.2.1 Aerosol inverse problems

It is well-known that the characteristics of the aerosol particle size, which can be represented as a size distribution function in a mathematical formalism, say $n(r)$, plays an important role in climate modelling due to its uncertainty (Houghton *et al.* 1966). So, the determination of the particle size distribution function becomes a basic task in aerosol research (Bockmann 2001; Bockmann and Kirsche 2006; Bohren and Huffman 1983; Davies 1974; Mccartney 1976; Twomey 1977). Since the relationship between the size of atmospheric aerosol particles and the wavelength dependence of the extinction coefficient was first suggested by Angström (Ångström 1929), the size distribution has begun to be retrieved by extinction measurements.

For the sun-photometer, the attenuation of the aerosols can be written as the integral equation of the first kind

$$\tau_{aero}(\lambda) = \int_0^\infty \pi r^2 Q_{ext}(r, \lambda, \eta) n(r) \mathrm{d}r + \varrho(\lambda), \tag{12.3}$$

where r is the particle radius; $n(r)$ is the columnar aerosol size distribution (i.e. the number of particles per unit area per unit radius interval in a vertical column through the atmosphere); η is the complex refractive index of the aerosol particles; λ is the wavelength; $\varrho(\lambda)$ is the error/noise; and $Q_{ext}(r, \lambda, \eta)$ is the extinction efficiency factor from Mie theory. Since aerosol optical thickness (AOT) can be obtained from the measurements of the solar flux density with sun-photometers, one can retrieve the size distribution by the inversion of AOT measurements through the above equation. This type of method is called extinction spectrometry, which is not only the earliest method to apply remote sensing to determine atmospheric aerosol size characteristics, but also the most mature method thus far.

A common feature for all particle size distribution measurement systems is that the relation between noiseless observations and the size distribution function can be expressed as a first kind Fredholm integral equation, e.g., see Nguyen and Cox (1989), Voutilainenand and Kaipio (2000), Wang (2007), Wang and Yang (2008) and Wang (2008). For the aerosol attenuation problem (12.3), let us rewrite (12.3) in the form of the abstract operator equation

$$\begin{aligned} &K : \mathcal{X} \longrightarrow \mathcal{Y}, \\ &(Kn)(\lambda) + \varrho(\lambda) = \int_0^\infty k(r, \lambda, \eta) n(r) \mathrm{d}r + \varrho(\lambda) = o(\lambda) + \varrho(\lambda) = d(\lambda), \end{aligned} \tag{12.4}$$

where $k(r, \lambda, \eta) = \pi r^2 Q_{ext}(r, \lambda, \eta)$; $\mathcal{X}$ denotes the function space of aerosol size distributions; and $\mathcal{Y}$ denotes the observation space. Both $\mathcal{X}$ and $\mathcal{Y}$ are considered to be the separable Hilbert space. Note that τ_{aero} in Equation (12.3) is the measured term, it inevitably induces noise/errors. Hence, $d(\lambda)$ is actually a perturbed right-hand side. Keeping in mind the operator symbol, Equation (12.4) can be written as

$$Kn + \varrho = o + \varrho = d. \tag{12.5}$$

12.2.2 Land surface parameter retrieval problem

As is well-known, the anisotropy of the land surface can be best described by the Bidirectional Reflectance Distribution Function (BRDF). With the progress of the multiangular remote sensing, it seems that the BRDF models can be inverted to estimate structural parameters and spectral component signatures of Earth surface cover type (see Strahler *et al.* (1994) and Roujean *et al.* (1992)). The state-of-the-art of BRDF is the use of the linear kernel-driven models, mathematically described as the linear combination of the isotropic kernel, volume scattering kernel and geometric optics kernel. The information extraction on the terrestrial biosphere and other problems for retrieval of land surface albedos from satellite remote sensing have been considered by many authors in recent years, see for instance the survey papers on the kernel-based bidirectional reflectance distribution function (BRDF) models by Pokrovsky and Roujean (2002, 2003) and Pokrovsky *et al.* (2003) and references therein. The computational stability is characterized by the algebraic operator spectrum of the kernel-matrix and the observation errors. Therefore, the retrieval of the model coefficients is of great importance for computation of the land surface albedos. Other than observation errors, a limited or insufficient number of observations is one of the most severe obstacles to the estimation of BRDF. Therefore, it is very desirable to develop new techniques for the robust estimation of the BRDF model parameters because of the scarcity of the number of observations.

The linear kernel-based BRDF model can be described as follows (Roujean *et al.* 1992):

$$f_{iso} + k_{vol}(t_i, t_v, \phi) f_{vol} + k_{geo}(t_i, t_v, \phi) f_{geo} = r(t_i, t_v, \phi), \tag{12.6}$$

where r is the bidirectional reflectance; the kernels k_{vol} and k_{geo} are so-called kernels, that is, known functions of illumination and of viewing geometry that describe volume and geometric scattering respectively; t_i and t_v are the zenith angle of the solar direction and the zenith angle of the view direction respectively; ϕ is the relative azimuth of the Sun and view direction; and f_{iso}, f_{vol} and f_{geo} are three unknown parameters to be adjusted to fit the observations. Theoretically, f_{iso}, f_{vol} and f_{geo} are closely related to the biomass such as leaf area index (LAI), Lambertian reflectance, sunlit crown reflectance, and viewing and solar angles. The vital task then is to retrieve appropriate values of the three parameters.

Generally speaking, the BRDF model includes kernels of many types. However, it was demonstrated that the combination of RossThick (k_{vol}) and LiSparse (k_{geo}) kernels had the best overall ability to fit BRDF measurements and to extrapolate BRDF and albedo (Hu *et al.* 1997; Li *et al.* 1999; Privette *et al.* 1997; Wanner *et al.* 1995). A suitable expression for the RossThick kernel k_{vol} was derived by Roujean *et al.* (1992). It is reported that the LiTransit kernel k_{Transit}, instead of the kernel k_{geo}, is more robust and stable than the LiSparse nonreciprocal kernel and the reciprocal LiSparse kernel k_{sparse} (LiSparseR) where the LiTransit kernel and the LiSparse kernel are related by

$$k_{\text{Transit}} = \begin{cases} k_{\text{sparse}}, & B \le 2, \\ \frac{2}{B} k_{\text{sparse}}, & B > 2, \end{cases}$$

and B is given by $B := B(t_i, t_v, \phi) = -O(t_i, t_v, \phi) + \sec t_i' + \sec t_v'$ in Li *et al.* (2000). More detailed explanation about O and t' in the definition of k_{Transit} can be found in Wanner *et al.* (1995).

In order to use the combined linear kernel model, a key issue is to solve the inverse model numerically in a stable way. However, it is difficult to do so in practical applications due to the ill-posed nature of the inverse problem. So far, statistical methods have been developed for solving the inverse problem, which is based on expression of *a priori* knowledge of a model parameter such as a joint probability density $p_s(s_1, s_2, \cdots, s_k)$; and the expression for *a priori* knowledge of the model accuracy and measurement noise should be a conditional joint probability density $p_D(d_{\text{obs}}, S)$: $p_D(d_{\text{obs}}, S) = p_D(d_{\text{obs}}|S)p_s(S)$, where d_{obs} refers to the observed data. From the Bayesian inference formula, we have $p(S|d_{\text{obs}}) = p_D(d_{\text{obs}}|S)p_s(S)/p_D(d_{\text{obs}})$, where $p_D(d_{\text{obs}}) = \int_S p_D(d_{\text{obs}}|S)p_s(S)dV_s$ and dV_s is the volume differential elements in the parameter space. The $p_D(d_{\text{obs}}|S)$ can be interpreted as the prior knowledge of the model prediction of d_{obs} giving parameters in parameter space, $p_s(S)$ as the prior knowledge of the parameters, and $p_D(d_{\text{obs}})$ as prior knowledge of the marginal density of observation. In Li *et al.* (2001), the authors applied a least squares fitting method for recovering the interesting parameters. In Pokrovsky and Roujean (2002, 2003), the authors utilized the algebraic QR decomposition technique and also suggested using the singular value decomposition for the inversion of the BRDF model. An optimal design scheme for the angular sampling is addressed. Later in Pokrovsky *et al.* (2003), they compared several inversion techniques and uncertainty in albedo estimates from the SEVIRI/MSG observing system by using POLDER BRDF measurements. Though the solution method is algebraic, their description of the problem is still quite statistical.

12.3 Well-posedness and ill-posedness

From this section to the end of the chapter, unless specified, we will denote the operator equation as

$$K(x) = y, \tag{12.7}$$

which is an appropriate expression for an observing system, with K the response function (linear or nonlinear), x the unknown input and y the observed data. In particular, if K is linear mapping, we will denote the response system as

$$Kx = y, \tag{12.8}$$

which is clearly a special case of (12.7). We will also sometimes use K as an operator in infinite spaces and sometimes as a matrix. We assume that the readers can readily recognize them.

The Problem (12.7) is said to be properly posed or well-posed in the sense that it has the following three properties.

(C_1) There exists a solution of the problem, i.e., existence.

(C_2) There is at most one solution of the problem, i.e., uniqueness.

(C_3) The solution depends continuously on the variations of the right hand side (data), i.e., stability.

The condition (C_1) can be easily fulfilled if we enlarge the solution space of the problem (12.7). The condition (C_2) is seldom satisfied for many indirectly measurement problems.

This means that more than one solution may be found for problem (12.7) and the information about the model is missing. In this case, *a priori* knowledge about the solution must be incorporated and built into the model. The requirement of stability is the most important one. If problem (12.7) lacks the property of stability, then the computed solution has nothing to do with the true solution since the practically computed solution is contaminated by unavailable errors. Therefore, there is no way to overcome this difficulty unless additional information about the solution is available. Again, *a priori* knowledge about the solution should be involved.

The precise mathematical definition of a well-posed problems is as follows:

Definition 12.3.1 *Let $K : \mathcal{X} \to \mathcal{Y}$ be a linear or nonlinear mapping, $\mathcal{X}$ and $\mathcal{Y}$ are two normed spaces. The operator equation (12.7) is said to be well-posed provided that the following holds.*

(1) For each $y \in \mathcal{Y}$ there exists (at least one) $x \in \mathcal{X}$, called a solution, for which (12.7) holds.

(2) For each $y \in \mathcal{Y}$ there exists at most one $x \in \mathcal{X}$ such that (12.7) holds.

(3) The solution is stable with respect to perturbations in y, i.e., for each sequence $\{x_n\} \subset \mathcal{X}$ with $y_n = Kx_n \to y = Kx$ as $n \to \infty$, it follows that $x_n \to x$ as $n \to \infty$.

Equations for which at least one of the three properties is violated are called improperly posed or ill-posed.

If Equation (12.7) is well-posed, then K has a well-defined, continuous inverse operator K^{-1}. In particular, $K^{-1}(K(x)) = x$ for any $x \in \mathcal{X}$ and Range$(K) = \mathcal{Y}$. In this case both the algebraic nature of the spaces and the topologies of the spaces are ready to be employed.

The particle size distribution model (12.5) is a linear model in infinite spaces. The ill-posedness is self-evident because at least one of the three requirements for well-posed problems is violated. Note that (12.6) is a linear model in finite spaces, therefore it is easy to rewrite it in a finite rank operator equation

$$K\mathbf{x} = \mathbf{y}, \tag{12.9}$$

by setting $\mathbf{x} = [f_{iso}, f_{vol}, f_{geo}]^{\top}$ and $\mathbf{y} = [y_j]$ with the entries $y_j = r_j(t_i, t_v, \phi)$, where $\mathbf{y}$ is the measurement data. The inverse problem is how to recover the model parameters $\mathbf{x}$ given the measurement data $\mathbf{y}$. The discrete ill-posedness arises because the linear kernel driven BRDF model is usually underdetermined if there are too few looks or poor directional range, or the observations are highly linearly dependent and noisy. For example, a single angular observation may lead to an under-determined system whose solutions are infinite (the null space of the kernel contains non-zero vectors) or the system has no solution (the rank of the coefficient matrix is not equal to the augmented matrix). In practice, random uncertainty in the reflectances sampled translates into uncertainty in the BRDF and albedo. We note that noise inflation depends on the sampling geometry alone. For example, for MODIS and MISR sampling, they vary with latitude and time of year; but for kernel-based models, they do not depend on wavelength or the type of BRDF viewed. Therefore, the random noise in the observation (BRDF) and the small singular values of K control the error propagation.

12.4 Regularization

12.4.1 Imposing *a priori* constraints on the solution

For effective inversion of the ill-posed kernel driven model, we have to impose an *a priori* constraint to the interested parameters. This leads to solving a constrained least squares error problem

$$\min J(\mathbf{x}), \quad \text{s.t.} \quad K\mathbf{x} = \mathbf{y}, \quad \Delta_1 \leq c(\mathbf{x}) \leq \Delta_2, \tag{12.10}$$

where $J(\mathbf{x})$ denotes an object functional, which is a function of $\mathbf{x}$, $c(\mathbf{x})$ is the constraint to the solution $\mathbf{x}$, Δ_1 and Δ_2 are two constants that specify the bounds of $c(\mathbf{x})$. Usually, $J(\mathbf{x})$ is chosen as the norm of $\mathbf{x}$ with a different scale. If the parameter $\mathbf{x}$ comes from a smooth function, then $J(\mathbf{x})$ can be chosen as a smooth function, otherwise, $J(\mathbf{x})$ can be non-smooth.

The constraint $c(\mathbf{x})$ can be smooth or non-smooth. A generically used constraint is the smoothness. It assumes that physical properties in a neighbourhood of space or in an interval of time present some coherence and generally do not change abruptly. In practice, we can always find regularities of a physical phenomenon with respect to certain properties over a short period of time, say problems in image vision (Horn and Schunck 1981). The smoothness *a priori* has been one of the most popular *a priori* assumptions in applications. The general framework is the so-called regularization, which will be explained in the next section.

12.4.2 Tikhonov variational regularization

Most inverse problems in the real environment are generally ill-posed. Regularization methods are widely-used to solve such ill-posed problems. The complete theory for regularization was developed by Tikhonov and his colleagues (Tikhonov and Arsenin 1977). For the discrete model (12.9), we suppose that $\mathbf{y}$ is the true right-hand side, and denote $\mathbf{y_n}$ as the measurements with noise that represents the bidirectional reflectance. The Tikhonov regularization method is to solve a regularized minimization problem

$$J^{\alpha}(\mathbf{x}) := \|K\mathbf{x} - \mathbf{y_n}\|_2^2 + \alpha\|D^{1/2}\mathbf{x}\|_2^2 \longrightarrow \min \tag{12.11}$$

instead of solving

$$J(\mathbf{x}) = \|K\mathbf{x} - \mathbf{y_n}\|_2^2 \longrightarrow \min. \tag{12.12}$$

In (12.11), α is the regularization parameter and D is a positively (semi-)definite operator. By a variational process, the minimizer of (12.11) satisfies

$$K^{\top}K\mathbf{x} + \alpha D\mathbf{x} = K^{\top}\mathbf{y_n}. \tag{12.13}$$

The operator D is a scale matrix that imposes a smoothness constraint to the solution $\mathbf{x}$. The scale operator D and the regularization parameter α can be considered to be some kind of *a priori* information, which will be discussed next.

Choices of the scale operator D

To regularize the ill-posed problem discussed in the previous subsection, the choice of the scale operator D has great impact on the performance of the regularization. Note that the matrix D plays the role of imposing a smoothness constraint on the parameters and in improving the condition of the spectrum of the adjoint operator $K^\top K$. Therefore, it should be positively definite or at least positively semi-definite. One may readily see that there may be a choice in the identity. However this choice does not fully employ the assumption about the continuity of the parameters.

In Wang *et al.* (2007a), we assume that the operator equation (12.9) is the discretized version of a continuous physical model

$$K(x(\tau)) = y(\tau) \tag{12.14}$$

with K the linear/nonlinear operator, $x(\tau)$ the complete parameters describing the land surfaces and y the observation. Most of the kernel model methods reported in the literature may have the above formulation. Hence instead of establishing regularization for the operator equation (12.9) in the Euclidean space, it is more convenient to perform the regularization to the operator equation (16.8) on an abstract space. So from *a priori* considerations we suppose that the parameters x is a smooth function, in the sense that x is continuous on $[a, b]$, is differentiable almost everywhere and its derivative is square-integrable on $[a, b]$. By Sobolev's imbedding theorem (see, e.g., Tikhonov and Arsenin (1977) and Xiao *et al.* (2003)), the continuous differentiable function x in $W^{1,2}$ space embeds into integrable continuous function space L_2 automatically. Here, the L_2 space is defined as the set of functions that are square-integrable, i.e., $L_2(\Omega) := \{x(t) : \int_\Omega |x(t)|^2 dt < \infty\}$; the Sobolev $W^{1,2}$ space is defined as the set of functions that are continuous and differentiable with the bounded norms of itself and its generalized derivatives in L_2, i.e., $W^{1,2}(\Omega) := \{x(t) : x(t) \in C(\Omega), x(t) \in L_2(\Omega), \frac{dx}{dt} \in L_2(\Omega), \int_\Omega (x^2 + \sum_{i=1}^n (\frac{dx}{dt_i})^2) dt_1 dt_2 \cdots dt_n < \infty\}$, where $C(\Omega)$ denotes the space of the continuous functions on Ω. The inner product of two functions $x(\tau)$ and $y(\tau)$ in $W^{1,2}$ space is defined by

$$(x(\tau), y(\tau))_{W^{1,2}} := \int_\Omega \left(x(\tau)y(\tau) + \sum_{i=1}^n \frac{\partial x}{\partial \tau_i} \frac{\partial y}{\partial \tau_j} \right) \mathrm{d}\tau_1 \mathrm{d}\tau_2 \cdots \mathrm{d}\tau_n, \tag{12.15}$$

where Ω is the assigned interval of the definition.

Now we construct a regularizing algorithm that has an approximate solution $x^\alpha \in W^{1,2}[a, b]$ that converges, as the error level approaches zero, to the actual parameters in the norm of space $W^{1,2}[a, b]$; precisely we construct the functional

$$J^\alpha(x) = \rho_F[Kx, y] + \alpha L(x), \tag{12.16}$$

where $\rho_F[Kx, y] = \frac{1}{2}\|Kx - y\|_{L_2}^2$, $L(x) = \frac{1}{2}\|x\|_{W^{1,2}}^2$.

Assume that the variation of $x(\tau)$ is flat near the boundary of the integral interval $[a, b]$. In this case, the derivatives of x are zeros at the boundary of $[a, b]$. Let h_r be the step size of the grids in $[a, b]$, which could be equidistant or adaptive. Then after discretization of $L(x)$,

D is a tridiagonal matrix of the form

$$D := D_1 = \begin{bmatrix} 1+\frac{1}{h_r^2} & -\frac{1}{h_r^2} & 0 & \cdots & 0 \\ -\frac{1}{h_r^2} & 1+\frac{2}{h_r^2} & -\frac{1}{h_r^2} & \cdots & 0 \\ \vdots & \ddots & \ddots & \ddots & \vdots \\ 0 & \cdots & -\frac{1}{h_r^2} & 1+\frac{2}{h_r^2} & -\frac{1}{h_r^2} \\ 0 & \cdots & 0 & -\frac{1}{h_r^2} & 1+\frac{1}{h_r^2} \end{bmatrix}.$$

For the linear model (12.6), after the kernel normalization, we may consider $[a, b] = [-1, 1]$. Thus, D has the above form with $h_r = 2/(N-1)$.

There are many kinds of techniques for choosing the scale matrix D appropriately. In Phillips–Twomey's formulation of regularization (see, e.g., Wang *et al.* (2006a)), the matrix D is created by the norm of the second differences, $\sum_{i=2}^{N-1}(x_{i-1} - 2x_i + x_{i+1})^2$, which leads to the following form of matrix D

$$D := D_2 = \begin{bmatrix} 1 & -2 & 1 & 0 & 0 & 0 & \cdots & 0 & 0 & 0 & 0 \\ -2 & 5 & -4 & 1 & 0 & 0 & \cdots & 0 & 0 & 0 & 0 \\ 1 & -4 & 6 & -4 & 1 & 0 & \cdots & 0 & 0 & 0 & 0 \\ 0 & 1 & -4 & 6 & -4 & 1 & \cdots & 0 & 0 & 0 & 0 \\ \vdots & \vdots & \ddots & \ddots & \ddots & \ddots & \ddots & \vdots & \vdots & \vdots & \vdots \\ 0 & 0 & 0 & \cdots & 0 & 1 & -4 & 6 & -4 & 1 & 0 \\ 0 & 0 & 0 & \cdots & 0 & 0 & 1 & -4 & 6 & -4 & 1 \\ 0 & 0 & 0 & \cdots & 0 & 0 & 0 & 1 & -4 & 5 & -2 \\ 0 & 0 & 0 & \cdots & 0 & 0 & 0 & 0 & 1 & -2 & 1 \end{bmatrix}.$$

However, the matrix D is badly conditioned and thus the solution to minimize the functional $J^\alpha[x]$ with D as the smooth constraint is observed to have some oscillations (Wang *et al.* 2006a). Another option is the negative Laplacian (see, e.g., (Wang and Yuan 2003; Wang 2007)): $Lx := -\sum_{i=1}^{n} \frac{\partial^2 x}{\partial \tau_i^2}$, for which the scale matrix D for the discrete form of the negative Laplacian Lx is

$$D := D_3 = \begin{bmatrix} 1 & -1 & 0 & \cdots & 0 & 0 \\ -1 & 2 & -1 & \cdots & 0 & 0 \\ \vdots & \vdots & \vdots & \cdots & \vdots & \vdots \\ 0 & 0 & 0 & -1 & 2 & -1 \\ 0 & 0 & 0 & \cdots & -1 & 1 \end{bmatrix}.$$

where we assume that the discretization step length is 1. The scale matrix D_3 is positive semi-definite but not positive definite and hence the minimization problem may not work efficiently for severely ill-posed inverse problems. Another option of the scale matrix D is the identity, i.e., $D := D_4 = \text{diag}(e)$, where e is the components of all the ones, however this scale matrix is too conservative and may lead to over-regularization.

Regularization parameter selection methods

As noted above, the choice of the regularization parameter α is important to tackle the ill-posedness. *A priori* choice of the parameter α allows $0 < \alpha < 1$. However the *a priori* choice of the parameter does not reflect the degree of approximation that may lead to either over-estimate or under-estimate of the regularizer.

We will use the widely-used discrepancy principle (see Tikhonov and Arsenin (1977), Tikhonov *et al.* (1995) and Xiao *et al.* (2003)) to find an optimal regularization parameter. In fact, the optimal parameter α^* is a root of the nonlinear function

$$\Psi(\alpha) = \|K\mathbf{x}_\alpha - \mathbf{y_n}\|^2 - \delta^2, \tag{12.17}$$

where δ is the error level to specify the approximate degree of the observation to the true noiseless data, $\mathbf{x}_\alpha$ denotes the solution of the problem in Equation (12.13) corresponding to the value α of the related parameter. Noting that $\Psi(\alpha)$ is differentiable, fast algorithms for solving the optimal parameter α^* can be implemented. In this paper we will use the cubic convergent algorithm developed in Wang and Xiao (2001):

$$\alpha_{k+1} = \alpha_k - \frac{2\Psi(\alpha_k)}{\Psi'(\alpha_k) + (\Psi'(\alpha_k)^2 - 2\Psi(\alpha_k)\Psi''(\alpha_k))^{\frac{1}{2}}}. \tag{12.18}$$

In the above cubic convergent algorithm, the functions $\Psi'(\alpha)$ and $\Psi''(\alpha)$ have the following explicit expressions:

$$\begin{aligned}\Psi'(\alpha) &= -\alpha\beta'(\alpha),\\ \Psi''(\alpha) &= -\beta'(\alpha) - 2\alpha\left[\left\|\frac{\mathrm{d}\mathbf{x}_\alpha}{\mathrm{d}\alpha}\right\|^2 + \left(\mathbf{x}_\alpha, \frac{\mathrm{d}^2\mathbf{x}_\alpha}{\mathrm{d}\alpha^2}\right)\right],\end{aligned}$$

where $\beta(\alpha) = \|\mathbf{x}_\alpha\|^2$, $\beta'(\alpha) = 2(\mathrm{d}\mathbf{x}_\alpha/d\alpha, \mathbf{x}_\alpha)$, and $\mathbf{x}_\alpha$, $\mathrm{d}\mathbf{x}_\alpha/\mathrm{d}\alpha$ and $\mathrm{d}^2\mathbf{x}_\alpha/\mathrm{d}\alpha^2$ can be obtained by solving the following equations:

$$(K^\top K + \alpha D)\mathbf{x}_\alpha = K^\top \mathbf{y_n}, \tag{12.19}$$

$$(K^\top K + \alpha D)\frac{\mathrm{d}\mathbf{x}_\alpha}{\mathrm{d}\alpha} = -D\mathbf{x}_\alpha, \tag{12.20}$$

$$(K^\top K + \alpha D)\frac{\mathrm{d}^2\mathbf{x}_\alpha}{\mathrm{d}\alpha^2} = -2D\frac{\mathrm{d}\mathbf{x}_\alpha}{\mathrm{d}\alpha}. \tag{12.21}$$

To solve the linear matrix–vector Equations (12.19)–(12.21), we use the Cholesky (square root) decomposition method. A remarkable characteristic of the solution of (12.19)–(12.21) is that the Cholesky decomposition of the coefficient matrix $K^\top K + \alpha D$ needs only be performed once, then the three vectors $\mathbf{x}_\alpha$, $\mathrm{d}\mathbf{x}_\alpha/\mathrm{d}\alpha$, $\mathrm{d}^2\mathbf{x}_\alpha/\mathrm{d}\alpha^2$ can be obtained cheaply.

12.4.3 Direct regularization

Instead of Tikhonov regularization, our goal in this subsection is to solve an equality constrained l_2 problem

$$\|\mathbf{x}\|_2 \longrightarrow \min, \ \text{s.t.}\ \tilde{K}\mathbf{x} + \mathbf{n} = \mathbf{y_n}, \tag{12.22}$$

where $\tilde{K} \in \mathbb{R}^{M\times N}$ is a perturbation of K (i.e., if we regard K as an accurate operator, then $\tilde{K}$ is an approximation to K which may contain error or noise), $\mathbf{x} \in \mathbb{R}^N$, $\mathbf{n}$, $\mathbf{y_n} \in \mathbb{R}^M$.

In Pokrovsky and Roujean (2002), the authors suggested using the singular value decomposition for solving an ill-conditioned linear system $Ax = b$, but no strict interpretation is given in their paper. As mentioned above, the ill-posedness is largely due to the small singular values of the linear operator. Let us denote the singular value decomposition of $\tilde{K}$ as

$$\tilde{K} = U_{M\times N}\Sigma_{N\times N}V_{N\times N}^{\top} = \sum_{i=1}^{N}\sigma_i u_i v_i^{\top},$$

where both $U = [u_i]$ and $V = [v_i]$ are orthonormal matrices, that is, the products of U with its transpose and V with its transpose are both identity matrices; Σ is a diagonal matrix whose non-zero entries consist of the singular values of $\tilde{K}$. The traditional least-squares error (LSE) solution $\mathbf{x}_{lse}$ of the constrained optimization system (12.22) can be expressed by the singular values and singular vectors in the form

$$\mathbf{x}_{lse} = \sum_{i=1}^{N}\frac{1}{\sigma_i}(u_i^{\top}\mathbf{y_n})v_i. \tag{12.23}$$

If the rank of $\tilde{K}$ is $p \le \min\{M, N\}$, then the above solution form inevitably encounters numerical difficulties, since the denominator contains numerically infinitesimal values. Therefore, to solve the problem by the SVD, we must impose *a priori* information. As we have noted, Tikhonov regularization solves a variation problem by incorporating *a priori* information into the solution. In this subsection, we consider another way of incorporating *a priori* information to the solution. The idea is quite simple: instead of filtering the small singular values by replacing the small singular values with small positive numbers, we just make a truncation of the summation, that is, the terms containing small singular values are replaced by zeroes. In this way, we obtain a regularized solution of the least squares problem (12.22) of minimal norm

$$\mathbf{x}_{lse}^{trunc} = \sum_{i=1}^{p}\frac{1}{\sigma_i}(u_i^{\top}\mathbf{y_n})v_i, \tag{12.24}$$

and

$$\min_{\mathbf{x}}\left\|\tilde{K}\mathbf{x} - \mathbf{y_n}\right\|_2^2 = \sum_{i=p+1,\cdots}\left|u_i^{\top}\mathbf{y_n}\right|^2. \tag{12.25}$$

We wish to examine the truncated singular value decomposition more. Note that, in practice, $\tilde{K}$ may not be exactly rank deficient, but may instead be numerically rank deficient, that is, it has one or more small but non-zero singular values such that $p_\delta < \text{rank}(\tilde{K})$. Here, p_δ refers to the numerical δ-rank of a matrix, see, e.g., Wang *et al.* (2006b) for details. It is clear from Equation (12.24) that the small singular values inevitably give rise to difficulties. The regularization technique for SVD means that some of the small singular values are truncated when in computation, and is hence called the numerically truncated singular value decomposition (NTSVD). Now assume that K is corrupted by the error matrix B_δ. Then, we replace K by a matrix $K_{\tilde{p}}$ that is close to K and mathematically rank deficient. Our choice of $K_{\tilde{p}}$ is obtained by replacing the small non-zero singular values $\sigma_{\tilde{p}+1}$, $\sigma_{\tilde{p}+2}, \ldots$ with exact zeros, that is,

$$K_{\tilde{p}} = \sum_{i=1}^{\tilde{p}} \sigma_i u_i v_i^\top, \tag{12.26}$$

where $\tilde{p}$ is usually chosen as p_δ. We call (12.26) the numerically truncated singular value decomposition of K. Now, we use (12.26) as the linear kernel to compute the least squares solutions. Actually, we solve the problem $\min_{\mathbf{x}} \|K_{\tilde{p}}\mathbf{x} - \mathbf{y_n}\|_2$, and obtain the approximate solution $\mathbf{x}_{lse}^{appr}$ of the minimal-norm

$$\mathbf{x}_{lse}^{appr} = K_{\tilde{p}}^{\dagger}\mathbf{y_n} = \sum_{i=1}^{\tilde{p}} \frac{1}{\sigma_i}(u_i^\top \mathbf{y_n})v_i, \tag{12.27}$$

where $K_{\tilde{p}}^{\dagger}$ denotes the Moore–Penrose generalized inverse.

Let us explain in more details the NTSVD for the underdetermined linear system. In this case, the number of independent variables is more than the number of observations, that is, $M < N$. Assume that the δ-rank of $\tilde{K}$ is $\tilde{p} \leq \min\{M, N\}$. It is easy to augment $\tilde{K}$ to be an $N \times N$ square matrix $\tilde{K}_{aug}$ by padding zeros underneath its M non-zero rows. Similarly, we can augment the right-hand side vector $\mathbf{y_n}$ with zeros. The singular decomposition of $\tilde{K}$ can be rewritten as

$$\tilde{K}_{aug} = U\Sigma V^\top$$

with

$$U = [u_1\, u_2\, \cdots\, u_N]_{N\times N}, \quad V = [v_1\, v_2\, \cdots, v_N]_{N\times N},$$

$$\Sigma = \text{diag}(\sigma_1, \sigma_2, \cdots, \sigma_{\tilde{p}}, 0, \cdots, 0).$$

From this decomposition, we find that there are $N - \tilde{p}$ theoretical zero singular values of the diagonal matrix Σ. These $N - \tilde{p}$ zero singular values will inevitably induce high numerical instability.

12.4.4 Statistical regularization

Bayesian statistics provides a conceptually simple process for updating uncertainty in the light of evidence. Initial beliefs about some unknown quantity are represented by *a priori* distribution. Information in the data is expressed by the likelihood function $L(\mathbf{x}|\mathbf{y})$. The *a priori* distribution $p(\mathbf{x})$ and the likelihood function are then combined to obtain the *a posteriori* distribution for the quantity of interest. The *a posteriori* distribution expresses our revised uncertainty in the light of the data, in other words, an organized appraisal in the consideration of previous experience.

The role of Bayesian statistics is very similar to the role of regularization. Now, we establish the relationship between the Bayesian estimation and the regularization. A continuous random vector $\mathbf{f}$ is said to have a Gaussian distribution if its joint probability distribution function has the form

$$p_x(\mathbf{x}; \mu, C) = \frac{1}{\sqrt{(2\pi)^N \det(C)}} \exp\left(-\frac{1}{2}(\mathbf{x}-\mu)^\top C^{-1}(\mathbf{x}-\mu)\right), \tag{12.28}$$

where $\mathbf{x},\ \mu \in \mathbb{R}^N$, C is an n-by-n symmetric positive definite matrix, and $\det(\cdot)$ denotes the matrix determinant. The mean is given by $E(\mathbf{x}) = \mu$ and the covariance matrix is $cov(\mathbf{x}) = C$.

Suppose $\mathbf{y} = K\mathbf{x} + \mathbf{n}$ is a Gaussian distribution with mean $K\mathbf{x}$ and covariance $C_\mathbf{n}$, where $C_\mathbf{n}$ is the noise covariance of the observation noise and model inaccuracy. Then by (12.28) we obtain

$$p(\mathbf{y}|\mathbf{x}) = \frac{1}{\sqrt{(2\pi)^M \det(C_\mathbf{n})}} \exp\left(-\frac{1}{2}(\mathbf{y}-K\mathbf{x})^\top C_\mathbf{n}^{-1}(\mathbf{y}-K\mathbf{x})\right). \tag{12.29}$$

From (12.28), the prior probability distribution is given by $p(\mathbf{x}) = \exp(-\frac{1}{2}\mathbf{x}^\top \mathbf{C}_\mathbf{x}^{-1}\mathbf{x})/\sqrt{(2\pi)^N \det(C_\mathbf{x})}$. By Bayesian statistical inference and the above two equations, we obtain an *a posteriori* log likelihood function

$$L(\mathbf{x}|\mathbf{y}) = \log p(\mathbf{x}|\mathbf{y}) = -\frac{1}{2}(\mathbf{y}-K\mathbf{x})^\top C_\mathbf{n}^{-1}(\mathbf{y}-K\mathbf{x}) - \frac{1}{2}\mathbf{x}^\top C_\mathbf{x}^{-1}\mathbf{x} + \zeta, \tag{12.30}$$

where ζ is constant with respect to $\mathbf{x}$. The maximum a posteriori estimation is obtained by maximizing (12.30) with respect to $\mathbf{x}$,

$$\mathbf{x} = (K^\top C_\mathbf{n}^{-1} K + C_\mathbf{x}^{-1})^{-1} K^\top C_\mathbf{n}^{-1}\mathbf{y}. \tag{12.31}$$

The easiest way of choosing $C_\mathbf{n}$ and $C_\mathbf{x}$ is by letting $C_\mathbf{n} = \sigma_\mathbf{n}^2 I_M,\ C_\mathbf{x} = \sigma_\mathbf{x}^2 I_N$, and then (12.31) becomes

$$\mathbf{x} = (K^\top K + \xi I_M)^{-1} K^\top \mathbf{y}, \tag{12.32}$$

where $\xi = \sigma_\mathbf{n}^2/\sigma_\mathbf{x}^2$, which is the noise-to-signal ratio. It is clear that the solution obtained by maximum *a posteriori* estimation has the same form as the solution of the Tikhonov regularization.

12.5 Optimization techniques

12.5.1 Sparse inversion in l_1 space

It deserves attention that the ill-posedness is the intrinsic feature of the inverse problems. Unless some additional information/knowledge such as monotonicity, smoothness, boundedness or the error bound of the raw data are imposed, the difficulty is hardly to be solved. Generally speaking, the kernel-driven BRDF model is semiempirical, the retrieved parameters $\mathbf{x}$ are mostly considered as a kind of weight function though it is a function of Leaf Area Index (LAI), Lambertian reflectance, sunlit crown reflectance, and viewing and solar angles. Therefore, $\mathbf{x}$ is not necessarily positive. However, since it is a weight function, an appropriate arrangement of the components of $\mathbf{x}$ can yield the same results. That is to say, $\mathbf{x}$ can be 'made' to be non-negative. The problem remaining is to develop some proper methods to solve the 'artificial' problem. Our new meaning to the solution $\mathbf{f}^*$ is related to the l_1 norm problem

$$\min_{\mathbf{x}} \ \|\mathbf{x}\|_1, \ \text{s.t.} \ K\mathbf{x} = \mathbf{y}, \ \mathbf{x} \geq 0, \tag{12.33}$$

which automatically imposes *a priori* information by considering the solution in l_1 space. Because of the limitations of the observation system, one may readily see that the recovered land surface parameters are discrete and sparse. Therefore, if an inversion algorithm is not robust, the outliers far from the true solution may occur. In this situation, the *a priori* constrained l_1 minimization may work better than the conventional regularization techniques. The model (12.33) can be reduced to a linear programming problem (Wang *et al.* 2005; Ye 1997; Yuan 2001), hence linear programming methods can be used for solving the inverse problem.

The l_1 norm solution method is looking for a feasible solution within the feasible set $S = \{\mathbf{x} : K\mathbf{x} = \mathbf{y}, \ \mathbf{x} \geq 0\}$. So it is actually searching for an interior point within the feasible set S, hence it is called the interior point method. The dual standard form of (12.33) is in the form

$$\max \ \mathbf{y}^\top \mathbf{g}, \ \text{s.t.} \ \mathbf{s} = \mathbf{e} - K^\top \mathbf{g} \geq 0, \tag{12.34}$$

where $\mathbf{e}$ is a vector with all components equaling 1. Therefore, the optimality conditions for $(\mathbf{x}, \mathbf{g}, \mathbf{s})$ to be a primal-dual solution triplet are that

$$K\mathbf{x} = \mathbf{y}, \ K^\top \mathbf{g} + \mathbf{s} = \mathbf{e}, \ \tilde{S}\tilde{F}\mathbf{e} = 0, \ \mathbf{x} \geq 0, \ \mathbf{s} \geq 0, \tag{12.35}$$

where $\tilde{S} = \text{diag}(s_1, s_2, \cdots, s_N)$, $\tilde{F} = \text{diag}(x_1, x_2, \cdots, x_N)$ and s_i, x_i are components of vectors $\mathbf{s}$ and $\mathbf{x}$ respectively. The notation $\text{diag}(\cdot)$ denotes the diagonal matrix whose only non-zero components are the main diagonal line.

The interior point method generates iterates $\{\mathbf{x}_k, \mathbf{g}_k, \mathbf{s}_k\}$ such that $\mathbf{x}_k > 0$ and $\mathbf{s}_k > 0$. As the iteration index k approaches infinity, the equality-constraint violations $\|\mathbf{y} - K\mathbf{x}\|$ and $\|K^\top \mathbf{g}_k + \mathbf{s}_k - \mathbf{e}\|$ and the duality gap $\mathbf{x}_k^\top \mathbf{s}_k$ are driven to zero, yielding a limiting point that solves the primal and dual linear problems. For the implementation procedures and examples about using the algorithm, please refer to Wang *et al.* (2007b, 2009) for details.

12.5.2 Optimization methods for l_2 minimization model

The conventional Tikhonov regularization method is equivalent to the constrained l_2 minimization problem

$$\min_{\mathbf{x}} \ \|\mathbf{x}\|_2, \quad \text{s.t. } K\mathbf{x} = \mathbf{y}. \tag{12.36}$$

This reduces to solve an unconstrained optimization problem

$$\mathbf{x} = \mathrm{argmin}_{\mathbf{x}} J^{\alpha}(\mathbf{x}), \quad J^{\alpha}(\mathbf{x}) = \frac{1}{2}\left\|K\mathbf{x} - \mathbf{y}\right\|_2^2 + \frac{\alpha}{2}\left\|x\right\|_2^2. \tag{12.37}$$

There are different solution methods developed in optimization research fields for solving a minimization problem. Because our model is already in a regularization form, different optimization methods can be employed. Here we give a review of several particular useful methods that may be used by users.

The gradient is and Hessian of $J^{\alpha}(\mathbf{x})$ are given by

$$\mathrm{grad}_{\mathbf{x}}[J^{\alpha}(\mathbf{x})] = (K^{\top}K + \alpha I)^{-1}\mathbf{x} - K^{\top}\mathbf{y} \text{ and } \mathrm{Hess}_{\mathbf{x}}[J^{\alpha}(\mathbf{x})] = K^{\top}K + \alpha I,$$

respectively. Hence at the kth iterative step, the gradient and Hessian of $J^{\alpha}(\mathbf{x}_k)$ can be expressed as $\mathrm{grad}_k[J^{\alpha}]$ and $\mathrm{Hess}_k[J^{\alpha}]$, which are evaluated by $\mathrm{grad}_{\mathbf{x}_k}\ [J^{\alpha}(\mathbf{x}_k)]$ and $\mathrm{Hess}_{\mathbf{x}_k}\ [J^{\alpha}(\mathbf{x}_k)]$ respectively.

Newton-type methods

Newton-type methods are based on the Gauss–Newton method and its various variations. We only supply the algorithm for the Gauss–Newton method in this subsection. The Gauss–Newton method is an extension of Newton method in one-dimensional space to higher dimensional space. The iteration formula reads as

$$\mathbf{x}_{k+1} = \mathbf{x}_k - \tau_k(\mathrm{Hess}_k[J^{\alpha}])^{-1}\mathrm{grad}_k[J^{\alpha}], \tag{12.38}$$

where τ_k, a damping parameter, that can be solved by a line search technique, is used to control the direction $(\mathrm{Hess}_k[J^{\alpha}])^{-1}\mathrm{grad}_k[J^{\alpha}]$. One may also apply a more popular technique, called the trust region technique, to control the direction $(\mathrm{Hess}_k[J^{\alpha}])^{-1}\mathrm{grad}_k[J^{\alpha}]$ within a reliable generalized ball in every iteration (Wang and Yuan 2005; Wang 2007). We recall that the inverse of $\mathrm{Hess}_k[J^{\alpha}]$ should be avoided for saving the amount of computation. Instead, linear algebraic decomposition methods can be applied to solve $(\mathrm{Hess}_k[J^{\alpha}])^{-1}\mathrm{grad}_k[J^{\alpha}]$.

There are different variations of the Gauss–Newton method, which are based on the approximation of the explicit Hessian matrix $\mathrm{Hess}_{\mathbf{x}}[J^{\alpha}]$. For these extensions to well-posed and ill-posed problems, please refer to Dennis and Schnable (1983), Kelley (1999), Nocedal (1980), Wang and Yuan (2005), Wang (2007) and Yuan (1994, 1993) for details.

Gradient-type methods

Let us consider the linear operator equation

$$Kx = y, \tag{12.39}$$

where K, $\mathbf{x}$ and $\mathbf{y}$ have the same meaning as before. The functional $J(\mathbf{x})$ to be minimized is given by

$$J(\mathbf{x}) = \frac{1}{2}\left\| K\mathbf{x} - \mathbf{y} \right\|_2^2. \tag{12.40}$$

The gradient of $J(\mathbf{x})$ is given by $\text{grad}_{\mathbf{x}}[J(\mathbf{x})] = (K^\top K)^{-1}\mathbf{x} - K^\top \mathbf{y}$. At the kth iterative step, the gradient of $J(\mathbf{x}_k)$ can be expressed as $\text{grad}_k[J]$, which is evaluated by $\text{grad}_{\mathbf{x}_k}[J(\mathbf{x}_k)]$. Note that for ill-posed inversion, the direct inverse of $K^\top K$ should be avoided, otherwise, significant perturbation of error will occur (Wang *et al.* 2007a).

Unlike Newton-type methods, the gradient method does not need the Hessian information. We first recall the well-known fixed point iteration method in the standard mathematical textbook: the fixed point iteration formula for solving (12.39) is

$$\mathbf{x}_{k+1} = \mathbf{x}_k + \tau(y - K\mathbf{x}_k), \tag{12.41}$$

where $\tau \in (0, 2/\|K\|)$ and K is linear, bounded and non-negative. The method can be considered as a special case of the gradient method. One may readily see that this method is very similar to the method of successive-approximations, where a very simple way to introduce the method is as follows. Consider the operator $T(\mathbf{x}) = \mathbf{x} + \tau(\mathbf{y} - K\mathbf{x})$, where τ is the so-called relaxation parameter. Any solution of (12.39) is equivalent to finding a fixed point of the operator T, i.e., solve for $\mathbf{x}$ from $\mathbf{x} = T(\mathbf{x})$. Assuming that T is a contraction mapping, then by the method of successive approximations, we obtain the following iterative scheme $\mathbf{x}_{k+1} = T(\mathbf{x}_k)$, i.e., iterative formula (12.41). The method converges if and only if $K\mathbf{x} = \mathbf{y}$ has a solution.

Now we introduce a very simple gradient method, the steepest descent method, the iteration formula reads as

$$\mathbf{x}_{k+1} = \mathbf{x}_k + \tau_k \cdot K^\top(y - K\mathbf{x}_k), \tag{12.42}$$

where τ_k is obtained by line search, i.e., $\tau_k = \text{argmin}_{\tau>0} J(\mathbf{x}_k - \tau \text{grad}_k[J])$. If we restrict stepsize τ_k to be fixed in $(0, 2/\|K^\top K\|)$, then the steepest descent method reduces to the famous Landweber–Fridman iteration method. More extensions include the non-monotone gradient method and different applications in applied science and can be found in, e.g., Barzilai and Borwein (1988), Brakhage (1987), Fletcher (2005), Wang (2007) and Wang and Ma (2007).

Finally we want to mention that for underdetermined ill-posed problems, regularization constraints or *a priori* constraints should be incorporated into the minimization model (12.40) first, e.g., $J(\mathbf{x})$ in (12.40) should be replaced by $J^\alpha(\mathbf{x})$ in (12.37), then we may apply the aforementioned gradient methods. Application examples on aerosol particle size distribution function retrieval problems and non-monotone gradient method are include in Wang (2008).

12.6 Kernel-based BRDF model inversion

12.6.1 Inversion by NTSVD

Consider the linear combination of three kernels k_{geo}, k_{vol} and the isotropic kernel

$$\hat{f}_{iso} + \hat{f}_{geo}k_{geo}(t_i, t_v, \phi) + \hat{f}_{vol}k_{vol}(t_i, t_v, \phi) = \hat{r}$$

for each observation. Considering the smoothing technique technique in l_2 space, we solve the following constrained optimization problem

$$\min \|[\hat{f}_{iso}, \hat{f}_{geo}, \hat{f}_{vol}]^\top\|_2, \text{ s.t. } \hat{f}_{iso} + \hat{f}_{geo}k_{geo} + \hat{f}_{vol}k_{vol} = \hat{r}. \tag{12.43}$$

Let us just consider an extreme example for the kernel-based BRDF model: that is, if only a single observation is available at one time, then it is clear that the above equation has infinitely many solutions. If we denote

$$K = [1 \quad k_{geo}(t_i, t_v, \phi) \quad k_{vol}(t_i, t_v, \phi)]_{1\times 3},$$

then the singular decomposition of the zero augmented matrix K_{aug} leads to $K_{aug} = U_{3\times3}\ \Sigma_{3\times3}\ V_{3\times3}^\top$ with $U = [u_1\ u_2\ u_3]$, $\Sigma = \text{diag}(\sigma_1, \sigma_2, \sigma_3)$, $V = [v_1\ v_2\ v_3]$, where each $u_i,\ v_i,\ i = 1, 2, 3$, are the 3-by-1 columns. Our *a priori* information is based on searching for a minimal norm solution within the infinite set of solutions, that is, the solution $f^* = [\hat{f}^*_{iso}, \hat{f}^*_{geo}, \hat{f}^*_{vol}]^\top$ satisfies $\hat{f}^*_{iso} + \hat{f}^*_{geo}k_{geo}(t_i, t_v, \phi) + \hat{f}^*_{vol}k_{vol}(t_i, t_v, \phi) = \hat{r}$ and at the same time $\|f^*\| \longrightarrow$ minimum.

12.6.2 Tikhonov regularized solution

Denote by M the number of measurements in the kernel-based models. Then the operator Equation (12.9) can be rewritten in the following matrix–vector form

$$K\mathbf{x} = \mathbf{y}, \tag{12.44}$$

where

$$K = \begin{bmatrix} 1 & k_{geo}(1) & k_{vol}(1) \\ 1 & k_{geo}(2) & k_{vol}(2) \\ \vdots & \vdots & \vdots \\ 1 & k_{geo}(M) & k_{vol}(M) \end{bmatrix}, \quad \mathbf{x} = \begin{bmatrix} f_{iso} \\ f_{geo} \\ f_{vol} \end{bmatrix}, \quad \mathbf{y} = \begin{bmatrix} r_1 \\ r_2 \\ \vdots \\ r_M \end{bmatrix}.$$

In which, $k_{geo}(k)$ and $k_{vol}(k)$ represent the values of kernel functions $k_{geo}(t_i, t_v, \phi)$ and $k_{vol}(t_i, t_v, \phi)$ corresponding to the kth measurement for $k = 1, 2, \cdots$; r_k represents the kth observation for $k = 1, 2, \cdots$.

By Tikhonov Regularization, we solve for a regularized solution $\mathbf{x}^{\alpha}$ by minimizing the functional

$$J(\mathbf{x}, \alpha) = \frac{1}{2}\|K\mathbf{x} - \mathbf{y}\|^2 + \frac{1}{2}\alpha\|D\mathbf{x}\|^2. \tag{12.45}$$

Choices of the parameter α and the scale operator D are discussed in Section 12.4.2.

12.6.3 Land surface parameter retrieval results

We use the combination of the RossThick kernel and LiTransit kernel in the numerical tests. In practice, the coefficient matrix K cannot be determined accurately, and a perturbed version $\tilde{K}$ is obtained instead. Also instead of the true measurement $\mathbf{y}$, the observed measurement $\mathbf{y_n} = \mathbf{y} + \mathbf{n}$ is the addition of the true measurement $\mathbf{y}$ and the noise $\mathbf{n}$, which for simplicity is assumed to be additive Gaussian random noise. Therefore it suffices to solve the following operator equation with perturbation

$$\tilde{K}\mathbf{x} = \mathbf{y_n},$$

where $\tilde{K} := K + \delta B$ for some perturbation matrix B and δ denotes the noise level (upper bound) of $\mathbf{n}$ in (0,1). In our numerical simulation, we assume that B is a Gaussian random matrix, and also that

$$\|\mathbf{y_n} - \mathbf{y}\| \leq \delta < \|\mathbf{y_n}\|. \tag{12.46}$$

The above assumption about the noise can be interpreted as that the Signal-to-Noise Ratio (SNR) should be greater than 1. We make such an assumption as we believe that observations (BRDF) are not trustworthy otherwise. It is clear that (12.44) is an underdetermined system if $M \leq 2$ and an overdetermined system if $M > 3$. Note that for satellite remote sensing, because of the restrictions in view and illumination geometries, $\tilde{K}^{\top}\tilde{K}$ needs not have a bounded inverse (Gao *et al.* 1998; Li *et al.* 2001; Verstraete *et al.* 1996). We believe that the proposed regularization method that can be employed to find an approximate solution $\mathbf{x}_{\alpha}^{\dagger}$ satisfies

$$\|\tilde{K}\mathbf{x}_{\alpha}^{\dagger} - \mathbf{y_n}\| \longrightarrow \min.$$

We choose the widely used 73 data sets referred in Li *et al.* (2001). Of the 73 sets of BRDF measurements, only 18 sets of field-measured BRDF data with detailed information about the experiments are known, including biophysical and instrumental information (Deering *et al.* 1992, 1994, 1995; Eck *et al.* 1992; Hu *et al.* 1997; Kimes *et al.* 1985, 1986). These data sets cover a large variety of vegetative cover types, and are fairly well representative of the natural and cultivated vegetation. Table 12.1 summarizes the basic properties of the data sets used in this chapter. For those selected field data, the observations are sufficient. If the kernel matrix $\tilde{K}$ is well-conditioned, the problem is well-posed, and the regularization is unnecessary, which can also be considered a regularization procedure with zero regularization parameter. Even for sufficient observations, most of the kernel matrix $\tilde{K}$ could be ill-conditioned, and hence the problem is discrete ill-posed. In that situation, one has to resort to regularization or preconditioning to make the proposed algorithm workable.

Table 12.1 Data sets used in the simulation

Data	Cover type	LAI
ranson_soy.827	Soy	2.9
kimes.orchgrass	Orchard grass	1
Parabola.1994.asp-ifc2	Aspen	5.5

We choose one or two observations as a limited number of observations and compare the retrieval results by different regularization methods (Wang *et al.* 2007a, 2008). Comparison results are given in Tables 12.2 and 12.3. In the two tables, these methods are denoted by: NTSVD, Tikh(a) ($\alpha = \delta^2$); Tikh(b) (D is in the form of D_1 and α is chosen by *a posteriori* method addressed in Section 12.4.2); Tikh(c) (D is in the form D_2 and α is chosen by *a posteriori* method addressed in Section 12.4.2); Tikh(d) (D is in the form D_3 and α is chosen by *a posteriori* method addressed in Section 12.4.2) and Tikh(e) (D is in the form D_4 and α is chosen by *a posteriori* method addressed in Section 12.4.2). The true White Sky Albedo (WSA) is calculated from well-posed situations using AMBRALS, i.e., full observation data. It deserves pointing out that the standard operational algorithm used in

Table 12.2 Comparison of computational values of the WSAs from data sets in Table 12.1 for single observation and for two observations with the true WSAs values (multiangular observations) for VisRed band. Reproduced by permission of American Geophysical Union; ©2008 AGU

	Methods	Single observation	Two observations	True WSAs
	NTSVD	0.0449047	0.0442712	
	Tikh(a)	0.0638937	0.0386419	
	Tikh(b)	0.0401528	0.0560726	
ranson_soy.827				0.0405936
	Tikh(c)	0.0633967	0.0590594	
	Tikh(d)	−0.0009147	0.0539707	
	Tikh(e)	0.0476311	0.0583683	
	NTSVD	0.1082957	0.1058740	
	Tikh(a)	0.0397185	0.0860485	
	Tikh(b)	0.0753925	0.1214918	
kimes.orchgrass				0.0783379
	Tikh(c)	0.26211583	0.4365220	
	Tikh(d)	−0.0018020	0.0383555	
	Tikh(e)	0.1137684	0.1707774	
	NTSVD	0.0364620	0.0389198	
	Tikh(a)	0.0447834	−0.0040831	
	Tikh(b)	0.0262501	0.0102457	
Parabola.1994.asp-ifc2				0.0227972
	Tikh(c)	0.0798633	−0.0874783	
	Tikh(d)	−0.0006110	−0.0401510	
	Tikh(e)	0.0375009	0.0547068	

Table 12.3 Comparison of computational values of the WSAs from the data sets in Table 12.1 for single observation and for two observations with the true WSAs values (multiangular observations) for Nir band. Reproduced by permission of American Geophysical Union; ©2008 AGU

	Methods	Single observation	Two observations	True WSAs
	NTSVD	0.4469763	0.4348320	
	Tikh(a)	0.6359822	0.4195730	
	Tikh(b)	0.3996775	0.5439493	
ranson_soy.827				0.3653728
	Tikh(c)	0.6310461	0.9247240	
	Tikh(d)	−0.0091045	−0.0098136	
	Tikh(e)	0.4741162	0.6277249	
	NTSVD	0.3890207	0.37216767	
	Tikh(a)	0.2048903	0.2945934	
	Tikh(b)	0.2708260	0.4458619	
kimes.orchgrass				0.2963261
	Tikh(c)	0.9415755	1.8140732	
	Tikh(d)	−0.0064732	0.1927318	
	Tikh(e)	0.4086801	0.6015300	
	NTSVD	0.5517209	0.5741842	
	Tikh(a)	0.6776356	−0.0617838	
	Tikh(b)	0.3972022	0.2398577	
Parabola.1994.asp-ifc2				0.4240376
	Tikh(c)	1.2084479	−0.8953630	
	Tikh(d)	−0.0092437	−0.4071125	
	Tikh(e)	0.5674424	0.8185223	

AMBRALS does not work for such severely ill-posed situations. If we regard WSA > 1 or WSA < 0 as failed inversion, it is clear that our proposed method works for all of the cases. It follows from the experiments that our method (Tikh(b)) (proposed in Wang *et al.* (2008)) works, for a single observation, and performs better than the NTSVD (proposed in Wang *et al.* (2007a)) and the standard Tikhonov regularization with *a priori* choice of the regularization parameter (Tikh(a)).

In the next, we use an atmospherically corrected moderate resolution imaging spectroradiometer (MODIS) 1B product acquired on a single day as an example of a single observation BRDF at certain viewing direction. Each pixel has a different view zenith angle and relative azimuth angle. The data MOD021KM.A2001135-150 with horizontal tile number (26) and vertical tile number (4) covers Shunyi county of Beijing, China. The three parameters are retrieved by using this 1B product. Figure 12.1 plots the reflectance for band 1 of a certain day DOY = 137. In the MODIS AMBRALS algorithm, when insufficient reflectances or a poorly representative sampling of high quality reflectances are available for a full inversion, a database of archetypal BRDF parameters is used to supplement the data and a magnitude inversion is performed (Strahler *et al.* 1999; Verstraete *et al.* 1996). We note that the standard MODIS AMBRALS algorithm cannot work for such an extreme case, even for MODIS magnitude inversion, since it is hard to obtain seasonal data associated with a dynamic land cover in a particular site. But our method still works for such an extreme case because that smoothness

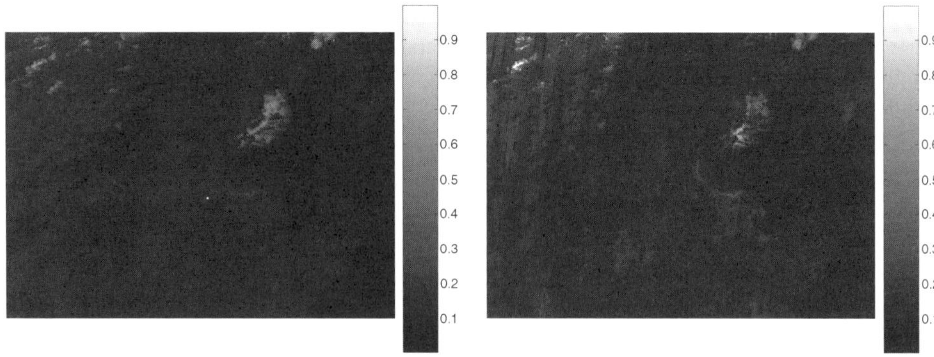

Figure 12.1 Reflectance for band 1 of MOD021KM.A2001137 (left), and the white-sky albedo retrieved by proposed Tikhonov regularization method (right). Reproduced by permission of American Geophysical Union; ©2008 AGU.

constraint is already implanted into the model. We plot the white-sky albedo (WSA) retrieved by Tikh(b) for band 1 of one observation (DOY=137) in Figure 12.1(right). We see that the albedo retrieved from insufficient observations can generate the general profile. We observe that most of the details are preserved, though the results are not perfect. The results are similar to that from the NTSVD method developed in Wang *et al.* (2007a). Hence, we conclude that both methods can be considered useful methods for the retrieval of land surface parameters and for computing land surface albedos. Thus both algorithms can be considered as supplement algorithms for the robust estimation of the land surface BRDF/albedos.

We also test our algorithms to the Landsat Thematic Mapper (TM) data measured in Shunyi county of Beijing, China. The TM sensor is an advanced, multi-spectral scanning, Earth resources instrument designed to achieve higher image resolution, sharper spectral separation, improved geometric fidelity, and greater radiometric accuracy and resolution. Figure 12.2 plots the reflectance for band 5 on May 17 2001. The spatial resolution for the TM sensor on band 5 is 30 metres. The white-sky albedo (WSA) retrieved by Tikh(b) for band 5 of one observation on May 17 2001 is plotted in Figure 12.2(right) The retrieved results show that our algorithms work for satellite data with high spatial resolutions.

We want to emphasize that our method can generate smoothing data to help the retrieval of parameters once sufficient observations are unavailable. As we have pointed out in Wang *et al.* (2007a, 2008), we do not suggest discarding the useful history information (e.g., data that is not too old) and the multiangular data. Instead, we should fully employ such information if it is available. The key to why our algorithm outperforms previous algorithms is that our algorithm is adaptive, accurate and very stable, which solves the kernel-based BRDF model of any order, which may be a supplement for the BRDF/albedo retrieval product.

For the remote sensor MODIS, which can generate a product by using 16 days different observations data, this is not a strict restriction for MODIS, since it aims at global exploration. For other sensors, the period for their detection of the same area will be longer than 20 days or more. Therefore, for vegetation in the growing season, the reflectance and albedos will change significantly. Hence robust algorithms to estimate BRDF and albedos in such cases are highly desired. Our algorithm is a proper choice, since it can generate retrieval results which quite

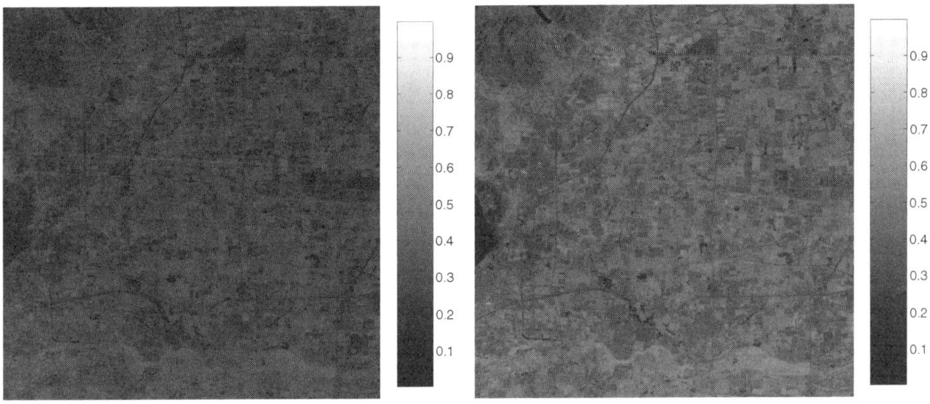

Figure 12.2 Reflectance for band 5 of Landsat Thematic Mapper Data (TM) on May 17 2001 (left) and the white-sky albedo retrieved by proposed Tikhonov regularization method (right). Reproduced by permission of American Geophysical Union; ©2008 AGU.

approximate the true values of different vegetation type of land surfaces by capturing just one time of observation.

Moreover, for some sensors with high spatial resolution, the quasi multiangular data are impossible to obtain. This is why there are not high resolution albedo products. But with our algorithm, we can achieve the results. This is urgently needed in real applications.

12.7 Aerosol particle size distribution function retrieval

We consider retrieving the aerosol particle size distribution function $n(r)$ from the attenuation equation (12.3). But it is an infinite dimensional problem with only a finite set of observations, so it is improbable to implement such a system by computer to get a continuous expression of the size distribution $n(r)$. Numerically, we solve the discrete problem of operator equation (12.5). Using collocation (Wang *et al.* 2006a), the infinite problem can be written in a finite dimensional form by sampling some grids $\{r_j\}_{j=1}^N$ in the interval of interests $[a, b]$.

Denoting by $K = (K_{ij})_{N\times N}$, $\mathbf{n}$, ϱ and $\mathbf{d}$ the corresponding vectors, we have

$$K\mathbf{n} + \varrho = \mathbf{d}. \tag{12.47}$$

This discrete form can be used for computer simulations.

Phillips–Twomey's regularization is based on solving the problem

$$\min_{\mathbf{n}} \; Q(\mathbf{n}), \;\; \text{s.t.} \; \|K\mathbf{n} - \mathbf{d}\| = \Delta, \tag{12.48}$$

where $Q(\mathbf{n}) = (D\mathbf{n}, \mathbf{n})$, where D is a pre-assigned scale matrix.

In Phillips–Twomey's formulation of regularization, the choice of the scale matrix is vital. They chose the form of the matrix D by the norm of the second differences,

$\sum_{i=2}^{N-1}(n_{i-1} - 2n_i + n_{i+1})^2$, which corresponds to the following form of matrix $D = D_2$. However, the matrix D is badly conditioned. For example, with $N = 200$, the largest singular value is 15.998012. The smallest singular value is 6.495571×10^{-17}. This indicates that the condition number of the matrix D defined by the ratio of the largest singular value to the smallest singular value equals 2.462911×10^{17}, which is worse. Hence, for small singular values of the discrete kernel matrix K, the scale matrix D cannot have them filtered even with a large Lagrangian multiplier μ. This numerical difficulty encourages us to study a more robust scale matrix D, which is formulated in the next section.

We consider the Tikhonov regularization in Sobolev $W^{1,2}$ space as mentioned in Section 13.4.2. By the variational process, we solve a regularized linear system of equations

$$K^\top K\mathbf{n} + \alpha H\mathbf{n} - K^\top \mathbf{d} = 0, \tag{12.49}$$

where H is a triangular matrix in the form of D_1. For choice of the regularization parameter, We consider the *a posteriori* approach mentioned in Section 12.4.2. Suppose we are interested in the particle size in the interval $[0.1, 4]$ μm, the step size is $h_r = 3.9/(N-1)$. Now choosing the discrete nodes $N = 200$, the largest singular value of H is $1.041482176501067 \times 10^4$ by double machine precision, and the smallest singular value of H is 0.99999999999953 by double machine precision. Compared with the scale matrix D of Phillips–Twomey's regularization, the condition number of H is $1.041482176501554 \times 10^4$, which is better than D in filtering small singular values of the discrete kernel K.

To perform the numerical computations, we apply the technique developed in King *et al.* (1978), that is, we assume that the actual aerosol particle size distribution function consists of the multiplication of two functions $h(r)$ and $f(r)$: $n(r) = h(r)f(r)$, where $h(r)$ is a rapidly varying function of r, while $f(r)$ is more slowly varying. In this way we have

$$\tau_{aero}(\lambda) = \int_a^b [k(r, \lambda, \eta)h(r)]f(r)dr, \tag{12.50}$$

where $k(r, \lambda, \eta) = \pi r^2 Q_{ex}(r, \lambda, \eta)$ and we denote $k(r, \lambda, \eta)h(r)$ as the new kernel function which corresponding to a new operator Ξ:

$$(\Xi f)(r) = \tau_{aero}(\lambda). \tag{12.51}$$

For simplicity of notation, the discretization of Ξ is again denoted by the matrix A.

The size distribution function $n_{true}(r) = 10.5r^{-3.5}\exp(-10^{-12}r^{-2})$ is used to generate synthetic data. The particle size radius interval of interest is $[0.1, 2]$ μm. This aerosol particle size distribution function can be written as $n_{true}(r) = h(r)f(r)$, where $h(r)$ is a rapidly varying function of r, while $f(r)$ is more slowly varying. Since most measurements of the continental aerosol particle size distribution reveal that these functions follow a Junge distribution (Junge 1955), $h(r) = r^{-(\nu^*+1)}$, where ν^* is a shaping constant with typical values in the range 2.0–4.0, therefore it is reasonable to use an $h(r)$ of Junge type as the weighting factor to $f(r)$. In this work, we choose $\nu^* = 3$ and $f(r) = 10.5r^{1/2}\exp(-10^{-12}r^{-2})$. The form of this size distribution function is similar to the one given by Twomey (1975), where a rapidly changing function $h(r) = Cr^{-3}$ can be identified, but it is more similar to a Junge distribution for $r \geq 0.1$ μm. One can also generate other particle number size distributions and compare the

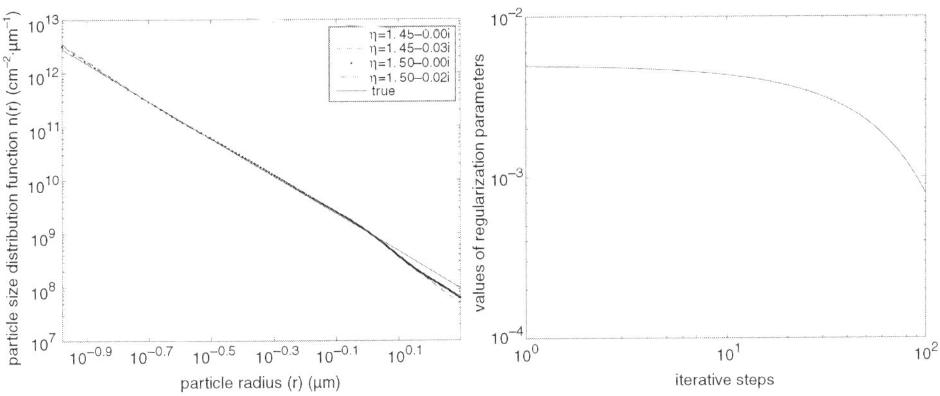

Figure 12.3 Left: Input and retrieved results with our inversion method in the case of error level $\delta = 0.05$ and different complex refractive indices. Right: iterative computational values of regularization parameters when the error level $\delta = 0.05$.

reconstruction with the input. In the first place, the complex refractive index η is assumed to be $1.45 - 0.00i$ and $1.50 - 0.00i$, respectively. Then we invert the same data, supposing that η has an imaginary part. The complex refractive index η is assumed to be $1.45 - 0.03i$ and $1.50 - 0.02i$, respectively. The precision of the approximation is characterized by the Root Mean-Square Error (RMSE)

$$\text{RMSE} = \sqrt{\frac{1}{m}\sum_{i=1}^{m}\frac{(\tau_{comp}(\lambda_i) - \tau_{meas}(\lambda_i))^2}{(\tau_{comp}(\lambda_i))^2}}, \tag{12.52}$$

which describes the average relative deviation of the retrieved signals from the true signals. In which, τ_{comp} refers to the retrieved signals, τ_{meas} refers to the measured signals. Numerical illustrations are plotted in Figure 12.3 with noise level $\delta = 0.05$ for different refractive indices. The behavior of the regularization parameter is plotted in Figure 12.3(right). The RMSEs for each case are shown in Table 12.4.

Table 12.4 The root mean-square errors for different noise levels

Noise levels	$\eta = 1.45 - 0.00i$	$\eta = 1.45 - 0.03i$	$\eta = 1.50 - 0.02i$
$\delta = 0.005$	1.6443×10^{-5}	1.2587×10^{-5}	2.2773×10^{-5}
$\delta = 0.01$	1.6493×10^{-5}	1.2720×10^{-5}	2.2847×10^{-5}
$\delta = 0.05$	1.6996×10^{-5}	1.3938×10^{-5}	2.3504×10^{-5}

12.8 Conclusion

In this chapter, we investigate the regularization and optimization methods for the solution of the kernel-based remotely sensed inverse problems. We reformulate the problem in functional space by introducing the first kind operator equations, then the solution methods in l_1 and l_2 spaces are considered. The regularization strategy and optimization solution techniques are fully described. The equivalence between the Tikhonov regularization and Baysian statistical inference is established. We want to emphasize that there are different ways that can be developed to impose *a priori* information (Wang 2007). For example, (P1) the unknowns **x** can be bounded. This method requires a good *a priori* upper bound for **x**; (P2) applying different weights to the components of **x**, then solve the kernel model under the constraint of the weights; (P3) imposing historical information on **x** provided that such historical information exists; (P4) simplifying the physical model by solving a l^p norm problem, which means the unknowns **x** can be obtained under the l^p scale (Wang *et al.* 2009).

Acknowledgments

We would like to thanks the referees' valuable suggestions which greatly help us improve the quality of the chapter. This research is supported by National "973" Key Basic Research Developments Program of China under grant numbers 2007CB714400 and 2005CB422104 and National Natural Science Foundation of China (NSFC) under grant number 10871191.

References

Ångström, A. (1929) On the atmospheric transmission of sun radiation and on dust in the air. *Geografiska Annaler*, **11**, 156–166.

Atzberger, C. (2004) Object-based retrieval of biophysical cannopy variables using artificial neural nets and radiative transfer models. *Remote Sensing of Environment*, **93**, 53–67.

Barzilai, J. and Borwein, J. (1988) Two-point step size gradient methods. *IMA Journal of Numerical Analysis*, **8**, 141–148.

Bockmann, C. (2001) Hybrid regularization method for the ill-posed inversion of multiwavelength lidar data in the retrieval of aerosol size distributions. *Applied Optics*, **40**, 1329–1342.

Bockmann, C. and Kirsche, A. (2006) Iterative regularization method for lidar remote sensing. *Computer Physics Communications*, **174**, 607–615.

Bohren, G.F. and Huffman, D.R. (1983) *Absorption and Scattering of Light by Small Particles*. New York, John Wiley & Sons.

Brakhage, H. (1987) On ill-posed problems and the method of conjugate gradients. In H.W. Engl and C.W. Groetsch (eds), *Inverse and Ill-Posed Problems*. Academic Press, Boston, 165–175.

Camps-Valls, G. (2008) New machine-learning paradigm provides advantages for remote sensing. *SPIE Newsroom*, DOI: 10.1117/2.1200806.1100.

Davies, C.N. (1974) Size sistribution of atmospheric aerosol. *Journal of Aerosol Science*, **5**, 293–300.

Deering, D.W., Eck, T.F. and Grier, T. (1992) Shinnery oak bidirectional reflectance properties and canopy model inversion. *IEEE Trans. Geosci. Remote Sensing*, **30**(2), 339–348.

Deering, D.W., Middleton, E.M. and Eck, T.F. (1994) Reflectance anisotropy for a spruce–hemlock forest canopy. *Remote Sens. of Environ.*, **47**, 242–260.

Deering, D.W., Shmad, S.P., Eck, T.F. and Banerjee, B.P. (1995) Temporal attributes of the bidirectional reflectance for three forest canopies. *International Geoscience and Remote Sensing Symposium (IGARSS'95)*, 1239–1241.

Dennis, J.E. and Schnable, R.B. (1983) *Numerical Methods for Unconstrained Optimization and Nonlinear Equations*. Englewood Cliffs, N.J., Prentice Hall.

Eck, T.F. and Deering, D.W. (1992) Spectral bidirectional and hemispherical reflectance chacteristics of selected sites in the streletskay steppe. *Proceedings of the (1992) International Geoscience and Remote Sensing Symposium*. IEEE Geosci. and Remote Sens. Soc., New Jersey, 1053–1055.

Engl, H.W., Hanke, M. and Neubauer, A. (1996) *Regularization of Inverse Problems*. Dordrecht, Kluwer.

Fletcher, R. (1987) *Practical Methods of Optimisation*. 2nd edn. Chichester, John Wiley & Sons.

Fletcher, R. (2005) On the Barzilai-Borwein method, *Optimization and control with applications, Appl. Optim.*, **96**, 235–256, New York, Springer.

Gao, F., Strahler, A.H., Lucht, W., Xia, Z. and Li, X. (1998) Retrieving albedo in small sample size. *IEEE Int. Geosci. Remote Sens. Symp. Proc. 1998*, **5**, 2411–2413.

Hansen, P.C. (1998) *Rank-deficient and Discrete Ill-posed Problems: Numerical Aspects of Linear Inversion*. SIAM, Philadelphia.

Horn, B.K.P. and Schunck, B.G. (1981) Determining optical flow. *Artificial Intelligence*, **17**, 185–203.

Houghton, J.T., Meira Filho, L.G., Callander, B.A., Harris, N., Kattenberg, A. and Maskell, K. (1966) *Climate Change 1995*. Published for the Intergovernmental Panel on Climate Change, Cambridge University Press.

Hu, B.X., Lucht, W., Li, X.W. and Strahler, A.H. (1997) Validation of kernel-driven semiempirical models for the surface bidirectional reflectance distribution function of land surfaces. *Remote Sensing of Environment*, **62**, 201–214.

Junge, C.E. (1955) The size distribution and aging of natural aerosols as determined from electrical and optical data on the atmosphere. *J. Meteor.*, **12**, 13–25.

Kelley, C.T. (1999) *Iterative Methods for Optimization*. SIAM in Applied Mathematics.

Kimes, D.S., Newcomb, W.W. and Tucker, C.J. (1985) Directional reflectance factor distributions for cover types of Northen Africa. *Remote Sensing of Environment*, **18**, 1–19.

Kimes, D.S., Newcomb, W.W. and Nelson, R.F. (1986) Directional reflectance distributions of a hardwood and a pine forest canopy. *IEEE Transactions on Geoscience and Remote Sensing*, **24**, 281–293.

King, M.D., Byrne, D.M., Herman, B.M. and Reagan, J.A. (1978) Aerosol size distributions obtained by inversion of spectral optical depth measurements. *Journal of Aerosol Science*, **35**, 2153–2167.

Li, X., Wang, J., Hu, B. and Strahler, A.H. (1998) On unitilization of *a priori* knowledge in inversion of remote sensing models. *Science in China D*, **41**, 580–585.

Li, X., Gao, F., Liu, Q., Wang, J.D. and Strahler, A.H. (2000) Validation of a new GO kernel and inversion of land surface albedo by kernel-driven model (1). *Journal of Remote Sensing*, **4** (appl.), 1–7.

Li, X., Gao, F., Wang, J. and Strahler, A.H. (2001) *A priori* knowledge accumulation and its application to linear BRDF model inversion. *J. Geophys. Res.*, **106**(D11), 11925–11935.

Li, X. and Wang, J. (1995) *Vegetation Optical Remote Sensing Model and Vegetation Structure Paramerization*. Beijing, Science Press.

Li, X., Wang, J. and Strahler, A.H. (1999) Apparent reciprocal failure in BRDF of structured surfaces. *Progress of Natural Sciences*, **9**, 747–750.

Liang, S.L. (2004) *Quantitiative Remote Sensing of Land Surfaces*. New Jersey, John Wiley & Sons.

Liang, S.L. (2008) *Advances in Land Remote Sensing: System, Modelling, Inversion and Application*. Springer.

Mccartney, G.J. (1976) *Optics of Atmosphere*. New York, John Wiley & Sons.

Nashed, M.Z. (1976) Perturbations and approximations for generalized inverses and linear operators equations. In M.Z. Nashed (ed.) *Generalized Inverses and Applications*, 325–396, New York, Academic Press.

Nguyen, T. and Cox, K. (1989) A method for the determination of aerosol particle distributions from light extinction data. In *Abstracts of the American Association for Aerosol Research Annual Meeting*. American Association of Aerosol Research, Cincinnati, Ohio, 330–330.

Nocedal, J. (1980) Updating quasi-Newton matrices with limited storage. *Mathematics of Computation*, **95**, 339–353.

Phillips, D.L. (1962) A technique for the numerical solution of certain integral equations of the first kind. *Journal of the Association for Computing Machinery*, **9**, 84–97.

Pokrovsky, O. and Roujean, J.L. (2002) Land surface albedo retrieval via kernel-based BRDF modelling: I. statistical inversion method and model comparison. *Remote Sensing of Environment*, **84**, 100–119.

Pokrovsky, O.M. and Roujean, J.L. (2003) Land surface albedo retrieval via kernel-based BRDF modelling: II. an optimal design scheme for the angular sampling. *Remote Sens. Environ.*, **84**, 120–142.

Pokrovsky, I.O., Pokrovsky, O.M. and Roujean, J.L. (2003). Development of an operational procedure to estimate surface albedo from the SEVIRI/MSG observing system by using POLDER BRDF measurements: II. Comparison of several inversion techniques and uncertainty in albedo estimates. *Remote Sensing of Environment*, **87** (2-3), 215–242.

Privette, J.L., Eck, T.F. and Deering, D.W. (1997) Estimating spectral albedo and nadir reflectance through inversion of simple bidirectional reflectance distribution models with AVHRR/MODIS-like data. *J. Geophys. Res.*, **102**, 29529–29542.

Roujean, J.L., Leroy, M. and Deschamps, P.Y. (1992) A bidirectional reflectance model of the Earth's surface for the correction of remote sensing data. *J. Geophys. Res.*, **97**, 20455–20468.

Strahler, A.H., Li, X.W., Liang, S., Muller, J.-P., Barnsley, M.J. and Lewis, P. (1994) *MODIS BRDF/Albedo product: Algorithm technical basis document*. NASA EOS-MODIS Doc., Vol. 2.1, p. 55.

Strahler, A.H., Lucht, W., Schaaf, C.B., Tsang, T., Gao, F., Li, X., Muller, J.P., Lewis, P. and Barnsley, M.J. (1999) *MODIS BRDF/albedo product: Algorithm theoretical basis document*. NASA EOS-MODIS Doc., V5.0, 53pp. (URL: http://modarch.gsfc.nasa.gov/MODIS/LAND/#albedo-BRDF).

Tarantola, A. (1987) *Inverse Problems Theory: Methods for Data Fitting and Model Parameter Estimation*. New York, Elsevier Science.

Tikhonov, A.N. and Arsenin, V.Y. (1977) *Solutions of Ill-posed Problems*. New York, John Wiley & Sons.

Tikhonov, A.N., Goncharsky, A.V., Stepanov, V.V. and Yagola, A.G. (1995) *Numerical Methods for the Solution of Ill-Posed Problems*. Dordrecht, Kluwer.

Twomey, S. (1963) On the numerical solution of Fredholm integral equations of the first kind by the inversion of the linear system produced by quadrature. *Journal of the Association for Computing Machinery*, **10**, 97–101.

Twomey, S. (1975) Comparison of constrained linear inversion and an iterative nonlinear algorithm applied to the indirect estimation of particle size distributions. *Journal of Computational Physics*, **18**, 188–200.

Twomey, S. (1977) *Atmospheric Aerosols*. Amsterdam, Elsevier Science.

Vapnik, V.N. (1995) *The Nature of Statistical Learning Theory*. Berlin, Spring-Verlag.

Verstraete, M.M., Pinty, B. and Myneny, R.B. (1996) Potential and limitations of information extraction on the terrestrial biosphere from satellite remote sensing. *Remote Sensing Environment*, **58**, 201–214.

Voutilainenand, A. and Kaipio, J.P. (2000) Statistical inversion of aerosol size distribution data. *Journal of Aerosol Science*, **31** (Suppl.1), 767–768.

Wang, Y.F. and Xiao, T.Y. (2001) Fast realization algorithms for determinig regularization parameters in linear inverse problems. *Inverse Problems*, **17**, 281–291.

Wang, Y.F. and Yuan, Y.X. (2003) A trust region algorithm for solving distributed parameter identification problem. *Journal of Computational Mathematics*, **21**(6), 759–772.

Wang, Y.F. and Yuan, Y.X. (2005) Convergence and regularity of trust region methods for nonlinear ill-posed inverse problems. *Inverse Problems*, **21**, 821–838.

Wang, Y.F., Li, X.W., Ma, S.Q., Yang, H., Nashed, Z. and Guan, Y.N. (2005) BRDF model inversion of multiangular remote sensing: ill-posedness and interior point solution method. *Proceedings of the 9th International Symposium on Physical Measurements and Signature in Remote Sensing (ISPMSRS)*, **XXXVI**, 328–330.

Wang, Y.F., Fan, S.F., Feng, X., Yan, G.J. and Guan, Y.N. (2006a) Regularized inversion method for retrieval of aerosol particle size distribution function in $W^{1,2}$ space. *Applied Optics*, **45**, 7456–7467.

Wang, Y.F., Wen, Z., Nashed, Z. and Sun, Q. (2006b) Direct fast method for time-limited signal reconstruction. *Applied Optics*, **45**, 3111–3126.

Wang, Y.F., Li, X.W., Nashed, Z., Zhao, F., Yang, H., Guan, Y.N. and Zhang, H. (2007a) Regularized kernel-based BRDF model inversion method for ill-posed land surface parameter retrieval. *Remote Sensing of Environment*, **111**, 36–50.

Wang, Y.F., Fan, S.F. and Feng, X. (2007b) Retrieval of the Aerosol Particle Size Distribution Function by Incorporating *a Priori* Information. *Journal of Aerosol Science*, **38**, 885–901.

Wang, Y.F. (2007) *Computational Methods for Inverse Problems and Their Applications*. Beijing, Higher Education Press.

Wang, Y.F. and Ma, S.Q. (2007) Projected Barzilai–Borwein methods for large scale non-negative image restorations. *Inverse Problems in Science and Engineering*, **15**, 559–583.

Wang, Y.F., Yang, C.C. and Li, X.W. (2008) A regularizing kernel-based BRDF model inversion method for ill-posed land surface parameter retrieval using smoothness constraint. *Journal of Geophysical Research*, **113**, D13101, doi:10.1029/2007JD009324.

Wang, Y.F., Ma, S.Q., Yang, H., Wang, J.D. and Li, X.W. (2009) On the effective inversion by imposing *a priori* information for retrieval of land surface parameters. *Science in China D*, **52**(3).

Wang, Y.F. and Yang, C.C. (2008) A regularizing active set method for retrieval of atmospheric aerosol particle size distribution function. *Journal of Optical Society of America A*, **25**, 348–356.

Wang, Y.F. (2008) An efficient gradient method for maximum entropy regularizing retrieval of atmospheric aerosol particle size distribution function. *Journal of Aerosol Science*, **39**, 305–322.

Wang, Y.F. Cao, J.J., Yuan, Y.X., Yang C.C. and Xiu N.H. (2009) Regularizing active set method for nonnegatively constrained ill-posed multichannel image restoration problem, *Applied Optics*, **48**, 1389–1401.

Wanner, W., Li, X. and Strahler, A.H. (1995) On the derivation of kernels for kernel driven models of bidirectional reflectance. *J. Geophys. Res.*, **100**, 21077–21090.

Xiao, T.Y., Yu, S.G. and Wang, Y.F. (2003) *Numerical Methods for the Solution of Inverse Problems*. Beijing, Science Press.

Ye, Y.Y. (1997) *Interior Point Algorithms: Theory and Analysis*. Chichester, John Wiley & Sons.

Yuan, Y.X. (1994) Nonlinear programming: trust region algorithms. In S.T. Xiao and F. Wu (eds), *Proceedings of Chinese SIAM Annual Meeting*. Beijing, Tsinghua University Press, 83–97.

Yuan, Y.X. (1993) *Numerical Methods for Nonliear Programming*. Shanghai, Shanghai Science and Technology Publication.

Yuan, Y.X. (2001) A scaled central path for linear programming. *Journal of Comp. Math.*, **19**, 35–40.

13

Land and sea surface temperature estimation by support vector regression

Gabriele Moser and Sebastiano B. Serpico

Dept. of Biophysical and Electronic Engineering (DIBE) and Interuniversity Research Center in Environmental Monitoring (CIMA) University of Genoa, Genoa, Italy

Land and sea surface temperatures are important quantities for many hydrological and meteorological models and satellite infrared remote sensing represents an effective way to map them on global and regional scales. A supervised approach, based on support vector regression, has recently been developed to estimate surface temperature from satellite images. Such a strategy requires the user to set several internal parameters. Moreover, in order to integrate the resulting estimates into hydrological or meteorological data-assimilation schemes, a further important input is the statistics of the regression error affecting each pixel. In this chapter, we propose a method to automatically set the input parameters and two techniques to model the statistics of pixelwise regression error. Parameter optimization is achieved by numerically minimizing a generalization-error bound that can be computed by using only the training set, and error modelling is performed by integrating a Bayesian support vector regression approach with maximum-likelihood or confidence-interval supervised parameter estimators.

Some parts of this chapter have been derived from papers (Moser and Serpico 2007, 2008, 2009) listed in the References (reproduced by permission of IEEE).

Kernel Methods for Remote Sensing Data Analysis Edited by Gustavo Camps-Valls and Lorenzo Bruzzone

13.1 Introduction

Land surface temperature (LST) and sea surface temperature (SST) play crucial roles in many environmental models, such as the energy and mass exchange between the atmosphere and surface, vegetation and soil moisture, numerical weather prediction, climatic variability and global ocean circulation (Dash *et al.* 2002). Measurements made by satellite sensors are a feasible way to systematically retrieve several Earth-surface parameters, including LST and SST, from surface-leaving radiation on both a regional and global scale, because satellites capture a repetitive and consistent view of the Earth's surface at different observation scales. Many algorithms have been proposed for retrieving LST and SST from space radiometry (Haines *et al.* 2007; Jiménez-Muñoz and Sobrino 2007; Mao *et al.* 2008; Peres and DaCamara 2004a). Most of these are derived from radiative transfer theory (Tsang *et al.* 2001) and require prior information about the state of the atmosphere (e.g., temperature and water vapour profiles) and surface (e.g., emissivity); without this information, accurate surface temperature estimation becomes more difficult (Barton 1992; Chan and Gao 2005; Peres and DaCamara 2004b, 2005; Pinheiro *et al.* 2004; Rodger *et al.* 2005; Sun and Pinker 2004). Estimating these surface and atmospheric parameters is often a difficult task in itself and this problem is expected to be solved prior to (or jointly with) surface temperature estimation (Mao *et al.* 2008; Peres and DaCamara 2004b; Sun and Pinker 2004).

Using a different perspective, oriented to pattern-recognition and machine-learning and not to radiative transfer, an alternative approach to LST and SST estimation has recently been proposed in Zortea *et al.* (2006) and Moser and Serpico (2007); this approach is based on support vector regression (SVR). Support vector machines (SVMs) represent a family of powerful supervised learning techniques, and their application to the temperature estimation problem allows for computing a nonparametric approximation of the relationship between satellite data and matching *in-situ* measurements. This supervised regression strategy is theoretically proven to exhibit very good generalization and robustness (see Vapnik (1998) and Chapter 2). It has also been experimentally demonstrated that such a strategy can generate more accurate estimates, as compared with traditional techniques, but at the expense of increased complexity and computational time (Zortea 2007; Zortea *et al.* 2006). Unlike traditional methods, this SVR approach is supervised and requires *in-situ* temperature measurements for training purposes. However, it does not involve an explicit characterization of surface and/or atmospheric parameters. From this viewpoint, such an approach is complementary to traditional methods. Radiometric and/or subsurface ground temperature measurements are regularly acquired by several networks of micrometeorological stations, often for flood prevention and environmental monitoring purposes. Sea temperature measurements are acquired on a regular basis by moored and drifting buoys. In the context of remote sensing, SVR has also recently been used for estimating several other geo/biophysical parameters (Bazi and Melgani 2007; Bruzzone and Melgani 2005; Camps-Valls *et al.* 2006a,b; Durbha *et al.* 2007; Yang *et al.* 2006).

In this chapter, we will not address the applicative aspects related to LST and SST retrieval from satellite data, but we will focus on two methodological problems regarding the use of SVR to map these geophysical variables (a more application-oriented analysis can be found in Zortea (2007) and in Zortea *et al.* (2006). First, note that the SVR technique in Zortea *et al.* (2006) involves tuning several internal parameters whose values affect the resulting approximation function and must be initially set by the user – typically via 'trial-and-error' grid searches for the lowest regression error (e.g., cross-validation or leave-one-out error). This process is often time consuming because of the need to repeat the training for each node in the grid, and it is also sensitive to the (usually empirical) choice of the related ranges and discretization steps of the parameters. In particular, the time requirement for grid searches can become critical when large *in-situ* data sets are available for training purposes. This may occur, for instance, when *in-situ* stations continuously provide temperature measurements over a long time period.

In the present chapter, a method is presented that automatically sets the SVR input parameters in order to minimize estimation errors. It is applied in the context of using the SVR approach for LST and SST retrieval and it is based on minimizing an analytical upper bound for the leave-one-out regression error (namely, the so-called 'span bound'), which is known to be strongly correlated with regression errors on test data and can be computed by using only the training set (Chang and Lin 2005). Powell's method is the algorithm chosen to address the minimization problem because it is applicable to non-differentiable functions (Press *et al.* 2002), such as the considered span bound. This removes the need to introduce regularized versions of the functional, which may also involve manually setting more parameters.

Furthermore, in order to integrate temperature estimates effectively into data-assimilation schemes for hydrological or meteorological modelling (e.g., flood-prevention applications), another critical piece of information that needs to be input is often the statistics of the temperature regression error. Estimates of the root mean square error (RMSE) or the variance of the regression error are required input data for both variational (Caparrini and Castelli 2003) and filtering-based (e.g., Kalman-like (Zhou *et al.* 2006)) assimilation techniques. However, classical error-assessment procedures, such as cross-validation or leave-one-out sampling, provide only global information; they characterize each map of LST/SST estimates by a single numerical error index (e.g., an RMSE value) without providing a pointwise characterization of the error in each pixel (Lin and Weng 2004; Vapnik 1998).

Here, two methods are introduced to optimize statistical modelling of the SVR error for each sample. The key idea of these techniques is to generalize the Bayesian method in (Gao *et al.* 2002) by integrating more accurate approaches to model the statistics of the error contribution related to intrinsic data variability, such as noise. The two algorithms introduced in this framework are based on: (1) a confidence-interval approach and (2) a case-specific formulation of the maximum likelihood (ML) technique, tailored to the non-stationary pixelwise distribution of the SVR error.

The chapter is organized as follows. Previous work on LST and SST estimation and on the two considered methodological problems is summarized in Section 13.2. The methods developed to address these problems are described in Section 13.3. Experimental results on MSG-SEVIRI and AVHRR data are presented in Section 13.4, and conclusions are drawn in Section 13.5.

13.2 Previous work

13.2.1 LST and SST estimation from satellite data

LST and SST are intuitively meant to represent the skin-temperature values measured at the surface–atmosphere interface (Dash *et al.* 2002). However, these quantities can rapidly vary with spatial position on the Earth's surface and are, in general, inhomogeneous inside the instantaneous field of view (IFOV) of a given satellite sensor (especially in the case of a low-resolution sensor). Therefore, from a remote-sensing perspective, LST and SST more precisely represent the surface radiometric temperatures corresponding to the IFOV of the sensor and resulting from the integrated effect of the ensemble of (possibly thermally inhomogeneous) bodies within the IFOV (Dash *et al.* 2002; Merchant and Borgne 2004). Their estimation from satellite remote sensing is based on both the impact on thermal infrared (TIR) passive measurements of the temperature of the observed surface according to Planck's law of black-body radiation, and the behavior of radiation in the atmosphere as predicted by the radiative transfer theory (Tsang *et al.* 2001). Furthermore, given the measured top-of-the-atmosphere (TOA) spectral radiance, a direct inversion of Planck's law yields only the so-called brightness temperature that is not, by itself, a correct LST or SST estimate since it does not take into account the effects of surface emissivity and of the transmittance along the atmospheric path between the surface and the satellite. In the infrared range commonly used for temperature retrieval (8–13 μm), this atmospheric effect is mainly due to the water-vapour and temperature vertical profiles (Dash *et al.* 2002). Traditional LST estimation methods assume the surface emissivity as an input datum derived from prior knowledge about the land cover (Pozo Vázquez *et al.* 1997) or estimated on the basis of satellite maps of the vegetation properties (e.g., through vegetation indices (Peres and DaCamara 2005)).

Single-channel approaches to LST or SST estimation are based on a single TIR channel whose brightness temperature is corrected for the atmospheric effect to compute an estimate of the desired surface temperature (Dash *et al.* 2002). An advantage of these approaches is that they are feasible for sensors endowed with just one TIR channel. On the other hand, they have quite strict requirements in terms of the accuracy of the input atmospheric temperature and the water-vapour vertical profiles. These profiles can be obtained by vertical sounding satellite instruments (e.g., TIROS) or from radiosondes, but the application of a single-channel LST estimation method to a TIR image may be critical when synchronous and co-located vertical profiles are not available or are affected by large uncertainties.

The split-window techniques (SWTs) are based on observations in multiple TIR channels and exploit the differential absorption between such channels in order to compensate for atmospheric effects. Basic SWT formulations express the estimate as a linear combination of the brightness temperatures in the TIR channels (Dash *et al.* 2002). More advanced SWTs also incorporate emissivity and water-vapour content estimates as further input data in quadratic formulations (Pozo Vázquez *et al.* 1997; Sun and Pinker 2004; Yu *et al.* 2008). The parameters of SWTs are typically calibrated (once and for all) for a given sensor (e.g., NOAA-AVHRR, MODIS and MSG-SEVIRI) by applying a least-squares fitting to predefined sets of values of surface and atmospheric data and the resulting simulated TOA brightness temperatures (Jiménez-Muñoz and Sobrino 2007; Li *et al.* 2001; Sun and Pinker 2004; Yu *et al.* 2008). These methods can then be applied to a satellite image in an unsupervised way because there is no need for training *in-situ* temperature data despite the fact that the above-mentioned surface and atmospheric data for the related geographical area are still required.

Further approaches that have been proposed for LST retrieval are the multiangle methods, which jointly exploit TIR measurements collected with different viewing angles (Dash *et al.* 2002), the two-temperature methods, which use TIR observations at different times to separate the impacts of emissivity and LST on the TOA radiance (Peres and DaCamara 2004a,b), or the combination of radiative transfer modelling and neural networks to jointly estimate LST and emissivity (Mao *et al.* 2008). In the case of SST, compositing methods have been proposed that combine SST estimates obtained from multiple swaths in order to increase the spatial coverage and to reduce the impact of possible cloud cover (Haines *et al.* 2007; Li *et al.* 2001) and multivariate regression have been used in Tandeo *et al.* (2009) to compensate for the effect of the solar zenith angle (and of other possible variables, such as wind speed or aerosol optical depth) on the SST estimation accuracy.

Thorough reviews of LST and SST estimation problems from satellite data can be found in Dash *et al.* (2002), Merchant and Borgne (2004) and Yu *et al.* (2008).

13.2.2 Parameter optimization and error modelling for SVR

Focusing first on automating parameter optimization for support vector techniques, we remark that this problem has been mainly addressed in the context of support vector classification (SVC) and formalized as the numerical minimization of upper bounds on the generalization error of the classifier (Chapelle *et al.* 2002). In Chapelle *et al.* (2002), several bounds (e.g., the 'radius-margin bound', RMB, and the 'span bound', SB), are introduced or revised, focusing on a SVC using a quadratic penalty on the slack variables (Vapnik 1998). Gradient-like minimization procedures are also developed for such bounds, provided that they are differentiable. This approach is extended in (Chung *et al.* 2003) to SVC using a linear penalty on the slack variables (Vapnik 1998) by combining a heuristic generalization of RMB and a quasi-Newton minimization algorithm. In Duan *et al.* (2003) several error bounds (i.e., the '$\xi\alpha$ bound' (Joachims 2000), the 'generalized approximate cross-validation' (Wahba *et al.* 1999), the 'approximate span bound' (Vapnik and Chapelle 1999) the 'Vapnik–Chervonenkis bound' (Burges 1998) and the above mentioned RMB) are revised and experimentally compared in the context of SVC with both linear and quadratic penalties on the slack variables. In Bartlett and Mendelson (2002) quantitative bounds on the error probability of general families of discriminant functions are formulated in terms of classification errors on the training samples and an analytical measure of the complexity of the classifier (i.e., the Rademacher complexity).

With regard to SVR, semiheuristic approaches to parameter setting have been proposed in Cherkassky and Ma (2003) and Kwok and Tsang (2003) that relate the parameters to the noise and input data statistics. Bayesian approaches to parameter estimation have been proposed in Chu *et al.* (2004) and Gao *et al.* (2002, 2003a,b), even though they focus on case-specific (e.g., non-ϵ-insensitive) SVR architectures. In Chang and Lin (2005), an approach based on gradient-like minimization of error bounds is extended to SVR: two parameter-optimization methods are developed by generalizing the above mentioned RMB and SB to regression and combining them with quasi-Newton procedures while employing SVR with quadratic penalty on the slack variables. In this approach, the RMB regression formulation exhibits a low correlation with actual regression errors on test samples, and the SB expression for regression is a non-differentiable function of the unknown parameters. To overcome the latter issue, a modified differentiable version of SB is used, even though this requires the introduction of another manually set smoothing parameter (Chang and Lin 2005). The Rademacher complexity bound holds in the case of regression as well, and a closed-form analytical formulation of the bound is available for a kernel-based functional approximator (such as those given by SVR), even though it does not have a bias term (Bartlett and Mendelson 2002; Shawe-Taylor *et al.* 2005).

We note that modelling the statistics of pointwise SVR error has been explored only recently in the literature. This may be related to the fact that pointwise estimates of regression uncertainty or confidence intervals are almost natural by-products of Bayesian approaches to regression, but SVR is an intrinsically nonbayesian technique. However, it has been proven in Gao *et al.* (2002) and Law and Kwok (2001) that SVR can be reformulated as a Bayesian 'maximum-*a-posteriori*' (MAP) regression with respect to suitably defined conditional likelihood and prior distributions. According to this formulation, the regression error on each sample can be expressed as the sum of two independent stochastic processes related to the SVR functional approximator and to the intrinsic uncertainty in the input data, respectively (Gao *et al.* 2002). A critical drawback of this method is that the statistics of the latter contribution are modelled only as a function of the SVR input parameters (Gao *et al.* 2002) whose values are optimized in order to generate accurate estimates and may not, in general, effectively characterize the input data distribution. Bayesian SVR has also been combined with an *ad hoc* (non-ϵ-insensitive) loss function (Chu *et al.* 2004) and mean-field-theoretic (Gao *et al.* 2003a) and variational (Gao *et al.* 2003b) formulations. The criticality remarked for (Gao *et al.* 2002) also holds for such methods.

13.3 Methodology

13.3.1 SVR for LST and SST estimation

Assuming the existence of a possibly complicated or unknown relationship between the surface temperature of a land or sea geographical area and the spectral radiance measured by a satellite sensor over this area, evaluating the surface temperature would require the inversion of this relation. From a learning-oriented perspective, this suggests using regression methods. Accordingly, a supervised estimation scheme based on SVR has been adopted and applied to passive sensor data in (Zortea *et al.* 2006) and (Moser and Serpico 2007).

Let $\mathbf{x}$ ($\mathbf{x} \in \mathbb{R}^d$) be a d-dimensional feature vector extracted from the acquired sensor data for a given pixel, y ($y \in \mathbb{R}$) be the corresponding Earth-surface physical quantity to be estimated (either LST or SST), and $f : \mathbb{R}^d \to \mathbb{R}$ be the unknown function relating the features with this physical quantity. SVR computes an estimate $\hat{f}$ by minimizing an upper bound on the probability that the regression error may be above a given threshold (Vapnik 1998). More precisely, given a set $\mathcal{D} = \{(\mathbf{x}_1, y_1), (\mathbf{x}_2, y_2), \ldots, (\mathbf{x}_\ell, y_\ell)\}$ of ℓ training samples, where $\mathbf{x}_i$ is the feature vector for the ith training sample and y_i is the corresponding *in-situ* measurement ($i = 1, 2, \ldots, \ell$), the resulting approximation is expressed as a linear combination of appropriate kernel functions centred on a subset of training samples (Vapnik 1998):

$$\hat{f}(\mathbf{x}) = \sum_{i \in \mathcal{S}} \beta_i^* K(\mathbf{x}_i, \mathbf{x}|\boldsymbol{\gamma}) + b^*, \tag{13.1}$$

where $\beta_1^*, \beta_2^*, \ldots, \beta_\ell^*$ are the weight coefficients of the linear combination, $K(\cdot, \cdot|\boldsymbol{\gamma})$ is a kernel function that is generally parameterized by a vector $\boldsymbol{\gamma}$ of parameters ($\boldsymbol{\gamma} \in \mathbb{R}^r$), b^* is a bias term, and $\mathcal{S} = \{i : \beta_i^* \neq 0\}$ (Vapnik 1998). If $i \in \mathcal{S}$ (i.e., $\beta_i^* \neq 0$) the training sample $(\mathbf{x}_i, y_i)$ is a support vector (SV). We assume that the vector $\boldsymbol{\gamma}$ of the internal parameters of the kernel is defined so that it can take on values in the whole space $\mathbb{R}^r$. The vector $\boldsymbol{\beta}^* = [\beta_1^*, \beta_2^*, \ldots, \beta_\ell^*]^\mathsf{T}$ of the coefficients characterizing the kernel expansion in Equation (13.1) is obtained by solving the following quadratic programming (QP) problem (Vapnik 1998):

$$\begin{cases} \min\limits_{\boldsymbol{\beta} \in \mathbb{R}^\ell} \left[\dfrac{1}{2} \boldsymbol{\beta}^\mathsf{T} \mathbf{K}(\boldsymbol{\gamma}, \mathcal{D}) \boldsymbol{\beta} - \mathbf{y}(\mathcal{D})^\mathsf{T} \boldsymbol{\beta} + \epsilon \|\boldsymbol{\beta}\|_1 \right] \\ \mathbf{1}^\mathsf{T} \boldsymbol{\beta} = 0, \quad \|\boldsymbol{\beta}\|_\infty \leq C, \end{cases} \tag{13.2}$$

where C and ϵ are SVR parameters, $\mathbf{K}(\boldsymbol{\gamma}, \mathcal{D})$ is the $\ell \times \ell$ matrix whose (i, j)th entry is given by $K(\mathbf{x}_i, \mathbf{x}_j|\boldsymbol{\gamma})$ $(i, j = 1, 2, \ldots, \ell)$, $\mathbf{y}(\mathcal{D}) = [y_1, y_2, \ldots, y_\ell]^\mathsf{T}$ is the vector of the ℓ *in-situ* measurements, and $\mathbf{1}$ is a vector with unitary components. Note that, unlike Chang and Lin (2005), we use SVR with linear penalty on the slack variables, since this usually ensures that the resulting functional approximator is more sparse, limiting the risk of overfitting and reducing the time required to generate maps of LST and SST estimates compared with using a quadratic penalty (Vapnik 1998).

The method's input parameters are C, ϵ and $\boldsymbol{\gamma}$. The parameter C $(C > 0)$ tunes the trade-off between the generalization capability of the functional approximator and the accuracy of the fitting on the training set: a large value of C allows a high regression accuracy on the training set but may encourage overfitting, thus causing high sensitivity to both noise and possible errors in the training data. A small value of C aims at optimizing the smoothness and generalization of the approximator, but it may result in a poorly fitted training set, causing a reduced accuracy of the approximation of the estimated function. If $0 < |\beta_i^*| < C$, the SV $(\mathbf{x}_i, y_i)$ is called an 'unbounded support vector' (USV) or 'free support vector' $(i = 1, 2, \ldots, \ell)$; $\mathcal{U} = \{i : 0 < |\beta_i^*| < C\}$ identifies the set of USVs. On the other hand, ϵ $(\epsilon > 0)$ represents the largest (absolute) error that is considered acceptable and not penalized by SVR (Vapnik 1998). The vector containing the SVR input parameters can be defined as $\boldsymbol{\theta} = (\log C, \log \epsilon, \boldsymbol{\gamma}) \in \mathbb{R}^m$, where $m = r + 2$ is the number of parameters, and the logarithm is applied to C and ϵ in order to let $\boldsymbol{\theta}$ take on values in the whole space $\mathbb{R}^m$, without an explicit positivity constraint.

Given the training set $\mathcal{D}$ and the parameter vector $\boldsymbol{\theta}$, the training consists of solving the QP problem in Equation (13.2). We stress the dependence on $\mathcal{D}$ and $\boldsymbol{\theta}$ by denoting as $\boldsymbol{\beta}^*(\boldsymbol{\theta}, \mathcal{D})$, $\mathcal{S}(\theta, \mathcal{D})$, $\mathcal{U}(\boldsymbol{\theta}, D)$ and $\hat{f}(\cdot|\boldsymbol{\theta}, \mathcal{D})$ the resulting weight vector of the linear combination, the related index sets of SVs and USVs, and the SVR functional approximator, respectively.

13.3.2 Automatic parameter optimization for SVR

The method presented here for automatically optimizing SVR parameters is based on searching for a parameter vector that minimizes an upper bound on the regression error. Specifically, one can prove under mild assumptions that the average 'leave-one-out' SVR error can be upper bounded by the following 'span-bound' (SB) functional (Chang and Lin 2005):

$$\mathcal{B}(\boldsymbol{\theta}|\mathcal{D}) = \epsilon + \frac{1}{\ell} \sum_{i \in \mathcal{S}(\boldsymbol{\theta}, \mathcal{D})} |\beta_i^*(\boldsymbol{\theta}, \mathcal{D})| s_i^2(\boldsymbol{\theta}, \mathcal{D}) +$$

$$+ \frac{\mathbf{y}(\mathcal{D})^\mathsf{T} \boldsymbol{\beta}^*(\boldsymbol{\theta}, \mathcal{D}) - \epsilon \|\boldsymbol{\beta}^*(\boldsymbol{\theta}, \mathcal{D})\|_1 - \boldsymbol{\beta}^*(\boldsymbol{\theta}, \mathcal{D})^\mathsf{T} \mathbf{K}(\boldsymbol{\gamma}, \mathcal{D}) \boldsymbol{\beta}^*(\boldsymbol{\theta}, \mathcal{D})}{C\ell}. \tag{13.3}$$

Here, $s_i(\boldsymbol{\theta}, \mathcal{D})$, named the 'span' of the SV $(\mathbf{x}_i, y_i)$, is a real coefficient computed by solving the following QP problem $(i \in \mathcal{S}(\boldsymbol{\theta}, \mathcal{D}))$:

$$s_i^2(\boldsymbol{\theta}, \mathcal{D}) = \min \left\{ \boldsymbol{\lambda}^\mathsf{T} \mathbf{K}_u(\theta, \mathcal{D}) \boldsymbol{\lambda} - 2\mathbf{K}_{ux}(\mathbf{x}_i|\boldsymbol{\theta}, \mathcal{D})^\mathsf{T} \boldsymbol{\lambda} + K(\mathbf{x}_i, \mathbf{x}_i|\boldsymbol{\gamma}) : \mathbf{1}^\mathsf{T} \boldsymbol{\lambda} = 1 \right\}, \tag{13.4}$$

where, denoting by $\ell_u(\boldsymbol{\theta}, \mathcal{D})$ the number of USVs, $\mathbf{K}_u(\boldsymbol{\theta}, \mathcal{D})$ is the $\ell_u(\boldsymbol{\theta}, \mathcal{D}) \times \ell_u(\boldsymbol{\theta}, \mathcal{D})$ matrix whose (j, h)th entry is given by $K(\mathbf{x}_j, \mathbf{x}_h|\boldsymbol{\gamma})$ $(j, h \in \mathcal{U}(\boldsymbol{\theta}, \mathcal{D}))$ and $\mathbf{K}_{ux}(\mathbf{x}|\boldsymbol{\theta}, \mathcal{D})$ $(\mathbf{x} \in \mathbb{R}^d)$ is the $\ell_u(\boldsymbol{\theta}, \mathcal{D})$-dimensional vector whose generic element is $K(\mathbf{x}, \mathbf{x}_j|\boldsymbol{\gamma})$ $(j \in \mathcal{U}(\boldsymbol{\theta}, \mathcal{D}))$.

Hence, given $\mathcal{D}$ and $\boldsymbol{\theta}$, the corresponding value $\mathcal{B}(\boldsymbol{\theta}|\mathcal{D})$ can be analytically computed by first solving the QP problem in Equation (13.2) – which yields the coefficients $\beta_i^*(\boldsymbol{\theta}, \mathcal{D})$ ($i = 1, 2, \ldots, \ell$) and identifies SVs and USVs – and then solving the QP problem in Equation (13.4) for each SV. Even though SB can be computed with only the training samples, it is known to be a very tight bound and it is strongly correlated with the regression error on test samples provided that the test and training samples are drawn from the same probability distribution (Chang and Lin 2005). A parameter vector $\boldsymbol{\theta}^* \in \mathbb{R}^m$ that minimizes $\mathcal{B}(\boldsymbol{\theta}|\mathcal{D})$ is searched for in $\mathbb{R}^m$. However, in general, $\mathcal{B}(\cdot|\mathcal{D})$ is a nondifferentiable function; this prevents the use of gradient-based minimization algorithms and requires methods that need only evaluate the objective function. Powell's method is adopted here, which is an iterative unconstrained minimization technique that does not require possible derivatives of the objective function.

The key idea of Powell's method is to emulate, without using gradient and Hessian information, the behavior of the conjugate gradient technique (which, by itself, is applicable to twice continuously differentiable functions) by identifying a set of m vectors in $\mathbb{R}^m$ that are almost conjugate. The minimum of the functional is computed by sequentially performing a line search[1] at each iteration for each 'pseudo-conjugate' direction (Powell 1964). Specifically, let $\boldsymbol{\theta}^t$ be the parameter vector computed at the tth iteration of the method, $\mathbf{u}^{i,t}$ be the ith pseudo-conjugate direction computed at the tth iteration, and $\bar{\boldsymbol{\theta}}^{i,t}$ be the parameter vector computed after the ith sequential line search of the tth iteration ($t = 0, 1, 2, \ldots; i = 1, 2, \ldots, m$). The method is initialized with a given parameter vector $\boldsymbol{\theta}^0$ and with the assumption that the initial directions $\mathbf{u}^{1,0}, \mathbf{u}^{2,0}, \ldots, \mathbf{u}^{m,0}$ are the canonical basis vectors in $\mathbb{R}^m$ (i.e., $\mathbf{u}^{i,0}$ has a unitary ith component, while all the other components are zero; $i = 1, 2, \ldots, m$). Then, for the tth iteration, the following operations are performed ($t = 0, 1, 2, \ldots$) (Powell 1964; Press *et al.* 2002):

1. set $\bar{\boldsymbol{\theta}}^{0,t} = \boldsymbol{\theta}^t$;

2. for each ith pseudo-conjugate direction $\mathbf{u}^{i,t}$, start from the current point $\bar{\boldsymbol{\theta}}^{i-1,t}$ and minimize $\mathcal{B}(\boldsymbol{\theta}|\mathcal{D})$ by a line search in the direction $\mathbf{u}^{i,t}$, by computing the scalar:

$$\rho^{i,t} = \arg\min_{\rho \in \mathbb{R}} \mathcal{B}(\bar{\boldsymbol{\theta}}^{i-1,t} + \rho \mathbf{u}^{i,t}|\mathcal{D}); \tag{13.5}$$

   then move to the resulting minimum point $\bar{\boldsymbol{\theta}}^{i,t} = \bar{\boldsymbol{\theta}}^{i-1,t} + \rho^{i,t}\mathbf{u}^{i,t}$ ($i = 1, 2, \ldots, m$);

3. update the parameter vector as the last point reached in the parameter space by the sequence of m line searches performed in step 2, i.e., set $\boldsymbol{\theta}^{t+1} = \bar{\boldsymbol{\theta}}^{m,t}$;

4. update the search directions by dropping the first direction $\mathbf{u}^{1,t}$ and introducing a new direction $(\boldsymbol{\theta}^{t+1} - \boldsymbol{\theta}^t)$, i.e., set $\mathbf{u}^{i,t+1} = \mathbf{u}^{i+1,t}$ for $i = 1, 2, \ldots, m-1$ and $\mathbf{u}^{m,t+1} = \boldsymbol{\theta}^{t+1} - \boldsymbol{\theta}^t$.

The method is iterated until the relative decrease in the functional between successive iterations falls below a given threshold, which is very close to the machine precision. This method is analytically proven to quadratically converge, under mild assumptions, at least to a local minimum (Powell 1964; Press *et al.* 2002). One drawback is that the pseudo-conjugate directions

[1] By 'line search' we mean the search for the minimum of a given function of m variables along a straight line in the related m-dimensional space.

may become linearly dependent during the iterations, thereby implicitly restricting the search space to a lower-dimensional subspace (Brent 1973). A simple strategy for resolving this problem while keeping quadratic convergence consists of reinitializing the pseudo-conjugate directions to the canonical basis of $\mathbb{R}^m$ after each cycle of $(m + 1)$ successive iterations (Press *et al.* 2002). Alternate solutions to this problem, either keeping or giving up the quadratic convergence property, are discussed in (Brent 1973) and (Press *et al.* 2002).

A technique that does not involve derivatives also has to be used for the line search in step 2. To achieve this end, Brent's method is chosen, which is based on the iterative combination of the 'inverse parabolic interpolation' and 'golden-section' approaches and is applicable to the minimization of single-variable, continuous (and possibly nondifferentiable) functions (Brent 1973). Given such a function, the golden-section method iteratively computes a sequence of bracketing triplets[2] that converges to a local minimum of the function (Press *et al.* 2002). The inverse parabolic interpolation method iteratively minimizes a parabolic interpolation of the input (single-variable) function computed according to the evaluations of the function at three distinct points. If the function is monomodal, this method is proven to converge to the desired local minimum point. Moreover, it typically converges faster than the golden-section approach. However, if the monomodality assumption fails, it may not converge (Press *et al.* 2002). Brent's method is based on the idea of iteratively exploiting the inverse parabolic interpolation approach when close to a monomodal valley of the function being minimized, while using the golden-section technique otherwise. Basically, during each iteration a parabolic interpolation step is attempted and a series of tests is performed to discriminate if the resulting solution is acceptable or if a golden-section step is required. More details about the resulting numerical procedure, which is not straightforward, can be found in Brent (1973) and in Press *et al.* (2002). The described automatic parameter-setting method will henceforth be referred to as the 'Powell-span bound' (PSB) technique.

13.3.3 Pointwise statistical modelling the SVR error

The problem of estimating the pixelwise probability distribution of the SVR error is addressed by extending the Bayesian approach by Gao, Gunn, and Harris (GGH for short) (Gao *et al.* 2002). Let the parameter vector $\boldsymbol{\theta}$ be fixed to a given value, for instance, by applying PSB. The GGH framework reformulates SVR in terms of a Bayesian MAP regression. Specifically, $f(\mathbf{x})$ $(\mathbf{x} \in \mathbb{R}^d)$ is formalized as a d-dimensional zero-mean Gaussian process and the relationship between the data sample $\mathbf{x}$ and the corresponding variable y is modelled by the process $f(\mathbf{x})$ up to a zero-mean additive noise-like stationary component $n(\mathbf{x})$, independent of $f(\mathbf{x})$ (Gao *et al.* 2002):

$$y = f(\mathbf{x}) + n(\mathbf{x}). \tag{13.6}$$

As a consequence, the SVR error $\eta(\mathbf{x}) = y - \hat{f}(\mathbf{x}|\boldsymbol{\theta}, \mathcal{D})$ in the generic sample $(\mathbf{x}, y) \in \mathbb{R}^{d+1}$ is decomposed as the sum of two independent stochastic processes:

$$\eta(\mathbf{x}) = s(\mathbf{x}) + n(\mathbf{x}), \tag{13.7}$$

[2] Recall that given a continuous function $g : \mathbb{R} \to \mathbb{R}$, a 'bracketing triplet' is a triplet $(a, b, c) \in \mathbb{R}^3$ such that $a < b < c$ and $g(b) \leq \min\{g(a), g(c)\}$. Thanks to the continuity of g, this condition ensures that a local minimum point of g exists in (a, c).

where $s(\mathbf{x}) = f(\mathbf{x}) - \hat{f}(\mathbf{x}|\boldsymbol{\theta}, \mathcal{D})$ is a non-stationary process formalizing the error contribution due to the SVR functional approximation and $n(\mathbf{x})$ represents the error contribution related to intrinsic data variability (e.g., noise) (Gao *et al.* 2002). When conditioned to the training set $\mathcal{D}$, $s(\mathbf{x})$ is shown to be approximately zero-mean Gaussian $\mathcal{N}[0, \sigma_s^2(\mathbf{x}|\boldsymbol{\theta}, \mathcal{D})]$ with a variance:

$$\sigma_s^2(\mathbf{x}|\boldsymbol{\theta}, \mathcal{D}) = K(\mathbf{x}, \mathbf{x}|\boldsymbol{\gamma}) - \mathbf{K}_{ux}(\mathbf{x}|\boldsymbol{\theta}, \mathcal{D})^{\mathsf{T}} \mathbf{K}_u(\boldsymbol{\theta}, \mathcal{D})^{-1} \mathbf{K}_{ux}(\mathbf{x}|\boldsymbol{\theta}, \mathcal{D}). \tag{13.8}$$

On the other hand, $n(\mathbf{x})$ is a zero-mean process, independent of $\mathcal{D}$, with variance:

$$\sigma_n^2(\boldsymbol{\theta}) = \epsilon^2 \left[\frac{2}{(\epsilon C)^2} + \frac{\epsilon C + 3}{3(\epsilon C + 1)} \right]. \tag{13.9}$$

Thus, given $\mathcal{D}$, the regression error $\eta(\mathbf{x})$ is zero-mean with variance equal to:

$$\sigma_\eta^2(\mathbf{x}|\boldsymbol{\theta}, \mathcal{D}) = \sigma_s^2(\mathbf{x}|\boldsymbol{\theta}, \mathcal{D}) + \sigma_n^2(\boldsymbol{\theta}). \tag{13.10}$$

Note that this model was developed in Gao *et al.* (2002) by assuming that $f(\mathbf{x})$ is a zero-mean process and by focusing on SVR without the bias term b^* in Equation (13.1). These restrictions can be partially removed by assuming that $f(\mathbf{x})$ has a constant mean independent of $\mathbf{x}$ – but not necessarily zero – and by assuming that b^* is equal to this mean value. Under these working hypotheses (which we assume hold in order to ensure the analytical tractability of the problem), it is straightforward to replicate the arguments given in Gao *et al.* (2002) to prove Equations (13.7)–(13.9).

Thanks to Equations (13.8)–(13.10), GGH provides a closed-form expression for the point-wise SVR error variance as a function of (C, ϵ) and of the set of USVs. Since the USVs are selected by the training process, this approach allows SVR to analytically compute an estimate of the standard deviation of the regression error occurring in each sample. However, a drawback is that the stationary contribution σ_n is expressed as an algebraic function of C and ϵ, whose values are usually optimized in order to minimize a global measure of the regression error (see Sections 13.2 and 13.3.2). The resulting optimal values for C and ϵ are not chosen to model the data variability accurately. This may yield large errors in the estimated standard deviation values. This criticality is also emphasized by the fact that σ_n is an unbounded function of (C, ϵ) and diverges when the product (ϵC) vanishes (see Equation (13.9)).

The methods presented here aim to remove this drawback and extend GGH by integrating it with more accurate estimators of the variance contribution that depends on data uncertainty. The key idea is to keep the decomposition of the regression error in Equation (13.7), while modelling $n(\mathbf{x})$ as a generic $\mathcal{N}(0, \sigma_n^2)$ stationary process and estimating σ_n by a supervised approach. A Gaussian assumption for $n(\mathbf{x})$ is usually accepted for the data noise in the case of passive sensors. Two estimation methods are introduced in this framework.

The first technique is a formulation of the ML approach for the non-stationary error distribution of the training samples. Thanks to the Gaussian assumption for $n(\mathbf{x})$, the distribution of the error $\eta_i = y_i - \hat{f}(\mathbf{x}_i|\boldsymbol{\theta}, \mathcal{D})$ on each ith training sample, when conditioned to $\mathcal{D}$, is Gaussian $\mathcal{N}[0, \sigma_s^2(\mathbf{x}_i|\boldsymbol{\theta}, \mathcal{D}) + \sigma_n^2]$ $(i = 1, 2, \ldots, \ell)$. Assuming $\eta_1, \eta_2, \ldots, \eta_\ell$ are independent (which is a simplifying assumption usually accepted in non-contextual image analysis), the log-likelihood of the vector $\boldsymbol{\eta} = [\eta_1, \eta_2, \ldots, \eta_\ell]^{\mathsf{T}}$ of the errors in the training samples is (up to additive or multiplicative constants):

$$\mathcal{L}(\sigma_n|\boldsymbol{\theta},\mathcal{D}) = -\sum_{i=1}^{\ell}\left\{\frac{\eta_i^2}{\sigma_s^2(\mathbf{x}_i|\boldsymbol{\theta},\mathcal{D})+\sigma_n^2} + \log\left[\sigma_s^2(\mathbf{x}_i|\boldsymbol{\theta},\mathcal{D})+\sigma_n^2\right]\right\}. \tag{13.11}$$

The numerical maximization of this log-likelihood function yields the estimate $\hat{\sigma}_n$ of σ_n computed by the first method. Since $\mathcal{L}(\cdot|\boldsymbol{\theta},\mathcal{D})$ is twice continuously differentiable, the Newton–Raphson algorithm is applied to this maximization task (Press *et al.* 2002). The resulting technique is called 'Maximum-Likelihood Generalized GGH' (ML-GGGH).

The second method computes a value of σ_n that yields an optimal fit between the model-based probability that the regression error lies within given confidence intervals and the empirical percentages of the training-set errors in the same intervals. Specifically, for a generic value of σ_n, for any $p \in [0, 1]$, and for the ith training sample, the non-negative number $\alpha_i(p, \sigma_n|\boldsymbol{\theta},\mathcal{D})$ given by:

$$P\{|\eta_i| \leq \alpha_i(p,\sigma_n|\boldsymbol{\theta},\mathcal{D})|\boldsymbol{\theta},\mathcal{D}\} = p \tag{13.12}$$

can be computed analytically according to the modelled Gaussian distribution $\mathcal{N}[0, \sigma_s^2(\mathbf{x}_i|\boldsymbol{\theta},\mathcal{D})+\sigma_n^2]$ of η_i, conditioned to $\mathcal{D}$ $(i = 1, 2, \ldots, \ell)$. Then, $\alpha_i(p,\sigma_n|\boldsymbol{\theta},\mathcal{D})$ is the unique solution of the equation:

$$\operatorname{erf}\left[\frac{\alpha_i(p,\sigma_n|\boldsymbol{\theta},\mathcal{D})}{\sqrt{2\sigma_s^2(\mathbf{x}_i|\boldsymbol{\theta},\mathcal{D})+2\sigma_n^2}}\right] = p, \tag{13.13}$$

where erf$(\cdot)$ is the well-known error function (Press *et al.* 2002). For the model $\mathcal{N}[0, \sigma_s^2(\mathbf{x}_i|\boldsymbol{\theta},\mathcal{D})+\sigma_n^2]$ to accurately fit the observed error distribution of the training-set, it is desired that, for any $p \in [0, 1]$ the fraction $r(p,\sigma_n|\boldsymbol{\theta},\mathcal{D})$ of training samples $(\mathbf{x}_i, y_i)$ such that $|\eta_i| \leq \alpha_i(p,\sigma_n|\boldsymbol{\theta},\mathcal{D})$, is close to p. The second technique aims at ensuring this property by computing the value $\hat{\sigma}_n$ of σ_n that minimizes the following average distance between p and $r(p,\sigma_n|\boldsymbol{\theta},\mathcal{D})$:

$$\mathcal{J}(\sigma_n|\boldsymbol{\theta},\mathcal{D}) = \int_0^1 |r(p,\sigma_n|\boldsymbol{\theta},\mathcal{D}) - p|\,\mathrm{d}p. \tag{13.14}$$

Since $r(\cdot,\cdot|\boldsymbol{\theta},\mathcal{D})$ is, in general, a discontinuous function, the Brent's algorithm is chosen again to address the numerical minimization of $\mathcal{J}(\cdot|\boldsymbol{\theta},\mathcal{D})$. The resulting method is called 'confidence-interval generalized GGH' (CI-GGGH).

13.4 Experimental results

13.4.1 Data sets and experimental set-up

The methods described in Section 13.3 were experimentally validated for both LST and SST retrieval from satellite data. The LST data sets are composed of images acquired over Italy between 7:00 a.m. and 6:00 p.m. local time by the SEVIRI instrument on board Meteosat Second Generation (MSG) between August 12 2005, and September 19 2005. These data were received by the station at the Interuniversity Research Centre in Environmental Monitoring (CIMA) in Savona (Italy). The corresponding *in-situ* data are radiometric ground temperature measurements acquired by thirteen micrometeorological stations from the ARSIA ('*Agenzia regionale per lo sviluppo e l'innovazione nel settore agricolo e forestale*') network in the Italian Tuscany region. The samples acquired by seven of the thirteen stations were used for training purposes; the samples acquired by the other six stations were employed for testing. The resulting training and test sets were spatially disjoint. Samples affected by cloud cover were removed from the data set.

The experiments were separately performed during four different time slots (7–8 a.m., 9–12 a.m., 1–4 p.m. and 5–6 p.m.) in order to take into account the different relationships between LST and satellite observations at different times of the day. Accordingly, four LST data sets were established (one per time slot). To apply SVR, three features were considered for each sample: the measured brightness temperatures of the 10.82 μm SEVIRI channel; the difference between the brightness temperatures in the 10.82 μm and 12 μm channels; and the 'normalized difference vegetation index' (NDVI). One of the two brightness temperatures and the difference between them were used instead of two brightness temperatures because the difference feature is known to be important for temperature estimation purposes, thanks to its correlation with atmospheric effects on sensor radiance (Sobrino *et al.* 2004).

For the SST case, a data set was constructed from the publicly available multi-year multi-satellite 'AVHRR Pathfinder Matchup Database' of approximately co-temporal and co-located *in-situ* SSTs (acquired from both moored and drifting buoys) and AVHRR images. Details on this database can be found in Evans and Podestá (1998). Samples collected on two days (January 1 and 2, 1999) from globally distributed buoys, were used in the experiments. As in Evans and Podestá (1998) and Li *et al.* (2001), four features were considered for each sample: the secant of the satellite zenith angle, the measured brightness temperatures of AVHRR channels 4 and 3B, and the difference between the brightness temperatures of channels 4 and 5. The mid-wave infrared channel-3B brightness temperature was used for its correlation with night-time SST (Li *et al.* 2001) and because the samples in this data set were collected during two entire days, including both day-time and night-time acquisitions. In the day-time, channel 3B is also affected by solar radiance. Thus, from the viewpoint of SST estimation, this feature is expected to be noisier for samples collected during day-time than night-time. However, this effect will not be critical because of the high correlation coefficient with SST *in-situ* measurements (around 89%, irrespective of acquisition time) and of the robustness of SVR to noise.

For all of the data sets the features of the training and test samples and the temperature values were normalized to the interval [0, 1] and SVR was applied with a Gaussian radial basis function kernel, i.e. ($\mathbf{x}, \bar{\mathbf{x}} \in \mathbb{R}^d$):

$$K(\mathbf{x}, \bar{\mathbf{x}}|\delta) = \exp\left(-\frac{\|\mathbf{x} - \bar{\mathbf{x}}\|^2}{2\delta^2}\right), \tag{13.15}$$

which is parameterized by the standard deviation δ ($\delta > 0$).[3] This kernel has been employed frequently in remote-sensing applications, and the experimental analysis described in Zortea *et al.* (2006) confirms that it generates accurate temperature estimates. The resulting parameter vector is defined as $\boldsymbol{\theta} = (\log C, \log \epsilon, \log \delta)$, so that all three components of $\boldsymbol{\theta}$ can belong anywhere in an entire real line.

[3]The symbol δ is used instead of the more customary symbol σ in order to avoid confusion with the standard-deviation variables introduced for GGH, ML-GGGH and CI-GGGH.

13.4.2 Parameter-optimization results

PSB was initialized with $C^0 = 1$, $\epsilon^0 = 0.01$ and $\delta^0 = 0.5$ and applied to each data set. At most, seven iterations were sufficient for PSB to converge. The results were compared with those obtained by a hold-out grid-search: for each node in a predefined grid, SVR was trained on the training set; then the parameter vector yielding the smallest mean absolute error (MAE) for the test set was located in the grid. This represents an ideal case in which the true *in-situ* temperatures are assumed to be known on the test samples; this is not a realistic parameter-optimization strategy, but it represents a benchmark result. The grid was defined by the following sets of values: $C \in \{10^{-3}, 10^{-2}, \ldots, 10^4\}$, $\epsilon \in \{10^{-4}, 10^{-3}, \ldots, 10^{-1}\}$ and $\delta \in \{0.01, 0.02, \ldots, 1\}$.

The accuracies obtained for the test set by using SVR with the parameter configurations computed by PSB and the grid search are presented in Table 13.1. Both parameter-optimization strategies generated quite accurate temperature estimates. Specifically, MAE $\simeq$ 0.5K was obtained for SST, while the LST-estimation accuracy depended on the time slot, with MAE $\simeq$ 1.4–1.5K for the 7–8 a.m. and 5–6 p.m. slots, and MAE $\simeq$ 2.7–3.1K for 9–12 a.m. and 1–4 p.m.. This is consistent with the usual behavior of LST estimates from satellite data, which are typically more accurate in the first and last hours of day time and less accurate in the central hours (Ulivieri *et al.* 1994; Zortea 2007). This behavior is explained by a lower correlation of the satellite infrared observations with the *in-situ* measurements during the central hours

Table 13.1 MAE and RMSE for the test set and computation times of the training phases for functional approximators obtained by the PSB and hold-out grid-search, when applied to all considered data sets. The related parameter values and the training and test sample sizes (ℓ and ℓ_{test}, respectively) are also reported (partly reproduced by permission of IEEE; the 'LST 7–8 a.m.' and 'SST' subtables are © 2008 IEEE

	LST 7–8 a.m.		LST 9–12 a.m.		LST 1–4 p.m.	
	PSB	Hold-out	PSB	Hold-out	PSB	Hold-out
MAE [K]	1.51	1.40	2.72	2.16	3.11	2.80
RMSE [K]	1.93	1.86	3.41	2.86	3.86	3.52
time	47s	1h 55m 30s	1m 32s	5h 13m 1s	19s	56m 14s
ℓ	149		263		120	
ℓ_{test}	104		209		86	
C	4.976	0.1	2.666	0.1	1.733	0.1
ϵ	0.002	0.001	0.002	0.001	0.008	0.0001
δ	0.491	0.42	0.500	0.49	0.507	0.5

	LST 5–6 p.m.		SST	
	PSB	Hold-out	PSB	Hold-out
MAE [K]	1.44	1.43	0.54	0.46
RMSE [K]	1.79	1.81	0.73	0.67
time	18s	1h 21m 4s	2m 16s	4m 21s
ℓ	121		109	
ℓ_{test}	104		106	
C	1.739	0.100	3.457	10
ϵ	0.0003	0.001	0.002	0.01
δ	0.656	0.31	0.500	1

than the correlations during the first and last hours of day time. For instance, the correlation coefficients of the *in-situ* temperatures with the SEVIRI brightness temperature channels were about 76%, 70%, 62% and 85% for the 7–8 a.m., 9–12 a.m., 1–4 p.m. and 5–6 p.m. data sets, respectively (irrespective of training and test subdivision). For the SST data set, which gave smaller errors, this correlation coefficient was around 95%.

The optimal parameter vectors identified by PSB and by the hold-out grid search were quite different from each other. However, a very similar performance was exhibited by the two approaches, with slightly smaller regression errors for the hold-out grid search. This result is expected as this approach also exploits the true *in-situ* temperatures of the test samples. However, the differences between the MAE values for the two approaches were just 0.07K for the SST data set, and 0.11, 0.5, 0.3 and 0.01K for the LST data sets from the 7–8 a.m., 9–12 a.m., 1–4 p.m. and 5–6 p.m. time slots, respectively. The similarity in the regression performances can be explained by the strong correlation between MAE (on test set) and SB (computed on the training set) (Chang and Lin 2005). Figures 13.1 and 13.2 show the maps of LST estimates obtained by applying the functional approximators trained on the LST data

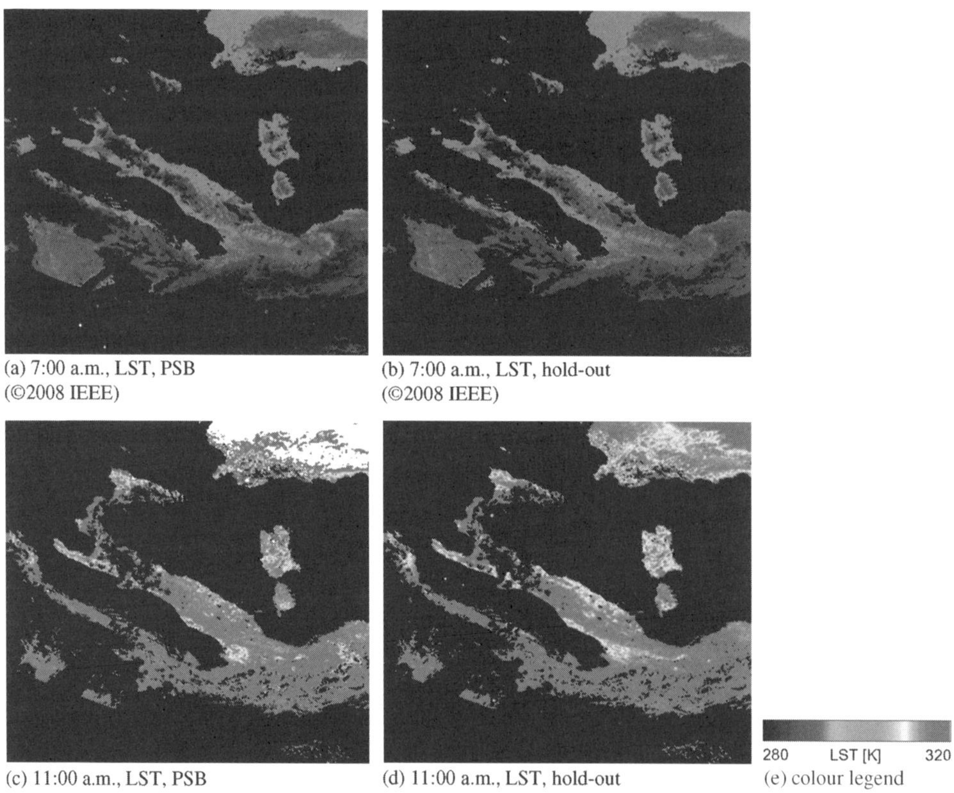

Figure 13.1 Maps of LST estimates generated by SVR when applied with parameter values optimized by PSB and the hold-out grid-search to MSG-SEVIRI images acquired over Italy on September 15 1999, at 7:00 a.m. and 11:00 a.m. local time. Cloud-covered and sea areas are masked in black (partly reproduced by permission of IEEE). (See plate 7)

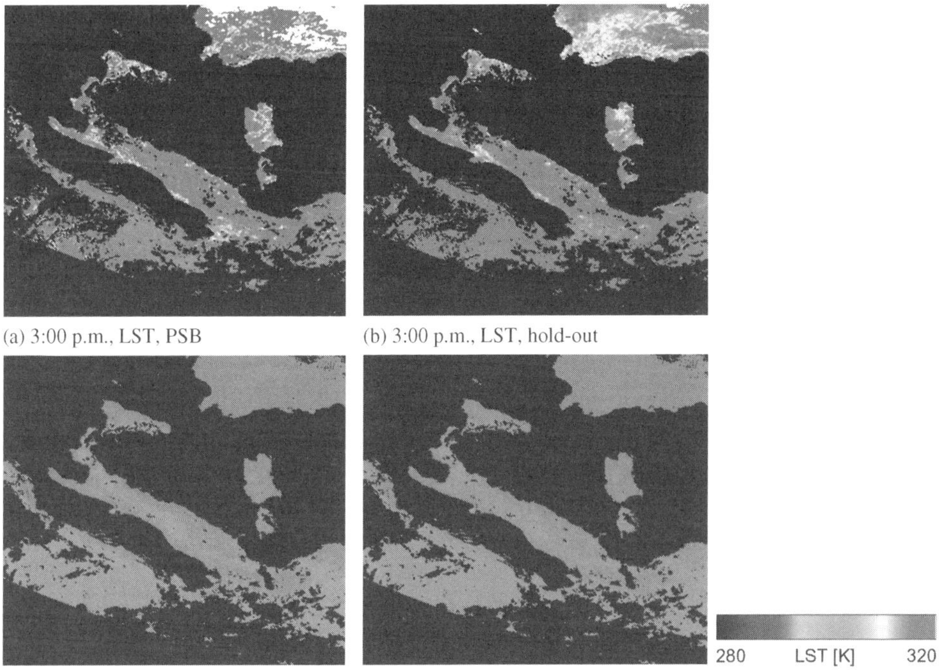

(a) 3:00 p.m., LST, PSB (b) 3:00 p.m., LST, hold-out

(c) 5:00 p.m., LST, PSB (d) 5:00 p.m., LST, hold-out (e) colour legend

Figure 13.2 Maps of LST estimates generated by SVR when applied with parameter values optimized by PSB and the hold-out grid-search to MSG-SEVIRI images acquired over Italy on September 15 1999, at 3:00 p.m. and 5:00 p.m. local time. Cloud-covered and sea areas are masked in black (partly reproduced by permission of IEEE). (See plate 8)

sets (with the parameter vectors computed by PSB and by the hold-out grid search) to MSG images acquired over Italy on September 15 1999, at 7:00 a.m., 11:00 a.m., 3:00 p.m. and 5:00 p.m.. A visual comparison further confirms the similarity between the PSB and grid-search results.

On the other hand, there are dramatic differences between the computation times of PSB and of the hold-out grid-search for LST estimation (all times in Table 13.1 refer to the same hardware configuration, i.e., a 3.4 GHz CPU with 1 GB RAM). The grid search required a few hours to identify optimal parameter vectors and train the related SVR approximators – due to the need to repeat the QP solution process at each node in the grid – whereas PSB always took less than 2 minutes. This is explained by the fact that the grid search performs a complete exploration of the parameter space (up to the adopted search ranges and quantization steps), whereas PSB performs a sequence of moves (in the parameter space) that approaches a locally optimum solution. From this viewpoint, PSB is a suboptimal approach, even though it still identifies parameter configurations affected by regression errors very close to those incurred by the grid search. The difference in computation time is less significant for the SST data set because the number of available training samples is smaller than for the LST data sets, and the time required by each QP numerical solution is correspondingly much shorter (Vapnik 1998).

Table 13.2 Experiments on the LST 7–8 a.m. and SST data sets: MAE and RMSE on the test set, computation times and optimal parameter values for the CV grid search; means and standard deviations (stdev) of the MAE and RMSE values obtained by PSB on the test sets, when varying the initial parameter vector in a predefined grid of 250 points; square roots of the integrated bias, variance and MSE computed on the test set of each of the considered data sets, according to the LST/SST estimators obtained by running PSB over 100 randomly selected training subsets (reproduced by permission of IEEE: © 2008 IEEE)

			LST 7–8 a.m.	SST
	MAE [K]		1.53	0.46
	RMSE [K]		1.96	0.67
Cross validation	time		2h 22m 33s	4m 21s
grid search	C		10	10
	ϵ		0.001	0.01
	δ		0.46	1
	MAE [K]	mean	1.53	0.57
Initialization		stdev	0.09	0.09
analysis	RMSE [K]	mean	1.97	0.80
		stdev	0.14	0.16
Bias-variance	Root integrated bias [K]		1.92	0.72
analysis	Root integrated variance [K]		0.76	0.55
	Root integrated MSE [K]		2.07	0.91

In order to thoroughly investigate the properties of PSB, three more specific experiments were carried out on the SST and 7–8 a.m. LST data sets. First, an experimental comparison was performed with the results given by the grid search for the parameter vector yielding the smallest cross-validation (CV) MAE computed on the training samples. Unlike the above mentioned hold-out approach, this is a feasible and popular parameter optimization technique. We used three-fold CV with a random subdivision of the training set into three disjoint almost equal-sized subsets and the same grid as in the hold-out case. Then, in order to assess the sensitivity of PSB to initialization conditions, we varied the initial parameter values in the ranges of $C^0 \in [0.1; 10]$, $\epsilon^0 \in [0.001; 0.1]$ and $\delta^0 \in [0.1; 1]$ by defining a grid of 250 points over these ranges and by initializing PSB to each node in this grid. Finally a bias-variance analysis was performed in order to investigate possible overfitting issues. The results of these specific experiments are summarized in Table 13.2.

The difference between the regression errors from PSB and the CV grid search were even smaller than those for the hold-out case: slightly smaller MAE and RMSE values were obtained by CV and PSB when applied to the SST and LST data sets, respectively. On the other hand, the computation time for CV was even longer than that for the hold-out search; this result is expected due to the three-fold approach.

As the PSB initial parameter values were varied in the above ranges the method converged, as expected, to different local minima of the SB functional. However, we note a remarkable stability of the resulting test-set errors. For both data sets, the MAE and RMSE standard deviations were much smaller than the corresponding average values, which were very close to the values reported in Table 13.1 for the specific case $C^0 = 1$, $\epsilon^0 = 0.01$ and $\delta^0 = 0.5$. The average times taken by PSB for the 250 runs were 1 m 17 s and 3 m 38 s for the LST and SST data sets, respectively. Such times were shorter than the corresponding times required by CV or hold-out grid searches (see Tables 13.1 and 13.2). However, the choice of the initial parameter vector affected the time required by PSB to converge to a local minimum for each run. With regard to the LST data set, the computation times for all PSB runs were much shorter than those for the grid searches. On the other hand, for the SST data set, PSB took a longer time than CV in 57 out of 250 runs. In fact, due to the small training-sample size of this data set, CV required a short time. Such results reconfirm that PSB, as compared with grid searches, yields a relevant reduction in the computational burden when many training samples are available, whereas this advantage may not be significant in small-sample size cases, where the execution time for grid searches is already non-critical.

In regard to the bias-variance analysis, we recall that given the regression problem formalized in Section 13.3.1, the MSE on an unknown test sample $\mathbf{x}$ ($\mathbf{x} \in \mathbb{R}^d$) can be decomposed into the sum of three contributions (Geman *et al.* 1992):

$$\mathrm{MSE}(\mathbf{x}, \boldsymbol{\theta}) = E\{[y - \hat{f}(\mathbf{x}|\boldsymbol{\theta}, \mathcal{D})]^2|\mathbf{x}\} = \mathrm{Var}\{y|\mathbf{x}\} + \\ + \left[E\{\hat{f}(\mathbf{x}|\boldsymbol{\theta}, \mathcal{D})|\mathbf{x}\} - E\{y|\mathbf{x}\}\right]^2 + \mathrm{Var}\{\hat{f}(\mathbf{x}|\boldsymbol{\theta}, \mathcal{D})|\mathbf{x}\}. \tag{13.16}$$

The first term does not depend on the adopted estimator and is related to the intrinsic data variability of the target variable y. The second contribution represents a measure of the average bias of the adopted estimator $\hat{f}(\mathbf{x}|\boldsymbol{\theta}, \mathcal{D})$ with respect to the conditional mean $E\{y|\mathbf{x}\}$ – the optimal estimator in the MSE sense (Papoulis and Pillai 2002), which is usually unknown. The third contribution is a measure of the variance of the estimator $\hat{f}(\mathbf{x}|\boldsymbol{\theta}, D)$ as a function of the random training samples in $\mathcal{D}$. There is often a trade-off (namely, the so-called 'bias-variance dilemma') between the bias and variance terms (Geman *et al.* 1992). For instance, sharply reducing the bias may cause overfitting, which is likely to yield an increase in the variance.

For both considered data sets, 100 subsets each composed of around 50 samples were randomly sampled from the available training set. PSB was initialized with $C^0 = 1$, $\epsilon^0 = 0.01$ and $\delta^0 = 0.5$ and run until convergence over each training subset, yielding a distinct trained functional approximator. Then, the integrated bias, integrated variance, and integrated MSE were computed for the test set (see Table 13.2), which represent estimates of the average bias, variance and MSE values. This relies on the assumption that the MSE contribution related to intrinsic data variability in Equation (13.16) is negligible when compared with the other two terms. In particular, the resulting integrated MSE is equal to the average of the 100 MSE values obtained from the test set by the functional approximators trained on the 100 subsets. Further details about such estimates can be found in Geman *et al.* (1992). The root integrated MSE values in Table 13.2 were slightly higher than the RMSE values in Table 13.1. This is probably due to the much smaller size of the training subsets involved in this experiment as compared with the original training-sample sizes. For both data sets (especially the LST data), the integrated biases were significantly larger than the corresponding integrated variances. The integrated bias was equal to 86% and 63% of the corresponding integrated MSE for the LST and SST data sets, respectively. This suggests that the test-set errors can be ascribed to the overall difference between the functional approximator generated by PSB and the unknown conditional mean of the target LST/SST given the feature values – not because of large fluctuations in the approximator as the training samples are varied. This behavior is expected to limit the risks of overfitting.

13.4.3 Results on the estimation of regression-error variance

CI-GGGH and ML-GGGH were applied to all of the data sets and experimentally compared with GGH and with a noiseless estimator that presents only the non-stationary error component. In all of these experiments, the input SVR parameters were set to the values identified by PSB. The results were quantitatively assessed for each method and test sample $(\bar{\mathbf{x}}_i, \bar{y}_i)$ by computing the error-variance estimate and analytically deriving the interval $[-\bar{\alpha}_{pi}, \bar{\alpha}_{pi}]$ such that the probability that the regression error on $(\bar{\mathbf{x}}_i, \bar{y}_i)$ belongs to $[-\bar{\alpha}_{pi}, \bar{\alpha}_{pi}]$ (conditioned to $\mathcal{D}$) equals a given value $p \in [0, 1]$ $(i = 1, 2, \ldots, \ell_{test})$; the derivation is based on modelled Gaussian error and is the same as in Equation (13.13)). Table 13.3 shows the percentages of

Table 13.3 Experimental results of the ML-GGGH (ML), CI-GGGH (CI), GGH and noiseless estimators: estimates $\hat{\sigma}_n$ and percentages of test samples $(\bar{\mathbf{x}}_i, \bar{y}_i)$ whose regression errors belong to the ranges $[-\bar{\alpha}_{pi}, \bar{\alpha}_{pi}]$ $(i = 1, 2, \ldots, \ell_{test})$ corresponding to several probability levels p (partly reproduced by permission of IEEE)

	LST 7–8 a.m. (© 2009 IEEE)					LST 9–12 a.m.			
p	noiseless	GGH	ML	CI	p	noiseless	GGH	ML	CI
50%	8%	100%	57%	50%	50%	11%	100%	47%	50%
60%	10%	100%	65%	57%	60%	14%	100%	58%	60%
70%	13%	100%	70%	66%	70%	18%	100%	69%	71%
80%	15%	100%	78%	70%	80%	22%	100%	78%	80%
90%	20%	100%	85%	79%	90%	26%	100%	88%	91%
$\hat{\sigma}_n$ [K]	0	11.4	1.8	1.5	$\hat{\sigma}_n$ [K]	0	21.2	3.2	3.2
	LST 1–4 p.m.					LST 5–6 p.m.			
p	noiseless	GGH	ML	CI	p	noiseless	GGH	ML	CI
50%	5%	100%	22%	31%	50%	9%	100%	50%	47%
60%	10%	100%	30%	38%	60%	15%	100%	63%	61%
70%	12%	100%	35%	58%	70%	20%	100%	70%	67%
80%	15%	100%	49%	64%	80%	24%	100%	78%	77%
90%	22%	100%	67%	77%	90%	31%	100%	88%	84%
$\hat{\sigma}_n$ [K]	0	32.6	2.8	2.4	$\hat{\sigma}_n$ [K]	0	32.5	1.7	1.6
	SST (© 2009 IEEE)								
p	noiseless	GGH	ML	CI					
50%	68%	100%	69%	68%					
60%	74%	100%	74%	74%					
70%	77%	100%	77%	77%					
80%	83%	100%	83%	83%					
90%	90%	100%	91%	90%					
$\hat{\sigma}_n$ [K]	0	12.3	0.1	0.0					

test samples $(\bar{\mathbf{x}}_i, \bar{y}_i)$ whose actual errors (computed according to the true temperature values and to the SVR estimates) belong to the resulting ranges $[-\bar{\alpha}_{pi}, \bar{\alpha}_{pi}]$ for several probability levels $p \in [0, 1]$. We note that the Gaussian shape of the SVR error is assumed to hold in ML-GGGH and CI-GGGH. For GGH $n(\mathbf{x})$ has a zero-mean symmetric monomodal non-gaussian distribution (Gao *et al.* 2002). Here, we compute the intervals $[-\bar{\alpha}_{pi}, \bar{\alpha}_{pi}]$ according to a Gaussian density for GGH as a simplifying approximation.

Focusing first on the LST data sets, the noiseless model underestimated the true probability levels due to a related underestimation of σ_n. This confirms the importance of the stationary error contribution and of an accurate estimation of its variance. On the other hand, GGH sharply overestimated σ_n computing $\hat{\sigma}_n = 10$–30K. Such results were not consistent with the observed test-set errors. For instance, for the 7–8 a.m. time slot the errors on the test samples belonged to the range $[-3.47; 5.10]$ K; this makes the GGH estimate $\hat{\sigma}_n = 11.4$K unrealistic – since the stationary and non-stationary components of the error variance are non-negative, the overall estimated error variance predicted by GGH on each pixel was always larger than $\hat{\sigma}_n^2$. Similar comments hold for the other time slots as well.

By contrast, the estimates $\hat{\sigma}_n$ given by ML-GGGH and CI-GGGH (which were quite close to each other) were below 2K and correspond to percentages of test samples that well approximated the related probability levels, especially for the 7–8 a.m., 9–12 a.m. and 5–6 p.m. time slots; slightly more accurate results were given by ML-GGGH than by CI-GGGH. For the 1–4 p.m. time slot, ML-GGGH and CI-GGGH slightly underestimated the SVR error variance. Again this can be interpreted as a consequence of the lower correlation between satellite observations and *in-situ* measurements during the central hours of day time, which affects the accuracy of the two supervised estimation algorithms.

Figure 13.3 shows, as examples, maps of estimates of the standard deviation of the SVR error computed by CI-GGGH for the MSG acquisitions described in Section 13.4.2; the maps given by ML-GGGH are similar. The standard deviation estimates in the Italian territory are around 1.25K during the 7–8 a.m. and 5–6 p.m. time slots, while larger values (around 2–4K) are obtained during the other two time slots. This is consistent with the overall test-set errors of the LST estimates generated by SVR (see Table 13.1). Note that such standard deviation maps were generated by processing only training data with no need for further test or validation samples. This is an important property of ML-GGGH and CI-GGGH, since they endow the SVR approach to LST and SST estimation with the capability to provide an intrinsic evaluation of the accuracy of the temperature estimates without requiring additional labelled data. For instance, larger standard deviations are found for the North Africa area included in the image; this suggests that less reliable LST estimates were computed by SVR in the corresponding image region, probably due to a lack of training samples in the area.

When applied to the SST data set, the noiseless estimator yields quite a good fit between the modelled and empirical distributions of the test-set errors, which may be interpreted as a limited presence of noise in this data set, with a slight overestimation of the probability levels. ML-GGGH and CI-GGGH correctly identified this situation by automatically computing almost negligible values for $\hat{\sigma}_n$, thus exhibiting an estimation performance almost identical to that of the noiseless estimator. However, since the stationary variance contribution is always non-negative, the slight overestimation introduced by the noiseless estimator could not be corrected by ML-GGGH and CI-GGGH. As in the LST experiments, GGH strongly overestimated σ_n, again producing an unrealistic error variance estimate as compared with test-set errors, which were in the range $[-2.63; 2.11]$K.

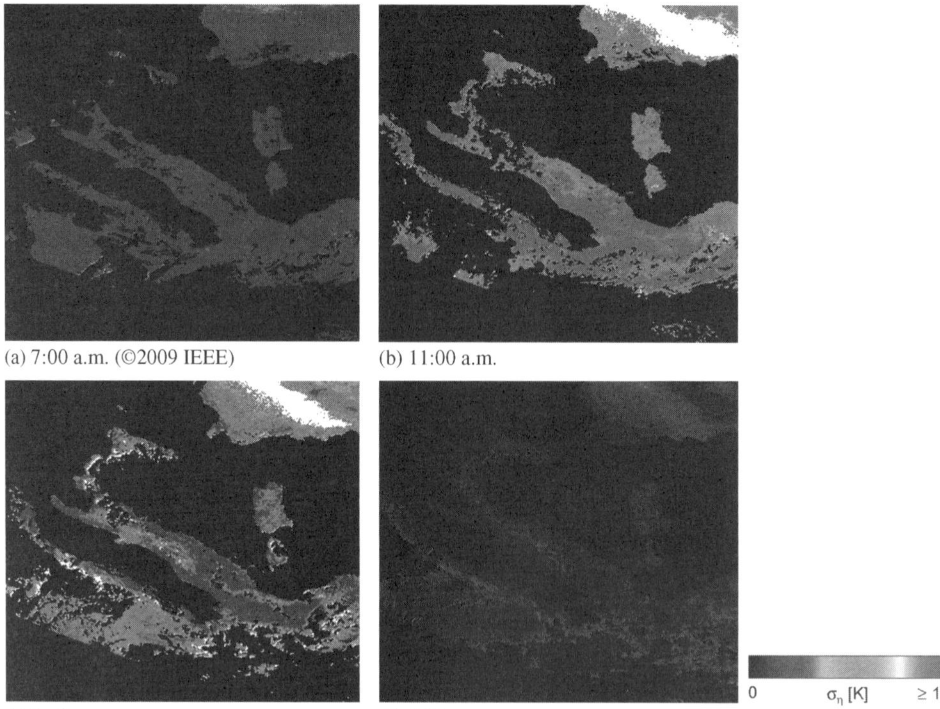

(a) 7:00 a.m. (©2009 IEEE) (b) 11:00 a.m. (c) 3:00 p.m. (d) 5:00 p.m. (e) colour legend

Figure 13.3 Maps of estimates of the standard deviations σ_η of the SVR error generated by the application of CI-GGGH to the same satellite acquisitions as in Figures 13.1 and 13.2. Cloud-covered and sea areas are masked in black (partly reproduced by permission of IEEE). (See plate 9)

13.5 Conclusions

This chapter presented methods in the SVR framework to automate the parameter-optimization processes and estimate pointwise regression error variance. These methods have been validated in the contexts of LST and SST retrieval from satellite infrared data.

In regard to the PSB technique for automatic parameter optimization, experimental results for both LST and SST data sets turned out to be very similar to those obtained by classical grid searches. On the other hand, the computation time for PSB applied to LST data sets turned out to be much shorter than the execution times of the grid searches. In the SST experiments, this difference in computation time was lower, due to the smaller number of training samples. This demonstrates the capability of PSB to exhibit accuracies very close to those achieved by grid searches with a much smaller computational burden. Note that the training sets employed here for LST experiments were made of a few hundred samples. Even in such cases, the grid searches required several hours to complete the training process; this makes them critical from a computational viewpoint when applied to much larger training sets. In addition, a grid search preliminary requires the user to manually predefine a search range and quantization step for each parameter. From this perspective, PSB is a feasible tool for fully automatic and fast training with essentially equivalent accuracies and no need for user interaction. This is important for using the SVR approach to estimate LST/SST; it speeds up updates of the functional approximator, which is regularly used to generate maps of surface temperature estimates. Thus, it is worth investigating the integration of PSB with incremental learning (Wang 2005).

The initial parameter configuration for PSB turned out not to be critical. It influenced the specific local minimum to which the algorithm converged without significantly affecting the test-set errors. The initial parameter vector had a stronger impact on the time taken to reach convergence. Regardless, in the LST experiments, this time was always significantly shorter than that required by grid searches. In the SST experiments, the time for the cross-validation grid search and the average time of PSB were comparable, because of the above-mentioned smaller training-sample size in this data set. Further analysis of the behavior of PSB from the viewpoint of the bias-variance dilemma was also performed by iteratively running PSB on randomly selected training subsets and then estimating the integrated biases and variances of the resulting estimators. The experimental results suggest that the integrated bias contributions were dominant with respect to the integrated variances for the three data sets, showing a limited sensitivity to the specific choice of the training samples – provided that they are drawn from the same probability distribution.

With regard to the problem of modelling space-variant statistics of the regression error, two fully automatic methods have been presented within the framework of the Bayesian reformulation of SVR. The experiments suggest that the two techniques accurately model the pixelwise probability distribution of the SVR errors on test samples by combining the capabilities to take into account both the stationary error component (by supervised parameter estimation) and the non-stationary error contribution (by Bayesian SVR). Slightly more accurate results were obtained for the ML estimation method than for the confidence-interval approach. In particular, both techniques significantly outperformed a previous Bayesian SVR technique that models the stationary contribution only as a function of the input SVR parameters; this technique gave quite poor results, even generating unrealistic error-variance estimates.

In particular, the standard deviation values computed by the two methods for the LST experiments were remarkably consistent with the temporal behavior of the overall regression accuracy during the day time. Similar to traditional methods (e.g., split-window techniques), SVR provides more accurate LST estimates when applied to images acquired in the first and last hours of day time than in the middle hours of the day time. The temporal behavior of the standard deviation estimates highlights these different situations: larger deviation values were obtained for the middle hours of the day than in the first and last hours. This further shows the ability of these methods to automatically provide a user/operator with an intrinsic evaluation of the expected regression accuracy.

The experiments also showed that, when the variance of the non-stationary error component is already overestimated by Bayesian SVR, the two techniques we presented do not amplify this overestimation, as they automatically compute negligible values for the stationary contribution. However, one drawback is that they cannot correct this overestimation, since the stationary and non-stationary variance terms are additive and non-negative. This possible overestimation issue did not turn out to be critical, at least, for the considered data sets. A future extension of this work could be a modification of the Bayesian approach adopted to compute the variance of the non-stationary term, aiming at tuning this component according to the values of the regression errors for the training samples. Future studies could also involve the integration of non-gaussian estimators allowing higher order modelling (e.g., skewness, kurtosis) of the error distribution.

Even though all of the methods described above have been tested in the context of surface-temperature retrieval, they are not application specific. Both the generalization-error bound concepts used by PSB and the Bayesian and parameter-estimation approaches for pixelwise error modelling are general. Experiments for different regression problems (e.g., estimation of other bio/geophysical parameters of the Earth's surface) are required to appreciate the effectiveness of these methods in other fields. It might also be worth integrating these algorithms with hydrological or meteorological data-assimilation techniques (e.g., for flood prevention and monitoring) (Boni *et al.* 2001; Caparrini and Castelli 2003).

Acknowledgments

This research was carried out within the framework of the PROSCENIO 2005–2008 programme funded by the Italian Department of Civil Protection. The support is gratefully acknowledged. The authors would also like to thank CIMA for providing the MSG images used for the experiments, ARSIA for acquiring temperature measurements in Tuscany (Italy), Professor Fabio Castelli from the University of Florence (Italy) for providing such measurements, the AVHRR Pathfinder Oceans group of the University of Miami (USA) for allowing free web access to the AVHRR Pathfinder Matchup Database of SST measurements, Dr Chih Chung Chang and Professor Chih Jen Lin from the University of Taipei (Taiwan) for freeware providing the LIBSVM software package, Dr Maciel Zortea for his assistance in data preparation and pre-processing, and Mr Alessio Agneessens for his help with the experiments.

References

Bartlett, P. and Mendelson, S. (2002) Rademacher and Gaussian complexities: risk bounds and structural result. *Journal of Machine Learning Research*, **3**, 463–484.

Barton, I.J. (1992) Satellite-derived sea surface temperature – a comparison between operational, theoretical, and experimental algorithms. *J. Applied Meteorology*, **31**, 432–442.

Bazi, Y. and Melgani, F. (2007) Semi-supervised PSO-SVM regression for biophysical parameter estimation. *IEEE Trans. Geosci. Remote Sensing*, **45**(6), 1887–1895.

Boni, G., Castelli, F. and Entekhabi, D. (2001) Sampling strategies and assimilation of ground temperature for the estimation of surface energy balance components. *IEEE Trans. Geosci. Remote Sensing*, **39**(1), 165–172.

Brent, R.P. (1973) *Algorithms for Minimization without Derivatives*. Prentice-Hall.

Bruzzone, L. and Melgani, F. (2005) Robust multiple estimator systems for the analysis of biophysical parameters from remotely sensed data. *IEEE Trans. Geosci. Remote Sensing*, **43**(1), 159–174.

Burges, C. (1998) *A Tutorial on Support Vector Machines for Pattern Recognition*. Kluwer Academic Publishers, Boston, MA, USA.

Camps-Valls, G., Bruzzone, L., Rojo-Álvarez, J.L. and Melgani, F. (2006a) Robust support vector regression for biophysical variable estimation from remotely sensed images. *IEEE Trans. Geosci. Remote Sensing*, **44**(3), 339–343.

Camps-Valls, G., Gómez-Chova, L., Muñoz-Marí, J., Vila-Francés, J., Amorós-López, J. and Calpe-Maravilla, J. (2006b) Retrieval of oceanic chlorophyll concentration with relevance vector machines. *Remote Sens. Environ.*, **105**(1), 23–33.

Caparrini, F. and Castelli, F. (2003) Mapping of land–atmosphere heat fluxes and surface parameters with remote sensing data. *Boundary-Layer Meteorology*, **107**, 605–633.

Chan, P.K. and Gao, B.C. (2005) A comparison of MODIS, NCEP and TMI sea surface temperature datasets. *IEEE Geosci. Remote Sensing Letters*, **2**(3), 270–274.

Chang, M.W. and Lin, C.J. (2005) Leave-one-out bounds for support vector regression model selection. *Neural Comp.*, **17**, 1188–1222.

Chapelle, O., Vapnik, V., Bousquet, O. and Mukherjee, S. (2002) Choosing multiple parameters for support vector machines. *Mach. Learn.*, **46**, 131–159.

Cherkassky, V. and Ma, Y. (2003) Practical selection of SVM parameters and noise estimation for SVM regression. *Neural Networks*, **17**, 113–126.

Chu, W., Keerthi, S. and Ong, C.J. (2004) Bayesian support vector regression with a unified loss function. *IEEE Trans. Neural Networks*, **15**, 29–44.

Chung, K.M., Kao, W.C., Sun, T., Wang, L.L. and Lin, C.J. (2003) Radius margin bounds for support vector machines with the RBF kernel. *Neural Comp.*, **15**, 2643–2681.

Dash, P., Gottsche, F.M., Olesen, F.S. and Fischer, H. (2002) Land surface temperature and emissivity estimation from passive sensor data: theory and pratice – current trends. *Int. J. Remote Sens.*, **23**(13), 2563–2594.

Duan, K., Keerthi, S.S. and Poo, A.N. (2003) Evaluation of simple performance measures for tuning SVM hyperparameters. *Neurocomputing*, **51**, 41–59.

Durbha, S.S., King, R.L. and Younan, N.H. (2007) Support vector machines regression for retrieval of leaf area index from multiangle imaging spectroradiometer. *Remote Sens. Environ.*, **107**(1–2), 348–361.

Evans, R. and Podestá, G. (1998) NOAA/NASA AVHRR oceans pathfinder sea surface temperature data set user's reference manual. Technical report, NOAA/NASA. Online: http://www.rsmas.miami.edu/groups/rrsl/pathfinder/.

Gao, J.B., Gunn, S.R. and Harris, C.J. (2002) A probabilistic framework for SVM regression and error bar estimation. *Machine Learning*, **46**, 71–89.

Gao, J.B., Gunn, S.R. and Harris, C.J. (2003a) Mean field method for the support vector machine regression. *Neurocomp.*, **50**, 391–405.

Gao, J.B., Gunn, S.R. and Harris, C.J. (2003b) SVM regression through variational methods and its sequential implementation. *Neurocomp.*, **55**, 151–167.

Geman, S., Bienenstock, E. and Doursat, R. (1992) Neural networks and the bias/variance dilemma. *Neural Computation*, **4**, 1–58.

Haines, S.L., Jedlovec, G.J. and Lazarus, S.M. (2007) A MODIS sea surface temperature composite for regional applications. *IEEE Trans. Geosci. Remote Sensing*, **45**(9), 2919–2927.

Jiménez-Muñoz, J.C. and Sobrino, J.A. (2007) Feasibility of retrieving land-surface temperature from ASTER TIR bands using two-channel algorithms: A case study of agricultural areas. *IEEE Geosci. Remote Sensing Letters*, **4**(1), 60–64.

Joachims, T. (2000) Estimating the generalization performance of a SVM efficiently. *Proc. of the International Conference on Machine Learning*.

Kwok, J.T. and Tsang, I.T. (2003) Linear dependency between ϵ and the input noise in ϵ-support vector regression. *IEEE Trans. Neural Networks*, **14**(3), 544–553.

Law, M.H. and Kwok, J.T. (2001) Bayesian support vector regression. *Proc. of AISTATS-2001, Key West, Florida, USA*, pp. 239–244.

Li, X., Pichel, W., Maturi, E., Clemente-Colon, P. and Sapper, J. (2001) Deriving the operational nonlinear multichannel sea surface temperature algorithm coefficients for NOAA-15 AVHRR/3. *Int. J. Remote Sensing*, **22**, 699–704.

Lin, C.J. and Weng, R.C. (2004) Simple probabilistic predictions for support vector regression. Technical report, National Taiwan University. Online: http://www.csie.ntu.edu.tw/ cjlin/papers/svrprob.pdf.

Mao, K., Shi, J., Tang, H., Li, Z.L., Wang, X. and Chen, K.S. (2008) A neural network technique for separating land surface emissivity and temperature from ASTER imagery. *IEEE Trans. Geosci. Remote Sensing*, **46**(1), 200–208.

Merchant, C.J. and Borgne, P.L. (2004) Retrieval of sea surface temperature from space, based on modelling of infrared radiative transfer: Capabilities and limitations. *J. Atmos. Ocean. Tech.*, **21**, 1734–1746.

Moser, G. and Serpico, S.B. (2007) Automatic land and sea surface temperature estimation from remote sensing data. *Proc. of SPIE-ISPRS-2007, Florence, Italy, 17-20 September 2007.*

Moser, G. and Serpico, S.B. (2008) Modelling the error statistics in support vector regression of surface temperature from infrared data. *Proc. of IGARSS-2008, Boston, USA, 6–11 July (2008) (in print).*

Moser, G. and Serpico, S.B. (2009) Automatic land and sea surface temperature estimation from remote sensing data. *IEEE Trans. Geosci. Remote Sensing (in press).*

Papoulis, A. and Pillai, S.U. (2002) *Probability, random variables, and stochastic processes.* McGraw-Hill.

Peres, L.F. and DaCamara, C.C. (2004a) Improving two-temperature method retrievals based on a nonlinear optimization approach. *IEEE Geosci. Remote Sensing Letters*, **2**(3), 232–236.

Peres, L.F. and DaCamara, C.C. (2004b) Inverse problems theory and application: analysis of the two-temperature method for land-surface temperature and emissivity estimation. *IEEE Geosci. Remote Sensing Letters*, **1**(3), 206–210.

Peres, L.F. and DaCamara, C.C. (2005) Emissivity maps to retrieve land-surface temperature from MSG/SEVIRI. *IEEE Trans. Geosci. Remote Sensing*, **43**(8), 1834–1844.

Pinheiro, C.T., Privette, J.L., Mahoney, R. and Tucker, C.J. (2004) Directional effects in a daily AVHRR land surface temperature dataset over Africa. *IEEE Trans. Geosci. Remote Sensing*, **42**(9), 1941–1954.

Powell, M.J.D. (1964) An efficient method for finding the minimum of a function of several variables without calculating derivatives. *Comp. J.*, **7**, 155–162.

Pozo Vázquez, D., Reyes, F.J.O. and Arboledas, L.A. (1997) Comparative study of algorithms for estimating land surface temperature from AVHRR data. *Rem. Sens. Environ.*, **62**(3), 215–222.

Press, W.H., Teukolsky, S.A., Wetterling, W.T. and Flannery, B.P. (2002) *Numerical Recipes in C.* Cambridge University Press, Cambridge, UK.

Rodger, A.P., Balick, L.K. and Clodius, W.B. (2005) The performance of the multi-spectral thermal imager (MTI) surface temperature retrieval algorithm at three sites. *IEEE Trans. Geosci. Remote Sensing*, **43**(3), 658–665.

Shawe-Taylor, J., Williams, C.K.I., Cristianini, N. and Kandola, J. (2005) On the eigenspectrum of the Gram matrix and the generalization error of kernel-PCA. *IEEE Trans. Information Theory*, **51**(7), 2510–2522.

Sobrino, J.A., Jiménez-Muñoz, J.C., El-Kharraz-Gomez, M., Romaguera, M. and Soria, G. (2004) Single-channel and two-channel methods for land surface temperature retrieval from DAIS data and its application to the Barrax site. *Int. J. Remote Sens.*, **25**(1), 215–230.

Sun, D. and Pinker, R.T. (2004) Case study of soil moisture effect on land surface temperature retrieval. *IEEE Geosci. Remote Sensing Letters*, **1**(2), 127–130.

Tandeo, P., Autret, E., Piollé, J.-F., Tournadre, J. and Ailliot, P. (2009) A multivariate regression approach to adjust AATSR sea surface temperature to in situ measurements. *IEEE Geosci. Remote Sensing Letters*, **6**(1), 8–12.

Tsang, L., Kong, J.A. and Ding, K.H. (2001) *Scattering of Electromagnetic Waves*, Vol. 1 Theories and applications, Wiley Interscience.

Ulivieri, C., Castronuovo, M.M., Francioni, R. and Cardillo, A. (1994) A split window algorithm for estimating land surface temperature from satellites. *Advances in Space Research*, **14**, 59–65.

Vapnik, V. and Chapelle, O. (1999) *Bounds on error expectation for support vector machine*. In *Advances in Large Margin Classifiers*, A.J. Smola, P. Bartlett, B. Scholkopf (eds), Schuurmans MIT Press.

Vapnik, V.N. (1998) *Statistical Learning Theory*. Wiley Interscience.

Wahba, G., Lin, Y. and Zhang, H. (1999) GACV for support vector machines. In *Advances in Large Margin Classifiers* A.J. Smola, B. Bartlett and B. Schölkopf (eds), Schuurmans MIT Press.

Wang, W. (2005) An incremental learning strategy for support vector regression. *Neural Processing Letters*, **21**, 175–188.

Yang, F., White, M.A., Michaelis, A.R., Ichii, K., Hashimoto, H., Votava, P., Zhu, A.X. and Nemani, R.R. (2006) Prediction of continental-scale evapotranspiration by combining MODIS and AmeriFlux data through support vector machine. *IEEE Trans. Geosci. Remote Sensing*, **44**(11), 3452–3461.

Yu, Y., Privette, J.L. and Pinheiro, A.C. (2008) Evaluation of split-window land surface temperature algorithms for generating climate data records. *IEEE Trans. Geosci. Remote Sensing*, **46**(1), 179–192.

Zhou, Y.H., McLaughlin, D. and Entekhabi, D. (2006) Assessing the performance of the ensemble Kalman filter for land surface data assimilation. *Mon. Weather. Review*, **134**(8), 2128–2142.

Zortea, M. (2007) Advanced pattern recognition techniques for environmental information extraction from remotely sensed data. Ph.D. thesis, University of Genoa.

Zortea, M., De Martino, M., Moser, G. and Serpico, S.B. (2006) Land surface temperature estimation from infrared satellite data using support vector machines *Proc. of IGARSS-2006, Denver, USA, (31) July – (4) August 2006*, pp. 2109–2112.

V

Kernel-based feature extraction

14

Kernel multivariate analysis in remote sensing feature extraction

Jerónimo Arenas-García[1] and Kaare Brandt Petersen[2]
[1] *Dept. Signal Theory and Communications, Univ. Carlos III de Madrid, Spain*
[2] *Epital ApS, Denmark*

Feature extraction has become an important topic in remote sensing due to the usually very high dimensionality of data, as well as the high redundancy among spectral bands. This chapter reviews different multivariate analysis techniques for feature extraction, providing a uniform treatment of Principal Component Analysis (PCA), Partial Least Squares (PLS), Canonical Correlation Analysis (CCA) and Orthonormalized PLS (OPLS). Nonlinear extensions are derived by means of kernel methods. Special attention is paid to a sparse version of Kernel OPLS, which can significantly reduce the computational burden both in training and operational phases. The discriminative power of these MVA methods has been studied on a classification scenario, showing the superiority of kernel over linear versions, and the convenience of using the target data for training the feature extractors.

14.1 Introduction

Most data analysis methods are able to deal efficiently with data in one or a few dimensions, but real world applications often involve patterns of much larger dimension than that. In Guyon and Elisseff (2003), for example, the listed benchmark data sets are of dimensions between 10^4 and 10^5. In remote sensing image processing, stacking spatial, spectral, temporal and multiangular information may lead to a high dimensionality too. In addition, new superresolution sensors are producing a high number of spectral channels. For that reason, areas like

Kernel Methods for Remote Sensing Data Analysis Edited by Gustavo Camps-Valls and Lorenzo Bruzzone

feature extraction and dimensionality reduction have become highly important independent subtopics of signal processing and data analysis, and the challenges have motivated the development of both completely new methods and nonlinear versions of well-known methods. A family of methods that has been successful in several scientific areas, and has recently gathered increasing interest in the remote sensing community, consists of the kernel extensions of multivariate techniques such as principal component analysis (PCA) and partial least squares (PLS).

Principal component analysis, also known as the *Hotelling transform* or the *Karhunen–Loeve transform*, is not only a very old multivariate technique (Pearson 1901), but also an extremely widespread method for dimensionality reduction. It consists of projecting the input data set onto the directions of largest input variance. Thus, PCA only considers the input data and does not take into account any target data set.

On the other hand, Partial Least Squares (PLS) is, in its most general form, a family of techniques for analysing relations between data sets by latent variables. It is a basic assumption that the information is over-represented in the data sets, and that these therefore can be reduced in dimensionality by the latent variables. How these latent variables are exactly found, and how the data are projected, vary within the approach, but they are often maximizing the covariance of two projected expressions. Some of the appealing properties of PLS that have made it popular are that it can handle data sets with more dimensions than samples and with massive collinearity between the variables.

The basic PLS algorithm considers two data sets $\mathbf{X}$ and $\mathbf{Y}$, where samples are arranged in rows, and consists of finding latent variables, which accounts for the covariance $\mathbf{X}^\top\mathbf{Y}$ between the data sets. This is done either as an iterative procedure or as an eigenvalue problem. Given the latent variables, the data sets $\mathbf{X}$ and $\mathbf{Y}$ are then transformed in a process that subtracts the information contained in the latent variables. This process, which is often referred to as *deflation*, can be done in a number of ways and these different approaches define the many variants of PLS described in more detail below.

The PLS method is closely related to other multivariate analysis (MVA) techniques, such as Canonical Correlation Analysis (CCA) and Principal Component Analysis (PCA) . In CCA, the aim is to find linear projections of the data sets $\mathbf{X}$ and $\mathbf{Y}$ to maximize *correlation* between the projected data sets. Thus, in contrast to PLS, CCA accounts for correlation rather than covariance, and this makes CCA a special case of PLS with its own characteristic properties (see Wegelin (2000) for further details). Principal component analysis, on the other hand, finds the linear projections for the data set $\mathbf{X}$ that maximize the covariance $\mathbf{X}^\top\mathbf{X}$ of these data. Thus, PCA is, in contrast to PLS, not considering the second data set $\mathbf{Y}$ at all, and is therefore not concerned about the relation *between* the data sets $\mathbf{X}$ and $\mathbf{Y}$. To the extent the data set $\mathbf{Y}$ can be regarded as target or output data, this property makes PCA an unsupervised approach and, in that sense, clearly different from PLS.

Since canonical correlation analysis is a special case of the general PLS algorithm, some may argue that the earliest trace of the ideas behind PLS is the paper introducing CCA from the mid-1930s by Hotelling (Hotelling 1936), a period in which linear multivariate analysis was founded. It is more accurate and in line with tradition, however, to accredit the early nonlinear iterative partial least squares (NIPALS) from 1966 presented in Wold (1966a,b) by Herman Wold as the historical starting point of PLS. In those papers, PLS was introduced as a general framework within the statistical community but have since then had success in applied scientific areas such as psychology, econometrics and chemometrics; see Geladi (1988) for a discussion of the early history of PLS.

The different needs, applications, and lines of development of PLS, have spanned a number of different PLS variants. In one variant of PLS, sometimes referred to as the 'canonical PLS' variant or 'PLS Mode A', the data sets **X** and **Y** are treated entirely equal (Wold 1975) and the deflation done by projecting the data into the space orthogonal to the latent variables. Because of the symmetry between **X** and **Y**, this approach favours an interpretation of PLS in the line of modelling data rather than regression from one data set to the other. The variant of PLS that has become particularly popular, however, is the algorithm presented in Wold *et al.* (1984) and studied in further details in Hoskuldsson (1988). The algorithm described in these, will in this paper be referred to as PLS2, and is based on the following two assumptions: first that the latent variables of **X** are good predictors of **Y**, and second that there is a linear relation between the latent variables of **X** and of **Y**. This linear relation implies a certain deflation scheme, where the latent variable of **X** is used to deflate also the **Y** data set. Several other variants of PLS exists such as Orthonormalized PLS (Worsley *et al.* 1998) and PLS-SB (Sampson *et al.* 1989); see Wegelin (2000) for a technical description, and Rosipal and Kramer (2006) for a well-written contemporary overview.

No matter how refined the various early developments of MVA become, they are still linear projections. Therefore, in those cases where the variables of the input and output spaces are not linearly related, the challenge of the data is still poorly handled. To counter this, different nonlinear versions of these methods have been developed and these can be categorized into two fundamentally different approaches: (1) The modified variants in which the linear relation between the latent variables are substituted by a nonlinear relation; and (2) the kernel variants in which MVA methods are reformulated to fit a kernel approach. In the second approach, the input data are mapped by a nonlinear function into a high-dimensional space in which ordinary linear MVA is performed on the transformed data. A central property of this kernel approach is the exploitation of the so-called kernel trick, i.e., that only the inner products in the transformed space are necessary and not the explicit nonlinear mapping; see Schöelkopf and Smola (2002) and Shawe-Taylor and Cristianini (2004) for excellent introductions to kernel techniques. The kernel approach can be regarded as a generalization of a linear property and many well-known linear methods for feature extraction have been *kernelized*, meaning reformulated in a kernel framework, e.g., Principal Component Analysis (Schölkopf *et al.* 1998), Fisher discriminants (Mika *et al.* 1999), Linear Discriminant Analysis (Baudat and Anouar 2000), and Canonical Correlation Analysis (Lai and Fyfe 2000). Among them, in this paper we will focus on kernel versions of MVA algorithms, which can be applied to both classification and regression problems.

The first paper to apply kernel methods on PLS was Rannar *et al.* (1994), but in this paper the authors applied a linear kernel framework to reduce the computational complexity in the input space rather than assuming a nonlinear mapping for flexibility and expressive power. Instead, it was Rosipal and Trejo who in 2001 first presented a *nonlinear* kernel variant of PLS (Rosipal and Trejo 2001). In that paper, the kernel matrix and the **Y** matrix are deflated in the same way, and the PLS variant is thus more in line with the PLS2 variant than with the traditional algorithm from 1975 (PLS Mode A). In this chapter the nonlinear kernel PLS by Rosipal and Trejo is referred to simply as KPLS2, although many details could advocate a more detailed nomenclature.

The appealing property of kernel algorithms in general is that one can obtain the flexibility of nonlinear expressions while still solving only linear equations. For this reason, kernel methods have been applied in a wide variety of fields, including remote sensing data analysis (see, among others, Arenas-García and Camps-Valls (2008), Banerjee *et al.* (2008),

Camps-Valls and Bruzzone (2005) and Muñoz-Marí *et al.* (2007)). The downside is that, for a data set of l samples, the kernel matrices to be handled are $l \times l$, which for even a moderate number of samples, quickly becomes a problem with respect to both memory and computing time. This problem is present not only in the training phase, but also when predicting the output given some large training data set: evaluating thousands of kernels for every new input vector is, in most applications, not acceptable. Furthermore, there is, for these so-called *dense solutions* in multivariate analysis, also the problem of overfitting. To counter the impractical dense solutions in kernel PLS, a few solutions have been proposed: In Hoegaerts *et al.* (2004), the feature mapping is approximated directly following the Nyström method and in Momma and Bennett (2003) the underlying cost function is modified to impose sparsity.

In this chapter we provide an overview of some of the most commonly used multivariate analysis methods, namely PCA, PLS and CCA, as well as a fourth method not so frequently used in the literature: the Orthonormalized PLS (OPLS). After reviewing the equations that define both the linear and kernel versions of these methods, we will also introduce a recently proposed variant of the kernel OPLS method (Arenas-García *et al.* 2007), which can efficiently handle feature extraction in large data sets with many dimensions, such as those that are typically encountered in remote sensing applications. The so-called Reduced Kernel OPLS (rKOPLS) algorithm consists of two parts: an orthonormalized variant of kernel PLS called KOPLS, and a sparse approximation for large scale data sets. Compared with related approaches like Rosipal *et al.* (2003), the KOPLS transforms only the input data, and keeps them orthonormal at two stages: the images in feature space and the projections in feature space. The sparse approximation is along the lines of Lee and Mangasarian (2001), that is, we are representing the reduced kernel matrix as an outer product of a reduced and a full feature mapping, and thus keeping more information than changing the cost function or doing simple subsampling.

The chapter is structured as follows. In the next section we review the principles of several multivariate analysis methods: PCA, PLS, CCA and OPLS. After describing their nonlinear extensions into the kernel framework in Section 14.3, the rKOPLS method is presented in Section 14.4, explaining some of its more relevant advantages. Section 14.5 presents experimental results in remote sensing data feature extraction for classification tasks, illustrating the properties of the different methods.

14.2 Multivariate analysis methods

Multivariate Analysis (MVA) refers to a family of methods for feature extraction that extract highly correlated projections of data representations in input and output spaces. Before providing a description of the most characteristic MVA methods, we start by giving the motivation for these feature extraction methods, and by introducing some of the notation that will be used throughout the chapter.

Consider we are given a set of data pairs $\{\mathbf{x}_i, \mathbf{y}_i\}_{i=1}^{l}$, with $\mathbf{x}_i \in \mathbb{R}^d$, and $\mathbf{y}_i \in \mathbb{R}^m$, or, using matrix notation, $\mathbf{X} = [\mathbf{x}_1, \ldots, \mathbf{x}_l]^\top$ and $\mathbf{Y} = [\mathbf{y}_1, \ldots, \mathbf{y}_l]^\top$, where superscript $^\top$ denotes matrix or vector transposition. Although not strictly necessary for the derivation of MVA algorithms, it is customary to centre the data (see, e.g., Shawe-Taylor and Cristianini (2004) p. 147). We will denote by $\tilde{\mathbf{X}}$ and $\tilde{\mathbf{Y}}$ the centred versions of the data matrices, so that $\mathbf{C}_{xx} = \frac{1}{l}\tilde{\mathbf{X}}^\top\tilde{\mathbf{X}}$ and $\mathbf{C}_{yy} = \frac{1}{l}\tilde{\mathbf{Y}}^\top\tilde{\mathbf{Y}}$ are the sample estimations of the covariance matrices of the input and output data, respectively, and $\mathbf{C}_{xy} = \frac{1}{l}\tilde{\mathbf{X}}^\top\tilde{\mathbf{Y}}$ represents the covariance between the input and output data.

The objective of a standard linear multiregression problem is to adjust a linear model for predicting the output variables from the input features, $\hat{\tilde{\mathbf{y}}} = \mathbf{W}^\top\tilde{\mathbf{x}}$. Normally, the mean square error is accepted as the objective criterion. It is well known that the solution to this problem can be easily obtained as:

$$\mathbf{W} = \tilde{\mathbf{X}}^\dagger\tilde{\mathbf{Y}}, \tag{14.1}$$

where $\tilde{\mathbf{X}}^\dagger$ is the Moore–Penrose pseudoinverse of $\tilde{\mathbf{X}}$, $\tilde{\mathbf{X}}^\dagger = (\tilde{\mathbf{X}}^\top\tilde{\mathbf{X}})^{-1}\tilde{\mathbf{X}}^\top$. However, this problem is not well-posed if linear dependencies exist among the components of the input vectors, since in that case $\mathbf{X}^\top\mathbf{X}$ becomes rank-deficient (and thus, non-invertible). This problem can easily arise in remote sensing applications when working with hyperspectral images because of the high collinearity between adjacent spatial features. The same problem is encountered in the small-sample-size problem (i.e., when $l < d$). Note that this is often the case when using Kernel Methods, since then input data lies in a space of very high (even infinite) dimensionality.

The solution suggested by MVA to the above singularity problem consists in projecting the input data into a subspace that preserves the most relevant information for the regression problem, using a projection matrix of size $d \times n_p$, $\mathbf{U} = [\mathbf{u}_1, \mathbf{u}_2, \cdots, \mathbf{u}_{n_p}]$, where $\mathbf{u}_i$ is the ith projection vector and n_p the dimension of the projection space. Effectively, if we denote by $\tilde{\mathbf{X}}' = \tilde{\mathbf{X}}\mathbf{U}$ a matrix containing the n_p extracted features of the original input data, the new solution to the regression problem is given by

$$\mathbf{W} = \tilde{\mathbf{X}}'^\dagger\tilde{\mathbf{Y}} = (\tilde{\mathbf{X}}'^\top\tilde{\mathbf{X}}')^{-1}\tilde{\mathbf{X}}'^\top\tilde{\mathbf{Y}}. \tag{14.2}$$

Note that now it is necessary only to invert an $n_p \times n_p$ matrix (normally $n_p \ll d$). Furthermore, if $\mathbf{U}$ is properly designed, $\tilde{\mathbf{X}}'^\top\tilde{\mathbf{X}}'$ will be full rank, making the solution unique.

In certain cases, feature extraction should also be considered for the objective (dependent) data, e.g. when output variables are heterogeneous and/or redundant or, in classification problems, when the number of classes is large. In these cases, output feature extraction can be expressed in similar terms as the process of projecting $\tilde{\mathbf{Y}}$ using the $m \times n_p$ matrix projection operator $\mathbf{V} = [\mathbf{v}_1, \mathbf{v}_2, \cdots, \mathbf{v}_{n_p}]$, i.e., $\tilde{\mathbf{Y}}' = \tilde{\mathbf{Y}}\mathbf{V}$.

The different MVA methods differ in how the projection matrices $\mathbf{U}$ and $\mathbf{V}$ are obtained, their common goal being computing informative projections, in the sense that the projected input and output data are maximally aligned. In the rest of the section we will review the particularities of several such methods: PCA, PLS, CCA and OPLS.

14.2.1 Principal component analysis (PCA)

One of the most simple schemes to reduce the dimensionality of data is Principal Component Analysis (PCA) (Jollife 1986), which only pays attention to the representation of the data in the original space. It is in this sense that we refer to PCA as a unsupervised feature extraction method.

By exploiting the covariance between the different variables (i.e., the columns of $\tilde{\mathbf{X}}$), PCA extracts the directions of the input space that maximize the variance of the projected data.

Traditionally, PCA is stated as the solution to an eigenvalue equation of the input covariance matrix

$$\mathbf{C}_{xx}\mathbf{u}_i = \lambda_i \mathbf{u}_i,$$

where the eigenvalues $\{\lambda_i\}_{i=1}^{n_p}$, with $\lambda_i \geq \lambda_{i+1}$, are the n_p largest eigenvalues of $\mathbf{C}_{xx}$, and $\{\mathbf{u}_i\}_{i=1}^{n_p}$ their corresponding eigenvectors, which define the projection vectors. This can also be expressed more compactly as:

$$\begin{aligned} \text{PCA:} \quad & \mathbf{U} = \arg\max_{\mathbf{U}} \ \mathrm{Tr}\{\mathbf{U}^\top \mathbf{C}_{xx}\mathbf{U}\} \\ & \text{subject to: } \mathbf{U}^\top\mathbf{U} = \mathbf{I}, \end{aligned} \tag{14.3}$$

where $\mathbf{I}$ is the identity matrix of size n_p.

Using Lagrange multipliers, it can be shown that the solution to (14.3) is obtained when the columns of the projection matrix $\mathbf{U}$ are the eigenvectors of the symmetric matrix $\mathbf{C}_{xx}$ associated with its n_p largest eigenvectors, or by a different set of vectors that form an orthonormal basis of the same space.

Typically, the PCA algorithm is implemented by extracting the projection vectors one by one, using a sequential method consisting of the following two phases.

1. The leading vector of the covariance matrix, and its corresponding eigenvalue, $\{\lambda_i, \mathbf{u}_i\}$, are extracted. Different methods can be used to extract the largest eigenvector of a symmetric matrix. For instance, in Table 14.1 we provide the pseudocode of the well-known power method.

2. The covariance matrix is deflated to remove the eigenvector obtained in the first step:

$$\mathbf{C}_{xx} \leftarrow \mathbf{C}_{xx} - \lambda_i \mathbf{u}_i \mathbf{u}_i^\top.$$

   This deflation process amounts to projecting the data matrix onto the orthogonal complement of the direction given by $\mathbf{u}_i$, and can also be expressed as

$$\tilde{\mathbf{X}} \leftarrow \tilde{\mathbf{X}}[\mathbf{I} - \mathbf{u}_i\mathbf{u}_i^\top].$$

Table 14.1 Iterative power method for extracting the largest eigenvector of $\mathbf{C}_{xx}$

1 - Initialize $\mathbf{u}$.*
2 - Iterate until convergence
    $\mathbf{u} \leftarrow \mathbf{C}_{xx}\mathbf{u}$
    $\mathbf{u} \leftarrow \mathbf{u}/\|\mathbf{u}\|_2$
3 - $\mathbf{u}$ is the leading eigenvector, and $\lambda = \frac{\|\mathbf{C}_{xx}\mathbf{u}\|_2}{\|\mathbf{u}\|_2}$ its associated eigenvalue

* To make sure that $\mathbf{u}$ does not lie in the null space of $\mathbf{C}_{xx}$, we can initialize it as any of the columns of the matrix.

By following this two-step process, we are not only guaranteeing that matrix $\mathbf{U}$ is a solution to (14.3), but also that the solution achieved at each iteration is optimal (according to PCA criterion) for the current number of projections.

An important property of the projections given by the PCA method is that they can provide the best approximation of the original data in the mean square error sense. In other words, and noting that that the expression of the projected data in the original space is given by $\tilde{\mathbf{X}}\mathbf{U}\mathbf{U}^\top$, the PCA problem can alternatively be reformulated as

$$\text{PCA(2):} \quad \mathbf{U} = \arg\min_{\mathbf{U}} \ \|\tilde{\mathbf{X}} - \tilde{\mathbf{X}}\mathbf{U}\mathbf{U}^\top\|_F^2, \tag{14.4}$$

where $\|\cdot\|_F$ is the Frobenius norm of a matrix. Indeed, this optimality criterion is the main motivation underlying the PCA method. However, when the extracted projections are to be used in a supervised learning task (either classification or regression), it is clear that the most useful projections will be those that are highly aligned with the objective data, and these are not necessarily those containing the highest variance in input space. For this reason, a better performance can be expected from supervised MVA methods that make use of the output data for the optimization of the projection matrix.

14.2.2 Partial least squares

Partial Least Squares (PLS), developed by Herman Wold in 1966 (Wold *et al.* 1984), is probably one of the simplest methods for supervised MVA. Among its many advantages is its ability to deal with highly correlated data, which has justified its very extensive use in many different fields, such as chemometrics or remote sensing. In these applications, data of interest are usually obtained from the exploration of a range of spectral frequencies, and the variables obtained in such a way frequently suffer from massive collinearity.

The underlying assumption of a PLS model is that the system of interest is driven by a few latent variables (also called factors or components), that are *linear* combinations of observed explanatory variables (i.e., spectral channels or bands). The central idea of PLS is to find a few eigenvectors of spectral matrices that will produce score values that both summarize the variance of spectral reflectance well and highly correlate with response variables (i.e. material class or corresponding biophysical parameter).

In order to do so, PLS looks for the projection vectors that maximize the covariance between the projected input and output data:

$$\text{PLS:} \quad \mathbf{u}, \mathbf{v} = \arg\max_{\mathbf{u},\mathbf{v}} \ \frac{\mathbf{u}^\top \mathbf{C}_{xy} \mathbf{v}}{\mathbf{u}^\top \mathbf{u} \, \mathbf{v}^\top \mathbf{v}}, \tag{14.5}$$

which can also be expressed in terms of the complete projection matrices as

$$\begin{aligned} \text{PLS(2):} \quad & \mathbf{U}, \mathbf{V} = \arg\max_{\mathbf{U},\mathbf{V}} \ \mathrm{Tr}\{\mathbf{U}^\top \mathbf{C}_{xy} \mathbf{V}\} \\ & \text{subject to:} \ \mathbf{U}^\top\mathbf{U} = \mathbf{V}^\top\mathbf{V} = \mathbf{I}. \end{aligned} \tag{14.6}$$

The solution to such a problem is given by the singular value decomposition (SVD) of $\mathbf{C}_{xy}$. As with PCA, practical implementations are usually based on a sequential two-steps scheme, in which projection vectors $\mathbf{u}_i$ and $\mathbf{v}_i$ are first obtained as the singular vectors associated with the largest singular value of $\mathbf{C}_{xy}$. The power method can also be used here, by noting that $\mathbf{u}_i$ is the first eigenvector of $\mathbf{C}_{xy}\mathbf{C}_{xy}^\top$ and $\mathbf{v}_i$ is proportional to $\mathbf{C}_{xy}^\top\mathbf{u}_i$. After obtaining the projection vectors, the cross-covariance matrix has to be deflated, and different PLS variants are described in the literature, which basically differ on how this deflation step is carried out.

For instance, PLS-SB (Sampson *et al.* 1989) proceeds in a PCA-like way, projecting data matrices $\tilde{\mathbf{X}}$ and $\tilde{\mathbf{Y}}$ into the orthogonal complements of the directions given by $\mathbf{u}_i$ and $\mathbf{v}_i$, respectively:

$$\begin{aligned}\tilde{\mathbf{X}} &\leftarrow \tilde{\mathbf{X}}[\mathbf{I} - \mathbf{u}_i\mathbf{u}_i^\top]\\ \tilde{\mathbf{Y}} &\leftarrow \tilde{\mathbf{Y}}[\mathbf{I} - \mathbf{v}_i\mathbf{v}_i^\top].\end{aligned} \tag{14.7}$$

This process is equivalent to the deflation of the cross-covariance matrix according to

$$\mathbf{C}_{xy} \leftarrow \mathbf{C}_{xy} - \sigma_i\mathbf{u}_i\mathbf{v}_i^\top,$$

where σ_i is the singular value associated with the pair $\{\mathbf{u}_i, \mathbf{v}_i\}$. By proceeding in such a way, it can be seen that PLS-SB achieves the same projection vectors that would be obtained from a block solution based on the SVD of the original covariance matrix $\mathbf{C}_{xy}$, and that the maximum number of projections that can be extracted is given by the rank of such a matrix.

A PLS method that is especially well-suited for regression is the algorithm presented in Wold *et al.* (1984), which we will refer to here as PLS2, in accordance with Wegelin (2000). Since the values that are going to be used for the regression of the output data are given at each iteration by $\tilde{\mathbf{X}}\mathbf{u}_i$, PLS2 introduces a deflation scheme consisting in projecting the columns of $\tilde{\mathbf{X}}$ and $\tilde{\mathbf{Y}}$ into the orthogonal complement of $\tilde{\mathbf{X}}\mathbf{u}_i$:

$$\tilde{\mathbf{X}} \leftarrow \left(\mathbf{I} - \frac{\tilde{\mathbf{X}}\mathbf{u}_i\mathbf{u}_i^\top\tilde{\mathbf{X}}^\top}{\|\tilde{\mathbf{X}}\mathbf{u}_i\|_2^2}\right)\tilde{\mathbf{X}} \tag{14.8}$$

$$\tilde{\mathbf{Y}} \leftarrow \left(\mathbf{I} - \frac{\tilde{\mathbf{X}}\mathbf{u}_i\mathbf{u}_i^\top\tilde{\mathbf{X}}^\top}{\|\tilde{\mathbf{X}}\mathbf{u}_i\|_2^2}\right)\tilde{\mathbf{Y}}. \tag{14.9}$$

It can be shown that the deflation of $\tilde{\mathbf{Y}}$ at each step is not necessary, and does not affect the projection vectors extracted by PLS2.

Since this deflation process does not decrease the rank of $\tilde{\mathbf{Y}}$, PLS2 can in general extract as many projections as the original dimensionality of the input data, in contrast to a maximum of rank$(\mathbf{C}_{xy})$ projections for PLS-SB. Another difference between the two algorithms is that in PLS2 the columns of the projected input data matrix, $\{\tilde{\mathbf{X}}\mathbf{u}_i\}_{i=1}^{n_p}$, are orthogonal, while for PLS-SB they are not. Regarding the projection vectors, it can be shown (Shawe-Taylor and Cristianini 2004, Subsection 6.7.1) that $\{\mathbf{u}_i\}_{i=1}^{n_p}$ constitutes a set of orthonormal vectors for both schemes, while the output projection vectors $\{\mathbf{v}_i\}_{i=1}^{n_p}$ are orthogonal only for PLS-SB.

14.2.3 Canonical correlation analysis

Canonical Correlation Analysis (CCA) was first proposed by H. Hotelling in 1936 (Hotelling 1936). In contrast to the maximization of covariance pursued by PLS, CCA searches for the directions of maximum correlation between input and output data. In this way, CCA can more conveniently deal with directions of the input (output) space that present very high variance, and that would therefore be over-emphasized by PLS, no matter whether they are very correlated with the output (input) space representation or not.

The maximization of correlation carried out by CCA allows one to state this method as the following maximization problem:

$$\text{CCA:} \qquad \mathbf{u}, \mathbf{v} = \arg\max_{\mathbf{u},\mathbf{v}} \frac{(\mathbf{u}^\top \mathbf{C}_{xy}\mathbf{v})^2}{\mathbf{u}^\top \mathbf{C}_{xx}\mathbf{u}\ \mathbf{v}^\top \mathbf{C}_{yy}\mathbf{v}}. \tag{14.10}$$

Now, noting that the above quotient is invariant to a scaling of the projection vectors **u** and **v**, this problem can be alternatively rewritten as

$$\begin{aligned} \text{CCA(2):} \qquad & \mathbf{u}, \mathbf{v} = \arg\max_{\mathbf{u},\mathbf{v}} \ \mathbf{u}^\top \mathbf{C}_{xy}\mathbf{v} \\ & \text{subject to: } \mathbf{u}^\top \mathbf{C}_{xx}\mathbf{u} = \mathbf{v}^\top \mathbf{C}_{yy}\mathbf{v} = 1, \end{aligned} \tag{14.11}$$

where the removal of the square constrains the solution to pairs $\{\mathbf{u}, \mathbf{v}\}$ resulting in a positive value of $\mathbf{u}^\top \mathbf{C}_{xx}\mathbf{v}$. Note that this is not a critical constraint since we are just interested in the directions defined by the projection vectors. Another important property of CCA is that the projection vectors are not affected by a scaling of the original data, as it can easily be seen from the maximization problem (14.10).

We can also express the CCA problem in terms of the complete projection matrices **U** and **V** as

$$\begin{aligned} \text{CCA(3):} \qquad & \mathbf{U}, \mathbf{V} = \arg\max_{\mathbf{U},\mathbf{V}} \ \mathrm{Tr}\{\mathbf{U}^\top \mathbf{C}_{xy}\mathbf{V}\} \\ & \text{subject to: } \mathbf{U}^\top \mathbf{C}_{xx}\mathbf{U} = \mathbf{V}^\top \mathbf{C}_{yy}\mathbf{V} = \mathbf{I}. \end{aligned} \tag{14.12}$$

Introducing Lagrange multipliers, and after some simple algebraic manipulations, it can be shown that CCA projection vectors are defined by the solutions of the following generalized eigenvalue problem:

$$\begin{pmatrix} \mathbf{0} & \mathbf{C}_{xy} \\ \mathbf{C}_{xy}^\top & \mathbf{0} \end{pmatrix} \begin{pmatrix} \mathbf{u} \\ \mathbf{v} \end{pmatrix} = \lambda \begin{pmatrix} \mathbf{C}_{xx} & \mathbf{0} \\ \mathbf{0} & \mathbf{C}_{yy} \end{pmatrix} \begin{pmatrix} \mathbf{u} \\ \mathbf{v} \end{pmatrix}, \tag{14.13}$$

where **0** represents a matrix of zeros of appropriate dimensions. The above equation can be reformulated as a standard eigenvalue problem, as explained in Hardoon *et al.* (2003).

Another interesting aspect of CCA is its equivalency to the minimization of the following least squares cost function:

$$\begin{aligned} \text{CCA(4):} \quad & \mathbf{U}, \mathbf{V} = \underset{\mathbf{U},\mathbf{V}}{\arg\min} \ \ \|\tilde{\mathbf{Y}}\mathbf{V} - \tilde{\mathbf{X}}\mathbf{U}\|_F^2 \\ & \text{subject to:} \ \mathbf{U}^\top \mathbf{C}_{xx}\mathbf{U} = \mathbf{V}^\top \mathbf{C}_{yy}\mathbf{V} = \mathbf{I}. \end{aligned} \tag{14.14}$$

In other words, (14.14) shows that CCA searches for projections of the input and output data matrices such that each column of $\tilde{\mathbf{Y}}'$ can be reconstructed from the corresponding column of $\tilde{\mathbf{X}}'$ with minimum square loss. Note, however, that in general we will be more interested in approximating the original label matrix, $\tilde{\mathbf{Y}}$, rather than a certain projection of it, and it is in this sense that OPLS becomes optimal.

14.2.4 Orthonormalized partial least squares

In this subsection we review Orthonormalized Partial Least Squares (OPLS) (Worsley *et al.* 1998), which is defined by the following maximization problem:

$$\begin{aligned} \text{OPLS:} \quad & \mathbf{U} = \underset{\mathbf{U}}{\arg\max} \ \ \text{Tr}\{\mathbf{U}^\top \mathbf{C}_{xy}\mathbf{C}_{xy}^\top \mathbf{U}\} \\ & \text{subject to:} \ \mathbf{U}^\top \mathbf{C}_{xx}\mathbf{U} = \mathbf{I}. \end{aligned} \tag{14.15}$$

Note that, unlike PLS and CCA, OPLS only performs dimensionality reduction in the input space. According to our discussion at the beginning of the section, when tackling a linear multiregression problem, as in many other supervised learning situations, we are primarily concerned with the extraction of good features in the input space.

As with CCA, the solution to the OPLS optimization problem leads to a generalized eigenvalue problem

$$\mathbf{C}_{xy}\mathbf{C}_{xy}^\top \mathbf{u} = \lambda \mathbf{C}_{xx}\mathbf{u}. \tag{14.16}$$

The above equation can be rewritten as a standard eigenvalue problem by premultiplying both sides of the equality by $\mathbf{C}_{xx}^{-1}$. A more convenient formulation can be obtained if we define $\mathbf{w} = \mathbf{C}_{xx}^{1/2}\mathbf{u}$. After premultiplying (14.16) by $\mathbf{C}_{xx}^{-1/2}$, we get

$$\mathbf{C}_{xx}^{-1/2}\mathbf{C}_{xy}\mathbf{C}_{xy}^\top \mathbf{C}_{xx}^{-1/2}\mathbf{w} = \lambda \mathbf{w}, \tag{14.17}$$

so that now vectors $\mathbf{w}$ are obtained as the eigenvectors of a symmetric matrix. Finally, OPLS projection vectors can be recovered from $\mathbf{w}$ as $\mathbf{u} = \mathbf{C}_{xx}^{-1/2}\mathbf{w}$.

It is important to note that (14.17) is only a valid expression when $\mathbf{C}_{xx}$ is a full rank matrix. When this is not the case, we could proceed by first removing the directions of the input space where the data do not present any variance (i.e. by projecting the input data into the subspace spanned by the eigenvectors of $\mathbf{C}_{xx}$ with non-zero eigenvalues).

The main argument in favour of OPLS with respect to other MVA methods is its optimality, in the mean square error sense, for performing linear multiregression of $\tilde{\mathbf{Y}}$ on the projected input data (Arenas-García and Camps-Valls 2008; Roweis and Brody 1999). In other words,

Table 14.2 Some relevant properties of the linear MVA methods reviewed in Section 14.2

	PCA	PLS-SB	PLS2	CCA	OPLS
Max. problem	$\mathbf{u}^\top \mathbf{C}_{xx}\mathbf{u}$	$\mathbf{u}^\top \mathbf{C}_{xy}\mathbf{v}$	$\mathbf{u}^\top \mathbf{C}_{xy}\mathbf{v}$	$\mathbf{u}^\top \mathbf{C}_{xy}\mathbf{v}$	$\mathbf{u}^\top \mathbf{C}_{xy}\mathbf{C}_{xy}^\top\mathbf{u}$
Constraints	$\mathbf{u}^\top \mathbf{u} = 1$	$\mathbf{u}^\top \mathbf{u} = 1$ $\mathbf{v}^\top \mathbf{v} = 1$	$\mathbf{u}^\top \mathbf{u} = 1$ $\mathbf{v}^\top \mathbf{v} = 1$	$\mathbf{u}^\top \mathbf{C}_{xx}\mathbf{u} = 1$ $\mathbf{v}^\top \mathbf{C}_{yy}\mathbf{v} = 1$	$\mathbf{u}^\top \mathbf{C}_{xx}\mathbf{u} = 1$
Max. n_p	$r(\tilde{\mathbf{X}})$	$r(\mathbf{C}_{xy})$	$r(\tilde{\mathbf{X}})$	$r(\mathbf{C}_{xy})$	$r(\mathbf{C}_{xy})$
$\{\tilde{\mathbf{X}}\mathbf{u}_i\}$ Orthogonal?	No	No	Yes	Yes	Yes

$r(\cdot)$ denotes the rank of the matrix inside the parentheses.

while CCA and PLS are optimal in the sense of maximizing correlation and covariance, respectively, OPLS becomes optimal with respect to the following problem:

$$\text{OPLS(2):} \qquad \mathbf{U} = \arg\min_{\mathbf{U}} \ \|\tilde{\mathbf{Y}} - \tilde{\mathbf{X}}\mathbf{U}\mathbf{W}\|_F^2, \tag{14.18}$$

where $\mathbf{W} = \tilde{\mathbf{X}}'^{\dagger}\tilde{\mathbf{Y}}$ is the optimal regression matrix. Note that, if one is interested in extracting features of the input data for a subsequent approximation of the label matrix, this is a much more sensible criterion than either (14.4) or (14.14).

As a summary, in Table 14.2 we review some of the most relevant properties of the linear MVA methods that we have analysed in this section.

14.3 Kernel multivariate analysis

All the algorithms reviewed in Section 14.2 can be used to extract linear projections of the input and output data. However, the quality of the extracted features will seriously degrade when the input and output spaces are not linearly related. To increase the expressiveness of MVA projections, several authors have proposed nonlinear extensions of these algorithms. In particular, the kernel framework has been applied to obtain nonlinear versions of all four algorithms reviewed in the previous section: Kernel PCA (Schölkopf *et al.* 1998), Kernel PLS (Rosipal and Trejo 2001), Kernel CCA (Lai and Fyfe 2000) and Kernel OPLS (Arenas-García *et al.* 2007).

As explained in previous chapters, Kernel Methods work by first projecting the input data into a high (possibly infinite) dimensional Reproducing Kernel Hilbert Space, where a standard linear algorithm is applied. This results in a nonlinear method under the perspective of the input space.

To introduce Kernel MVA (KMVA) algorithms, we will first consider a mapping function $\boldsymbol{\phi}(\cdot) : \mathbb{R}^d \rightarrow \mathcal{F}$ that projects input data into feature space. In this way, the new training data set is $\{\boldsymbol{\phi}(\mathbf{x}_i), \mathbf{y}_i\}_{i=1}^l$, where $\boldsymbol{\phi}(\mathbf{x}_i) \in \mathcal{F}$ and $\mathbf{y}_i \in \mathbb{R}^m$, or, using matrix notation, $\boldsymbol{\Phi} = [\boldsymbol{\phi}(\mathbf{x}_1), \cdots, \boldsymbol{\phi}(\mathbf{x}_l)]^\top$, $\mathbf{Y} = [\mathbf{y}_1, \cdots, \mathbf{y}_l]^\top$. As before, we will consider centred versions of these matrices, and denote them by $\tilde{\boldsymbol{\Phi}}$ and $\tilde{\mathbf{Y}}$.

To obtain the kernel version of a given linear algorithm, we have to replace the original input data matrix by $\tilde{\boldsymbol{\Phi}}$. By doing so, the projections of the input and output data will be given by $\tilde{\mathbf{\Phi}}' = \tilde{\mathbf{\Phi}}\mathbf{U}$ and $\tilde{\mathbf{Y}}' = \tilde{\mathbf{Y}}\mathbf{V}$, respectively, where the projection matrix $\mathbf{U}$ is now of size $\dim(\mathcal{F}) \times n_p$. However, the direct application of this idea suffers from some serious practical limitations when the dimension of $\mathcal{F}$ is very large (which is usually the case). Note, for instance, that the input covariance matrix, $\mathbf{C}_{\phi\phi} = \frac{1}{l}\tilde{\mathbf{\Phi}}^\top\tilde{\mathbf{\Phi}}$, which is usually needed by the different MVA methods, becomes of size $\dim(\mathcal{F}) \times \dim(\mathcal{F})$ under this approach.

To implement realizable kernel MVA algorithms we will have to rewrite the equations presented in Section 14.2 in terms of inner products in $\mathcal{F}$ only. Therefore, we will rely during our derivations on the availability of a kernel matrix $\mathbf{K}_x = \tilde{\mathbf{\Phi}}\tilde{\mathbf{\Phi}}^\top$ of dimensions $l \times l$, which contains the inner products of any two training points $[\mathbf{K}_x]_{ij} = \tilde{K}(\mathbf{x}_i, \mathbf{x}_j) = \langle\tilde{\boldsymbol{\phi}}(\mathbf{x}_i), \tilde{\boldsymbol{\phi}}(\mathbf{x}_j)\rangle$.

In the following subsections, we will see how the application of the Representer's Theorem (Schöelkopf and Smola 2002) (see also Chapter 2) can also be used here to rewrite linear MVA algorithms in terms of inner products only, thus allowing their extension to the kernel framework.

14.3.1 Kernel PCA

As in the linear case, the goal of KPCA is to find the projections that maximize the variance of the input data in the feature space. By simply replacing $\tilde{\mathbf{X}}$ by $\tilde{\mathbf{\Phi}}$ in (14.3), KPCA can be analytically formulated in the following way:

$$\text{KPCA:} \quad \mathbf{U} = \arg\max_{\mathbf{U}} \ \mathrm{Tr}\{\mathbf{U}^\top\tilde{\mathbf{\Phi}}^\top\tilde{\mathbf{\Phi}}\mathbf{U}\} \qquad \text{subject to: } \mathbf{U}^\top\mathbf{U} = \mathbf{I}. \tag{14.19}$$

To rewrite the above equations in terms of inner products only, we can make use of the Representer's Theorem (Schöelkopf and Smola 2002), which states that the projection vectors can be expressed as a linear combination of the training data in $\mathcal{F}$. Using matrix notation, this can be expressed as $\mathbf{U} = \tilde{\mathbf{\Phi}}^\top\mathbf{A}$, where $\mathbf{A} = [\boldsymbol{\alpha}_1, \cdots, \boldsymbol{\alpha}_{n_p}]^\top$, with vectors $\boldsymbol{\alpha}_i$ containing the coefficients corresponding to the ith projection vector. Now, KPCA can be reformulated as follows:

$$\text{KPCA(2):} \quad \mathbf{A} = \arg\max_{\mathbf{A}} \ \mathrm{Tr}\{\mathbf{A}^\top\mathbf{K}_x\mathbf{K}_x\mathbf{A}\} \qquad \text{subject to: } \mathbf{A}^\top\mathbf{K}_x\mathbf{A} = \mathbf{I}. \tag{14.20}$$

Therefore, the solution will now be given in terms of the $l \times n_p$ matrix $\mathbf{A}$, which defines the projection operator via $\mathbf{U} = \tilde{\mathbf{\Phi}}^\top\mathbf{A}$.

As with the linear MVA methods, it is relatively straightforward to obtain the generalized eigenvalue problem that is satisfied by the solution of (14.20):

$$\mathbf{K}_x\mathbf{K}_x\boldsymbol{\alpha} = \lambda\mathbf{K}_x\boldsymbol{\alpha}.$$

It is also possible to show (Schölkopf *et al.* 1998) that all the solutions of interest of the problem above also satisfy the following standard eigenvalue problem:

$$\mathbf{K}_x\boldsymbol{\alpha} = \lambda\boldsymbol{\alpha},$$

so that vectors $\{\boldsymbol{\alpha}_i\}$ are given by the eigenvectors associated with the largest eigenvalues of the kernel matrix, and their extraction can be implemented in very much the same way as the two-step sequential method that was described for linear PCA:

1. Extract the leading eigenvector of $\mathbf{K}_x$, and its corresponding eigenvalue: $\{\lambda_i, \boldsymbol{\alpha}_i\}$.
2. Deflate the kernel matrix according to $\mathbf{K}_x \leftarrow \mathbf{K}_x - \lambda_i \boldsymbol{\alpha}_i \boldsymbol{\alpha}_i^\top$.

Additionally, although the above procedure determines the projection directions of KPCA, (14.19) requires all projection vectors $\mathbf{u}_i$ to be of unitary norm. This can be implemented by a simple rescaling of the optimal coefficients:

$$\boldsymbol{\alpha}_i \leftarrow \frac{\boldsymbol{\alpha}_i}{\sqrt{\boldsymbol{\alpha}_i^\top \mathbf{K}_x \boldsymbol{\alpha}_i}}.$$

Regarding the optimality of KPCA in the least squares error sense, it is clear that property (14.4) will still hold for KPCA in feature space. In other words, KPCA will extract the features that allow for the best approximation (in the least squares error sense) of the input data in feature space, $\tilde{\boldsymbol{\Phi}}$.

Finally, it is important to mention that in many cases the projection matrix $\mathbf{U}$ cannot be explicitly calculated. However, note that we are actually not interested on this matrix itself, but on the projections of input data along the directions determined by the columns of $\mathbf{U}$. When we want to extract the features corresponding to a new pattern $\mathbf{z}$, we can do so using

$$\tilde{\boldsymbol{\phi}}'(\mathbf{z}) = \tilde{\boldsymbol{\phi}}(\mathbf{z})\mathbf{U} = \tilde{\boldsymbol{\phi}}(\mathbf{z})\tilde{\boldsymbol{\Phi}}^\top \mathbf{A} = \begin{pmatrix} \tilde{K}(\mathbf{x}_1, \mathbf{z}) \\ \vdots \\ \tilde{K}(\mathbf{x}_l, \mathbf{z}) \end{pmatrix} \mathbf{A},$$

which again requires only the computation of inner products (kernels) in $\mathcal{F}$.

14.3.2 Kernel PLS

Some recent developments based on kernel methods have lead to nonlinear PLS algorithms (Rosipal and Trejo 2001), which are based on maximizing the variance between the input data in $\mathcal{F}$ and the objective data matrix $\tilde{\mathbf{Y}}$:

$$\text{KPLS:} \quad \mathbf{U}, \mathbf{V} = \underset{\mathbf{U},\mathbf{V}}{\arg\max} \ \mathrm{Tr}\{\mathbf{U}^\top \tilde{\boldsymbol{\Phi}}^\top \tilde{\mathbf{Y}} \mathbf{V}\} \qquad \text{subject to: } \mathbf{U}^\top\mathbf{U} = \mathbf{V}^\top\mathbf{V} = \mathbf{I}. \tag{14.21}$$

When $\tilde{\boldsymbol{\Phi}}$ cannot be explicitly computed, it is necessary to rewrite the maximization problem in terms of inner products. To do so, we will again rely on the Representer's Theorem that states that $\mathbf{U} = \tilde{\mathbf{\Phi}}^\top \mathbf{A}$, so that KPLS can be reformulated as

$$\begin{aligned} \text{KPLS(2):} \qquad \mathbf{A}, \mathbf{V} &= \arg\max_{\mathbf{A},\mathbf{V}} \ \text{Tr}\{\mathbf{A}^\top \mathbf{K}_x \tilde{\mathbf{Y}} \mathbf{V}\} \\ \text{subject to: } & \mathbf{A}^\top \mathbf{K}_x \mathbf{A} = \mathbf{V}^\top \mathbf{V} = \mathbf{I}. \end{aligned} \tag{14.22}$$

Again, the solutions to this problem can be obtained from the singular value decomposition of $\mathbf{K}_x \tilde{\mathbf{Y}}$.

As in the linear case, it is possible to implement the sequential extraction of KPLS projections, with different deflation schemes leading to different KPLS versions. For instance, to derive the deflation scheme associated with KPLS-SB we will simply have to replace $\tilde{\mathbf{X}}$ and $\mathbf{u}_i$ in (14.7) by $\tilde{\mathbf{\Phi}}$ and $\tilde{\mathbf{\Phi}}^\top \boldsymbol{\alpha}_i$, respectively, arriving at

$$\begin{aligned} \tilde{\mathbf{\Phi}} &\leftarrow \tilde{\mathbf{\Phi}}[\mathbf{I} - \tilde{\mathbf{\Phi}}^\top \boldsymbol{\alpha}_i \boldsymbol{\alpha}_i^\top \tilde{\mathbf{\Phi}}] = \tilde{\mathbf{\Phi}} - \mathbf{K}_x \boldsymbol{\alpha}_i \boldsymbol{\alpha}_i^\top \tilde{\mathbf{\Phi}} \\ \tilde{\mathbf{Y}} &\leftarrow \tilde{\mathbf{Y}}[\mathbf{I} - \mathbf{v}_i \mathbf{v}_i^\top]. \end{aligned} \tag{14.23}$$

Note that implementing the deflation of $\tilde{\mathbf{\Phi}}$ could be unfeasible for certain selections of the mapping function $\boldsymbol{\phi}(\mathbf{x})$. Nevertheless, the eigenvalue problem to be solved involves only $\mathbf{K}_x$, which can still be deflated in those cases according to

$$\mathbf{K}_x \leftarrow [\tilde{\mathbf{\Phi}} - \mathbf{K}_x \boldsymbol{\alpha}_i \boldsymbol{\alpha}_i^\top \tilde{\mathbf{\Phi}}][\tilde{\mathbf{\Phi}}^\top - \tilde{\mathbf{\Phi}}^\top \boldsymbol{\alpha}_i \boldsymbol{\alpha}_i^\top \mathbf{K}_x] = \mathbf{K}_x - \mathbf{K}_x \boldsymbol{\alpha}_i \boldsymbol{\alpha}_i^\top \mathbf{K}_x. \tag{14.24}$$

We can proceed in a similar way to obtain the deflation equations for KPLS2. A direct kernel extension of (14.8) results in

$$\tilde{\mathbf{\Phi}} \leftarrow \left(\mathbf{I} - \frac{\mathbf{K}_x \boldsymbol{\alpha} \boldsymbol{\alpha}^\top \mathbf{K}_x}{\|\mathbf{K}_x \boldsymbol{\alpha}\|_2^2} \right) \tilde{\mathbf{\Phi}},$$

so that the kernel matrix can be deflated using

$$\mathbf{K}_x \leftarrow \left[\mathbf{I} - \frac{\mathbf{K}_x \boldsymbol{\alpha}_i \boldsymbol{\alpha}_i^\top \mathbf{K}_x}{\|\mathbf{K}_x \boldsymbol{\alpha}\|_2^2} \right] \mathbf{K}_x \left[\mathbf{I} - \frac{\mathbf{K}_x \boldsymbol{\alpha}_i \boldsymbol{\alpha}_i^\top \mathbf{K}_x}{\|\mathbf{K}_x \boldsymbol{\alpha}\|_2^2} \right], \tag{14.25}$$

which again does not require the explicit calculation of the input data in $\mathcal{F}$. For a more detailed description of KPLS2, as well as implementation details of this kernel MVA method, the reader is referred to Shawe-Taylor and Cristianini (2004).

14.3.3 Kernel CCA

In this subsection we consider a kernel extension of CCA. Unlike some previous works (Lai and Fyfe 2000), and in order to provide a similar treatment of all KMVA methods, we will consider that only the input data are mapped into a feature space. Furthermore, it has also been found that applying nonlinear projections to both input and output data can result in serious overfitting problems (Hardoon *et al.* 2003; Shawe-Taylor and Cristianini 2004).

Replacing $\tilde{\mathbf{X}}$ by $\tilde{\boldsymbol{\Phi}}$ in (14.12), and by virtue of the Representer's Theorem $\mathbf{U} = \tilde{\boldsymbol{\Phi}}^\top \mathbf{A}$, one can formulate Kernel CCA (KCCA) as

$$\text{KCCA:} \quad \mathbf{A}, \mathbf{V} = \underset{\mathbf{A},\mathbf{V}}{\arg\max} \quad \text{Tr}\{\mathbf{A}^\top \mathbf{K}_x \tilde{\mathbf{Y}} \mathbf{V}\}$$
$$\text{subject to: } \mathbf{A}^\top \mathbf{K}_x^2 \mathbf{A} = \mathbf{V}^\top \mathbf{C}_{yy} \mathbf{V} = \mathbf{I}. \tag{14.26}$$

Now, we can proceed in a very similar way to the linear case to derive the following generalized eigenvalue problem which characterizes KCCA:

$$\begin{pmatrix} \mathbf{0} & \mathbf{K}_x \tilde{\mathbf{Y}} \\ \tilde{\mathbf{Y}}^\top \mathbf{K}_x & \mathbf{0} \end{pmatrix} \begin{pmatrix} \boldsymbol{\alpha} \\ \mathbf{v} \end{pmatrix} = \lambda \begin{pmatrix} \mathbf{K}_x^2 & \mathbf{0} \\ \mathbf{0} & \mathbf{C}_{yy} \end{pmatrix} \begin{pmatrix} \boldsymbol{\alpha} \\ \mathbf{v} \end{pmatrix}, \tag{14.27}$$

where $\mathbf{0}$ represents a matrix of zeros of appropriate dimensions.

When using KCCA we are selecting the directions in feature space that have a maximum correlation with a certain projection of the target data (i.e., the kernel extension of the previously presented result (14.14)).

14.3.4 Kernel OPLS

As occuring with linear OPLS, and in comparison with other Kernel MVA methods, Kernel OPLS (KOPLS) presents the advantage of extracting the directions in feature space $\mathcal{F}$ that minimize the residuals of a multiregression scheme that approximates the label matrix (see (14.18)).

To obtain the maximization problem that defines KOPLS we will proceed as usual. Replacing $\tilde{\mathbf{X}}$ by $\tilde{\boldsymbol{\Phi}}$ in (14.15) and using the Representer's Theorem to rewrite the projection operator as $\mathbf{U} = \tilde{\boldsymbol{\Phi}}^\top \mathbf{A}$, we get

$$\text{KOPLS:} \quad \mathbf{A} = \underset{\mathbf{A}}{\arg\max} \quad \text{Tr}\{\mathbf{A}^\top \mathbf{K}_x \mathbf{K}_y \mathbf{K}_x \mathbf{A}\}$$
$$\text{subject to: } \mathbf{A}^\top \mathbf{K}_x^2 \mathbf{A} = \mathbf{I}. \tag{14.28}$$

Applying Lagrange multipliers, it can be shown that the columns of $\mathbf{A}$ are given by the solutions to the following generalized eigenvalue problem:

$$\mathbf{K}_x \mathbf{K}_y \mathbf{K}_x \boldsymbol{\alpha} = \lambda \mathbf{K}_x \mathbf{K}_x \boldsymbol{\alpha}. \tag{14.29}$$

To solve this problem, we can also use an iterative procedure consisting of calculating the leading pair $\{\lambda_i, \boldsymbol{\alpha}_i\}$, and deflating the matrices. The deflation equation for KOPLS is:

$$\mathbf{K}_x \mathbf{K}_y \mathbf{K}_x \leftarrow \mathbf{K}_x \mathbf{K}_y \mathbf{K}_x - \lambda_i \mathbf{K}_x \mathbf{K}_x \boldsymbol{\alpha}_i \boldsymbol{\alpha}_i^\top \mathbf{K}_x \mathbf{K}_x.$$

To understand such a deflation procedure, note that it is equivalent to

$$\tilde{\mathbf{Y}} \leftarrow \tilde{\mathbf{Y}} - \sqrt{\lambda_i} \mathbf{K}_x \boldsymbol{\alpha}_i,$$

Table 14.3 Properties of Kernel MVA methods.

	KPCA	KPLS-SB	KPLS2	KCCA	KOPLS
Max. problem	$\boldsymbol{\alpha}^\top \mathbf{K}_x \mathbf{K}_x \boldsymbol{\alpha}$	$\boldsymbol{\alpha}^\top \mathbf{K}_x \tilde{\mathbf{Y}} \mathbf{v}$	$\boldsymbol{\alpha}^\top \mathbf{K}_x \tilde{\mathbf{Y}} \mathbf{v}$	$\boldsymbol{\alpha}^\top \mathbf{K}_x \tilde{\mathbf{Y}} \mathbf{v}$	$\boldsymbol{\alpha}^\top \mathbf{K}_x \mathbf{K}_y \mathbf{K}_x \boldsymbol{\alpha}$
Constraints	$\boldsymbol{\alpha}^\top \mathbf{K}_x \boldsymbol{\alpha} = 1$	$\boldsymbol{\alpha}^\top \mathbf{K}_x \boldsymbol{\alpha} = 1$ $\mathbf{v}^\top \mathbf{v} = 1$	$\boldsymbol{\alpha}^\top \mathbf{K}_x \boldsymbol{\alpha} = 1$ $\mathbf{v}^\top \mathbf{v} = 1$	$\boldsymbol{\alpha}^\top \mathbf{K}_x^2 \boldsymbol{\alpha} = 1$ $\mathbf{v}^\top \mathbf{C}_{yy} \mathbf{v} = 1$	$\boldsymbol{\alpha}^\top \mathbf{K}_x^2 \boldsymbol{\alpha} = 1$
Max. n_p	$r(\mathbf{K}_x)$	$r(\mathbf{K}_x \tilde{\mathbf{Y}})$	$r(\mathbf{K}_x)$	$r(\mathbf{K}_x \tilde{\mathbf{Y}})$	$r(\mathbf{K}_x \tilde{\mathbf{Y}})$

$r(\cdot)$ denotes the rank of the matrix inside the parentheses. Note that $r(\mathbf{K}_x \tilde{\mathbf{Y}}) \leq \min\{r(\mathbf{K}_x), r(\tilde{\mathbf{Y}})\}$, and $r(\mathbf{K}_x) \leq l$.
To compute projections for new data, all methods require l kernel computations per pattern.

i.e., at each step we are removing from the label matrix the best approximation based on the newly extracted projections of the input data: $\mathbf{K}_x \boldsymbol{\alpha}_i$.

Since deflation of the label matrix decreases by 1 the rank of $\tilde{\mathbf{Y}}$, the rank of $\mathbf{K}_x \mathbf{K}_y \mathbf{K}_x$ also decreases by 1. Accordingly, the maximum number of features that can be extracted with KOPLS is rank$(\mathbf{K}_x \tilde{\mathbf{Y}})$, which for most mapping functions will be $\min\{l, m\}$.

14.3.5 Some considerations about Kernel MVA methods

In this section we have introduced the formulation of some of the most frequently used Kernel MVA algorithms. All these methods require the computation of the kernel matrix of the input data, $\mathbf{K}_x$, and this is precisely their main limitation when dealing with large data sets, since both computation time and storage requirements grow with the square of the number of samples in the training set. Another limitation of Kernel MVA is that matrix $\mathbf{A}$ will in general be dense, making the extraction of features for new data a computationally demanding task, requiring l kernel computations per pattern. In the next section, we will present a recent extension of KOPLS that both obtains sparse solutions and significantly decreases the required computation for the extraction of the projection vectors.

As a summary of this section, Table 14.3 compares some of the most important properties of the methods described.

14.4 Sparse Kernel OPLS

To obtain a more practical scheme, a simple approach is to impose sparsity in the projection vectors representation. Rather than expressing each projection vector as a combination of all training data in feature space, we can use the approximation $\mathbf{U} = \mathbf{\Phi}_R^\top \mathbf{B}$, where $\mathbf{\Phi}_R$ is a subset of the training data containing only R patterns ($R < l$) and $\mathbf{B} = [\boldsymbol{\beta}_1, \cdots, \boldsymbol{\beta}_{n_p}]$ contains the coefficients of the compact model. When doing so, projection of new data is now given by $\tilde{\boldsymbol{\phi}}'(\mathbf{z}) = \tilde{\boldsymbol{\phi}}(\mathbf{z}) \mathbf{\Phi}_R^\top \mathbf{B}$, so that only R kernels per pattern are needed. Note, however, that this approach is not equivalent to simple subsampling of the training data, since all training patterns will still be used in the maximization problem that defines the optimal matrix $\mathbf{B}$.

Regarding the selection of the data points to be included in $\mathbf{\Phi}_R$, several strategies can be adopted to obtain a reduced data set that is as rich as possible. Nevertheless, for not too small R, good results are obtained by a random selection, very much in line with sparse greedy approximation for SVM (Lee and Mangasarian 2001). Therefore, in the following we will rely on this very simple selection strategy.

Next, we explain how this approximation can be exploited to derive a recently proposed method for efficient Kernel OPLS feature extraction (rKOPLS) (Arenas-García *et al.* 2007). Replacing $\tilde{\mathbf{X}}$ and $\mathbf{U}$ in (14.15) by $\tilde{\mathbf{\Phi}}$ and $\mathbf{U} = \mathbf{\Phi}_R^\top \mathbf{B}$, respectively, leads to

$$\begin{aligned} \text{rKOPLS:} \quad & \mathbf{B} = \arg\max_{\mathbf{B}} \ \mathrm{Tr}\{\mathbf{B}^\top \mathbf{K}_R \mathbf{K}_y \mathbf{K}_R^\top \mathbf{B}\} \\ & \text{subject to: } \mathbf{B}^\top \mathbf{K}_R \mathbf{K}_R^\top \mathbf{B} = \mathbf{I}, \end{aligned} \tag{14.30}$$

where we have defined the reduced kernel matrix $\mathbf{K}_R = \mathbf{\Phi}_R \tilde{\mathbf{\Phi}}^\top$ of size $R \times l$. Similarly to the standard KOPLS algorithm, the solution of rKOPLS is obtained by solving

$$\mathbf{K}_R \mathbf{K}_y \mathbf{K}_R^\top \boldsymbol{\beta} = \lambda \mathbf{K}_R \mathbf{K}_R^\top \boldsymbol{\beta}. \tag{14.31}$$

A procedure for the sequential extraction of rKOPLS projections can be given by the iterative application of the two following steps.

1. Obtain the largest eigenvalue of the generalized eigenvalue problem, λ_i, and its corresponding eigenvector, $\boldsymbol{\beta}_i$. Normalize $\boldsymbol{\beta}_i$ to satisfy the constraints in rKOPLS:

$$\boldsymbol{\beta}_i \leftarrow \frac{\boldsymbol{\beta}_i}{\sqrt{\boldsymbol{\beta}_i^\top \mathbf{K}_R \mathbf{K}_R^\top \boldsymbol{\beta}_i}}.$$

2. Deflate the following matrix

$$\mathbf{K}_R \mathbf{K}_y \mathbf{K}_R^\top \leftarrow \mathbf{K}_R \mathbf{K}_y \mathbf{K}_R^\top - \lambda_i \mathbf{K}_R \mathbf{K}_R^\top \boldsymbol{\beta}_i \boldsymbol{\beta}_i^\top \mathbf{K}_R \mathbf{K}_R^\top.$$

   which, as for standard KOPLS, accounts for removing from $\tilde{\mathbf{Y}}$ the best approximation based on the current projections.

Table 14.4 compares the main properties of standard KOPLS and rKOPLS, and illustrates the main advantages that are obtained from using the sparsity constraint for $\mathbf{U}$. These advantages can be summarized as follows:

- rKOPLS only requires the computation of $R \times l$ kernels for the training phase (with $R \ll l$ normally) in contrast to l^2 kernel evaluations for standard kernel MVA methods;
- only R kernel evaluations per pattern are required to extract projections for new data, while l kernel evaluations per pattern are needed for standard kernel MVA methods;
- in spite of $\mathbf{K}_R$ being of size $R \times l$, rKOPLS only requires the explicit calculation of two smaller matrices, $\mathbf{K}_R \mathbf{K}_Y \mathbf{K}_R^\top$ and $\mathbf{K}_R \mathbf{K}_R^\top$, both of size $R \times R$;
- parameter R acts as a sort of regularizer, so that matrix $\mathbf{K}_R \mathbf{K}_R^\top$ is better conditioned than the equivalent $\mathbf{K}_x^2$ for standard KOPLS.

Table 14.4 Comparison of standard KOPLS and rKOPLS algorithms.

	KOPLS	rKOPLS
Max. problem	$\boldsymbol{\alpha}^\top \mathbf{K}_x \mathbf{K}_y \mathbf{K}_x \boldsymbol{\alpha}$	$\boldsymbol{\beta}^\top \mathbf{K}_R \mathbf{K}_y \mathbf{K}_R^\top \boldsymbol{\beta}$
Constraints	$\boldsymbol{\alpha}^\top \mathbf{K}_x^2 \boldsymbol{\alpha} = 1$	$\boldsymbol{\beta}^\top \mathbf{K}_R \mathbf{K}_R^\top \boldsymbol{\beta} = 1$
Kernel matrix size	$l \times l$	$R \times l$
Storage	$O(l^2)$	$O(R^2)$
No. of nodes	l	R
Max. n_p	$r(\mathbf{K}_x \tilde{\mathbf{Y}})$	$r(\mathbf{K}_R \tilde{\mathbf{Y}})$

$r(\cdot)$ denotes the rank of the matrix inside the parentheses. Note that $r(\mathbf{K}_R \tilde{\mathbf{Y}}) \leq \min\{r(\mathbf{K}_R), r(\tilde{\mathbf{Y}})\}$, and $r(\mathbf{K}_R) \leq R$.

Finally, we would like to mention that imposing sparsity in the same way that we have considered for rKOPLS is also a valid approach for most MVA methods. As an important exception, algorithm KPLS2 cannot benefit from this possibility, given that the deflation scheme would still require the evaluation of the complete kernel matrix.

14.5 Experiments: pixel-based hyperspectral image classification

14.5.1 Data set description and experimental setup

In this section, we analyse the performance of different linear and nonlinear MVA methods for feature extraction in remote sensing classification tasks, paying a special attention to the rKOPLS algorithm that we have just described. In particular, we focus on a typical case of high dimensional image classification in remote sensing: pixel-based hyperspectral image classification. In such a case, the input space is normally redundant because of the collinearity introduced by the (locally stationary) spatial features.

For our experiments, we have used the standard AVIRIS image taken over NW Indiana's Indian Pine test site in June 1992. We removed 20 noisy bands covering the region of water absorption, and finally worked with 200 spectral bands. The high number of narrow spectral bands induce a high collinearity between them. Discriminating among the major crops in the area can be very difficult (in particular, given the moderate spatial resolution of 20 metres), which has made the scene a challenging benchmark to validate classification accuracy of hyperspectral imaging algorithms. The calibrated data is available online (along with detailed ground-truth information) from http://dynamo.ecn.purdue.edu/~biehl/MultiSpec. In all our experiments we used the whole scene, consisting of the full 145×145 pixels, which contains 16 classes, ranging in size from 20 to 2468 pixels, and thus constituting a very difficult situation. Among the 10 366 pixels for which labels are available, 20% of the data is used for training the feature extractors, keeping the remaining 80% to test the discriminative power of the extracted features.

The suitability of MVA methods in general, and of the rKOPLS method in particular, to classification setups is guaranteed if an appropriate labelling scheme (e.g. 1-of-C encoding) is used to encode class membership into the label matrix. When doing so, features revealed to be useful to approximate $\mathbf{Y}$ can also be expected to provide useful information for predicting the data class. See, e.g., Rifkin and Klautau (2004) for a discussion about the application of square loss to classification problems.

To keep things simple, the discriminative power of all (linear and nonlinear) extracted features was tested using a simple classifier consisting of a linear least squares model followed by a 'winner-takes-all' activation function (Bishop 1996). To be more specific, we first obtain the optimal regression matrix for the least squares problem, i.e., $\mathbf{W} = \tilde{\mathbf{X}}'^{\dagger}\tilde{\mathbf{Y}}$ and $\mathbf{W} = \tilde{\mathbf{\Phi}}'^{\dagger}\tilde{\mathbf{Y}}$ for the linear and kernel methods, respectively, with $\mathbf{Y}$ being the target matrix using 1-of-C encoding scheme. After obtaining $\mathbf{W}$ with the training data, the estimated class of new patterns is estimated as $\hat{\mathbf{y}} = \text{w.t.a.}[\mathbf{W}^\top\tilde{\mathbf{x}}]$ for the linearly extracted features and $\hat{\mathbf{y}} = \text{w.t.a.}[\mathbf{W}^\top\tilde{\boldsymbol{\phi}}(\mathbf{x})]$ for the kernel methods.

14.5.2 Results description

Figure 14.1 shows the overall classification accuracy and kappa statistic when using the features extracted using the linear MVA methods, for a varying number of features n_p. First, note that the overall accuracy and kappa statistic curves exhibit very similar behaviors, which suggests that the obtained models are not unbiased in terms of accuracy. Note also how OPLS outperforms all other methods for any number of extracted features, confirming the superior discriminative power of the projections extracted using this method. The only other method that somehow gets close in behavior to OPLS is CCA. However, note that OPLS projections are easier to compute, since CCA comes as the solution to a more complex generalized eigenvalue problem. When the maximum number of projections are used, all methods except PLS-SB achieve the same error, but PCA and PLS2 require 200 features (i.e., the dimensionality of the input space), while 15 projections suffice for CCA and OPLS.

Figure 14.2 displays the classification maps using the features extracted with linear PCA (for $n_p = 10$ and $n_p = 100$), linear OPLS (for $n_p = 10$), and its kernel version rKOPLS. At first glance, it may appear that using PCA with 100 projections provides somehow more noisy results than when using only the ten first features extracted by PCA. Studying the

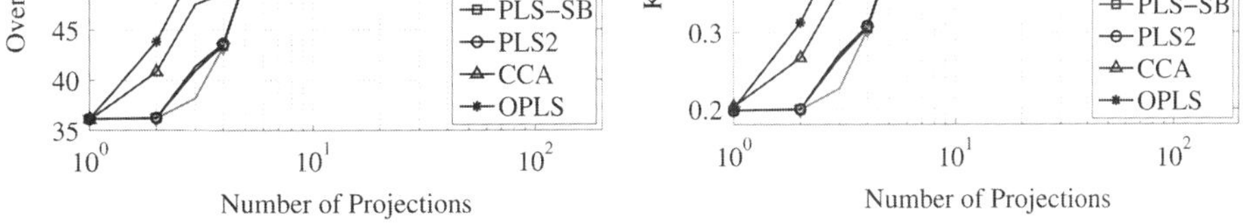

Figure 14.1 Average classification accuracy (%) (left) and kappa statistic (right) as a function of the number of projections extracted by the linear MVA methods.

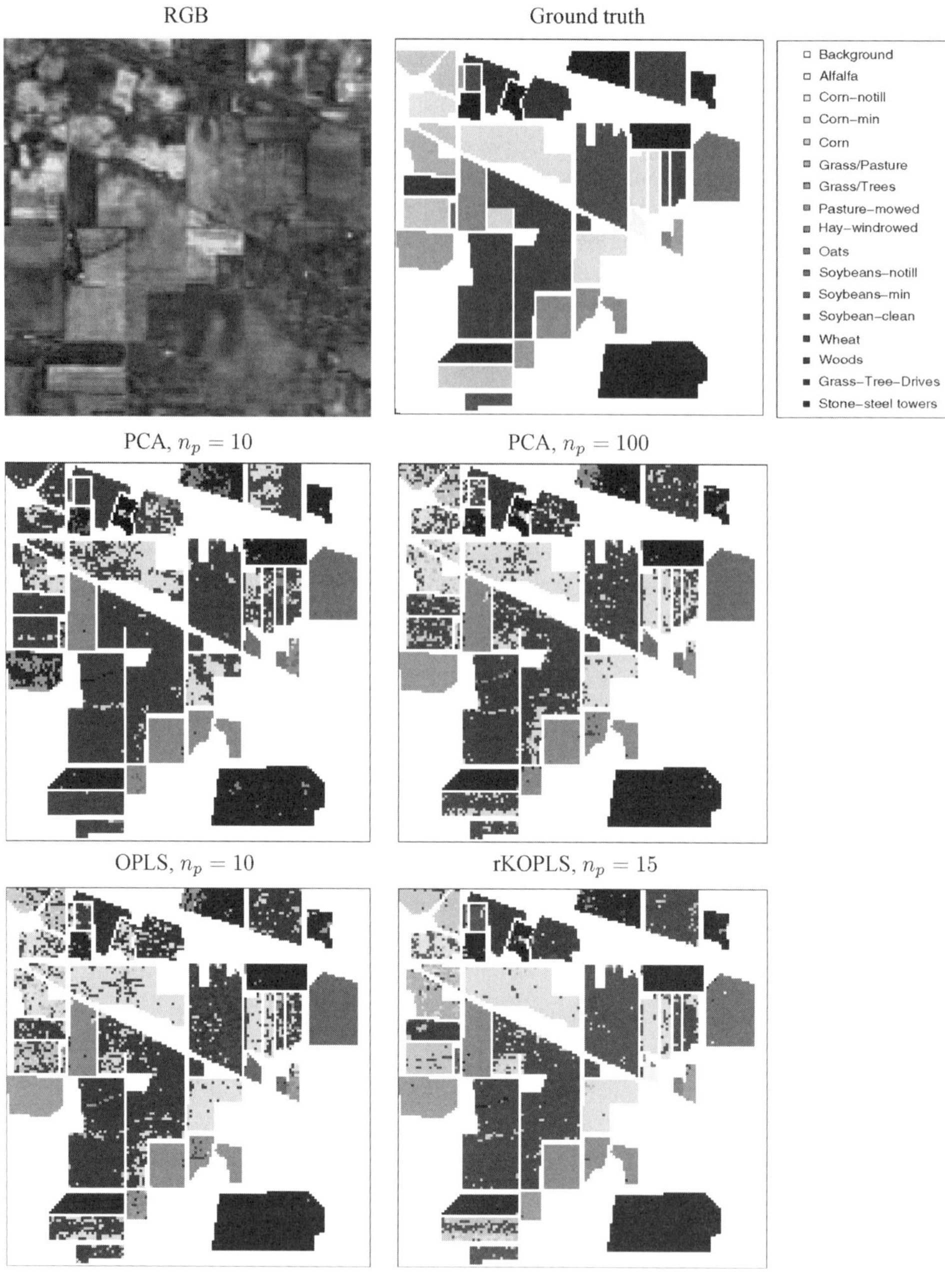

Figure 14.2 RGB composition and ground truth, along with the classification maps for Indiana's Indian Pine image, using the first n_p features extracted by different MVA methods: linear PCA, linear OPLS and rKOPLS.

images in more detail and comparing them with the true label information, it is evident that PCA with $n_p = 10$ lacks expressive power, and leads to incorrect classification in many of the crops that constitute the image. OPLS with $n_p = 10$ obtains similar results than PCA with $n_p = 100$, which illustrates that OPLS features are able to better summarize the most discriminative information contained in the input variables. The images also suggest that it would be beneficial to exploit spatial context in order to further improve the accuracy of the classifier, for instance by adding a postprocessing stage where such spatial information is taken into account.

For the nonlinear case, we have considered KPCA, KPLS2 and KOPLS with Gaussian kernel $k(\mathbf{x}_i, \mathbf{x}_j) = \exp(-\|\mathbf{x}_i - \mathbf{x}_j\|_2^2/2\sigma^2)$, using a ten-fold cross-validation procedure on the training set to select σ. Figure 14.3 displays the classification accuracy for the three methods. The sparse rKOPLS method (with $R = 1000$) was used instead of the more complex standard KOPLS. The same conclusions obtained for the linear case also apply to MVA methods in feature space. Therefore, we see that rKOPLS achieves a better performance than the other two methods, which require a much larger number of projections to achieve comparable performance. KPLS2 also outperforms KPCA, confirming the convenience of using the target matrix when feature extraction is considered for supervised learning. In the limit of n_p, we should stress that the behavior of the three methods would be very similar. However, we can obtain much better results from rKOPLS with a lower computational and memory burden.

When comparing Figures 14.1 and 14.3, we can also see that nonlinear methods obtain much higher recognition rates than linear MVA techniques (72 % vs. 85 %, approximately), demonstrating the superior expressive power of the kernel extensions. The superiority of nonlinear techniques is also clear when comparing in Figure 14.2 the class predictions of OPLS and rKOPLS for the different pixels in the image.

To finish the section, we have also studied the influence of parameter R on rKOPLS performance. Figure 14.4 shows rKOPLS recognition rates for $n_p = 15$ and for varying R. For this case, we have considered that rKOPLS features were used as the inputs of two different classifiers: (1) a simple linear model trained with the pseudoinverse of the projections (as illustrated before); and (2) a linear SVM with parameter C selected using ten-fold cross-validation. The

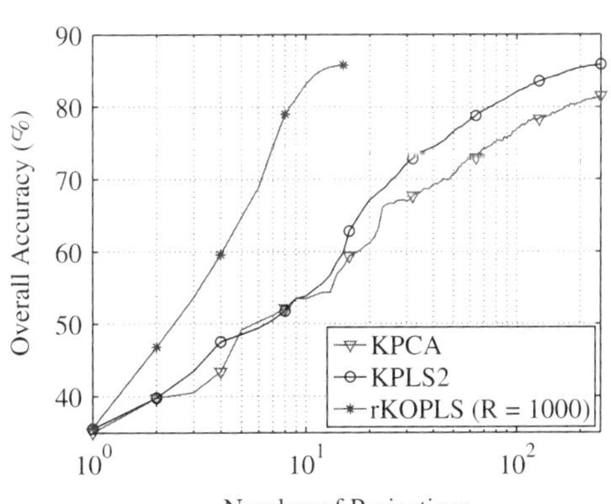

Figure 14.3 Overall accuracy as a function of the number of extracted features for different MVA schemes: KPCA, KPLS2 and the sparse KOPLS method (rKOPLS) with $R = 1000$.

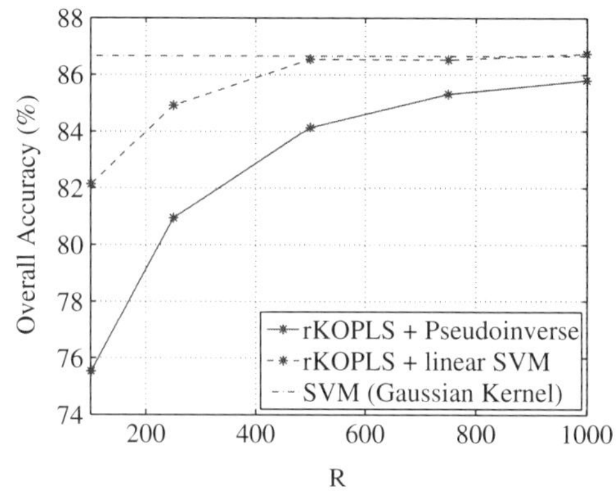

Figure 14.4 Overall accuracy of two different linear classification models using rKOPLS features as input variables. Performance is shown as a function of parameter R. The accuracy of an RBF-SVM classifier is also indicated for reference purposes.

accuracy of an RBF-SVM is also displayed in the figure as a reference for comparison. It can be seen that the offered (linear) solution rapidly approximates the state-of-the-art (nonlinear) SVM classifier as R is increased. The RBF-SVM solution needed 1295 support vectors to reach these results. We would like to emphasize that rKOPLS imposes sparsity in a much more effective way than a simple subsampling, since matrix $\mathbf{K}_R$ still retains information about all training points.

14.6 Conclusions

In this chapter we have reviewed several MVA techniques for feature extraction, providing a uniform treatment of PCA, PLS, CCA and OPLS. All these methods can be formulated as the solution to generalized eigenvalue or singular value decomposition problems, and can therefore be solved with standard techniques for matrix analysis and decomposition. In spite of its popularity, PCA presents the limitation of not using the target data, thus performing suboptimally in supervised problems. Several results and optimality conditions have been presented concerning all four methods. In particular, OPLS enjoys the important property of minimizing the approximation error when performing multiregression of the label matrix.

The main drawback of the previous algorithms is that they are constrained to linear projections. However, their expressivity can easily be extended by applying kernel methods. After reviewing KPCA, KPLS, KCCA and KOPLS, we have described a sparse version of KOPLS, rKOPLS, which provides good approximations of the original solution, while significantly reducing the computational burden of the training phase, and providing more compact machines.

The discriminative power of these MVA methods has been studied on a classification setup, where the better expressive power of the linear methods over the nonlinear ones was shown, as well as the convenience of using the target data for training the projection vectors. This suggests that (K)PLS, and especially (K)CCA and (K)OPLS, should be preferred to (K)PCA in supervised problems.

Acknowledgments

The work of J. Arenas-García was partly supported by the Spanish Ministry of Science and Innovation under grant TEC-2008-02473 and by Madrid Community grant S-505/TIC/0223.

References

Arenas-García, J. and Camps-Valls, G. (2008) Efficient kernel orthonormalized PLS for remote sensing applications. *IEEE Trans. Geoscience and Remote Sensing*, **46**, 2872–2881.

Arenas-García, J., Petersen, K.B. and Hansen, L.K. (2007) Sparse kernel orthonormalized PLS for feature extraction in large data sets *Advances in Neural Information Processing Systems, 19*. MIT Press.

Banerjee, A., Burlina, P. and Diehl, C.P. (2008) A support vector method for anomaly detection in hyperspectral images. *IEEE Trans. Geoscience and Remote Sensing*, **44**, 2282–2291.

Baudat, G. and Anouar, F. (2000) Generalized discriminant analysis using a kernel approach. *Neural Computation*, (12), 2385–2404.

Bishop, C.M. (1996) *Neural Networks for Pattern Recognition*. Oxford University Press.

Camps-Valls, G. and Bruzzone, L. (2005) Kernel-based methods for hyperspectral image classification. *IEEE Trans. Geoscience and Remote Sensing*, **43**, 1351–1362.

Geladi, P. (1988) Notes on the history and nature of partial least squares (PLS) modelling. *Journal of Chemometrics*, **2**, 231–246.

Guyon, I. and Elisseff, A. (2003) An introduction to variable and feature selection. *Journal of Machine Learning Research*, **3**, 1157–1182.

Hardoon, D.R., Szedmak, S. and Shawe-Taylor, J. (2003) Canonical correlation analysis: An overview with application to learning methods. Technical report.

Hoegaerts, L., Suykens, J.A.K., Vanderwalle, J., and Moor, B.D. (2004) Primal space sparse kernel partial least squares regression for large problems *Proc. Intl. Joint Conference on Neural Networks (IJCNN)*, pp. 561–566.

Hoskuldsson, A. (1988) PLS regression methods. *Journal of Chemometrics*, **2**, 211–228.

Hotelling, H. (1936) Relations between two sets of variates. *Biometrika*, **28**, 321–377.

Jollife, I.T. (1986) *Principal Component Analysis*. Springer-Verlag.

Lai, P.L. and Fyfe, C. (2000) Kernel and nonlinear canonical correlation analysis. *International Journal of Neural Systems*, **10**(5), 365–377.

Lee, Y.J. and Mangasarian, O.L. (2001) RSVM: Reduced support vector machines. *Data Mining Institute Technical Report 00-07, July 2000. CD Proceedings of the SIAM International Conference on Data Mining, Chicago, April 5–7.*

Mika, S., Ratsch, G., Weston, J., Schölkopf, B. and Muller, K.R. (1999) Fisher discriminant analysis with kernels *Proc. Neural Networks for Signal Processing*, pp. 41–48.

Momma, M. and Bennett, K. (2003) Sparse kernel partial least squares regression *Proc. Conference on Learning Theory (COLT)*, pp. 216–230.

Muñoz-Marí, J., Bruzzone, L. and Camps-Valls, G. (2007) A support vector domain description approach to supervised classification of remote sensing images. *IEEE Trans. Geoscience and Remote Sensing*, **45**, 2683–2692.

Pearson, K. (1901) On lines and planes of closest fit to systems of points in space. *Philosophical Magazine*, **2**(6), 559–572.

Rannar, S., Lindgren, F., Geladi, P. and Wold, S. (1994) A PLS kernel algorithm for data sets with many variables and fewer objects. Part 1: Theory and algorithm. *Journal of Chemometrics*, **8**, 111–125.

Rifkin, R. and Klautau, A. (2004) In defense of one-vs-all classification. *Journal of Machine Learning Research*, **5**, 101–141.

Rosipal, R. and Kramer, N. (2006) Overview and recent advances in partial least squares *Subspace, Latent Structure and Feature Selection Techniques*, Springer.

Rosipal, R. and Trejo, L.J. (2001) Kernel partial least squares regression in reproducing kernel Hilbert space. *Journal of Machine Learning Research*, **2**, 97–123.

Rosipal, R., Trejo, L.J. and Matthews, B. (2003) Kernel PLS-SVC for linear and nonlinear classification *Proc. Intl. Conf. on Machine Learning (ICML)*, pp. 640–647.

Roweis, S. and Brody, C. (1999) Linear heteroencoders. Technical report.

Sampson, P.D., Streissguth, A.P., Barr, H.M. and Bookstein, F.L. (1989) Neurobehavioral effects of prenatal alcohol: Part II. Partial Least Squares analysis. *Neurotoxicology and teratology*, **11**, 477–491.

Schölkopf, B. and Smola, A. (2002) *Learning with Kernels*. MIT Press.

Schölkopf, B., Smola, A. and Muller, K.R. (1998) Non linear component analysis as kernel eigenvalue problem. *Neural Computation*, **10**, 1299–1319.

Shawe-Taylor, J. and Cristianini, N. (2004) *Kernel Methods for Pattern Analysis*. Cambridge University Press.

Wegelin, J.A. (2000) A survey of partial least squares (PLS) methods, with emphasis on the two-block case. Technical report.

Wold, H. (1966a) *Research Papers in Statistics*, Wiley, NY p. 411.

Wold, H. (1966b) *Multivariate Analysis*, Academic Press p. 391.

Wold, H. (1975) Path models with latent variables: the NIPALS approach. *Quatitative Sociology: International Perspectives on Mathematical and Statistical Model Building*, pp. 307–357.

Wold, S., Albano, C., Dunn, W.J., Edlund, U., Esbensen, K., Geladi, P., Hellberg, S., Johansson, E., Lindberg, W. and Sjoström, M. (1984) Multivariate data analysis in chemistry *Chemometrics, Mathematics and Statistics in Chemistry*, Reidel Publishing Company, p. 17.

Worsley, K., Poline, J., Friston, K. and Evans, A. (1998) Characterizing the response of pet and fMRI data using multivariate linear models (MLM). *Neuroimage*, **6**, 305–319.

15

KPCA algorithm for hyperspectral target/anomaly detection

Yanfeng Gu
College of Electronics and Information Engineering, Harbin Institute of Technology, Harbin, P. R. China

Target detection and anomaly detection are two important applications of hyperspectral imagery. High dimensionality and complex (nonlinear) correlation between spectral bands make difficult target/anomaly detection in hyperspectral imagery. In addition, changing imaging conditions also make the same materials have potentially different spectral observations. Kernel methods (KM) provide powerful approaches to solve these problems. In this chapter we will introduce important applications of KM to target/anomaly detection, in order to construct effective algorithms for feature extraction. By means of these algorithms, the target detection and anomaly detection performance is greatly improved.

15.1 Introduction

Recently, hyperspectral images have been widely applied in many civilian and military fields, such as resource surveys, prospecting, environment inspection, object detection, etc. Accordingly, more and more researchers are attracted into research into hyperspectral image processing and applications that include dimensionality reduction, feature extraction and selection, classification and detection. Target detection and anomaly detection are two important applications of hyperspectral imagery. The main objective of target detection is to search the

Kernel Methods for Remote Sensing Data Analysis Edited by Gustavo Camps-Valls and Lorenzo Bruzzone

pixels in the hypercube for the presence of a specific material (a 'target'). Unlike target detection, anomaly detection attempts to locate anything that looks different spatially or spectrally from its surroundings. In spectral anomaly detectors, no *prior* knowledge of the target spectral signature is utilized or assumed.

High dimensionality and high (possibly nonlinear) correlation between spectral bands make target detection and anomaly detection difficult in hyperspectral imagery. In addition, changing image conditions can also make the same materials potentially have different spectral observations. Thus, feature extraction and dimensionality reduction become key procedures before detection or in the detection process itself. Kernel Methods (KM) currently provide powerful approaches to solving these problems.

In this chapter, we introduce important applications of KM to target detection and anomaly detection in hyperspectral imagery, and focus on constructing effective kernel algorithms for feature extraction and dimensionality reduction for the real tasks of target and anomaly detection. After using these algorithms for feature extraction, the performance of target detection and anomaly detection is greatly improved.

The remainder of this chapter is organized as follows. In Section 15.2, the motivation for introducing KM into hyperspectral target/anomaly detection is given. In Sections 15.3–15.5, three important aspects of hyperspectral images processing are described, including feature extraction for dimensionality reduction, target detection and anomaly detection. In these sections, we describe concrete applications of KM in hyperspectral images, and provide kernel-based algorithms for feature extraction, target detection and anomaly detection. Finally, conclusions are drawn.

15.2 Motivation

15.2.1 Feature extraction of hyperspectral images

In most cases, hyperspectral sensors oversample the *spectral* signal to ensure that any narrow features are adequately represented. In some cases, sensors may also oversample the *spatial* signal. An important function of hyperspectral signal processing is to eliminate redundancy in the spectral and spatial sample data while preserving the high-quality features required by detection, discrimination and classification. This dimensionality reduction is implemented in a scene-dependent (adaptive) manner and may be implemented as a distinct step in the processing or as an integral part of the overall algorithm. Regardless of how it is implemented, the dimensionality reduction algorithm must be designed to preserve the information of interest to downstream detection, classification, or spectral unmixing algorithms. Since dimensionality reduction with a small loss of information is always possible, the reduction can be achieved without significantly degrading detection performance or decreasing the separability between the different classes (classification performance). Dimensionality reduction leads to significant reductions in computational complexity and also reduces the number of pixels required to obtain statistical estimates of a given accuracy. Remember that the number of samples (pixels) required to obtain a statistical estimate with a given accuracy increases drastically with the dimensionality of the data. The most widely used algorithm for dimensionality reduction is principal component analysis (PCA) or, equivalently, Karhünen–Loéve transformation (Landgrebe 2002; Shaw 2002; Stein 2002; Manolakis 2002; Keshava 2002).

15.2.2 Introducing KM for hyperspectral image processing

Now let us observe correlation characteristics among spectral bands of hyperspectral images. Let I be the b-dimensional data space of hyperspectral images. We construct band vectors by using all pixels in the hyperspectral image of a band. For example, the band vector V_i, $V_i = [V_{i1}, V_{i2}, \ldots, V_{im}]$, where $m = l \times p$ is the total number of pixels resident in the hyperspectral image) consists of all pixels of the hyperspectral image, and V_{ik}, $k = 1, 2, \ldots, m$ represents the kth pixel. The correlation coefficient r_{ij} between the band vector V_i and the band vector V_j is defined as follows:

$$r_{ij} = \sum_{k=1}^{m} (V_{ik} - \bar{V}_i)(V_{jk} - \bar{V}_j) \Bigg/ \sqrt{\sum_{k=1}^{m} (V_{ik} - \bar{V}_i)^2 \sum_{k=1}^{m} (V_{jk} - \bar{V}_j)^2}, \qquad (15.1)$$

where $\bar{V}_i$ and $\bar{V}_j$ are the mean vectors of V_i and V_j, respectively.

According to Equation (15.1), we can calculate the correlation matrix of all bands of hyperspectral data, which consists of $b \times b$ elements in r_{ij}. The correlation matrix of a real hyperspectral image is illustrated in Figure 15.1. Here, the standard Indian Pine Test Site scene, Northwestern Indiana June 1992, is used; it contains a mixed agriculture/forestry landscape. The blocky aspect of the correlation matrix is evident.

The content mentioned above just concerns correlation characteristics between spectral bands. It combines two physical characteristics of hyperspectral data: hyperspectral data contain a lot of redundant information, and the correlation of neighbouring bands is higher than those of bands that are far from each other. In addition, the local correlation of neighbouring bands has approximate continuity and transmissibility. The complicated correlation characteristics between spectral bands make dimensionality reduction, feature extraction and

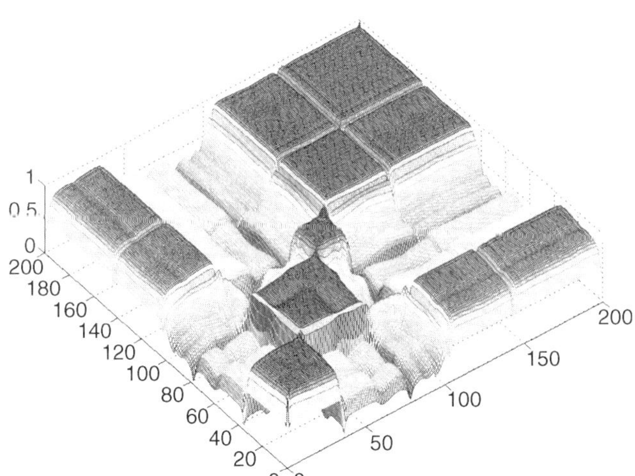

Figure 15.1 Correlation matrix of the hyperspectral Indian Pine Test Site scene, Northwestern Indiana showing an evident blocky characteristic.

downstream detection lead to a huge challenge. In particular, for the task of target detection and anomaly detection, important information about target or anomaly pixels is hidden or mixed in background clutters generally. Most linear methods, typically used in pattern analysis, however, cannot effective concentrate and extract useful features from high-dimensional data including a lot of redundant information, due to the nonlinearity nature of hyperspectral images. In contrast to linear algorithms, KM have shown huge potential in aspects of processing nonlinear problems. This is the motivation that we need to introduce KM into hyperspectral image processing.

15.2.3 Hyperspectral images for numerical experiments

In order to test the effectiveness of the corresponding kernel-based hyperspectral image processing methods, numerical experiments are performed on real AVIRIS data. After those bands that are corresponding to the water absorption regions are removed, low SNR and bad bands, 126 available bands remained in the 0.4–1.8μm wavelength range. A scene of 400×400 pixels in size was selected for our experiments, in which there are mainly nine classes of ground cover, most of which are very different in terms of their spectral characteristics. The ground truth map is available. The ground sampling distance of the hyperspectral images is 3.5 m. The sixth band of the hyperspectral images that includes the testing scenes is shown in Figure 15.2. From this scene, we select two regions one for target detection and one for anomaly detection. Figure 15.3 illustrates real hyperspectral images and the corresponding distribution of target/anomaly pixels from the hyperspectral images scenes for target detection and anomaly detection, respectively.

Figure 15.2 The 10th band from real hyperspectral images used in numerical experiments.

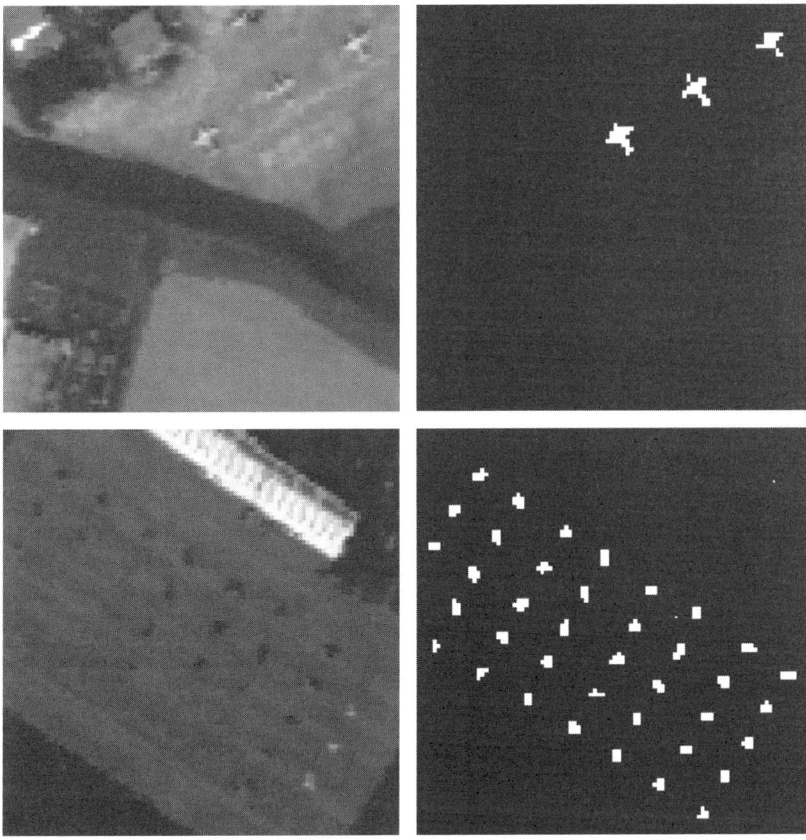

Figure 15.3 Real scene and truth distribution of panels for (top) target detection, (bottom) anomaly detection.

15.3 Kernel-based feature extraction in hyperspectral images

15.3.1 Principal component analysis

Dimensionality reduction of hyperspectral data usually starts with a correlation between hyperspectral bands. The conventional approach to deal with such a problem of information redundancy is principal component analysis (PCA), which is an orthogonal linear transformation. PCA has been applied to multi-spectral images for optimal band selection (Jia and Richards 1999). But the results from PCA are optimal only in the sense of minimum squared error (MSE) but definitely non-optimal in terms of class discrimination and separability. In addition, PCA in full hyperspectral data space has large computational loads (Gu *et al.* 2002).

Given a set of samples $\mathbf{z}_i$, $i = 1, 2, \ldots, N$, with $\mathbf{z}_i \in \mathbb{R}^d$, it is necessary to first calculate the mean μ_z and the covariance matrix Σ_z before performing PCA. The PCA transform can be written as follows

$$\mathbf{z}_p = \mathbf{T}_p^\top(\mathbf{z} - \mu_z), \tag{15.2}$$

where $\mathbf{z}_p$ is the transformed data, and $\mathbf{T}_p$ is the transform matrix. $\mathbf{T}_p$ is composed of the eigenvectors obtained by performing the eigenvalue decomposition on Σ_z, namely,

$$\lambda\boldsymbol{\Sigma}_z = \lambda\mathbf{v}, \tag{15.3}$$

where $\mathbf{v} = [\mathbf{v}^1, \mathbf{v}^2, \ldots, \mathbf{v}^N]$, $\mathbf{v}^i = [v_{i1}, v_{i2}, \ldots, v_{id}]^\top$ is the ith eigenvector of $\boldsymbol{\Sigma}_z$ that corresponds to the ith eigenvalue λ_i of $\boldsymbol{\Sigma}_z$.

15.3.2 Kernel mapping

In most cases, original hyperspectral image data have nonlinear characteristics. In order to better extract nonlinear features from hyperspectral data, KM can be effectively used. In this section, we introduce a method for construction of Mercer kernel in *Reproducing Kernel Hilbert Space* (RKHS).

According to statistical learning theory, in order to construct the optimal hyperplane in high dimensional space, only the dot product of support vectors in feature space is required, and one does not need to know the specific form of mapping $\boldsymbol{\phi}$, which maps samples from input space into feature space.

Now we consider an expression for the dot product in Hilbert space:

$$\langle\boldsymbol{\phi}(\mathbf{x}), \boldsymbol{\phi}(\mathbf{y})\rangle = K(\mathbf{x}, \mathbf{y}). \tag{15.4}$$

According to the Hilbert–Schmidt's theorem, $K(\mathbf{x}, \mathbf{y})$ is any symmetric function that satisfies Mercer's condition.

15.3.3 Kernel Principal Component Analysis (KPCA)

Given a set of centred samples $\mathbf{y}_k$, $k = 1, 2, \ldots, N$, $\mathbf{y}_k \in \mathbb{R}^d$, $\sum_{k=1}^{l} \mathbf{y}_k = 0$, and d is the dimensionality of the samples. First, the sample set $\{\mathbf{y}_k\}$ is mapped into a feature space $\mathbb{R}^X$ by $\boldsymbol{\phi} : \mathbb{R}^d \to \mathbb{R}^X$ and the covariance matrix $\boldsymbol{\Sigma}_{\boldsymbol{\phi}}$ is computed in the feature space $\mathbb{R}^X$,

$$\boldsymbol{\Sigma}_{\boldsymbol{\phi}} = \frac{1}{N}\sum_{k=1}^{N} \boldsymbol{\phi}(\mathbf{y}_k)\boldsymbol{\phi}(\mathbf{y}_k)^\top, \tag{15.5}$$

where $\boldsymbol{\phi}(\mathbf{y}_k)$ is a centred sample in the feature space.

Let $\mathbf{v} \neq 0$ be an eigenvector of $\boldsymbol{\Sigma}_{\boldsymbol{\phi}}$ that corresponds to a positive eigenvalue λ of $\boldsymbol{\Sigma}_{\boldsymbol{\phi}}$. So the eigenvector is in the space spanned by the mapped samples, i.e. $\mathbf{v} \in span\{\boldsymbol{\phi}(\mathbf{y}_1), \boldsymbol{\phi}(\mathbf{y}_2), \ldots, \boldsymbol{\phi}(\mathbf{y}_N)\}$. This can be described as

$$\mathbf{v} = \sum_{k=1}^{N} \beta_k\boldsymbol{\phi}(\mathbf{y}_k), \tag{15.6}$$

where β_k is the weighting coefficient of $\boldsymbol{\phi}(\mathbf{y}_k)$. Thus, eigenvalue decomposition can be written as

$$\lambda \mathbf{v} = \boldsymbol{\Sigma}_{\boldsymbol{\phi}} \mathbf{v}. \tag{15.7}$$

Furthermore, we multiply both sides of Equation (15.7) by $\boldsymbol{\phi}(\mathbf{y})$ and obtain the following equation:

$$\lambda(\boldsymbol{\phi}(\mathbf{y})\mathbf{v}) = (\boldsymbol{\phi}(\mathbf{y})\boldsymbol{\Sigma}_{\boldsymbol{\phi}}\mathbf{v}). \tag{15.8}$$

For all d eigenvectors, Equation (15.13) can be also written as

$$\lambda_i(\boldsymbol{\phi}(\mathbf{y})\mathbf{v}^i) = (\boldsymbol{\phi}(\mathbf{y})\boldsymbol{\Sigma}_{\boldsymbol{\phi}}\mathbf{v}^i), \tag{15.9}$$

where $i = 1, 2, \ldots, d$, $\mathbf{v}^i = \sum_{k=1}^{N} \beta_k^i \boldsymbol{\phi}(\mathbf{y}_k)$. According to the construction of the kernel function, define an $d \times d$ dimensionality of kernel matrix $\mathbf{K}$ in feature space as

$$(\mathbf{K})_{i,j} = (\boldsymbol{\phi}(\mathbf{y}_i) \cdot \boldsymbol{\phi}(\mathbf{y}_j)) = k(\mathbf{y}_i, \mathbf{y}_j) \tag{15.10}$$

and consider an eigenvalue decomposition for the expansion coefficients $\boldsymbol{\beta}^i = [\beta_{i1}, \beta_{i2}, \ldots, \beta_{id}]^\top$ by using kernel matrix $\mathbf{K}$ as

$$\lambda_i \boldsymbol{\beta}^i = \mathbf{K}\beta^i. \tag{15.11}$$

The solution obtained $(\lambda_i, \boldsymbol{\beta}^i)$ has to be normalized by imposing $\lambda_i(\beta^i \boldsymbol{\beta}^i) = 1$ and to be centred by substituting centred $\mathbf{K}_c$ for the $\mathbf{K}$. Matrix $\mathbf{K}_c$ is given by

$$\mathbf{K}_c = \mathbf{K} - \mathbf{1}_p \cdot \mathbf{K} - \mathbf{K} \cdot \mathbf{1}_p + \mathbf{1}_p \cdot \mathbf{K} \cdot \mathbf{1}_p, \tag{15.12}$$

where $\mathbf{1}_p$ is a $p \times p$ matrix with all elements equal to $1/p$. To obtain the principal components in feature space, the procedure, which projects the mapped sample $\boldsymbol{\phi}(\mathbf{y})$ on to vector $\mathbf{v}^i$, is necessary, namely,

$$\langle \mathbf{v}^i, \boldsymbol{\phi}(\mathbf{y}) \rangle = \sum_{k=1}^{N} \beta_k^i \langle \boldsymbol{\phi}(\mathbf{y}_k), \boldsymbol{\phi}(\mathbf{y}) \rangle = \sum_{k=1}^{N} \beta_k^i k(\mathbf{y}_k, \mathbf{y}). \tag{15.13}$$

The content given above is the basic theory of KPCA. The method has excellent ability to process nonlinear data (Schölkopf *et al.* 1998; Müller *et al.* 2001; Ruiz and Lopez-de-Teruel 2001). Figure 15.4 shows the excellent nonlinear performance of feature extraction of KPCA compared with PCA. Here, notice that the original hyperspectral images are represented by data cubes and hence the data have to be first converted into two-dimensional form before they are transformed. Each band of the HSI is viewed as a sample vector, and the number of total samples to be transformed is equal to the number of the original bands. Thus, after these samples are centred and normalized, KPCA is performed on them (Schölkopf *et al.*

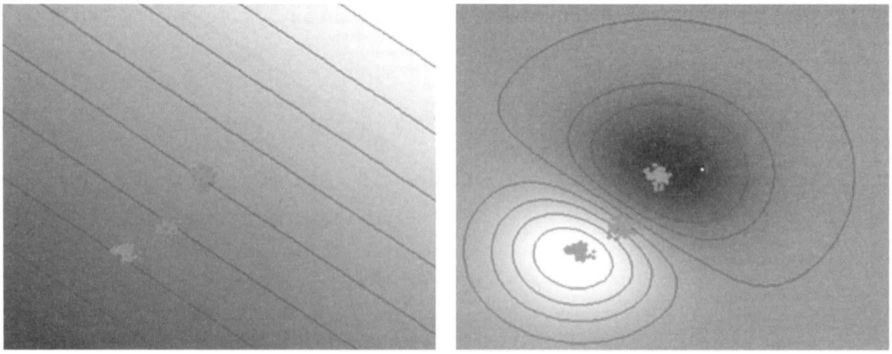

(a) (b)

Figure 15.4 The 1st linear and the 1st nonlinear features extracted by (a) PCA and (b) KPCA.

1998, 1999, 2000; Müller *et al.* 2001; Ruiz and Lopez-de-Teruel 2001). See also Chapters 2 and 14.

In practical applications, such as the subsequent target/anomaly detection, the selection of the kernel function is a key step for the full mining potential of KM in the application of hyperspectral images. In the proposed kernel-based feature extraction methods for target/anomaly detection in hyperspectral images, the Gaussian Radial Basis Function (RBF) as $(\mathbf{K})_{ij} = K(\mathbf{y}_i, \mathbf{y}_j) = \exp(-\|\mathbf{y}_i - \mathbf{y}_j\|^2)$ is adopted. Kwon *et al.* (2005) enumerates two main reasons why the Gaussian RBF kernel is chosen. The first reason is that this kernel is translation-invariant, and the second is that the Fourier transform of this kernel is also Gaussian. The hyperspectral target/anomaly detector is generally obtained under the Gaussian assumption, so the Gaussian RBF kernel is very suitable for the application of target/anomaly detection. The choice of the parameter is very critical. It should be carefully chosen so that the overall data variations can be fully exploited by the Gaussian RBF kernel. It is noticeable that the test hyperspectral images are first normalized by a constant before KPCA is performed in the subsequent target/anomaly detection. Generally, the constant is set to the maximum value obtained from all the spectral components of the spectral vectors in the corresponding test images. So the entries of the normalized pixel vectors fit into the interval of spectral values between zero and one. The rescaling of pixel vectors was mainly performed to utilize the dynamic range of the Gaussian RBF kernel effectively (Kwon 2005).

15.4 Kernel-based target detection in hyperspectral images

The process of detecting and identifying a target in hyperspectral imagery can be considered as consisting of two stages. The first stage is an anomaly detector that identifies spectral anomalies or a localized spectral difference. The second stage identifies whether or not the anomaly is a target or natural clutter. This stage can be achieved if the spectral signature of the target is known; this can be obtained from a spectral library or a spectral matched filter designed from a set of training data.

15.4.1 The concept of target detection

In target detection applications, the main objective is to search the pixels of an HSI data cube for the presence of a specific material (target). Conceptually, at least at a theoretical level, target detection can be viewed as a binary hypothesis testing problem, where each pixel is assigned a target or non-target label. However, there are some fundamental practical differences that have a great impact upon the development and evaluation of practical algorithms for detection versus classification applications. In surveillance applications, the size of the objects (targets) we are searching for constitutes a very small fraction of the total search area. Therefore, the target class will either be empty or sparsely populated. On the other hand, the general no-target class includes almost all pixels in the cube and is the union of the different specific background classes. We shall use the term background to refer to all non-target pixels of a scene. Usually, targets are man-made objects with spectra that differ from the spectra of natural background pixels (Healey and Slater 1999; Manolakis, *et al.* 2001; Thai and Healey 2002). More related content can be found in Chapters 7 and 8.

15.4.2 Invariant subpixel material detector

In hyperspectral target detection, the main task is to detect, discriminate and identify materials. The concrete processing includes two basic approaches: region-by-region and pixel-by-pixel. The pixel-by-pixel processing is necessary when the spatial resolution of images is low and mixed pixels exist in the hyperspectral images. Thus, identifying a target pixel by pixel is a dependable approach in most cases, and the detection algorithm tries to estimate whether a pixel is the target or it includes the target.

In the linear spectral mixture model (LSMM), the spectrum of a mixed pixel is represented as a linear combination of component spectra (endmembers). The weight of each endmember spectrum (abundance) is proportial to the fraction of the pixel area covered by the endmember (Manolakis 2000; Keshava 2002). Consider a hyperspectral image with N bands. Each pixel can be thus represented by an N-dimensional vector. Each pixel includes background and noise and possibly includes target. Thus, the linear spectral mixing of the pixel can be expressed by

$$\mathbf{y} = \alpha\mathbf{T} + \beta\mathbf{B} + \mathbf{n}, \tag{15.14}$$

where $\mathbf{T}$ and $\mathbf{B}$ represent the target and the background respectively, α, $\boldsymbol{\beta}$ represent the abundance of target and background respectively, and the $\mathbf{n}$ is noise fraction.

Using the linear mixture model, Bea Thai *et al.* proposed an invariant subpixel material detection (ISMD) algorithm (Thai 2002). In the ISMD algorithm, the target subspace and background subspace are obtained by performing singular value decomposition (SVD) on the matrix of training samples. ISMD provides a concrete expression of the generalized likelihood ratio as follows

$$\tilde{L}(y) = \frac{\mathbf{y}^{\top}(\mathbf{I} - \mathbf{B}\mathbf{B}^{\top})\mathbf{y}}{\mathbf{y}^{\top}(\mathbf{I} - \mathbf{A}\mathbf{A}^{\dagger})\mathbf{y}}, \tag{15.15}$$

where $\mathbf{A} = [\mathbf{B}\ \mathbf{T}]$, $\mathbf{A}^{\dagger}$ is the pseudo-inverse of $\mathbf{A}$, and $\mathbf{I}$ is the identity matrix. The detailed derivation of the Equation (15.15) can be found in the literature (Thai 2002).

The obtained subspace is, however, described by a group of linear basis vectors. Therefore, using KPCA to construct target and background subspaces is a better choice. In this way, we can obtain the kernel invariant subpixel detection algorithm (KISD) (Zhang and Gu 2004).

15.4.3 Kernel invariant subpixel detection

Figure 15.5 illustrates the processing chain of the KISD algorithm. Using the linear mixture model and substituting the kernel subspace of target and background for the original subspace, the hyperspectral target detection can be described as the following hypothesis test:

$$\mathbf{H}_0 : \mathbf{y} = \boldsymbol{\beta}_k \mathbf{B}_k + \mathbf{n} \quad \text{(target absent)} \tag{15.16}$$

$$\mathbf{H}_1 : \mathbf{y} = \alpha_k \mathbf{T}_k + \boldsymbol{\beta}_k \mathbf{B}_k + \mathbf{n} \quad \text{(target present)}, \tag{15.17}$$

where $\mathbf{T}_k$ and $\mathbf{B}_k$ represent the kernel subspace of the target and the background respectively, α_k, $\boldsymbol{\beta}_k$ represent the abundance of target and background respectively, $\mathbf{n}$ is Gaussian white noise fraction. $\mathbf{T}_k$ and $\mathbf{B}_k$ are obtained by performing KPCA on training samples of targets and background clutters, respectively. The dimensionality of the $\mathbf{T}_k$ and $\mathbf{B}_k$ are the numbers of

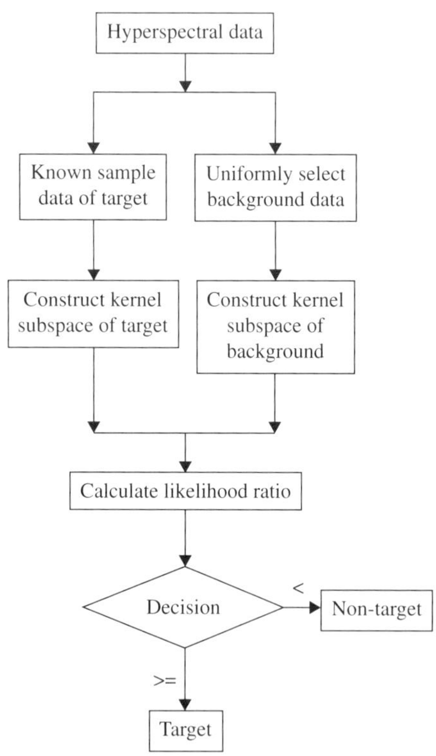

Figure 15.5 The processing flowchart of the KISD algorithm.

nonlinear kernel principal components remaining and it can be determined, based on energy ratio between the sum of the eigenvalues remaining and all eigenvalues by setting a desired threshold.

Thus the likelihood ratio is given by

$$L(y) = \frac{p_1(\mathbf{y}|H_1)}{p_0(\mathbf{y}|H_0)}, \tag{15.18}$$

where $p_0(\mathbf{y}|H_0)$ and $p_1(\mathbf{y}|H_1)$ are the value of the conditional probability density function of the observation y under two hypothesis tests H_1 and H_0, respectively. The corresponding formula of KISD is as follows:

$$\tilde{L}(\mathbf{y}) = \frac{\mathbf{y}^\top(\mathbf{I} - \mathbf{B}_k\mathbf{B}_k^\top)\mathbf{y}}{\mathbf{y}^\top(\mathbf{I} - \mathbf{A}_k\mathbf{A}_k^\dagger)\mathbf{y}}, \tag{15.19}$$

where $\mathbf{A}_k = [\mathbf{B}_k\ \mathbf{T}_k]$, $\mathbf{B}_k$ and $\mathbf{T}_k$ are kernel subspace of background samples and target samples constructed by KPCA respectively; $\mathbf{A}_k^\dagger$ is the pseudo-inverse of $\mathbf{A}_k$.

Using the Gram–Schmidt process, one can extract matrix $\mathbf{Q}_k$ from $\mathbf{A}_k$, of which orthonormal columns span the same subspace as the columns of $\mathbf{A}_k$. Thus expression (15.23) can be modified as follows:

$$\tilde{L}(\mathbf{y}) = \frac{\mathbf{y}^\top(\mathbf{I} - \mathbf{B}_k\mathbf{B}_k^\top)\mathbf{y}}{\mathbf{y}^\top(\mathbf{I} - \mathbf{Q}_k\mathbf{Q}_k^\top)\mathbf{y}}. \tag{15.20}$$

The expression (15.24) can be used as a discriminant to detect a hyperspectral subpixel target pixel by pixel. If the likelihood ratio in expression (15.24) exceeds the detection threshold in terms of a pixel, it can be concluded that the pixel includes the target. Otherwise, it can be considered that there is no target in the pixel. So the kernel-based invariant subspace detection (KISD) for hyperspectral target detection is performed.

In the experiments, in order to test the ability of the KISD algorithm in terms of spectral variability, three classes of different ground cover are randomly selected as the training samples of the background (non-target), and the training samples of target and background are obtained from a spectral library. Figure 15.6 provides the spectrum of these two target samples that are randomly selected from the target training set. By comparing the spectrum of two targets, it is clear that the spectral difference is significant. In order to test the ability of the proposed algorithm to process the sparsity of the target, the number of selected target samples is only one ninth of the background samples. In order to prove the effectiveness of the algorithm, ISMD, matched filtering detector (MFD) (Manolakis *et al.* 2001) and orthogonal subspace projection (OSP) (Harsanyi *et al.* 1994; Chang 2005) are realized in the experiment to be compared with the method. In the proposed kernel-based methods, the RBF kernel function is used, in which σ is set to 70 experimentally.

The detection results of the four algorithms are shown in Figure 15.7. The dimensionality of the target subspace in four algorithms is one and the dimensionality of the background subspace is selected as 1, 5 and 10 in each algorithm respectively. From the corresponding detection result, it can be noticed that the separability between targets and background of the proposed algorithm and ISMD is better than those from MFD and OSP. This indicates that the

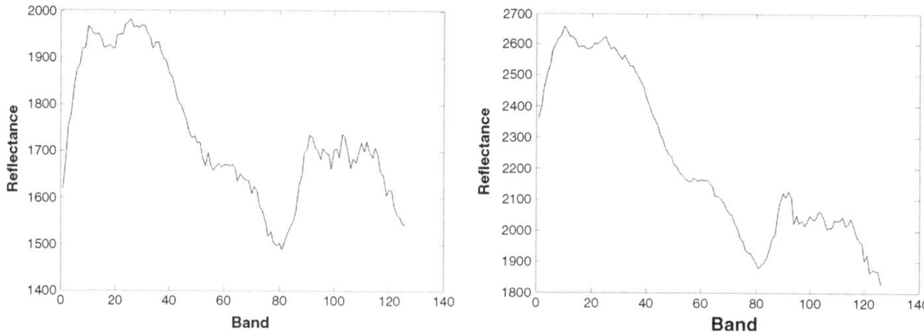

Figure 15.6 Spectral variability of targets from two different pixels.

detection ratios of the proposed algorithm and ISMD are higher than those from MFD and OSP. In addition, it can also be found that the detection result obtained by the proposed algorithm under different numbers of the background subspace is almost identical. This indicates that the construction of the background subspace in the proposed algorithm is more effective than the other algorithms.

To validate the effectiveness of the proposed algorithm, we evaluate the detection performance of the four methods by means of the receiver operating characteristics (ROC). The ROC represents the varying relationship of detection probability P_d and false alarm rate P_f, and can provide a quantitative comparison of the detection performance. Based on the ground truth, the coordinate ranges of those panels in the image scene was obtained. Anomalous targets and false alarms were identified by determining whether anomalous-like targets fell into the coordinate ranges of the panels or not. The identifications of targets and false alarms were performed on the scene, while each pixel is considered as a unit of the identifications. P_d and P_f are defined respectively as $P_d = N_d/N_T$ and $P_f = N_f/N_A$, where N_d is the number of the detected pixels, N_T is the total number of truth target pixels in the scene, N_f is the number of the false alarms, and N_A is the total number of image pixels.

Figure 15.8 shows the performance of these four methods for target detection in terms of the ROC. From this figure, it can be found that the proposed method outperforms the other three methods.

15.5 Kernel-based anomaly detection in hyperspectral images

15.5.1 The concept of anomaly detection

Anomaly detectors are pattern recognition schemes that are used to detect objects that might be of military and other field interest. Almost all anomaly detectors attempt to locate anything that looks different spatially or spectrally from its surroundings. In spectral anomaly detection algorithms, pixels (materials) that have a significantly different spectral signature from their neighbouring background clutter pixels are identified as spectral anomalies. Spectral anomaly detection algorithms can also use spectral signatures to detect anomalies embedded within

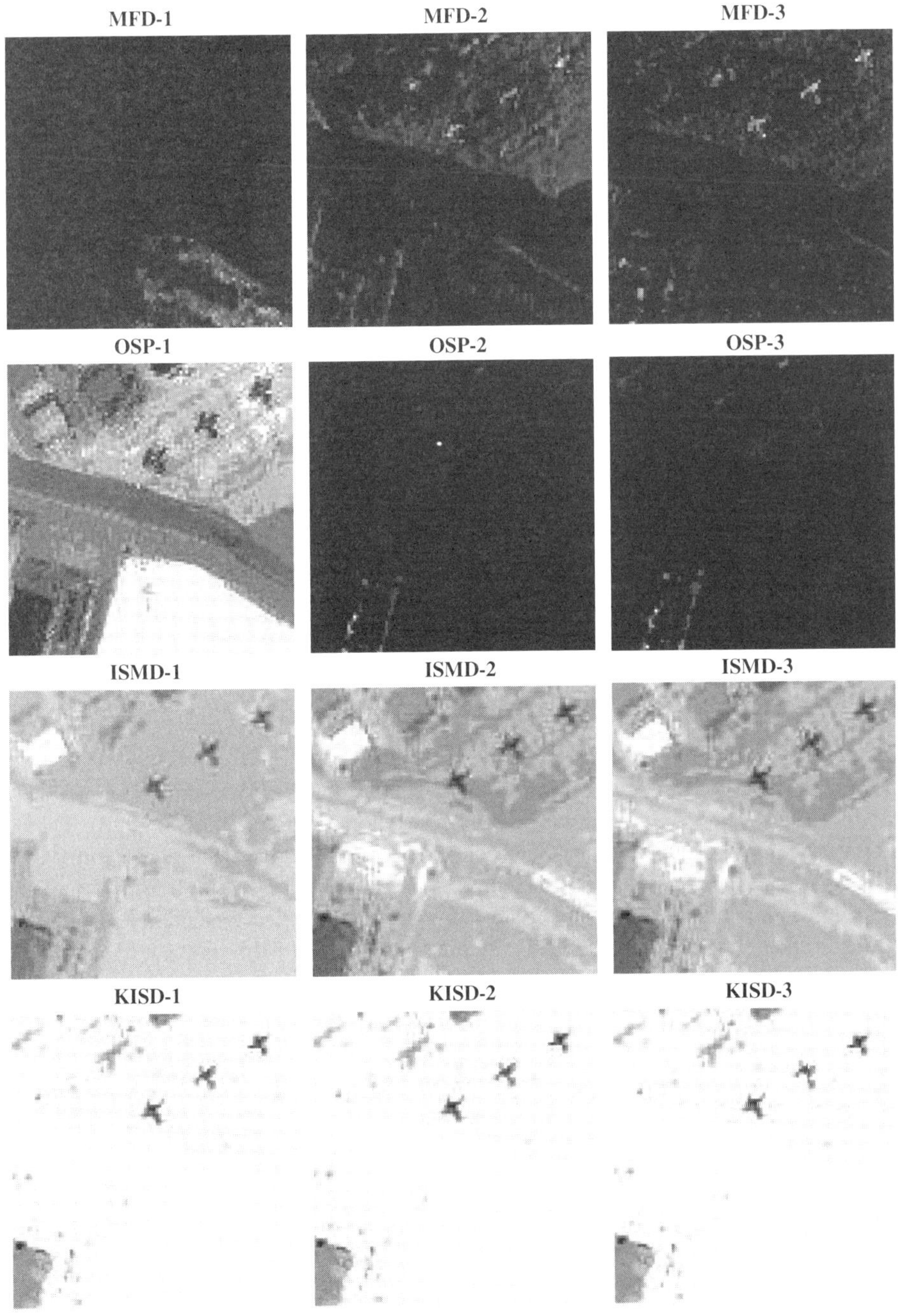

Figure 15.7 The detection result by using four algorithms. The dimensionality of the background subspace is 1, 5 and 10 respectively.

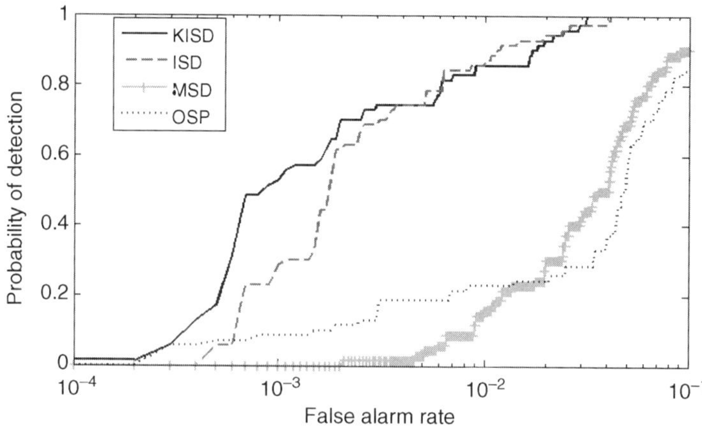

Figure 15.8 The ROC comparison of four methods for target detection.

background clutter with a very low signal-to-noise ratio (SNR). In spectral anomaly detectors, no *prior* knowledge of the target spectral signature is utilized or assumed (Stein *et al.* 2002; Chang and Chiang 2002; Kwon *et al.* 2005; Gu *et al.* 2006, 2008).

15.5.2 RX detector

A spectral anomaly detection algorithm was developed to detect targets with unknown spectral distribution against a background with unknown covariance (Reed and Yu 1990). The algorithm is now commonly referred to as the RX the detector, and it has been successfully applied to many multi-spectral/hyperspectral detection applications. The RX algorithm assumes that hyperspectral data follow the multivariate Gaussian distribution. Under this assumption, a generalized likelihood ratio test (GLRT) is constructed to test the hypotheses for estimating the existence of targets in each pixel of the scene.

Let each input spectral signal containing N_b spectral bands be denoted by $\mathbf{x}(n) = [x_1(n), x_2(n), \ldots, x_{N_b}(n)]^\top, n = 1, 2, \ldots, N_b$. Define $\mathbf{x}_B$ to be an $N_b \times m$ matrix of the m reference background clutter pixels. Namely, $m = n_1 \times n_2$, $n_1 \times n_2$ is the size of the sliding window for detection. The observed spectrum of each pixel is represented as a column vector in the sample matrix $\mathbf{x}_B$, $\mathbf{x}_B = [\mathbf{x}(1), \mathbf{x}(2), \ldots, \mathbf{x}(m)]$. The two competing hypotheses that the RX algorithm must distinguish are given by

$$\mathbf{H}_0 : \mathbf{x} = \mathbf{n} \quad \text{(Target absent)} \tag{15.21}$$

$$\mathbf{H}_1 : \mathbf{x} = a\mathbf{s} + \mathbf{n} \quad \text{(Target present)}, \tag{15.22}$$

where $a > 0$, $\mathbf{n}$ is a vector that represents the process of the background clutters and noise, and $\mathbf{s}$ is the spectral signature of the anomalous target given by $\mathbf{s} = [s_1, s_2, \ldots, s_{N_b}]^\top$. The signature s of the target and covariance $\boldsymbol{\Sigma}_B$ of background clutters are assumed to be unknown. Thus, the model assumes that the data come from two normal probability density functions with the same covariance matrix but different means. Under $\mathbf{H}_0$, the data (background clutters) are modelled as $N(0, \mathbf{\Sigma}_B)$, and under $\mathbf{H}_1$ the data are modelled as $\mathcal{N}(\mathbf{s}, \mathbf{\Sigma}_B)$. Based on the description given above, expression for the conventional RX detector is given by

$$\mathrm{RX}(\mathbf{x}) = (\mathbf{x} - \hat{\mu}_B)^{\top} \hat{\Sigma}_B^{-1} (\mathbf{x} - \hat{\mu}_B), \tag{15.23}$$

where $\hat{\Sigma}_B$ is the covariance matrix estimated from the reference data of the background clutters, and $\hat{\mu}_B$ is the estimated sample mean of the background clutters. Let η be a threshold, if $\mathrm{RX}(\mathbf{x}) \geq \eta$, then the assumption that a target present is adopted, otherwise the assumption is that the target is absent.

15.5.3 Selective KPCA Feature Extraction for Anomaly Detection

In this section, a selective kernel principal component analysis (SKPCA) (Gu *et al.* 2006, 2008) algorithm based on high-order statistics is introduced for anomaly detection in hyperspectral imagery. First, KPCA is performed on original hyperspectral data to mine fully a high-order correlation between spectral bands. Then average local singularity is defined based on high-order statistic in a local sliding window, which is used as a measure for selecting the most informative nonlinear components for anomaly detection. By selective KPCA the information of anomalous targets is extracted to a maximum extent and background clutters are well suppressed in the component selected. Finally, the selected component with maximum average local singularity is used as the input for anomaly detectors. Numerical experiments are conducted on real hyperspectral images collected by AVIRIS. The results strongly prove the effectiveness of the proposed algorithm.

After KPCA, research is focused on how to evaluate the information about anomaly targets present in nonlinear principal components based on the Gaussian assumption. The most effective component for anomaly detection is selected based on its amount of information about anomalies, which is equivalent to deciding whether its histogram was close to a Gaussian shape or not. As widely known, high-order statistics can measure how far a distribution is from Gaussian. Therefore, based on third- and fourth-order moments, i.e. skewness and kurtosis, local singularity (LS) is defined as a measure for selecting the most effective component for detection.

Given a random variable t let $\hat{m}$ and $\hat{\sigma}$ be the mean and variance estimations based on N observations t_i, $i = 1, 2, \ldots, N$, respectively. The *skewness* is defined as

$$\gamma_3 = E\{[t - E(t)]^3\} \Big/ (E\{[t - E(t)]^2\})^{3/2}, \tag{15.24}$$

where E is the expectation operator. An estimator of the skewness can be written as

$$\hat{\gamma}_3 = \sum_{i=1}^{N} (t_i - \hat{m})^3 \Big/ (N - 1)\hat{\sigma}^3. \tag{15.25}$$

The *kurtosis* is defined as

$$\gamma_4 = -3 + E\{[t - E(t)]^4\} \Big/ (E\{[t - E(t)]^2\})^2 \tag{15.26}$$

and its estimation is written as

$$\hat{\gamma}_4 = -3 + \sum_{i=1}^{N} (t_i - \hat{m})^4 \Big/ (N-1)\hat{\sigma}^4. \tag{15.27}$$

Generally, the skewness is used to measure the symmetry of the data distribution, and the kurtosis is used to measure trailing (Zhao *et al.* 1995). In anomaly detection, if data in the local windows follow a Gaussian distribution, then corresponding skewness and kurtosis both equal zero. If there are anomalies in the local windows, the Gaussian distribution is broken and the absolute values of the skewness and the kurtosis become large. When skewness and kurtosis are used to check Gaussian distribution of a region, we define two thresholds corresponding to skewness and kurtosis respectively as

$$T_s = \tau_s \theta_s \tag{15.28}$$

$$T_k = \tau_k \theta_k, \tag{15.29}$$

where θ_s and θ_k are positive constants and $\theta_s \gg 1, \theta_k \gg 1$, τ_s and τ_k are equal to $\sqrt{6/(n_1 n_2)}$ and $\sqrt{24/(n_1 n_2)}$ respectively, $n_1 n_2$ is the size of the local window. Apparently, when skewness and kurtosis are used to describe singularity in a local window for effective detection, the window size of the LS should be the same as the detection window in the RX detector, which depends on the size of the targets. If the absolute values of the skewness and the kurtosis are larger than the thresholds T_s andT_k respectively, then this means that the data distribution in the local window is not Gaussian.

To measure properly the singularity of each component transformed, the LS should be calculated in the local regions and the LS of the whole component should be an average value. The average LS of the whole component is denoted by the symbol $\bar{N}_{\text{ALS}}$. The detailed calculation of $\bar{N}_{\text{ALS}}$ corresponding to one component is as follows.

1. Construct a local sliding window with size $n_1 \times n_2$, let the moving interval be equal to the half width of the window and let C be the total sliding time.
2. Let $\bar{N}_{\text{ALS}} = 0$ and $j = 1$.
3. Calculate the absolute values A_s^j and A_k^j of the skewness and the kurtosis in the jth local window
4. Compare A_s^j and A_k^j with two thresholds T_s and T_k respectively; if both A_s^j and A_k^j are larger than T_s and T_k respectively, then $\bar{N}_{\text{ALS}} = \bar{N}_{\text{ALS}} + 1$; else hold $\bar{N}_{\text{ALS}}$ invariant.
5. If $j = C$, then the calculation is over and record the value of $\bar{N}_{\text{ALS}}$; else let $j = j + 1$ and return to (3).

After the above process is performed on each component, the nonlinear principal component with maximum $\bar{N}_{\text{ALS}}$ is selected as the input of the conventional RX detector.

The detailed process of the selective KPCA algorithm is as follows.

1. Convert the original hyperspectral data $\mathbf{y}$ from 3D-form to 2D-form, namely, orderly arrange all rows of the jth 2D-image into a 1D row vector $\mathbf{y}_j$, $j = 1, 2, \ldots, M$, and obtain 2D data $\mathbf{Y}$, where $\mathbf{y} = [\mathbf{y}_1, \mathbf{y}_2, \ldots, \mathbf{y}_M]^\top$.
2. Centre Y, consider $y_k (k = 1, 2, \ldots, M)$ as a set of centred random samples, designate a kernel function and calculate kernel matrix $\mathbf{K}$ according to Equation (15.11).
3. Calculate the centred kernel matrix $\mathbf{K}_c$ according to Equation (15.13).
4. Perform singular value decomposition on $\mathbf{K}_c$ and obtain those eigenvectors $\boldsymbol{\beta}^k$, $k = 1, 2, \ldots, q$ corresponding to non-zero eigenvalues in high-dimensional feature space.
5. Project $\mathbf{y}$ onto β^k, $k = 1, 2, \ldots, q$, obtain transformed nonlinear principal components.
6. Select the component X with the maximum average LS according to the procedure mentioned above.
7. Make a sliding local window on X, and calculate local mean and variance.
8. Realize an anomaly detection on the local window by using the RX detector in Equation (15.23).
9. Slide the window and repeat step (7) until the entire scene is covered.
10. Output detection result.

By the selective KPCA, information of anomalous targets is maximally extracted and background clutters are well suppressed in the selected component. Such feature extraction will be useful for improving performance of the conventional RX algorithm. Generally, all pixel vectors in the HSI are first normalized by a constant, which is the maximum value obtained from all the spectral components of spectral vectors in the corresponding test image before performing KPCA. In this chapter, the value of the Gaussian kernel width σ was determined experimentally and was set to 70. The first ten nonlinear principal components transformed on the original hyperspectral data by KPCA are shown in Figure 15.9. From those components, it is found that useful information is sufficiently extracted and well concentrated, and complete anomaly information is concentrated into the 4th component.

According to the defined average local singularity, the average local singularity of each component was calculated for selecting the maximum one. In both the calculation of the average local singularity and the subsequent anomaly detection, the size of the local window was set to 11×11, and both θ_s and θ_k are set to 11 experimentally. Figure 15.10 presents the curve of the average local singularity as a function of the nonlinear principal components. From the curve, it is easy to discover that the 4th component has the maximum average local singularity. From Figure 15.9, it is found that the 4th component includes almost all information about anomalous targets and is more effective for anomaly detection than the other components, so it is the chosen one for detection. This fact proves that the LS rule defined performs well.

To validate the effectiveness of the proposed algorithm, a comparison was made between the proposed algorithm and the feature extraction method based on KPCA with the energy rule. When the variance of noise is normalized to one, the energy rule becomes the SNR rule. The

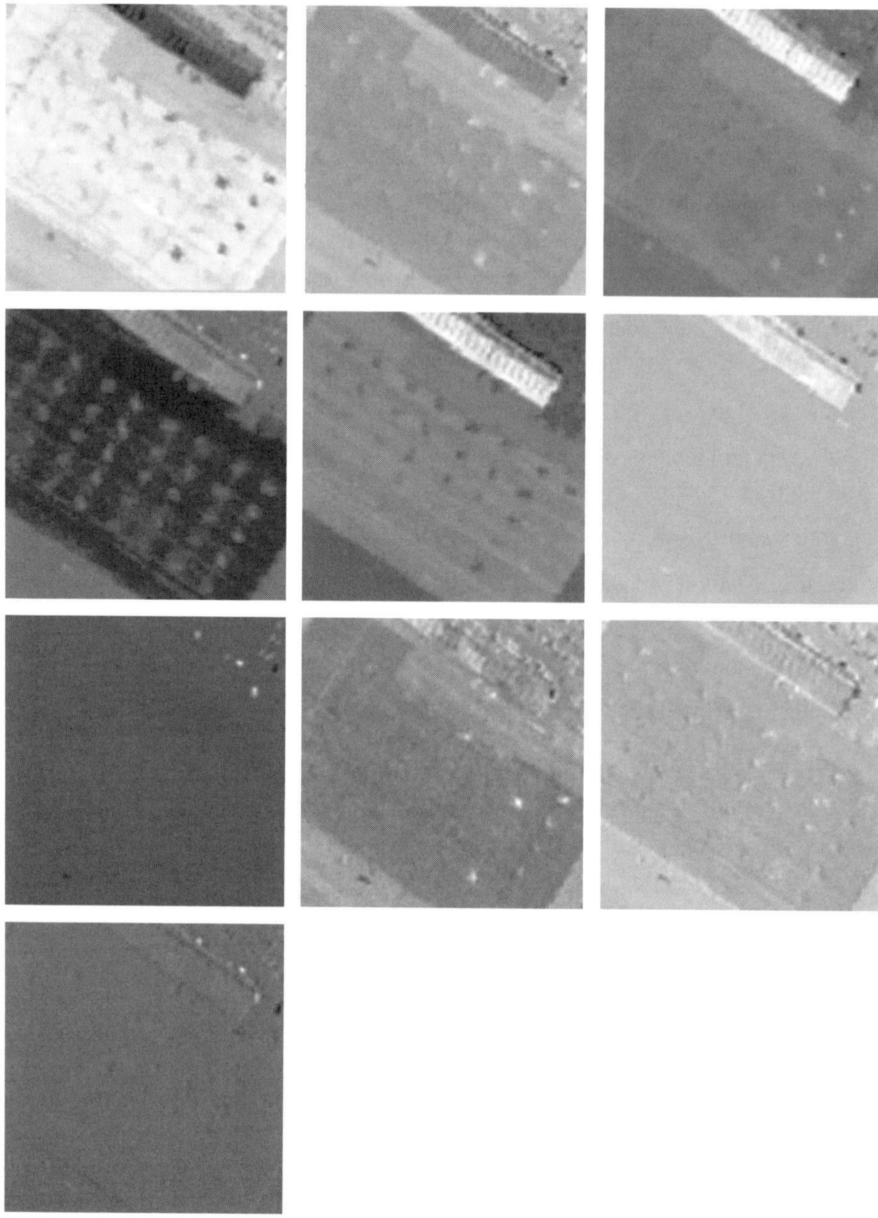

Figure 15.9 The first ten components transformed by KPCA from the first scene.

conventional RX algorithm is also carried out in this comparison. In addition, a comparison between KPCA with the most effective component and KPCA with multiple components was also conducted. In the method with multiple components, the first ten components with large eigenvalues were used. Figure 15.11 shows the detection results in the form of the ROC.

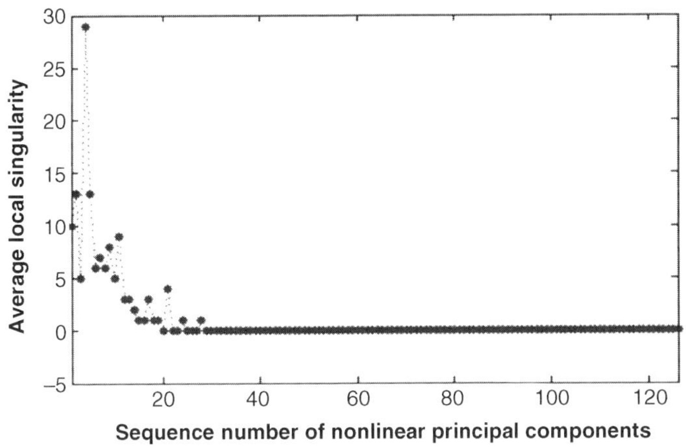

Figure 15.10 The curve of average local singularity accompanied by the sequence number of different nonlinear principal components.

From Figure 15.11(a), it can be seen that the detection performance of KPCA with the proposed LS rule is better than KPCA with the energy rule. Furthermore, the proposed selective KPCA algorithm greatly modifies the performance of the conventional RX detector. From Figure 15.11(b), it can be seen that the detection performance of the method using the chosen component with the most singular information is better than the method using multiple components with concentrated energy. Related content can be also found in Chapter 8.

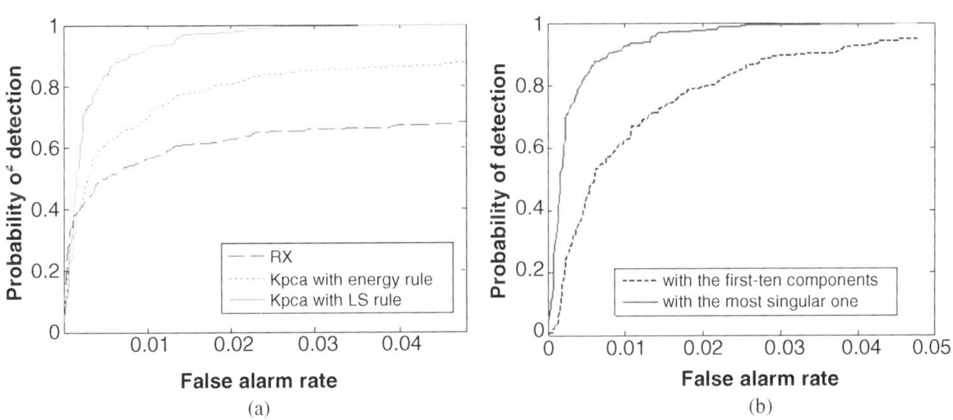

Figure 15.11 The comparison of ROCs in two cases: (a) between KPCA with LS rule, KPCA with energy rule, and the conventional RX detector; (b) between detection with the most singular component transformed (i.e. single) and with the first-ten components transformed (i.e. multiple).

15.6 Conclusions

In this chapter we first introduced the motivation for using kernel methods for target and anomaly detection in hyperspectral images. Then, dividing the importance of this chapter into three aspects concerning hyperspectral image processing, we described kernel-based feature extraction and its application for target and anomaly detection in hyperspectral images in detail. In these applications, we constructed effective methods for feature extraction using KM and greatly improved the performance of the existing algorithms for target and anomaly detection, respectively. The numerical experiments demonstrated the effectiveness of the proposed kernel-based methods. In summary, kernel methods have been demonstrated to be a powerful tool for detection applications in hyperspectral images.

Acknowledgments

This chapter builds on work undertaken over several years that has often benefited from colleagues, notably Ye Zhang (my Ph.D. supervisor), Junping Zhang and Ying Liu, and this is gratefully acknowledged. I am also grateful for the review comments received on the original version of this chapter. This work is supported by the Project of Natural Science Foundation of China No. 60402025.

References

Chang, C.-I. (2005) Orthogonal Subspace Projection (OSP) revisited: a comprehensive study and analysis, *IEEE Trans. Geosci. Remote Sensing*, **43**(3), 502–518.

Chang, C.-I. and Chiang, S.-S. (2002) Anomaly detection and classification for hyperspectral imagery, *IEEE Trans. Geosci. Remote Sensing*, **40**(6), 1314–1325.

Gu, Y., Liu, Y. and Zhang Y. (2006) A Selective Kernel PCA Algorithm for anomaly detection in hyperspectral imagery, *ICASSP 2006, IEEE International Conference on Acoustics, Speech and Signal Processing-Proceedings*, pp. II725–II728.

Gu, Y., Liu, Y. and Zhangm Y. (2008) A Selective KPCA Algorithm based on high-order statistics for anomaly detection in hyperspectral imagery, *IEEE Geoscience and Remote Sensing Letters*, **1**, 43–47.

Gu, Y., Zhang, Y. and Zhang, J. (2002) A Kernel Based Nonlinear subspace projection method for reduction of hyperspectral image dimensionality, *IEEE International Conference on Image Processing, ICIP'2002*, Rochester, NY, pp. II357–II360.

Harsanyi, J. and Chang, C.-I. (1994) Hyperspectral image classification and dimensionality reduction: an orthogonal subspace projection approach, *IEEE Trans. Geosci. Remote Sensing*, **32**(4), 779–785.

Healey, G. and Slater, D. (1999). Models and methods for automated material identification in hyperspectral imagery acquired under unknown illumination and atmospheric conditions, *IEEE Trans. Geosci. Remote Sensing*, **37**(6), 2706–2717.

Jia, X. and Richards, J.A. (1999) Segmented principal components transformation for efficient hyperspectral remote sensing image display and classification, *IEEE Trans. Geosci. Remote Sensing*, **37**, 538–542.

Keshava, N. and Mustard, J.F. (2002) Spectral unmixing, *IEEE Signal Process. Mag.*, **19**(1), 44–57.

Kwon, H. and Nasrabadi, N.M. (2005) Kernel RX-Algorithm: A nonlinear anomaly detector for hyperspectral imagery, *IEEE Trans. Geoscience and Remote Sensing*, **43**(2), 388–397.

Landgrebe, D. (2002) Hyperspectral image data analysis, *IEEE Signal Process. Mag.*, **1**, 17–28.

Manolakis, D., Siracusa, C. and Shaw, G. (2001) Hyperspectral subpixel target detection using the linear mixing model, *IEEE Trans. Geosci. Remote Sensing*, **39**(7), 29–43.

Manolakis, D. and Shaw, G. (2002) Detection algorithms for hyperspectral imaging applications, *IEEE Signal Process. Mag.*, **19**(1), 29–43.

Müller, K.-R., Mika, S., Rätsch, G. and Tsuda, K. (2001) An introduction to kernel-based learning algorithms, *IEEE Trans. Neural Networks*, **12**, 181–201.

Reed, S. and Yu, X. (1990). Adaptive multiple-band CFAR detection of an optical pattern with unknown spectral distribution, *IEEE Trans. Acoust., Speech, Signal Process*, **38**, 1760–1770.

Ruiz, A. and Lopez-de-Teruel, P.E. (2001) Nonlinear kernel-based statistical pattern analysis, *IEEE Trans. Neural Networks*, **12**, 16–32.

Schölkopf, B. (2000) The Kernel Trick for Distances, *Microsoft Research Technical Report*, MSR-TR-2000-51, May, pp. 1–10.

Schölkopf, B., Smola, A.J. and Müller, K.-R. (1998) Nonlinear component analysis as a kernel eigenvalue problem, *Neural Comput*, **10**, 1299–1319.

Schölkopf, B., Mika, S., Burges, C.J.C., Knirsch, P. and Müller, K.-R. (1999) Input space versus feature space in kernel-based methods, *IEEE Trans. Neural Networks*, **10**, 1000–1017.

Shaw, G. and Manolakis, D. (2002) Signal processing for hyperspectral image exploitation, *IEEE Signal Process. Mag.*, **1**, 12–16.

Stein, W.J., Scott, G.B., Beaven, G. and Lawrence, E.H. *et al.* (2002) Anomaly detection from hyperspectral imagery, *IEEE Signal Processing Magazine*, **19**(1), 58–69.

Thai, B. and Healey. G. (2002) Invariant subpixel material detection in hyperspectral imagery. *IEEE Trans. Geosci. Remote Sensing*, **40**(3), 599–608.

Zhang, Y. and Gu, Y. (2004) Kernel-based invariant subspace method for hyperspectral target detection, *ICASSP 2004, IEEE International Conference on Acoustics, Speech and Signal Processing-Proceedings*, pp. V801–V804.

Zhao, Y., Zhuang, X. and Ting, S. (1995) Gaussian mixture density modelling of non-gaussian source for autoregressive process, *IEEE Trans. Signal Processing*, **43**(4), 894–903.

16

Remote sensing data classification with kernel nonparametric feature extractions

Bor-Chen Kuo[1], Jinn-Min Yang[2] and Cheng-Hsuan Li[1]

[1]*Graduate Institute of Educational Measurement and Statistics, National Taichung University, Taiwan*

[2]*Department of Mathematics Education, National Taichung University, Taiwan*

In recent years, many studies have shown that kernel methods are computationally efficient, robust, and stable for pattern analysis. Many kernel-based classifiers have been designed and applied to classify remote-sensed data and some results show that kernel-based classifiers give satisfactory performances. Much research into hyperspectral image classification also shows that Nonparametric Weighted Feature Extraction (NWFE) is a powerful tool for extracting hyperspectral image features. But NWFE is still based on linear transformation. In this chapter, the kernel method is applied to extend NWFE to Kernel-based NWFE (KNWFE). The new KNWFE possesses the advantages of both linear and nonlinear transformation. In addition, Fuzzy Linear Feature Extraction (FLFE) and its Kernel-based version (KFFE) are introduced. The experimental results show that KNWFE and KFFE outperform KPCA and GDA.

Some parts of this chapter have been derived from the paper (Kuo, Li and Yang 2008) listed in the References (reproduced by permission of IEEE).

Kernel Methods for Remote Sensing Data Analysis Edited by Gustavo Camps-Valls and Lorenzo Bruzzone

16.1 Introduction

Remote sensing data such as hyperspectral data generally have hundreds of measured bands, which potentially provide more accurate and detailed information for classification. However, collecting the ground-truth of a hyperspectral image can be very difficult and expensive. Hence, the classifiers often suffer from the Hughes phenomenon (Hughes 1968), also known as the curse of dimensionality. Feature extraction and selection are powerful methods of alleviating the problem. The feature selection method is defined as the process of selecting a subset from a set of original features. The most important issue for feature selection is to find an efficient search strategy for obtaining a suitable feature subset for classification, particularly for high-dimensional data. Feature extraction uses all the features to construct a transformation that maps the original data to a low-dimensional subspace. An illustration is shown in Figure 16.1. The two distributions projected on any of the original axes overlap. However, after feature extraction, they can be separated completely on the transformed axis.

In recent years, many studies (Schölkopf *et al.* 1998; Shawe-Taylor and Cristianini 2004) show that kernel methods are computationally efficient, robust, and stable for pattern analysis. Many kernel-based classifiers have been designed and applied to classify remote sensing data and results show that kernel-based classifiers give sound performances (Camps-Valls and Bruzzone 2005; Camps-Valls *et al.* 2008; Melgani and Bruzzone 2004). The main idea of kernel methods is to map the input data from the original space to a so-called feature space by a nonlinear mapping. A learning algorithm designed to discover linear patterns in the feature space and the algorithms can be implemented in terms of inner products in feature space, and the inner products can be calculated directly from the original data by employing a kernel function. Importantly, a linear relationship can be found in the feature space, which is equivalent to seeking the nonlinear relationship in the original space.

Some studies (Benediktsson *et al.* 2005; Dundar and Landgrebe 2004; Hsieh *et al.* 2006; Landgrebe 2005; Richards, 2005; Song *et al.* 2005) have also shown that nonparametric feature extractions (NFEs) are powerful in reducing the dimensionality of hyperspectral image data. However, most of the NFEs are still based on the linear transformation, which may fail in handling nonlinear problems. Thus, nonlinear feature extraction methods are necessary.

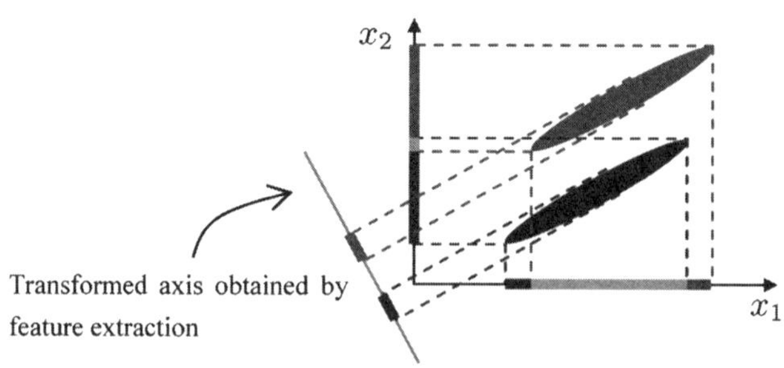

Figure 16.1 The feature extraction method can obtain complete separation on the transformed axis.

In this chapter, two kernel-based nonparametric feature extractions are introduced and applied to hyperspectral data classification in a supervised sense. One is called Kernel-based Nonparametric Weighted Feature Extraction (KNWFE) and the other is Kernel-based Fuzzy Feature Extraction (KFFE), along with experiments to compare performances on hyperspectral image classification. This chapter is organized as follows. The reviews of linear discriminant analysis (LDA), the kernel-based version of LDA, i.e., generalized discriminant analysis (GDA), NWFE, and fuzzy linear feature extraction (FLFE) are introduced in Section 16.2. Section 16.3 contains the main ideas and theoretical development of KNWFE and KFFE. Theoretical development is discussed from three aspects: the definition of scatter matrices, the regularization method, and the eigen-decomposition technique. The details of the last two aspects are discussed in Section 16.4. The performances of Kernel-based Principal Component Analysis (KPCA), GDA, KNWFE and KFFE are compared in Section 16.5 based on real hyperspectral images. Section 16.6 contains comments and conclusions.

16.2 Related feature extractions

In this section, four feature extraction methods, Fisher's Linear Discriminant Analysis (LDA) (Fukunaga 1990), the kernel version of LDA (Baudat and Anouar 2000), Nonparametric Weighted Feature Extraction (NWFE) (Kuo and Landgrebe 2004) and Fuzzy Linear Feature Extraction (FLFE) (Yang *et al.* 2006) are introduced. The following are some notations for this chapter worth noting.

Suppose a multiclass problem

$$\{\mathbf{x}_1, \ldots, \mathbf{x}_n\} = \left\{\mathbf{x}_1^{(1)}, \ldots, \mathbf{x}_{n_1}^{(1)}, \ldots, \mathbf{x}_1^{(L)}, \ldots, \mathbf{x}_{n_L}^{(L)}\right\} \subset \mathbb{R}^{N_b},$$

where $\{\mathbf{x}_1^{(i)}, \ldots, \mathbf{x}_{n_i}^{(i)}\}$ are the samples in class i, $i = 1, \ldots, L$, and $n = n_1 + \cdots + n_L$. Let $X_i^T = [\boldsymbol{\phi}(\mathbf{x}_1^{(i)}), \ldots, \boldsymbol{\phi}(\mathbf{x}_{n_i}^{(i)})]$, and $X^T = [X_1^T, \ldots, X_L^T]$, then the kernel matrix $\mathbf{K} = K(\mathbf{x}_i, \mathbf{x}_j)$ for $1 \leq i, j \leq n$ on samples is XX^T, i.e., $\mathbf{K} = XX^T$.

16.2.1 Linear discriminant analysis

Linear Discriminant Analysis (LDA) is often used for dimension reduction in classification problems. It is also called the parametric feature extraction method (Fukunaga 1990), since LDA uses the mean vector and covariance matrix of each class. Usually within-class, between-class, and mixture scatter matrices are used to formulate the criterion of class separability. The goal of LDA is to find a transformation matrix A to maximize the tranformed class separability that is composed of between-class and within-class scatter matrices. A within-class scatter matrix for L classes is expressed by

$$S_w^{LDA} = \sum_{i=1}^{L} \sum_{j=1}^{n_i} \frac{1}{n} \left(\mathbf{x}_j^{(i)} - \mathbf{m}_i\right) \left(\mathbf{x}_j^{(i)} - \mathbf{m}_i\right)^T,$$

where $\mathbf{m}_i$ is the class mean. A between-class scatter matrix is expressed as

$$S_b^{LDA} = \sum_{i=1}^{L} \frac{n_i}{n} (\mathbf{m}_i - \mathbf{m}_0)(\mathbf{m}_i - \mathbf{m}_0)^T,$$

where $\mathbf{m}_0$ represents the expected vector of the mixture distribution and is given by

$$\mathbf{m}_0 = \sum_{i=1}^{L} \frac{n_i}{n} \mathbf{m}_i.$$

The optimal features are determined by maximizing the Fisher criterion given by

$$\mathbf{A} = \arg\max_{A} \operatorname{tr}\left(\left(A^T S_w^{LDA} A \right)^{-1} A^T S_b^{LDA} A \right).$$

To find A is equivalent to solving the following generalized eigenvalue problem

$$S_b^{LDA} \mathbf{v_h} = \lambda_h S_w^{LDA} \mathbf{v_h}, \quad h = 1, \ldots, p, \quad p \leq L - 1,$$

where p denotes the dimensionality of the reduced subspace, $(\lambda_h, \mathbf{v_h})$ represent the eigen-pair of $(S_w^{LDA})^{-1} S_b^{LDA}$ and $\lambda_1 \geq \lambda_2 \geq \cdots \geq \lambda_p$. Thus, the transformation matrix $\mathbf{A} = [v_1, \ldots, v_p]$ can be obtained.

16.2.2 Generalized discriminant analysis

A kernel-based LDA was proposed by Baudat and Anouar (2000) and called generalized discriminant analysis (GDA). The scatter matrices, S_b^{GDA}, S_w^{GDA} and S_t^{GDA} of GDA in the feature space $\mathcal{H}$ are defined as

$$S_b^{GDA} = \sum_{i=1}^{L} \frac{n_i}{n} (\mathbf{M}_i - \mathbf{M}_0)(\mathbf{M}_i - \mathbf{M}_0)^T,$$

$$S_w^{GDA} = \sum_{i=1}^{L} \sum_{j=1}^{n_i} \frac{1}{n} \left(\boldsymbol{\phi}\left(\mathbf{x}_j^{(i)}\right) - \mathbf{M}_i \right) \left(\boldsymbol{\phi}\left(\mathbf{x}_j^{(i)}\right) - \mathbf{M}_i \right)^T,$$

and

$$S_t^{GDA} = S_b^{GDA} + S_w^{GDA},$$

where

$$\mathbf{M}_i = \frac{1}{n_i}\sum_{j=1}^{n_i} \boldsymbol{\phi}\left(\mathbf{x}_j^{(i)}\right) \quad \text{and} \quad \mathbf{M}_0 = \sum_{i=1}^{L}\sum_{j=1}^{n_i} \frac{1}{n}\boldsymbol{\phi}\left(\mathbf{x}_j^{(i)}\right).$$

Similar to the LDA, the transformation matrix of GDA can be obtained by maximizing the Fisher criterion:

$$\mathbf{A} = \arg\max_{A} \left\{ \operatorname{tr}\left(\left(A^T S_t^{GDA} A\right)^{-1} A^T S_b^{GDA} A\right)\right\}.$$

This maximization is equivalent to the following generalized eigenvalue resolution: to find eigenvalues λ and eigenvectors $\mathbf{v}$ of

$$S_b^{GDA}\mathbf{v} = \lambda S_t^{GDA}\mathbf{v},$$

where $\mathbf{v}$ is a vector in the feature space $\mathcal{H}$.

Suppose the generalized eigenvector $\mathbf{v}$ is a linear combination of training samples in the feature space, i.e., there exists $\alpha_j^{(i)} \in \mathbb{R}$ such that

$$\mathbf{v} = \sum_{i=1}^{L}\sum_{j=1}^{n_i} \alpha_j^{(i)}\boldsymbol{\phi}\left(\mathbf{x}_j^{(i)}\right) = X^T\boldsymbol{\alpha}, \boldsymbol{\alpha} = \left[\alpha_1^{(1)}, \ldots, \alpha_{n_1}^{(1)}, \ldots, \alpha_1^{(L)}, \ldots, \alpha_{n_L}^{(L)}\right]^T.$$

This means that $\mathbf{v}$ can be solved for some suitable $\alpha_j^{(i)}$. By some block matrix computations, $\alpha_j^{(i)}$ can be found by solving the following generalized eigenvalue problem:

$$\mathbf{K}\mathbf{W}^{\mathbf{GDA}}\mathbf{K}\boldsymbol{\alpha} = \lambda\mathbf{K}\mathbf{K}\boldsymbol{\alpha},$$

where

$$\mathbf{W}^{\mathbf{GDA}} = \begin{bmatrix} W_1^{GDA} & & & 0 \\ & W_2^{GDA} & & \\ & & \ddots & \\ 0 & & & W_L^{GDA} \end{bmatrix} \quad \text{and} \quad W_i^{GDA} = \begin{bmatrix} \frac{1}{n_i} & \cdots & \frac{1}{n_i} \\ \vdots & \ddots & \vdots \\ \frac{1}{n_i} & \cdots & \frac{1}{n_i} \end{bmatrix},$$

where $i = 1, \ldots, L$.

Next, the eigen-decomposition of kernel matrix $\mathbf{K}$ is employed for stabilizing and improving the resolution in GDA. Let $\mathbf{K} = \mathbf{P}\boldsymbol{\Gamma}\mathbf{P}^T$ be the eigen-decomposition of $\mathbf{K}$, where $\mathbf{P}$ is a matrix with orthonormal columns and $\boldsymbol{\Gamma}$ is a diagonal matrix of nonzero eigenvalues with decreasing order. Assume that $\boldsymbol{\beta} = \boldsymbol{\Gamma}\mathbf{P}^T\boldsymbol{\alpha}$. Then the generalized eigenvalue problem can be converted to

$$\mathbf{P}^T\mathbf{W}^{\mathbf{GDA}}\mathbf{P}\boldsymbol{\beta} = \lambda\boldsymbol{\beta}.$$

After $\boldsymbol{\beta}$ is obtained, the $\boldsymbol{\alpha}$ can be computed by $\boldsymbol{\alpha} = \mathbf{P}\boldsymbol{\Gamma}^{-1}\boldsymbol{\beta}$ and divided by $\sqrt{\boldsymbol{\alpha}^T\mathbf{K}\boldsymbol{\alpha}}$ to get normalized vectors $\bar{\boldsymbol{\alpha}}$, i.e.,

$$\bar{\boldsymbol{\alpha}} = \left[\bar{\alpha}_1^{(1)}, \ldots, \bar{\alpha}_{n_1}^{(1)}, \ldots, \bar{\alpha}_1^{(L)}, \ldots, \bar{\alpha}_{n_L}^{(L)}\right]^T = \frac{\boldsymbol{\alpha}}{\sqrt{\boldsymbol{\alpha}^T\mathbf{K}\boldsymbol{\alpha}}}.$$

Finally, the projection component of an unknown mapped sample $\boldsymbol{\phi}(\mathbf{z})$ can be expressed as

$$\boldsymbol{\phi}(\mathbf{z})^T\mathbf{v} = \sum_{i=1}^{L}\sum_{j=1}^{n_i} \bar{\alpha}_j^{(i)} K\left(\mathbf{z}, \mathbf{x}_j^{(i)}\right).$$

Note that the rank of the between-scatter matrix of GDA is the same with LDA. Thus, only $L-1$ features can be extracted at most by GDA.

16.2.3 Nonparametric weighted feature extraction

There are three major disadvantages of LDA. One is that it works well only if the distributions of classes are normal-like distributions. When the distributions of classes are non-normal-like or multi-modal mixture distributions, the performance of LDA is unsatisfactory. The second disadvantage of LDA is that the rank of the between-scatter matrix is less than or equal to $L-1$. So only $L-1$ features can be extracted at most by using LDA. The third limitation is that if the within-class covariance is singular, which often occurs in high dimensional problems, LDA will have a poor performance on classification, Nonparametric Weighted Feature Extraction (NWFE) was proposed by Kuo and Landgrebe (2004) to impove these problems.

The main idea of NWFE is to put different weights on every sample to compute the 'weighted means' and defining a new separability critrion with new nonparametric between-class and within-class scatter matrices. The between-class scatter matrix S_b^{NW} and the within-class scatter matrix S_w^{NW} are defined as

$$S_b^{NW} = \sum_{i=1}^{L} P_i \sum_{\substack{j=1 \\ j \neq i}}^{L} \sum_{\ell=1}^{n_i} \frac{\lambda_\ell^{(i,j)}}{n_i} \left(\mathbf{x}_\ell^{(i)} - M_j\left(\mathbf{x}_\ell^{(i)}\right)\right)\left(\mathbf{x}_\ell^{(i)} - M_j\left(\mathbf{x}_\ell^{(i)}\right)\right)^T$$

and

$$S_w^{NW} = \sum_{i=1}^{L} P_i \sum_{\ell=1}^{n_i} \frac{\lambda_\ell^{(i,i)}}{n_i} \left(\mathbf{x}_\ell^{(i)} - M_i\left(\mathbf{x}_\ell^{(i)}\right)\right)\left(\mathbf{x}_\ell^{(i)} - M_i\left(\mathbf{x}_\ell^{(i)}\right)\right)^T,$$

where the scatter matrix weight $\lambda_\ell^{(i,j)}$ is defined by

$$\lambda_\ell^{(i,j)} = \frac{\|\mathbf{x}_\ell^{(i)} - M_j(\mathbf{x}_\ell^{(i)})\|^{-1}}{\sum_{t=1}^{n_i}\|\mathbf{x}_t^{(i)} - M_j(\mathbf{x}_t^{(i)})\|^{-1}}$$

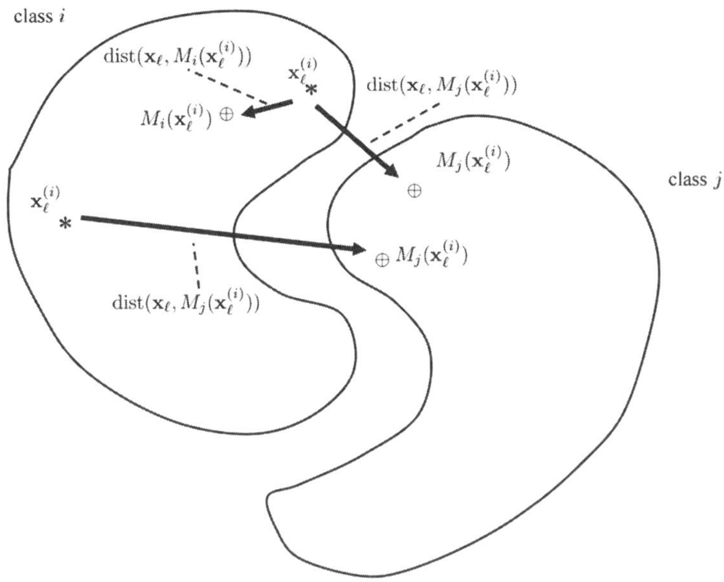

Figure 16.2 The relationship between sample points and their local means in NWFE.

and the

$$M_j(\mathbf{x}_\ell^{(i)}) = \sum_{k=1}^{n_j} w_{\ell k}^{(i,j)} \mathbf{x}_k^{(j)}, \quad w_{\ell k}^{(i,j)} = \frac{\|\mathbf{x}_\ell^{(i)} - \mathbf{x}_k^{(j)}\|^{-1}}{\sum_{t=1}^{n_j} \|\mathbf{x}_\ell^{(i)} - \mathbf{x}_t^{(j)}\|^{-1}}$$

denotes the weighted mean with respect to $\mathbf{x}_\ell^{(i)}$ in class j.

Figure 16.2 shows the ideas of NWFE. The scatter matrix weight $\lambda_\ell^{(i,j)}$ will be closed to 1 if the distance between $\mathbf{x}_\ell^{(i)}$ and $M_j(\mathbf{x}_\ell^{(i)})$ is small. Otherwise, $\lambda_\ell^{(i,j)}$ will be closed to 0. Similarly, the weight $w_{\ell k}^{(i,j)}$ for computing weighted means will be closed to 1 if the distance between $\mathbf{x}_\ell^{(i)}$ and $\mathbf{x}_k^{(j)}$ is small. Otherwise, $w_{\ell k}^{(i,j)}$ will be closed to 0. NWFE proposes the 'weighted mean' and using weighted between- and within-class vector. The separability of classes in a data set can be measured by S_b^{NW} and S_w^{NW}.

The goal of NWFE is to find a linear transformation $\mathbf{A} \in \mathbb{R}^{d \times p}$, $p \leq d$, that maximizes the between-class scatter and minimizes the within-class scatter. The columns of $\mathbf{A}$ are the optimal features by optimizing the Fisher criteria, i.e.,

$$\mathbf{A} = \arg\max_{\mathbf{A}} \operatorname{tr}\left(\left(\mathbf{A}^T S_w^{NW} \mathbf{A}\right)^{-1} \left(\mathbf{A}^T S_b^{NW} \mathbf{A}\right)\right).$$

This maximization is equivalent to find eigenvalues λ and eigenvectors $\mathbf{v}$ for the generalized eigenvalue problem

$$S_b^{NW} \mathbf{v} = \lambda S_w^{NW} \mathbf{v}.$$

To reduce the effect of the cross-products of within-class distances and prevent the singularity, some regularized techniques (Kuo *et al.* 2003), can be applied to within-class scatter matrix. In NWFE, within-class scatter matrix is regularized by

$$S_w^{NW} = 0.5S_w^{NW} + 0.5\text{diag}\left(S_w^{NW}\right),$$

where diag(S_w^{NW}) means diagonal matrix of S_w^{NW}.

16.2.4 Fuzzy linear feature extraction

The main idea of Fuzzy Linear Feature Extraction (FLFE) (Yang *et al.* 2006) originates from the non-parametric discriminant analysis (NDA) (Fukunaga, 1990), which exhibits an essential observation that the training samples approaching the class boundary should be emphasized and given larger weights. Unlike NDA and NWFE, which defined the separability by using Euclidean distance directly, we discover the fuzzification procedure of the fuzzy k-nearest neighbours algorithm (Keller *et al.* 1985) that can be employed to find the samples near the class boundary more intuitively than NDA and NWFE.

The fuzzification procedure of the fuzzy kNN algorithm is

$$\mu_j(\mathbf{x}_\ell^{(i)}) = \begin{cases} 0.51 + 0.49 \times s_j/k & \text{if } j = i \\ 0.49 \times s_j/k & \text{if } j \neq i \end{cases},$$

where $\mathbf{x}_\ell^{(i)}$ is the ℓth training sample in class i, k is a constant, which can be regarded as an index for collecting local information surrounding each training sample, and s_j is the number of these k samples that belongs to class j. If we look closer into the membership values of each sample calculated in light of $\mu_j(\mathbf{x}_\ell^{(i)})$, the nonzero membership values reveal the classes that the sample is close to, and vice versa. An example of two classes is illustrated in Figure 16.3, where the parameter k is set to 3. The membership values of some samples are marked and those circled samples can be regarded as samples near the class boundary.

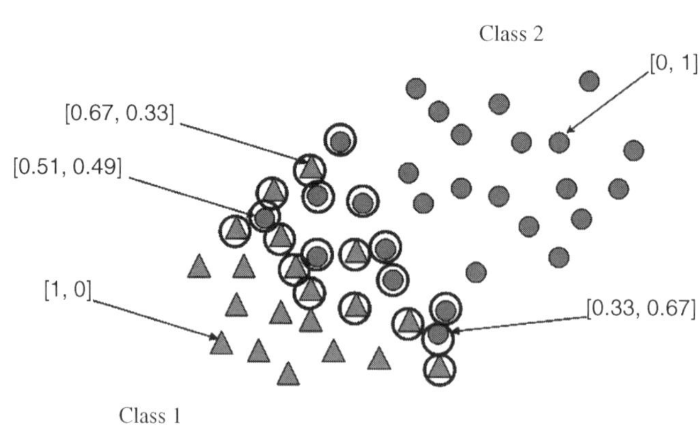

Figure 16.3 Illustration of finding the boundary samples based on the fuzzification scheme of the fuzzy kNN algorithm.

The fuzzy within-class scatter matrix S_w^{FL} and fuzzy between-class scatter matrix S_b^{FL} of FLFE are defined as follows.

$$S_w^{FL} = \sum_{i=1}^{L} P_i \sum_{\ell=1}^{n_i} a_\ell^{(i,i)} \left(\mathbf{x}_\ell^{(i)} - \mathbf{m}_i\right) \left(\mathbf{x}_\ell^{(i)} - \mathbf{m}_i\right)^T,$$

$$S_b^{FL} = \sum_{i=1}^{L} P_i \sum_{\substack{j=1 \\ j \neq i}}^{L} \sum_{\ell=1}^{n_i} b_\ell^{(i,j)} \left(\mathbf{x}_\ell^{(i)} - \mathbf{m}_j\right) \left(\mathbf{x}_\ell^{(i)} - \mathbf{m}_j\right)^T,$$

where P_i and $\mathbf{m}_i$ are the prior probability and the mean of class i, respectively. In addition, $a_\ell^{(i,i)}$ and $b_\ell^{(i,j)}$ are the weighting factors of sample $\mathbf{x}_\ell$ for within-class and between-class scatter matrices, respectively. They are defined as follows

$$a_\ell^{(i,i)} = 1 - \frac{\mu_i \left(\mathbf{x}_\ell^{(i)}\right)^{r_1}}{\sum_{\ell=1}^{n_i} \mu_i \left(\mathbf{x}_\ell^{(i)}\right)^{r_1}} \text{ and } b_\ell^{(i,j)} = \frac{\mu_j \left(\mathbf{x}_\ell^{(i)}\right)^{r_2}}{\sum_{\ell=1}^{n_i} \mu_j \left(\mathbf{x}_\ell^{(i)}\right)^{r_2}},$$

where r_1 and r_2 are weighting exponents. The designs of $a_\ell^{(i,i)}$ and $b_\ell^{(i,j)}$ enable samples near the class boundary to gain more weights. Unlike LDA, FLFE is nonparametric and is capable of extracting more than $L-1$ features that are utilized for real data classification.

16.3 Kernel-based NWFE and FLFE

In this section, two kernel-based feature extraction methods are introduced. The first is the kernel-based NWFE and the other is kernel-based FLFE.

16.3.1 Kernel-based NWFE

The kernel-based NWFE was proposed by Kuo *et al.*, (2008). The between-class scatter matrix S_b^{KNW} and the within-class scatter matrix S_w^{KNW} of KNWFE in the feature space $\mathcal{H}$ are defined as

$$S_b^{KNW} = \sum_{i=1}^{L} P_i \sum_{\substack{j=1 \\ j \neq i}}^{L} \sum_{\ell=1}^{n_i} \frac{\lambda_\ell^{(i,j)}}{n_i} \left(\boldsymbol{\phi}\left(\mathbf{x}_\ell^{(i)}\right) - M_j\left(\boldsymbol{\phi}\left(\mathbf{x}_\ell^{(i)}\right)\right)\right) \left(\boldsymbol{\phi}\left(\mathbf{x}_\ell^{(i)}\right) - M_j\left(\boldsymbol{\phi}\left(\mathbf{x}_\ell^{(i)}\right)\right)\right)^T$$

and

$$S_w^{KNW} = \sum_{i=1}^{L} P_i \sum_{\ell=1}^{n_i} \frac{\lambda_\ell^{(i,i)}}{n_i} \left(\boldsymbol{\phi}\left(\mathbf{x}_\ell^{(i)}\right) - M_i\left(\boldsymbol{\phi}\left(\mathbf{x}_\ell^{(i)}\right)\right)\right) \left(\boldsymbol{\phi}\left(\mathbf{x}_\ell^{(i)}\right) - M_i\left(\boldsymbol{\phi}\left(\mathbf{x}_\ell^{(i)}\right)\right)\right)^T,$$

where the scatter matrix weight $\lambda_\ell^{(i,j)}$ is defined by

$$\lambda_\ell^{(i,j)} = \frac{\left\|\boldsymbol{\phi}\left(\mathbf{x}_\ell^{(i)}\right) - M_j\left(\boldsymbol{\phi}\left(\mathbf{x}_\ell^{(i)}\right)\right)\right\|^{-1}}{\sum_{t=1}^{n_i}\left\|\boldsymbol{\phi}\left(\mathbf{x}_t^{(i)}\right) - M_j\left(\boldsymbol{\phi}\left(\mathbf{x}_t^{(i)}\right)\right)\right\|^{-1}},$$

$M_j(\boldsymbol{\phi}(\mathbf{x}_\ell^{(i)})) = \sum_{k=1}^{n_j} w_{\ell k}^{(i,j)}\boldsymbol{\phi}(\mathbf{x}_k^{(j)})$ denotes the weighted mean with respect to $\boldsymbol{\phi}(\mathbf{x}_\ell^{(i)})$ in class j and

$$w_{\ell k}^{(i,j)} = \frac{\left\|\boldsymbol{\phi}\left(\mathbf{x}_\ell^{(i)}\right) - \boldsymbol{\phi}\left(\mathbf{x}_k^{(j)}\right)\right\|^{-1}}{\sum_{t=1}^{n_j}\left\|\boldsymbol{\phi}\left(\mathbf{x}_\ell^{(i)}\right) - \boldsymbol{\phi}\left(\mathbf{x}_t^{(j)}\right)\right\|^{-1}}.$$

The following lemmas and theorems show that every part in S_b^{KNW} and S_w^{KNW} can be evaluated by the elements in the kernel matrix $\mathbf{K}$ or the kernel function K. Owing to the length of this chapter, the proofs of the following lemmas, theorems and corollary will be omitted. One can find these proofs in Kuo *et al.* (2008).

Lemma 16.3.1 *The weighted mean in class j with respect to* $\boldsymbol{\phi}(\mathbf{x}_\ell^{(i)})$ *is*

$$M_j(\boldsymbol{\phi}(\mathbf{x}_\ell^{(i)})) = X_j^T \begin{bmatrix} w_{\ell 1}^{(i,j)} \\ \vdots \\ w_{\ell n_j}^{(i,j)} \end{bmatrix}.$$

Lemma 16.3.2 *Let* $\mathbf{K}^{(i,j)} = X_i X_j^T$, *the* (i, j) *block of the kernel matrix* $\mathbf{K}$, *and*

$$W^{(i,j)} = \begin{bmatrix} w_{11}^{(i,j)} & \cdots & w_{1n_j}^{(i,j)} \\ \vdots & \ddots & \vdots \\ w_{n_i 1}^{(i,j)} & \cdots & w_{n_i n_j}^{(i,j)} \end{bmatrix}.$$

The scatter matrix weight $\lambda_\ell^{(i,j)}$ *is*

$$\lambda_\ell^{(i,j)} = \frac{\left[\mathbf{K}_{\ell\ell}^{(i,i)} + (W^{(i,j)}\mathbf{K}^{(j,j)}(W^{(i,j)})^T)_{\ell\ell} - 2(\mathbf{K}^{(i,j)}(W^{(i,j)})^T)_{\ell\ell}\right]^{-1/2}}{\sum_{t=1}^{n_i}\left[\mathbf{K}_{tt}^{(i,i)} + (W^{(i,j)}\mathbf{K}^{(j,j)}(W^{(i,j)})^T)_{tt} - 2(\mathbf{K}^{(i,j)}(W^{(i,j)})^T)_{tt}\right]^{-1/2}}.$$

The following theorem shows that S_b^{KNW} and S_w^{KNW} can be evaluated by matrix multiplications.

Theorem 16.3.3 *Let*

$$\Lambda^{(i,j)} = \text{diag}\left\{\frac{\lambda_1^{(i,j)}}{n_i}, \ldots, \frac{\lambda_{n_i}^{(i,j)}}{n_i}\right\} \text{ and } W^{(i,j)} = \begin{bmatrix} w_{11}^{(i,j)} & \cdots & w_{1n_j}^{(i,j)} \\ \vdots & \ddots & \vdots \\ w_{n_i 1}^{(i,j)} & \cdots & w_{n_i n_j}^{(i,j)} \end{bmatrix}.$$

The within-class scatter matrix S_w^{KNW} *becomes*

$$S_w^{KNW} = X^T \mathbf{W}^{\mathbf{KNW}} X,$$

where $\mathbf{W}^{\mathbf{KNW}} = W_1 - W_2 - W_2^T + W_3$, *and*

$$W_1 = \text{diag}\left\{P_1 \Lambda^{(1,1)}, \ldots, P_L \Lambda^{(L,L)}\right\},$$

$$W_2 = \text{diag}\left\{P_1 \Lambda^{(1,1)} W^{(1,1)}, \ldots, P_L \Lambda^{(L,L)} W^{(L,L)}\right\}, \text{ and}$$

$$W_3 = \text{diag}\left\{P_1 (W^{(1,1)})^T \Lambda^{(1,1)} W^{(1,1)}, \ldots, P_L (W^{(L,L)})^T \Lambda^{(L,L)} W^{(L,L)}\right\}.$$

The between-class scatter matrix S_b^{KNW} *becomes*

$$S_b^{KNW} = X^T (\mathbf{B}^{\mathbf{KNW}} - \mathbf{W}^{\mathbf{KNW}}) X,$$

where $\mathbf{B}^{\mathbf{KNW}} = B_1 - B_2 - B_2^T + B_3$, *and*

$$B_1 = \text{diag}\left\{P_1 \sum_{j=1}^{L} \Lambda^{(1,j)}, \ldots, P_L \sum_{j=1}^{L} \Lambda^{(L,j)}\right\},$$

$$B_2 = \begin{bmatrix} P_1 \Lambda^{(1,1)} W^{(1,1)} & \cdots & P_1 \Lambda^{(1,L)} W^{(1,L)} \\ \vdots & \ddots & \vdots \\ P_L \Lambda^{(L,1)} W^{(L,1)} & \cdots & P_L \Lambda^{(L,L)} W^{(L,L)} \end{bmatrix} \text{ and}$$

$$B_3 = \sum_{i=1}^{L} P_i \text{diag}\left\{(W^{(i,1)})^T \Lambda^{(i,1)} W^{(i,1)}, \ldots, (W^{(i,L)})^T \Lambda^{(i,L)} W^{(i,L)}\right\}.$$

From the above theorem the linear transformation $\mathbf{A}$ (has p columns) of KNWFE in the feature space $\mathcal{H}$ can be obtained by solving the following problem:

$$\mathbf{A} = \arg\max_{\mathbf{A}} \text{tr}\left(\left(\mathbf{A}^T X^T \mathbf{W}^{\mathbf{KNW}} X \mathbf{A}\right)^{-1} \left(\mathbf{A}^T X^T \left(\mathbf{B}^{\mathbf{KNW}} - \mathbf{W}^{\mathbf{KNW}}\right) X \mathbf{A}\right)\right).$$

From the above theorem, we obtain the following corollary.

Corollary 16.3.4 *If the rank of X^T is N_b, then the Fisher criterion with respect to KNWFE with linear kernel is the same as NWFE.*

16.3.2 Kernel-based FLFE

Some key components must be taken into consideration before constructing the kernel-based fuzzy feature extraction (KFFE). The first is about the distance measure in feature space $\mathcal{H}$ because the membership values have to be calculated through

$$\mu_j\left(\boldsymbol{\phi}\left(x_\ell^{(i)}\right)\right) = \begin{cases} 0.51 + 0.49 \times \frac{s_j}{k} & \text{if } j = i \\ 0.49 \times \frac{s_j}{k} & \text{if } j \neq i. \end{cases} \tag{16.1}$$

The second is how the features are extracted by KFFE via the kernel trick. Finally, we have to exhibit the projection of an unknown sample in feature space $\mathcal{H}$. In the following, we primarily reformulate the fuzzy within-class and between-class scatter matrices in feature space $\mathcal{H}$.

The fuzzy within-class scatter matrix S_w^{KF} in feature space $\mathcal{H}$ is defined as

$$S_w^{KF} = \sum_{i=1}^{L} P_i \sum_{\ell=1}^{n_i} a_\ell^{(i,i)}(\boldsymbol{\phi}(\mathbf{x}_\ell^{(i)}) - \mathbf{M}_i)(\boldsymbol{\phi}(\mathbf{x}_\ell^{(i)}) - \mathbf{M}_i)^T,$$

where $\mathbf{M}_i = \sum_{\ell=1}^{n_i} \boldsymbol{\phi}(\mathbf{x}_\ell^{(i)})/n_i$ is the mean of class i in feature space $\mathcal{H}$ and

$$a_\ell^{(i,i)} = 1 - \left(\mu_i\left(\boldsymbol{\phi}\left(\mathbf{x}_\ell^{(i)}\right)\right)^{r_1} \Big/ \sum_{\ell=1}^{n_i} \mu_i\left(\boldsymbol{\phi}\left(\mathbf{x}_\ell^{(i)}\right)\right)^{r_1}\right).$$

The fuzzy between-class scatter matrix S_b^{KF} in feature space $\mathcal{H}$ is given by

$$S_b^{KF} = \sum_{i=1}^{L} P_i \sum_{\substack{j=1 \\ j\neq i}}^{L} \sum_{\ell=1}^{n_i} b_\ell^{(i,j)}(\boldsymbol{\phi}(\mathbf{x}_\ell^{(i)}) - \mathbf{M}_j)(\boldsymbol{\phi}(\mathbf{x}_\ell^{(i)}) - \mathbf{M}_j)^T,$$

where $b_\ell^{(i,j)} = \mu_j(\boldsymbol{\phi}(\mathbf{x}_\ell^{(i)}))^{r_2} / \sum_{\ell=1}^{n_i} \mu_j(\boldsymbol{\phi}(\mathbf{x}_\ell^{(i)}))^{r_2}$ and $\mathbf{M}_j = \sum_{\ell=1}^{n_j} \boldsymbol{\phi}(\mathbf{x}_\ell^{(j)})/n_j$.

In the following lemmas and theorems, we demonstrate how the scatter matrices S_b^{KF} and S_w^{KF} can be computed by the entries in the kernel matrix $\mathbf{K}$ or the kernel function K. The proofs of these lemmas and theorems are similar to those in KNWFE.

Lemma 16.3.5 *The mean of class j, $\mathbf{M}_j$, in feature space $\mathcal{H}$ is defined by*

$$\mathbf{M}_j = \frac{1}{n_j} \sum_{\ell=1}^{n_j} \boldsymbol{\phi}(x_\ell^{(j)}) = X_j^T 1_{n_j},$$

where $1_{n_j} = [\frac{1}{n_j}, \ldots, \frac{1}{n_j}]$ is an $1 \times n_j$ constant matrix.

Theorem 16.3.6 *Let*

$$\Lambda_w^{(i,i)} = \mathrm{diag}(a_1^{(i,i)}, \ldots, a_{n_i}^{(i,i)}).$$

The within-class scatter matrix can be expressed as

$$S_w^{KF} = X^T \mathbf{W^{KF}} X,$$

where $\mathbf{W^{KF}} = \bar{W}_1 - \bar{W}_2 - \bar{W}_2^T + \bar{W}_3$ *with*

$$\bar{W}_1 = \mathrm{diag}(P_1 \Lambda_w^{(1,1)}, \ldots, P_L \Lambda_w^{(L,L)}),$$
$$\bar{W}_2 = \mathrm{diag}(P_1 \Lambda_w^{(1,1)} \mathbf{1}_{n_1}, \ldots, P_L \Lambda_w^{(L,L)} \mathbf{1}_{n_L}), \text{ and}$$
$$\bar{W}_3 = \mathrm{diag}(P_1 \mathbf{1}_{n_1}^T \Lambda_w^{(1,1)} \mathbf{1}_{n_1}, \ldots, P_L \mathbf{1}_{n_L}^T \Lambda_w^{(L,L)} \mathbf{1}_{n_L}).$$

Theorem 16.3.7 *Let*

$$\Lambda_b^{(i,j)} = \mathrm{diag}(b_1^{(i,j)}, \ldots, b_{n_i}^{(i,j)}).$$

The between-class scatter matrix can be expressed as

$$S_b^{KF} = X^T(\mathbf{B^{KF}} - \mathbf{W_b^{KF}})X,$$

where $\mathbf{B^{KF}} = \bar{B}_1 - \bar{B}_2 - \bar{B}_2^T + \bar{B}_3$ *and* $\mathbf{W_b^{KF}} = \bar{W}_{b1} - \bar{W}_{b2} - \bar{W}_{b2}^T + \bar{W}_{b3}$ *with*

$$\bar{B}_1 = \mathrm{diag}(P_1 \sum_{j=1}^{L} \Lambda_b^{(1,j)}, \ldots, P_L \sum_{j=1}^{L} \Lambda_b^{(L,j)}),$$

$$\bar{B}_2 = \begin{bmatrix} P_1 \Lambda_b^{(1,1)} \mathbf{1}_{n_1 n_1} & \ldots & P_1 \Lambda_b^{(1,L)} \mathbf{1}_{n_1 n_L} \\ \vdots & \ddots & \vdots \\ P_L \Lambda_b^{(L,1)} \mathbf{1}_{n_L n_1} & \ldots & P_L \Lambda_b^{(L,L)} \mathbf{1}_{n_L n_L} \end{bmatrix},$$

$$\bar{B}_3 = \sum_{i=1}^{L} P_i \mathrm{diag}(\mathbf{1}_{n_i n_1}^T \Lambda_b^{(i,1)} \mathbf{1}_{n_i n_1}, \ldots, \mathbf{1}_{n_i n_L}^T \Lambda_b^{(i,L)} \mathbf{1}_{n_i n_L}),$$

$$\mathbf{1}_{n_i n_j} = \begin{bmatrix} \frac{1}{n_i} & \cdots & \frac{1}{n_i} \\ \vdots & \ddots & \vdots \\ \frac{1}{n_i} & \cdots & \frac{1}{n_i} \end{bmatrix}_{n_i \times n_j},$$

$$\bar{W}_{b1} = \mathit{diag}(P_1 \Lambda_b^{(1,1)}, \ldots, P_L \Lambda_b^{(L,L)}),$$
$$\bar{W}_{b2} = \mathit{diag}(P_1 \Lambda_b^{(1,1)} \mathbf{1}_{n_1}, \ldots, P_L \Lambda_b^{(L,L)} \mathbf{1}_{n_L}), \text{ and}$$
$$\bar{W}_{b3} = \mathit{diag}(P_1 \mathbf{1}_{n_1}^T \Lambda_b^{(1,1)} \mathbf{1}_{n_1}, \ldots, P_L \mathbf{1}_{n_L}^T \Lambda_b^{(L,L)} \mathbf{1}_{n_L}).$$

16.4 Eigenvalue resolution with regularization

There are two types of feature extractions, one applies scatter matrices S_w and S_b in original space $\mathbb{R}^{N_b}$ and the other uses these scatter matrices in the feature space $\mathcal{H}$. The transformation matrix $\mathbf{A}$ cannot be computed explicitly in the feature space because only the inner products in feature space can be obtained by the kernel function and the corresponding feature mapping is implicit. Fortunately, the concept of dual form can solve this problem. The eigenvalue decomposition method in the feature space will be introduced in this section.

Recall that the scatter matrices of KNWFE are

$$S_w^{KNW} = X^T \mathbf{W}^{\mathbf{KNW}} X \quad \text{and} \quad S_b^{KNW} = X^T(\mathbf{B}^{\mathbf{KNW}} - \mathbf{W}^{\mathbf{KNW}})X,$$

and the scatter matrices of KFFE are

$$S_w^{KF} = X^T \mathbf{W}^{\mathbf{KF}} X \quad \text{and} \quad S_b^{KF} = X^T(\mathbf{B}^{\mathbf{KF}} - \mathbf{W}_{\mathbf{b}}^{\mathbf{KF}})X.$$

These two methods have the same forms

$$S_w = X^T \mathbf{W} X \quad \text{and} \quad S_b = X^T(\mathbf{B} - \mathbf{W_b})X, \tag{16.2}$$

i.e., $\mathbf{W} = \mathbf{W_b} = \mathbf{W}^{\mathbf{KNW}}$, $\mathbf{B} = \mathbf{B}^{\mathbf{KNW}}$ in KNWFE and $\mathbf{W_b} = \mathbf{W}_{\mathbf{b}}^{\mathbf{KF}}$, $\mathbf{W} = \mathbf{W}^{\mathbf{KF}}$, $\mathbf{B} = \mathbf{B}^{\mathbf{KF}}$ in KFFE.

Our goal is to find the transformation matrix $\mathbf{A}$ that maximizes the criterion

$$\mathbf{A} = \arg\max_{\mathbf{A}} \operatorname{tr}((\mathbf{A}^T S_w \mathbf{A})^{-1}(\mathbf{A}^T S_b \mathbf{A})). \tag{16.3}$$

The maximization process is equivalent to solving the generalized eigenvalue resolution

$$S_b \mathbf{v}_h = \lambda_h S_w \mathbf{v}_h,$$

where $(\lambda_h, \mathbf{v}_h)$ is the eigen-pair of $(S_w)^{-1} S_b$ in the feature space $\mathcal{H}$ with $\lambda_1 \geq \lambda_2 \geq \cdots \geq \lambda_p$. Note that p is the dimension of the transformed space. Consequently, the transformation matrix $\mathbf{A}$ can be expressed as

$$\mathbf{A} = [\mathbf{v}_1, \ldots, \mathbf{v}_p].$$

Assume that the eigenvector $\mathbf{v}_h$ lies in the span of all training samples in feature space $\mathcal{H}$, i.e., it can be expressed as the dual form

$$\mathbf{v}_h = X^T \boldsymbol{\alpha}_h.$$

Thus, the transformation matrix $\mathbf{A}$ can be represented as

$$\mathbf{A} = [\mathbf{v}_1, \ldots, \mathbf{v}_p] = X^T[\boldsymbol{\alpha}_1, \ldots, \boldsymbol{\alpha}_p] = X^T \tilde{\mathbf{A}}, \tag{16.4}$$

where $\tilde{\mathbf{A}} = [\boldsymbol{\alpha}_1, \ldots, \boldsymbol{\alpha}_p]$.

Therefore, from (16.2) to (16.4), to find $\mathbf{A}$ is equivalent to finding $\tilde{\mathbf{A}}$ by

$$\tilde{\mathbf{A}} = \arg\max_{\tilde{\mathbf{A}}} \{\operatorname{tr}((\tilde{\mathbf{A}}\mathbf{KWK}\tilde{\mathbf{A}})^{-1}(\tilde{\mathbf{A}}^T(\mathbf{K}(\mathbf{B} - \mathbf{W_b})\mathbf{K}\tilde{\mathbf{A}}))\}.$$

Then $\boldsymbol{\alpha}_h$ can be solved through the generalized eigenvalue problem

$$\mathbf{K}(\mathbf{B} - \mathbf{W_b})\mathbf{K}\boldsymbol{\alpha}_h = \lambda \mathbf{KWK}\boldsymbol{\alpha}_h. \tag{16.5}$$

From (16.5), we achieve the second goal: that the feature $\boldsymbol{\alpha}_h$ can be solved via the kernel trick.

In the following, the same eigen-decomposition of the kernel matrix is employed as in GDA. Again, let $\mathbf{K} = \mathbf{P\Gamma P}^T$ and $\boldsymbol{\beta}_h = \mathbf{\Gamma P}^T\boldsymbol{\alpha}_h$, then $\boldsymbol{\beta}_h$ can be derived by solving the following generalized eigenvalue problem

$$\mathbf{P}^T(\mathbf{B} - \mathbf{W}_b)\mathbf{P}\boldsymbol{\beta}_h = \lambda(\mathbf{P}^T\mathbf{WP})\boldsymbol{\beta}_h. \tag{16.6}$$

Then $\boldsymbol{\beta}_h$ is the eigenvector of $(\mathbf{P}^T\mathbf{WP})^{-1}\mathbf{P}^T(\mathbf{B} - \mathbf{W_b})\mathbf{P}$. However, the estimation of $(\mathbf{P}^T\mathbf{WP})^{-1}$ may be singular or nearly singular. Hence, the following regularization is taken to circumvent the singularity problem

$$\mathbf{P}^T\mathbf{WP} = (1-t)\cdot(\mathbf{P}^T\mathbf{WP}) + t\cdot \mathrm{diag}(\mathbf{P}^T\mathbf{WP}), \tag{16.7}$$

where t is a regularization parameter.

After $\boldsymbol{\beta}_h$ is found, the $\boldsymbol{\alpha}_h$ can be computed by $\boldsymbol{\alpha}_h = \mathbf{P}\Gamma^{-1}\boldsymbol{\beta}_h$ and divided by

$$\sqrt{\boldsymbol{\alpha}_h^T\mathbf{K}\boldsymbol{\alpha}_h}$$

to get normalized vectors $\bar{\boldsymbol{\alpha}}_h$, i.e.,

$$\bar{\boldsymbol{\alpha}}_h = [\bar{\alpha}_{h_1}^{(1)}, \ldots, \bar{\alpha}_{h_{n_1}}^{(1)}, \ldots, \bar{\alpha}_{h_1}^{(L)}, \ldots, \bar{\alpha}_{h_{n_L}}^{(L)}]^T = \frac{\boldsymbol{\alpha}_h}{\sqrt{\boldsymbol{\alpha}_h^T\mathbf{K}\boldsymbol{\alpha}_h}}.$$

We then obtain the transformation matrix $\mathbf{A}$.

Finally, the projected components of an unknown mapped sample $\boldsymbol{\phi}(\mathbf{z})$ in $\mathcal{H}$ can be calculated from

$$\boldsymbol{\phi}(\mathbf{z})^T\mathbf{v}_h = \sum_{i=1}^{L}\sum_{\ell=1}^{n_i}\bar{\alpha}_{h_\ell}^{(l)}K(\mathbf{x}_\ell^{(i)}, \mathbf{z}), \text{ for } h = 1, \ldots, p. \tag{16.8}$$

Note that the projected components of an unknown sample are also computed by virtue of the kernel trick. The summaries of KNWFE and KFFE are described in Tables 16.1 and 16.2.

16.5 Experiments

16.5.1 Data sets

In this study, for investigating the influences of training sample sizes to the dimension, three distinct cases, $n_i = 20 < n < N_b$ (case 1), $n_i = 150 < N_b < n$ (case 2) and $N_b < n_i = 300 < n$ (case 3) will be discussed. Owing to these sample size constraints, some of the classes in selected hyperspectral images for the experiment are used. The MultiSpec© (Landgrebe

Table 16.1 KNWFE algorithm

(1) Compute the distances between each pair of sample points in the feature space and form the distance matrix.

(2) Compute $w_{ks}^{(i,j)}$ using the distance matrix and get the matrix $W^{(i,j)}$.

(3) Compute the scatter matrix weights $\lambda_{\ell}^{(i,j)}$ and get the matrix $\Lambda^{(i,j)}$.

(4) By Theorem 17.3.3, compute $\mathbf{B}^{\mathbf{KNW}}$, $\mathbf{W}^{\mathbf{KNW}}$ and, hence, $S_b^{KNW} = X^T(\mathbf{B}^{\mathbf{KNW}} - \mathbf{W}^{\mathbf{KNW}})X$ and $S_w^{KNW} = X^T\mathbf{W}^{\mathbf{KNW}}X$.

(5) Do eigen-decomposition for $\mathbf{K}$, i.e., $\mathbf{K} = \mathbf{P\Gamma P}^T$.

(6) Evaluate $\mathbf{P}^T\mathbf{W}^{\mathbf{KNW}}\mathbf{P}$ by (16.7), where t is set by 0.5, and then compute $\boldsymbol{\beta}_h$ by (16.6), finally compute $\boldsymbol{\alpha}_h$ and $\mathbf{v}_h$.

(7) Calculate the projection components of each sample from (16.8).

Table 16.2 KFFE algorithm

(1) Calculate the membership values of each training sample of each class in feature space $\mathcal{H}$ from (16.1).

(2) Compute $\Lambda_w^{(i,i)}$ and $\Lambda_b^{(i,j)}$.

(3) Compute matrices $\mathbf{B}^{\mathbf{KF}}$, $\mathbf{W}_{\mathbf{b}}^{\mathbf{KF}}$, $\mathbf{W}^{\mathbf{KF}}$ and kernel matrix $\mathbf{K}$.

(4) Decompose $\mathbf{K}$ using eigenvectors decomposition, i.e., $\mathbf{K} = \mathbf{P\Gamma P}^T$.

(5) Do eigen-decomposition for $\mathbf{K}$, i.e., $\mathbf{K} = \mathbf{P\Gamma P}^T$.

(6) Estimate $\mathbf{P}^T\mathbf{W}^{\mathbf{KF}}\mathbf{P}$ from (16.7) and compute $\boldsymbol{\beta}_h$ from (16.6), and $\boldsymbol{\alpha}_h$ and $\mathbf{v}_h$.

(7) Calculate the projection components of each sample from (16.8).

2003) was used to select training and test samples (100 test samples per class) in our experiments which uses the same method in (Benediktsson *et al.* 2005; Serpico and Moser 2007; Landgrebe 2003).

In our experiments, three real data sets are applied to compare the performances of KNWFE, KFFE and other kernel-based famous feature extraction methods. They are the Indian Pines: a mixed forest/agricultural site in Indiana (Landgrebe, 2003), the Kennedy Space Centre (KSC), Florida (Ham *et al.* 2005), and the Washington DC Mall hyperspectral image (Landgrebe, 2003) as an urban site. The first two of these data sets were gathered by a sensor known as the Airborne Visible/Infrared Imaging Spectrometer. The Indian Pines image, mounted from an aircraft flown at 65 000 ft altitude and operated by the NASA/Jet Propulsion Laboratory, with the size of 145 $\times$ 145 pixels has 220 spectral bands measuring approximately 20 m across on the ground. The simulated greyscale IR image and the ground truth map are shown in Figure 16.4(top), respectively. Since the size of the samples in some classes are too small to retain enough disjoint samples for training and testing, only eight classes, Corn-min, Corn-notill, Soybean-clean, Grass/Pasture, Soybeans-min, Hay-windrowed, Soybeans-notill, and Woods, were selected for the experiments.

The KSC data set was acquired over the KSC by the NASA AVIRIS instrument on March 23, 1996. AVIRIS acquires data in 224 bands of 10 nm width with centre wavelengths from

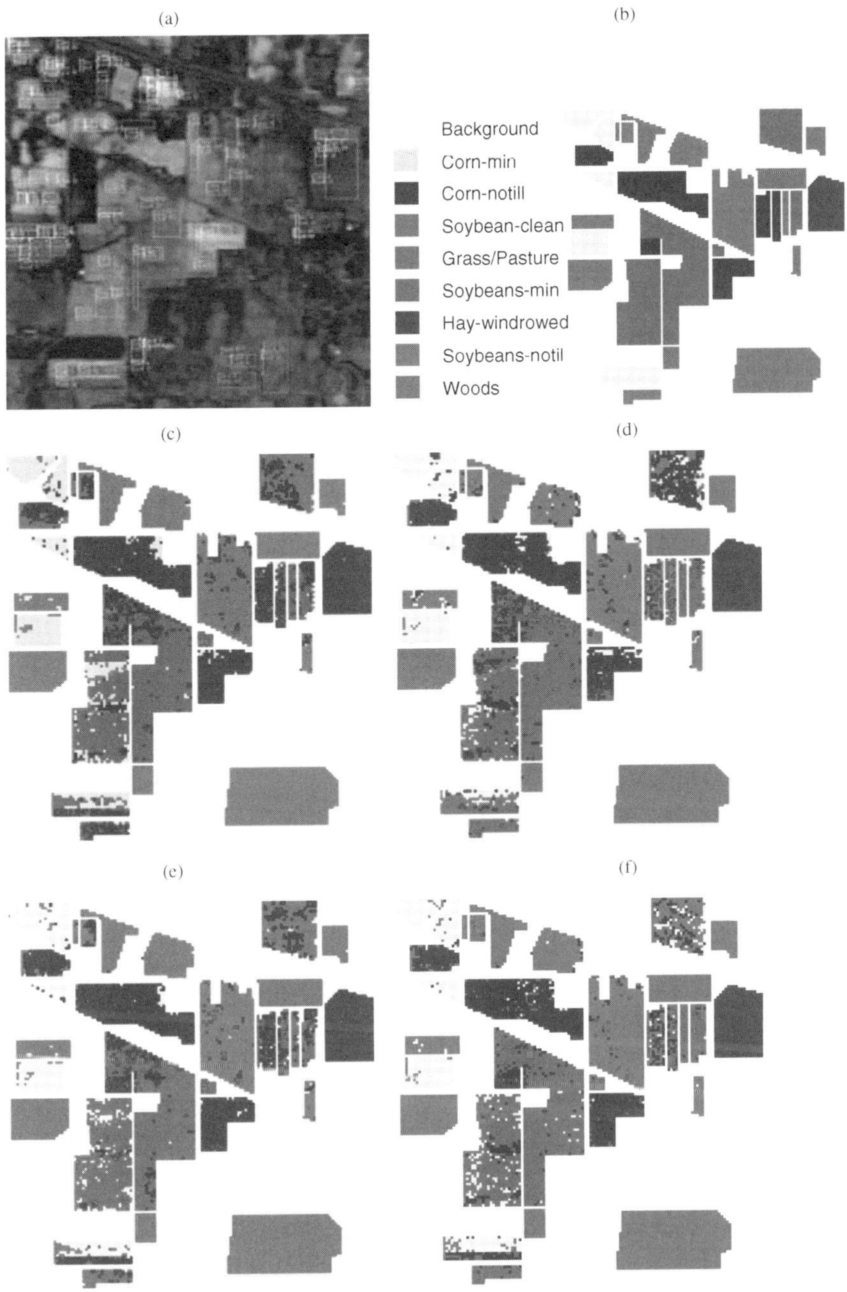

Figure 16.4 (a) Simulated greyscale IR image of the Indian Pines Site dataset. (b) Ground truth of the area with eight classes. (c) Classification map by KPCA-poly1 with ML classifier ($n_i = 300$, $p = 13$), (d) GDA-poly2 with SVM-poly1 classifier ($n_i = 300$, $p = 7$), (e) KNWFE-poly1 with ML classifier ($n_i = 300$, $p = 11$), and (f) KFFE-poly1 with 1NN classifier ($n_i = 300$, $p = 13$). (See plate 10)

400 to 2500 nm. The KSC data, acquired from an altitude of approximately 20 km, have a spatial resolution of 18 m. After removing water absorption and low SNR bands, 176 bands were used for the analysis (Ham *et al.* 2005). Owing to the sample size constraints, seven classes, Scrub, Graminoid marsh, Spartina marsh, Cattail marsh, Salt marsh, Mud flats, and Water, are selected.

The third data set, Washington DC Mall from an urban area, is a Hyperspectral Digital Imagery Collection Experiment (HYDICE) airborne hyperspectral data flightline over the Washington, DC Mall. Two hundred and ten bands were collected in the 0.4–2.4 μm region of the visible and infrared spectrum. Some water absorption channels are discarded, resulting in 191 channels. The dataset is available in the student CD-ROM of Landgrebe (2003). There are seven information classes in the Washington, DC data: roofs, roads, trails, grass, trees, water and shadows.

16.5.2 Experiment design

The purpose of this experiment is to compare the multiclass classification performances using ML, 1NN and soft-margin SVM classifiers with KPCA, GDA, KNWFE and KFFE features. These kernel-based feature extractions are applied with two types of kernels, polynomial kernels of degree d ($d = 1$ and 2) and the RBF kernel (width σ). In this study, classifiers ML and 1NN were implemented in PRTools (Duin *et al.* 2007) and soft-margin SVM in LIBSVM (Chang and Lin 2001). The multiclass SVM classification in LIBSVM is performed by one-against-one decomposition strategy (Hsu and Lin 2002). The other two kernel feature methods, KPCA and GDA, were implemented in STPRTool (Schlesinger *et al.* 2007). The free parameters used in KFLFE and SVM classifier are learned by grid-search on them by using five-fold cross-validation strategy. For example, in the soft-margin SVM with RBF kernel classifier, there are two parameters C and σ, which are used to control the trade-off between the margin and the size of the slack variables, and the flexibility of the kernel, respectively. Then, we use five-fold cross validation to find the best C within the given set $\{2^{-5}, 2^{-3}, \ldots, 2^{15}\}$ suggested by Chang and Lin, (2001) and σ within $\{2^5, 2^6, \ldots, 2^{20}\}$.

16.5.3 Experiment results

Tables 16.3–16.5 display the classification accuracies of testing data in cases 1, 2 and 3, respectively. In these tables, 'poly-d' indicates the polynomial kernel with degree d is used, and 'RBF' means that the RBF kernel is applied. Note that the best accuracy of each data set (in column) is highlighted in bold type. From Tables 16.3–16.5, we can find that

1. In the small sample size situation (case 1, $n_i = 20 < n < N_b$), the highest accuracies of all methods are 0.834 (KFFE-poly2 with ML classifier and KFFE-RBF with 1NN classifier), 0.943 (KFFE-RBF with 1NN and SVM-poly1 classifiers) and 0.840 (KNWFE-RBF with 1NN classifier) in Indian Pines, KSC and DC datasets, respectively.

2. In case 2 ($n_i = 150 < N_b < n$), the highest accuracies of all methods are 0.903 (KNWFE-poly1 with ML classifier and KFFE-poly2 with SVM-poly1 classifier), 0.976 (KFFE-RBF with 1NN classifier) and 0.851 (KNWFE-RBF with SVM-poly1 and SVM-RBF classifiers) in Indian Pines, KSC and DC datasets, respectively.

Table 16.3 The highest accuracies with respect to the dimensionality of the reduced subspace (in parentheses) for the three datasets in case 1 ($n_i = 20$). Partly reproduced from (Kuo *et al.* 2008) ©2008 IEEE

Feature extraction	Classifier	Indian Pines	KSC	DC
KPCA-poly1	ML	0.750(14)	0.914(8)	0.700(4)
	1NN	0.681(15)	0.853(11)	0.700(4)
	SVM-poly1	0.744(15)	0.797(7)	0.713(9)
	SVM-poly2	0.715(11)	0.799(7)	0.762(12)
	SVM-RBF	0.736(15)	0.894(14)	0.735(6)
KPCA-poly2	ML	0.766(14)	0.854(9)	0.709(12)
	1NN	0.663(14)	0.840(14)	0.757(12)
	SVM-poly1	0.783(15)	0.639(11)	0.681(11)
	SVM-poly2	0.759(12)	0.631(13)	0.734(11)
	SVM-RBF	0.775(9)	0.539(6)	0.704(12)
KPCA-RBF	ML	0.684(7)	0.920(12)	0.700(4)
	1NN	0.681(15)	0.840(13)	0.777(11)
	SVM-poly1	0.644(15)	0.793(12)	0.715(10)
	SVM-poly2	0.678(15)	0.804(14)	0.741(10)
	SVM-RBF	0.771(15)	0.903(12)	0.730(8)
GDA-poly1	ML	0.664(7)	0.931(5)	0.143(1)
	1NN	0.689(7)	0.933(6)	0.659(6)
	SVM-poly1	0.665(7)	0.893(6)	0.434(5)
	SVM-poly2	0.495(6)	0.911(6)	0.367(5)
	SVM-RBF	0.660(6)	0.904(6)	0.434(5)
GDA-poly2	ML	0.654(6)	0.524(5)	0.142(1)
	1NN	0.744(7)	0.884(6)	0.801(6)
	SVM-poly1	0.734(7)	0.866(6)	0.748(6)
	SVM-poly2	0.673(6)	0.841(5)	0.462(4)
	SVM-RBF	0.738(7)	0.880(6)	0.748(6)
GDA-RBF	ML	0.547(4)	0.664(6)	0.700(8)
	1NN	0.647(7)	0.930(6)	0.777(11)
	SVM-poly1	0.546(7)	0.840(6)	0.712(9)
	SVM-poly2	0.494(7)	0.926(6)	0.741(9)
	SVM-RBF	0.666(7)	0.840(6)	0.730(7)
KNWFE-poly1	ML	0.654(7)	0.920(5)	0.544(1)
	1NN	0.761(9)	0.937(15)	0.703(6)
	SVM-poly1	0.586(9)	0.903(14)	0.596(7)
	SVM-poly2	0.615(10)	0.844(7)	0.434(7)
	SVM-RBF	0.658(8)	0.763(7)	0.617(7)
KNWFE-poly2	ML	0.743(6)	0.843(11)	0.760(4)
	1NN	0.764(6)	0.894(12)	0.831(9)
	SVM-poly1	0.608(8)	0.917(12)	0.796(9)
	SVM-poly2	0.579(12)	0.823(15)	0.734(9)
	SVM-RBF	0.783(11)	0.724(12)	0.809(8)

(*Continued*)

Table 16.3 (*Continued*)

Feature extraction	Classifier	Indian Pines	KSC	DC
KNWFE-RBF	ML	0.695(6)	0.917(5)	0.689(5)
	1NN	0.778(10)	0.937(15)	**0.840(12)**
	SVM-poly1	0.649(14)	0.854(15)	0.833(8)
	SVM-poly2	0.735(10)	0.881(6)	0.831(9)
	SVM-RBF	0.800(12)	0.899(5)	0.731(14)
KFFE-poly1	ML	0.810(9)	0.924(7)	0.791(6)
	1NN	0.804(11)	0.933(6)	0.810(7)
	SVM-poly1	0.733(14)	0.924(9)	0.666(8)
	SVM-poly2	0.660(10)	0.927(15)	0.547(6)
	SVM-RBF	0.698(7)	0.923(5)	0.720(5)
KFFE-poly2	ML	**0.834(6)**	0.917(6)	0.807(6)
	1NN	0.831(14)	0.930(6)	0.806(3)
	SVM-poly1	0.814(14)	0.906(14)	0.823(6)
	SVM-poly2	0.708(14)	0.917(13)	0.776(8)
	SVM-RBF	0.783(11)	0.850(5)	0.814(4)
KFFE-RBF	ML	0.816(7)	0.941(5)	0.793(5)
	1NN	**0.834(14)**	**0.943(13)**	0.836(11)
	SVM-poly1	0.763(15)	**0.943(7)**	0.839(10)
	SVM-poly2	0.718(15)	0.923(15)	0.790(10)
	SVM-RBF	0.755(11)	0.896(5)	0.770(5)

Table 16.4 The highest accuracies with respect to the dimensionality of the reduced subspace (in parentheses) for the three datasets in case 2 ($n_i = 150$). Partly reproduced from (Kuo *et al.* 2008) ©2008 IEEE

Feature extraction	Classifier	Indian Pines	KSC	DC
KPCA-poly1	ML	0.879(7)	0.839(15)	0.770(13)
	1NN	0.750(15)	0.924(15)	0.768(13)
	SVM-poly1	0.759(15)	0.763(14)	0.720(12)
	SVM-poly2	0.823(13)	0.776(15)	0.788(11)
	SVM-RBF	0.754(14)	0.587(11)	0.727(10)
KPCA-poly2	ML	0.878(15)	0.527(14)	0.760(15)
	1NN	0.746(14)	0.787(15)	0.753(14)
	SVM-poly1	0.749(15)	0.564(15)	0.774(9)
	SVM-poly2	0.788(15)	0.609(15)	0.768(13)
	SVM-RBF	0.750(15)	0.467(15)	0.777(9)
KPCA-RBF	ML	0.816(15)	0.879(7)	0.770(13)
	1NN	0.734(15)	0.750(15)	0.775(12)
	SVM-poly1	0.746(15)	0.593(13)	0.702(11)
	SVM-poly2	0.774(15)	0.620(15)	0.721(8)
	SVM-RBF	0.835(14)	0.799(15)	0.788(12)

Table 16.4 (*Continued*)

Feature extraction	Classifier	Indian Pines	KSC	DC
GDA-poly1	ML	0.865(7)	0.964(6)	0.832(6)
	1NN	0.860(7)	0.967(6)	0.837(6)
	SVM-poly1	0.836(6)	0.936(6)	0.717(2)
	SVM-poly2	0.825(6)	0.939(6)	0.477(4)
	SVM-RBF	0.826(7)	0.944(6)	0.715(2)
GDA-poly2	ML	0.883(7)	0.943(6)	0.142(1)
	1NN	0.880(7)	0.960(6)	0.841(5)
	SVM-poly1	0.856(7)	0.954(6)	0.817(6)
	SVM-poly2	0.839(7)	0.956(6)	0.827(6)
	SVM-RBF	0.840(7)	0.954(6)	0.838(6)
GDA-RBF	ML	0.891(7)	0.946(6)	0.841(6)
	1NN	0.891(7)	0.953(6)	0.837(6)
	SVM-poly1	0.855(6)	0.851(6)	0.797(6)
	SVM-poly2	0.796(7)	0.881(6)	0.802(6)
	SVM-RBF	0.870(7)	0.924(6)	0.680(4)
KNWFE-poly1	ML	**0.903(15)**	0.846(6)	0.840(13)
	1NN	0.886(12)	0.954(15)	0.839(10)
	SVM-poly1	0.721(12)	0.881(15)	0.751(14)
	SVM-poly2	0.773(11)	0.833(15)	0.840(9)
	SVM-RBF	0.866(8)	0.947(15)	0.829(13)
KNWFE-poly2	ML	0.894(10)	0.614(15)	0.789(6)
	1NN	0.878(15)	0.881(15)	0.839(8)
	SVM-poly1	0.701(13)	0.789(15)	0.836(15)
	SVM-poly2	0.630(12)	0.524(14)	0.831(6)
	SVM-RBF	0.893(10)	0.841(11)	0.826(8)
KNWFE-RBF	ML	0.886(13)	0.960(9)	0.847(9)
	1NN	0.885(14)	0.971(15)	0.827(5)
	SVM-poly1	0.834(11)	0.931(6)	**0.851(6)**
	SVM-poly2	0.769(8)	0.934(6)	0.849(7)
	SVM-RBF	0.871(15)	0.961(8)	**0.851(12)**
KFFE-poly1	ML	0.883(7)	0.973(12)	0.821(3)
	1NN	0.871(9)	0.970(9)	0.834(5)
	SVM-poly1	0.840(15)	0.969(15)	0.843(13)
	SVM-poly2	0.775(8)	0.971(8)	0.841(11)
	SVM-RBF	0.841(13)	0.884(6)	0.819(13)
KFFE-poly2	ML	0.881(7)	0.941(6)	0.800(7)
	1NN	0.865(9)	0.973(15)	0.820(12)
	SVM-poly1	**0.903(13)**	0.930(6)	0.753(7)
	SVM-poly2	0.859(12)	0.939(6)	0.817(11)
	SVM-RBF	0.828(15)	0.921(13)	0.770(7)
KFFE-RBF	ML	0.876(9)	0.970(4)	0.824(4)
	1NN	0.860(11)	**0.976(10)**	0.830(11)
	SVM-poly1	0.849(11)	0.964(12)	0.847(14)
	SVM-poly2	0.861(7)	0.957(6)	0.844(12)
	SVM-RBF	0.826(11)	0.960(7)	0.797(11)

Table 16.5 The highest accuracies with respect to the dimensionality of the reduced subspace (in parentheses) for the three datasets in case 3 ($n_i = 300$). Partly reproduced from (Kuo *et al.* 2008) ©2008 IEEE

Feature extraction	Classifier	Indian Pines	KSC	DC
KPCA-poly1	ML	0.900(13)	0.530(8)	0.700(4)
	1NN	0.838(6)	0.814(15)	0.777(11)
	SVM-poly1	0.756(15)	0.703(14)	0.712(9)
	SVM-poly2	0.792(15)	0.726(14)	0.762(11)
	SVM-RBF	0.786(15)	0.634(14)	0.734(5)
KPCA-poly2	ML	0.900(15)	0.520(11)	0.708(12)
	1NN	0.837(15)	0.687(15)	0.757(12)
	SVM-poly1	0.772(14)	0.454(3)	0.684(12)
	SVM-poly2	0.801(15)	0.354(5)	0.734(9)
	SVM-RBF	0.799(14)	0.391(6)	0.704(12)
KPCA-RBF	ML	0.899(15)	0.886(10)	0.848(14)
	1NN	0.823(9)	0.929(14)	0.854(15)
	SVM-poly1	0.774(11)	0.773(14)	0.731(15)
	SVM-poly2	0.648(15)	0.769(6)	0.764(15)
	SVM-RBF	0.640(3)	0.943(15)	0.837(14)
GDA-poly1	ML	0.899(15)	0.886(10)	0.848(14)
	1NN	0.823(9)	0.929(14)	0.854(15)
	SVM-poly1	0.888(7)	0.973(6)	0.434(5)
	SVM-poly2	0.885(7)	0.971(6)	0.367(5)
	SVM-RBF	0.896(7)	0.980(6)	0.434(5)
GDA-poly2	ML	0.926(7)	0.956(6)	0.142(1)
	1NN	0.926(7)	0.966(6)	0.801(6)
	SVM-poly1	0.933(7)	0.949(6)	0.748(6)
	SVM-poly2	0.931(7)	0.947(6)	0.581(5)
	SVM-RBF	0.926(7)	0.957(6)	0.748(6)
GDA-RBF	ML	0.918(7)	0.974(6)	0.940(5)
	1NN	0.886(7)	0.973(6)	0.938(6)
	SVM-poly1	0.918(7)	0.937(6)	0.901(6)
	SVM-poly2	0.924(7)	0.976(6)	0.864(4)
	SVM-RBF	0.894(7)	0.930(6)	0.854(5)
KNWFE-poly1	ML	0.954(11)	0.800(14)	**0.959(15)**
	1NN	0.936(10)	0.970(9)	0.953(14)
	SVM-poly1	0.949(13)	0.897(9)	0.884(15)
	SVM-poly2	0.931(12)	0.877(15)	0.723(13)
	SVM-RBF	0.905(12)	0.941(15)	0.930(14)
KNWFE-poly2	ML	0.941(11)	0.550(11)	0.916(7)
	1NN	0.934(11)	0.861(15)	0.946(6)
	SVM-poly1	0.894(14)	0.707(11)	0.911(7)
	SVM-poly2	0.776(12)	0.593(15)	0.899(6)
	SVM-RBF	0.939(14)	0.884(15)	0.926(15)

Table 16.5 (*Continued*)

Feature extraction	Classifier	Indian Pines	KSC	DC
KNWFE-RBF	ML	0.944(11)	0.980(7)	0.951(15)
	1NN	0.941(14)	0.987(11)	0.947(15)
	SVM-poly1	0.916(15)	0.967(7)	0.923(9)
	SVM-poly2	0.891(8)	0.960(9)	0.916(8)
	SVM-RBF	0.929(14)	0.979(15)	0.923(9)
KFFE-poly1	ML	0.919(9)	0.984(12)	0.926(7)
	1NN	0.906(7)	0.980(7)	0.924(10)
	SVM-poly1	0.900(13)	0.980(13)	0.916(14)
	SVM-poly2	0.889(14)	0.976(10)	0.923(13)
	SVM-RBF	0.900(14)	0.986(10)	0.920(5)
KFFE-poly2	ML	0.936(7)	0.963(7)	0.911(5)
	1NN	**0.955(13)**	0.983(7)	0.920(9)
	SVM-poly1	0.923(15)	0.927(8)	0.933(8)
	SVM-poly2	0.911(11)	0.931(8)	0.924(9)
	SVM-RBF	0.921(7)	0.977(10)	0.941(14)
KFFE-RBF	ML	0.938(8)	**0.991(6)**	0.926(6)
	1NN	0.954(14)	**0.991(9)**	0.951(6)
	SVM-poly1	0.938(15)	0.981(7)	0.953(8)
	SVM-poly2	0.910(14)	0.979(8)	0.943(10)
	SVM-RBF	0.929(7)	0.990(8)	0.927(7)

3. In case 3 ($N_b < n_i = 300 < n$), the highest accuracies of all methods are 0.955 (KFFE-poly2 with 1NN classifier), 0.991 (KFFE-RBF with ML and 1NN classifiers) and 0.959 (KNWFE-poly1 with ML classifier) in Indian Pines, KSC and DC datasets, respectively.

4. We can observe that the best accuracies in each column all occur when applying KNWFE or KFFE. Therefore, KNWFE and KFFE, outperform other feature extraction under the best condition of all feature extractions.

5. For KSC dataset, KFFE-RBF with 1NN classifier is the best choice in the three cases. But for other two datasets, the best combination of feature extraction and classifier is not consistent.

Owing to the length restriction of this chapter, we choose the well-known Indian Pines Site image as an example and only some classified images are shown for comparison. The best classification mechanisms in case 3 ($n_i = 300$) and four feature extraction conditions (KPCA, GDA, KNWFE, and KFFE) are selected to generate the classified images.

Figure 16.4(top) shows the simulated Indian Pines Site image and the ground truth. Figure 16.4 also shows the classification maps resulting after applying KPCA with ML classifier, GDA with SVM classifier, KNWFE with ML classifier, and KFFE with 1NN classifier, which are the combinations with the highest classification accuracy. Here p is the number of features extracted by these methods with the highest accuracies in the corresponding methods. For instance, KNWFE-poly1 with ML classifier has the highest accuracy of all the combinations of KNWFE and classifiers and Figure 16.4(e) is the classification result of this combination. As indicated from Figures 16.4(c)–16.4(f), we can find that KNWFE and KFFE outperform the other feature extraction methods in 'Corn-min', 'Corn-notill', and 'Soybeans-notill' parts. In particular, KFFE has the best result in the 'Soybeans-min' part.

16.6 Comments and conclusions

In this chapter, two kernel-based nonparametric feature extraction methods, KNWFE and KFFE, were introduced and we have analysed and compared them and other kernel-based methods both theoretically and experimentally. From a theoretical point of view, NWFE and FLFE are special cases of KNWFE and KFFE with linear kernel, respectively. The experimental results of three hyperspectral images show that KNWFE and KFFE outperform KPCA and GDA under three training sample size conditions.

Finally, the performance of kernel-based methods largely depends on the choice of kernel functions. How to choose or create a kernel function that has the largest separability determined by KNWFE or KFFE is the next direction for research.

References

Baudat, G. and Anouar, F. (2000) Generalized discriminant analysis using a kernel approach. *Neural Computation*, **12**(10), 2385–2404.

Benediktsson, J. A., Palmason J. A. and Sveinsson J. R. (2005) Classification of hyperspectral data from urban areas based on extended morphological profiles. *IEEE Trans. Geosci. Remote Sens.*, **43**(3), 480–491.

Chang, C. C. and Lin, C. J. (2001) LIBSVM: a library for support vector machines. Software available at http://www.csie.ntu.edu.tw/ cjlin/libsvm.

Duin, R. P. W., Juszczak, P., Paclik, P., Pekalska, E., de Ridder, D., Tax, D. M. J. and Verzakov, S. (2007) *PRTools4.1, A Matlab Toolbox for Pattern Recognition*, Delft University of Technology.

Dundar, M. M. and Landgrebe, D. (2004) Toward an optimal supervised classifier for the analysis of hyperspectral Data. *IEEE Trans. Geosci. Remote Sens.*, **42**(1), 271–277.

Fukunaga, K. (1990) *Introduction to Sataistical Pattern Recognition*. San Diego, CA: Academic.

Camps-Valls, G. and Bruzzone, L. (2005) Kernel-based methods for hyperspectral image classification. *IEEE Trans. Geosci. Remote Sens.*, **43**(6), 1351–1362.

Camps-Valls, G., L. Gómez-Chova, J. Muñoz-Marí, J. L. Rojo-Álvarez and M. Martínez-Ramón (2008) Kernel-based framework for multi-temporal and multi-source remote sensing data classification and change detection. *IEEE Trans. Geosci. Remote Sens.*, **46**(6), 1822–1835.

Ham, J., Y. Chen, M. Crawford and J. Ghosh (2005) Investigation of the random forest framework for classification of hyperspectral data. *IEEE Trans. Geosci. Remote Sens.*, **43**(3), 492–501.

Hsieh, P. F., Wang D. S. and Hsu C. W. (2006) A linear feature extraction for multiclass classification problems based on class mean and covariance discriminant information extraction. *IEEE Trans. Pattern Anal. Machine Intell.*, **28**(2), 223–235.

Hsu, C.W. and Lin, C.J (2002) A comparison of methods for multi-class support vector machines. *IEEE Transactions on Neural Networks*, **13**, 415–425.

Hughes, G.F. (1968) On the mean accuracy of statistical pattern recognizers, *IEEE Trans. Inf. Theory*, **14**(1), 55–63.

Keller, J. M., Gray M. R. and Givens, J. A. Jr. (1985) A fuzzy K-nearest neighbour Algorithm. *IEEE Transactions on Systems, Man, and Cybernetics*, **15**(4), 580–58.

Kuo, B. C., Li C. H. and Yang J. M. (2008) Kernel nonparametric weighted feature extraction for hyperspectral image classification. Accepted by *IEEE Trans. Geosci. Remote Sens.*

Kuo, B. C. and Landgrebe, D. A. (2004) Nonparametric weighted feature extraction for classification. *IEEE Trans. Geosci. Remote Sens.*, **42**(5), 1096–1105.

Kuo, B. C., Landgrebe, D. A. Ko L. W. and Pai C. H. (2003) Regularized feature extractions for hyperspectral data classification. in *Proc. IGARSS*, Toulouse, France, July 21–25, 2003.

Landgrebe, D. (2003) *Signal Theory Methods in Multi-spectral Remote Sensing*. Hoboken, NJ: Chichester: John Wiley & Sons.

Landgrebe, D. (2005) Multi-spectral land sensing: Where from, where to? *IEEE Trans. Geosci. Remote Sens.*, **43**(3), 414–421.

Melgani, F. and Bruzzone, L. (2004) Classification of hyperspectral remote sensing images with support vector machines. *IEEE Trans. Geosci. Remote Sens.*, **42**(8), 1778–1790.

Mika, S., Ratsch, G. Scholkopf, B. Smola, A. Weston J. and Muller K. R. (1999) Invariant feature extraction and classification in kernel spaces. *Advances in Neural Information Processing Systems 12*. MIT Press: Cambridge, Mass.

Richards, J. A. (2005) Analysis of remotely sensed data: the formative decades and the future. *IEEE Trans. Geosci. Remote Sens.*, **43**(3), 422–432.

Schlesinger, M. I., Hlavac V. and Franc V. (2007). *Statistical Pattern Recognition Toolbox*. [Online]. Available: http://cmp.felk.cvut.cz/cmp/software/stprtool/index.html.

Schölkopf, B., Smola, A. and Müller K. R. (1998) Nonlinear component analysis as a kernel eigenvalue problem. *Neural Computation*, **10**, 1299–1319.

Serpico, S. B. and Moser, G. (2007) Extraction of spectral channels from hyperspectral images for classification purposes. *IEEE Trans. Geosci. Remote Sens.*, **45**(2), 484–495.

Shawe-Taylor, J. and Cristianini, N. (2004) *Kernel Methods for Pattern Analysis*. Cambridge University Press.

Song, X., Fan G. and Rao M. (2005) Automatic CRP mapping using nonparametric machine learning approaches. *IEEE Trans. Geosci. Remote Sens.*, **43**(4), 888–897.

Yang, J. M., Yu, P. T. Kuo B. C. and Hsieh T. Y. (2006) A novel fuzzy linear feature extraction for hyperspectral image classification. in *Proc. IGARSS*, 2006, pp. 3895–3898.

Index

Kernel Methods for Remote Sensing Data Analysis Edited by Gustavo Camps-Valls and Lorenzo Bruzzone